DEVELOPMENTS IN DAIRY CHEMISTRY—2

Lipids

CONTENTS OF VOLUME 1

Volume 1: Proteins

1. Chemistry of Milk Protein. HAROLD E. SWAISGOOD

2. Association of Caseins and Casein Micelle Structure. D. G. SCHMIDT

3. Inter-species Comparison of Milk Proteins. ROBERT JENNESS

4. Biosynthesis of Milk Proteins. T. B. MEPHAM, P. GAYE and J. C. MERCIER

5. The Enzymatic Coagulation of Milk. D. G. DALGLEISH

6. Heat-induced Coagulation of Milk. P. F. FOX

7. Age Gelation of Sterilized Milks. V. R. HARWALKAR

8. Changes in the Proteins of Raw Milk During Storage. ERNST H. REIMERDES

9. Nutritional Aspects of Milk Proteins. LEIF HAMBRAEUS

10. Manufacture of Casein, Caseinates and Co-precipitates. L. L. MULLER

11. Industrial Isolation of Milk Proteins: Whey Proteins. K. R. MARSHALL

12. Functional Properties of Milk Proteins and their Use as Food Ingredients. C. V. MORR

DEVELOPMENTS IN DAIRY CHEMISTRY—2

Lipids

Edited by

P. F. FOX

*Department of Dairy and Food Chemistry,
University College, Cork, Republic of Ireland*

APPLIED SCIENCE PUBLISHERS
LONDON and NEW YORK

APPLIED SCIENCE PUBLISHERS LTD
Ripple Road, Barking, Essex, England

Sole Distributor in the USA and Canada
ELSEVIER SCIENCE PUBLISHING CO., INC.
52 Vanderbilt Avenue, New York, NY 10017, USA

British Library Cataloguing in Publication Data

Developments in dairy chemistry.—(The developments series)
2: Lipids
1. Dairy products—Analysis and examination
I. Fox, P. F.   II. Series
636'.01'543   SF253

ISBN 0-85334-224-5

WITH 42 TABLES AND 88 ILLUSTRATIONS

© APPLIED SCIENCE PUBLISHERS LTD 1983

The selection and presentation of material and the opinions expressed in this publication are the sole responsibility of the authors concerned.

All rights reserved. No part of this publication may be reproduced, stored in a retrieval system, or transmitted in any form or by any means, electronic, mechanical, photocopying, recording, or otherwise, without the prior written permission of the copyright owner, Applied Science Publishers Ltd, Ripple Road, Barking, Essex, England

Typeset in Great Britain by Keyset Composition, Colchester
Printed in Great Britain by Galliard (Printers) Ltd, Great Yarmouth

PREFACE

Many of the desirable flavour and textural attributes of dairy products are due to their lipid components; consequently, milk lipids have, traditionally, been highly valued, in fact to the exclusion of other milk components in many cases. Today, milk is a major source of dietary lipids in western diets and although consumption of milk fat in the form of butter has declined in some countries, this has been offset in many cases by increasing consumption of cheese and fermented liquid dairy products.

This text on milk lipids is the second in a series entitled *Developments in Dairy Chemistry*, the first being devoted to milk proteins. The series is produced as a co-ordinated treatise on dairy chemistry with the objective of providing an authoritative reference source for lecturers, researchers and advanced students. The biosynthesis, chemical, physical and nutritional properties of milk lipids have been reviewed in eight chapters by world experts. However, space does not permit consideration of the more product-related aspects of milk lipids which play major functional roles in several dairy products, especially cheese, dehydrated milks and butter.

Arising from the mechanism of fatty acid biosynthesis and export of fat globules from the secretory cells, the fat of ruminant milks is particularly complex, containing members of all the major lipid classes and as many as 400 distinct fatty acids. The composition and structure of the lipids of bovine milk are described in Chapter 1, with limited comparison with non-bovine milk fats. Since the fatty acid profile of milk fat, especially in monogastric animals, may be modified by diet and other environmental factors, the biosynthesis of milk lipids is reviewed in Chapter 2 with the objective of indicating means by which the fatty acid profile, and hence the functional properties of the lipids, might be modified. Lipids in foods are normally present as an emulsion, stabilized by a layer of protein adsorbed at the oil–water interface. The fat in milk and cream exists as an oil-in-water emulsion with a unique stabilizing lipoprotein membrane, referred to as the milk fat globule membrane (MFGM). The inner layers of the MFGM are formed within the secretory cell and are relatively stable; however, the outer layers, which are acquired as the fat globule is exported through the apical membrane of the secretory cells, are unstable. Damage to the MFGM leads to chemical and physical instability of the fat phase in milk and hence the structure of the membrane has been the subject of considerable research, the results of which are reviewed in Chapter 3.

Lipids strongly influence, for good or evil, the flavour and texture of foods, especially high-fat products such as butter. The influence of various colloidal features of milk fat on the properties of milk and cream is considered in Chapter 4, while the crystallization of milk fat and how this may be controlled, modified and measured are reviewed in Chapter 5. Unfortunately, lipids are subject to chemical and enzymatic alterations which can cause flavour defects referred to as oxidative and hydrolytic rancidity, respectively. The storage stability of high-fat foods, especially mildly flavoured foods like milk, cream and butter, is strongly influenced by these changes which have been reviewed in Chapters 6 and 7.

Dietary lipids play many diverse nutritional roles, some of which are essential. However, dietary lipids, especially saturated lipids of animal origin, have been the subject of much controversy in recent years, particularly in regard to their possible role in atherosclerosis. Various aspects of the nutritional significance of lipids are discussed in Chapter 8.

Finally, I wish to thank sincerely the 14 authors who have contributed to this text and whose co-operation has made my task as editor a pleasure.

P. F. Fox

CONTENTS

Preface v

List of Contributors ix

1. The Composition and Structure of Milk Lipids . . . 1
   WILLIAM W. CHRISTIE

2. Influence of Nutritional Factors on the Yield, Composition and Physical Properties of Milk Fat 37
   J. C. HAWKE and M. W. TAYLOR

3. Origin of Milk Fat Globules and the Nature of the Milk Fat Globule Membrane 83
   T. W. KEENAN, DANIEL P. DYLEWSKI, TERRY A. WOODFORD and ROSEMARY H. FORD

4. Physical Chemistry of Milk Fat Globules 119
   P. WALSTRA

5. Physical Properties and Modification of Milk Fat . . . 159
   B. K. MORTENSEN

6. Lipolytic Enzymes and Hydrolytic Rancidity in Milk and Milk Products 195
   H. C. DEETH and C. H. FITZ-GERALD

7. Lipid Oxidation 241
   T. Richardson and M. Korycka-Dahl

8. The Nutritional Significance of Lipids 365
   M. I. Gurr

Index 419

LIST OF CONTRIBUTORS

WILLIAM W. CHRISTIE
  Department of Lipid Biochemistry and Enzymology, Hannah Research Institute, Ayr KA6 5HL, UK

H. C. DEETH
  Otto Madsen Dairy Research Laboratory, Queensland Department of Primary Industries, Hamilton, Queensland 4007, Australia

DANIEL P. DYLEWSKI
  Department of Biochemistry and Nutrition, Virginia Polytechnic Institute and State University, Blacksburg, Virginia 24061, USA

C. H. FITZ-GERALD
  Otto Madsen Research Laboratory, Queensland Department of Primary Industries, Hamilton, Queensland 4007, Australia

ROSEMARY H. FORD
  Department of Biochemistry and Nutrition, Virginia Polytechnic Institute and State University, Blacksburg, Virginia 24061, USA

M. I. GURR
  Nutrition Department, National Institute for Research in Dairying, University of Reading, Shinfield, Reading RG2 9AT, UK

J. C. Hawke
Department of Chemistry, Biochemistry and Biophysics, Massey University, Palmerston North, New Zealand

T. W. Keenan
Department of Biochemistry and Nutrition, Virginia Polytechnic Institute and State University, Blacksburg, Virginia 24061, USA

M. Korycka-Dahl
Department of Food Science, University of Wisconsin, Madison, Wisconsin 53706, USA

B. K. Mortensen
Danish Government Research Institute for Dairy Industry, Roskildevej 56, DK-3400 Hillerød, Denmark

T. Richardson
Department of Food Science, University of Wisconsin, Madison, Wisconsin 53706, USA

M. W. Taylor
Department of Food Technology, Massey University, Palmerston North, New Zealand

P. Walstra
Department of Food Science, Agricultural University, Wageningen, The Netherlands

Terry A. Woodford
Department of Biochemistry and Nutrition, Virginia Polytechnic Institute and State University, Blacksburg, Virginia 24061, USA

Chapter 1

THE COMPOSITION AND STRUCTURE OF MILK LIPIDS

WILLIAM W. CHRISTIE

Hannah Research Institute, Ayr, Scotland, UK

SUMMARY

The lipids of milk provide energy and many essential nutrients for the newborn animal. They also have distinctive physical properties that affect the processing of dairy products. In this review, the compositions and structures of both the major and minor components in milk lipids of many species are compared, and the data are discussed in terms of the nutrient requirements of the neonatal animal. The content and composition of milks from different species vary widely; presumably, these are evolutionary adaptations to differing environments. On the other hand, the distributions of lipid classes in milks are very similar. The fatty acid compositions of milks are usually complex and distinctive, depending on the nature of the fatty acids synthesized de novo *in the mammary gland and those received from the diet in each species. Characteristically, they contain short- and medium-chain fatty acids not found in other tissues. Triacylglycerols, the main lipid class, of milks from different species do, however, exhibit many structural similarities.*

1. INTRODUCTION

Lipids in milk provide a major source of energy and essential structural components for the cell membranes of the newborn in all mammalian species. They also confer distinctive properties on dairy foods that affect processing. In consequence, the composition, structure and chemistry of milk lipids have probably been studied more intensively than those from any other natural source. Milk lipids from the cow have received most attention because of their commercial importance, but human milk lipids have also been investigated in great detail, and those from many other species have been sampled for comparative biochemical purposes. The content and composition of lipids from milks from different species varies with such factors as diet, stage of lactation, number of lactations, breed and season, and these data are not always recorded by authors. As far as is possible, this review is concerned with milk from animals on an 'average' diet for the species, and in mid-lactation; other factors are discussed elsewhere. Published results can sometimes be misleading because of errors in sampling, sample handling or analytical methodology and the author has tended to ignore data that appear doubtful.

The lipid compositions of milks in general have been reviewed on a number of occasions[1-5] and reviews dealing with lipids of ruminant[6-8] and human[9-12] milk, in particular, have also appeared. The review by Morrison[1] is especially comprehensive and covers the literature until mid-1968. Although this chapter is intended to be a complete review of the subject within the space available, there has been some concentration on more recent publications and the reader is referred to Morrison's[1] review for a historical perspective and for the earlier literature. Most authors have tended to stress the complexity of milk lipids, but this is an aspect that can be exaggerated. A very high proportion (about 98%) of the lipids consist of a single lipid class, the triacylglycerols, with all other components being present at relatively low concentrations, although they may have appreciable nutritional or technological importance. Although many more fatty acids have been characterized as present in milk fat than in any other tissue, this is in part because no other lipid material has been investigated with such intensity. On the other hand, fatty acids with a wider range of chain lengths tend to be found in esterified form in milk fat than in other tissues of the same species.

2. THE CONTENT AND PHYSICAL FORM OF FAT IN MILK

Milk fat is secreted in the form of globules surrounded initially by a membrane, the milk fat globule membrane, which maintains the integrity of the globules and renders them compatible with their aqueous environment[5,13] (Chapter 3). The fat globules themselves consist almost entirely of triacylglycerols, whereas the membranes contain most of the complex lipids. After secretion, parts of the membrane detach and are found, together with their phospholipid constituents, with fragments of other mammary epithelial cell membranes in the skim-milk phase. In bovine milk, the skim-milk phospholipids are found to be identical in composition and structure to those of the milk fat globule membrane, confirming their common origin.[14] The same is likely to be true for other species, assuming similar mechanisms for milk fat secretion. The fat globules in bovine milk can vary somewhat in their size distribution but average 2–4 μm in diameter,[15] and the same has been found to be true for the human and for the few other species that have been examined.[16]

The ratio of fat to water in milk from different species can vary markedly, and is highly dependent on the stage of lactation. Some typical average values are listed in Table 1. Data for many more species are available in the review by Jenness and Sloan.[2] The content of fat in bovine milk is dependent on the breed and is highest in those from the Channel Islands, but 37 g litre^{-1} is a reasonable average value. The milks of other ruminants have a comparable or slightly higher content of fat, and the same is true of those of the human, other primates, pig and guinea-pig. Higher values have been recorded for some rodents and rabbits (100–200 g litre^{-1}), but the highest values of all (300–520 g litre^{-1}) have been found for marine mammals, especially those from colder waters such as seals and whales. Presumably this is an evolutionary adaptation which ensures that the young of these species can rapidly build up their stores of fat for energy and insulation to protect against the harsh environment. It has been pointed out that differences in the calorific values of milks from various species are due almost entirely to differences in the fat content.[2] The young of each species can vary markedly in their dependence on milk as a source of nutrients, for example from the human which is totally dependent to the hare which appears to have a very limited requirement, but no correlation between stage of maturity of the young and milk fat content has been observed.[2]

3. THE LIPID CLASS COMPOSITION OF MILKS

3.1. The Main Lipid Classes

In the milks of all species studied to date, triacylglycerols are by far the major lipid class, accounting for 97–98% of the total lipids, in many species (Table 2). The triacylglycerols are almost invariably accompanied by small

TABLE 1
THE FAT CONTENT OF MILKS FROM VARIOUS SPECIES

Species	Fat content (g litre^{-1})	Reference	Species	Fat content (g litre^{-1})	Reference
Cow	33–47	8	Marmoset	77	26
Buffalo	47	2	Rabbit	183	2
Sheep	40–99	8	Cottontail rabbit	336–352	27
Goat	41–45	8	Snowshoe hare	71	28
Musk-ox	109	17	Muskrat	110	29
Dall-sheep	32–206	18	Mink	134	30
Moose	39–105	2, 19	Chinchilla	117	31
Blackbuck antelope	93	20	Red kangaroo	9–119	2, 32
Elephant	85–190	21, 22	Bottle-nosed dolphin	62–330	2, 33
Human	38	2	Manatee	55–215	34
Squirrel monkey	10–51	2, 23	Pygmy sperm whale	153	35
Monkey (six species)	22–85	24	Harp seal	502–532	2, 36
Lemur (three species)	8–33	25	Horse	19	2
Pig	68	2	Polar bear	314–331	2, 37
Rat	103	2	Yeso brown bear	153	38
Guinea-pig	39	2	Bear (four species)	108–331	39

amounts of di- and mono-acylglycerols, free cholesterol and cholesterol esters (commonly in the ratio 10:1), unesterified (free) fatty acids and phospholipids (see Section 3.2). In addition to these, a number of minor simple lipids (see Section 3.3) and glycolipids (see Section 3.4) have been found. Comparatively high proportions of partial glycerides and unesterified fatty acids have been recorded in some published papers, but this usually indicates that faulty handling of the milk has led to some lipolysis. For example, the data for rat milk recorded in Table 2 must be regarded with some suspicion. In an analysis[44] in which the lipids of bovine milk were isolated with particular care, it was demonstrated that the small

TABLE 2

COMPOSITION OF INDIVIDUAL SIMPLE LIPIDS AND TOTAL PHOSPHOLIPIDS IN MILKS FROM VARIOUS SPECIES

Lipid class	Amount (wt% of the total lipids)					
	Cow	Buffalo	Human	Pig	Rat	Mink
Triacylglycerols	97·5	98·6	98·2	96·8	87·5	81·3
Diacylglycerols	0·36		0·7	0·7	2·9	1·7
Monoacylglycerols	0·027		T[a]	0·1	0·4	T[a]
Cholesterol esters	T[a]	0·1	T[a]	0·06		T[a]
Cholesterol	0·31	0·3	0·25	0·6	1·6	T[a]
Free fatty acids	0·027	0·5	0·4	0·2	3·1	1·3
Phospholipids	0·6	0·5	0·26	1·6	0·7	15·3
Reference	7	40	41	42	43	30

[a] T = Trace amount.

proportion of diacylglycerols present were mainly of the sn-1,2-configuration and were therefore probably intermediates in the biosynthesis of triacylglycerols rather than degradation products. Although relatively few species have been examined to date, the compositional data are sufficiently similar to suggest comparable mechanisms of synthesis and secretion. Phospholipids have been reported to comprise 15·3% of the total lipids of mink milk, a figure 10-fold higher than expected, but in this instance it has been suggested that there may have been contamination of the milk with disintegrating mammary tissue.[30] No lipids other than triacylglycerols were detected by thin-layer chromatography (TLC) in the milk fat of the bottle-nosed dolphin, but it would be surprising if small amounts of phospholipids are not present.[33]

3.2. The Phospholipids

Phospholipids are a small but important fraction of the lipids of milk and are found mainly in the milk fat globule membrane and other membranous material in the skim-milk phase. Their composition and biosynthesis have been reviewed.[45] The composition of the phospholipids from the milks of a number of species is shown in Table 3. Phosphatidylcholine, phosphatidylethanolamine and sphingomyelin are the main components in each instance with phosphatidylserine, phosphatidylinositol and lysophospholipids also being present. Although phosphatidic acid has occasionally been tentatively identified as a component of some milks, it was not found in a careful analysis of cows' milk.[49] Cardiolipin could not be detected in the milk of the cow[50] or buffalo[51] but was found in small amounts in sheep and goat milks.[51] There are marked similarities in the relative proportions of each of the phospholipids among species, and Morrison[1,46] has pointed out that the total concentration of the phosphocholine-containing components is fairly constant (52–60%), presumably because these perform the same structural function in each species. Again, the phospholipids of mink milk appear to be an exception.[30] Alkyl- and alk-1-enyl-ether forms of bovine milk phosphatidylcholine and phosphatidylethanolamine have been detected; 4% of the phosphatidylethanolamine and 1·3% of the phosphatidylcholine are in the plasmalogen forms.[52]

3.3. Minor Simple Lipid Constituents

A number of lipid- and fat-soluble compounds have been found in milks as relatively minor constituents. Although present in small amounts, they include several components of potential physiological importance to the newborn, including hormones and vitamins, as well as substances that contribute to the flavour and organoleptic properties of milk and dairy products. In addition, various foreign lipid-soluble substances may be found in milk, such as plasticizers or halogenated biphenyls. For obvious practical reasons, only the milks of the cow and human have been investigated in detail for the presence of any of these compounds.

3.3.1. Glycerolipids

Some of the minor lipid classes that have been found in milks are simply forms of the more common ones, but with distinctive fatty acid or other alkyl constituents that affect the chromatographic properties and permit separation. For example, triacylglycerols containing acetic acid, separable by TLC from those with only longer-chain fatty acids, have been found in trace amounts in milks from the cow, sheep and goat.[53] Triacylglycerols

TABLE 3
THE COMPOSITION OF THE PHOSPHOLIPIDS IN MILK FROM VARIOUS SPECIES

Amount (mol% of the total lipid phosphorus)

Species	Phosphatidyl-choline	Phosphatidyl-ethanolamine	Phosphatidyl-serine	Phosphatidyl-inositol	Sphingomyelin	Lysophospho-lipids[a]	Reference
Cow	34·5	31·8	3·1	4·7	25·2	0·8	46
Sheep	29·2	36·0	3·1	3·4	28·3		46
Buffalo	27·8	29·6	3·9	4·2	32·1	2·4	46
Goat	25·7	33·2	6·9	5·6[b]	27·9	0·5	47
Camel	24·0	35·9	4·9	5·9	28·3	1·0	46
Ass	26·3	32·1	3·7	3·8	34·1		46
Pig	21·6	36·8	3·4	3·3	34·9		46
Human	27·9	25·9	5·8	4·2	31·1	5·1	46
Cat	25·8	22·0	2·7	7·8[b]	37·9	3·4	47
Rat	38·0	31·6	3·2	4·9	19·2	3·1	43
Guinea-pig	35·7	38·0	3·2	7·1[b]	11·0	2·0	47
Rabbit	32·6	30·0	5·2	5·8[b]	24·9	0·4	47
Mouse[c]	32·8	39·8	10·8	3·6	12·5		48
Mink	52·8	10·0	3·6	6·6	15·3	8·3	30

[a] Mainly lysophosphatidylcholine but also lysophosphatidylethanolamine.
[b] Also contains lysophosphatidylethanolamine.
[c] Analysis of milk fat globule membrane phospholipids.

containing hydroxy-fatty acids which represent 0·61% of the total lipids of bovine milk, were isolated as the pyruvic acid–dinitrophenylhydrazone derivatives for characterization of the fatty acid components.[54] Similarly, triacylglycerols containing a β-keto fatty acid have been isolated by chromatographic procedures from butter fat (0·045% of the total) and characterized.[55] Alkyldiacylglycerols have been found in the neutral lipids of milk of the human (0·1%), cow (0·01%) and sheep (0·02%), together with trace amounts of 2-O-methoxy-substituted analogues and the corresponding phospholipid forms.[56,57] The same compounds are present in milks of the goat and pig, and somewhat greater amounts tend to be present in colostrum than in mid-term milk.[58] Alk-1-enyldiacylglycerols or neutral plasmalogens comprise 0·015% of the total lipids of bovine milk fat;[59] the aldehydes derived from these contain a relatively high proportion (30%) of branched-chain components.[60]

3.3.2. Sterols and Steroidal Hormones

By far the major sterol component (at least 95%) of most milks is cholesterol, but small amounts of other sterols have been found. In bovine milk, β-sitosterol, lanosterol, dihydrolanosterol,[61] Δ^4-cholesten-3-one, $\Delta^{3,5}$-cholestadiene-7-one,[62] and 7-dehydrocholesterol[63] have been isolated and adequately characterized, while campesterol, stigmasterol[64,65] and Δ^5-avenasterol[65] are almost certainly also present. A similar range of sterols is present in buffalo milk.[66,67] Human milk contains phytosterols in addition to cholesterol but the relative concentrations are dependent on the nature of the diet and the stage of lactation.[68,69] A number of steroidal hormones, especially progesterone, oestrogens and corticosteroids, have been found in milk. Although they may have some physiological importance for the newborn, most research interest has centred on monitoring progesterone levels for the prediction of oestrus and subsequently for the early diagnosis of pregnancy in cattle. All of these aspects of the occurrence of hormones in milk have been reviewed recently.[70,71]

3.3.3 Hydrocarbons

Squalene and carotene have long been recognized as minor constituents of milk fat. The distribution of the latter between the fat globule and its membrane has been a matter of controversy, but the most recent work has suggested that carotene is located mainly in the fat droplet.[72] Trace amounts of C_{17}–C_{48} odd- and even-numbered, normal and branched-chain, and possibly 1-cyclohexyl series of hydrocarbons have been detected in bovine milk.[73] In addition, pristane and phytane have been identified,[74] as have phyt-1-ene, phyt-2-ene, neophytadiene and many other related compounds.[75]

3.3.4. Lipid-Soluble Vitamins

The lipid-soluble vitamins, i.e. mainly vitamins A, D and E, in milks are of great nutritional importance for the newborn, and the compositions of these in the milks of various species[76–78] are listed in Table 4. The nature of the biologically active form of vitamin D in human milk, in particular, has been a matter for debate. Recent evidence suggests that 25-hydroxyvitamin D_3 accounts for 75% of the activity with vitamins D_2 and D_3 comprising most of the remainder.[79] Vitamin D sulphate either is not

TABLE 4
THE COMPOSITION OF LIPID-SOLUBLE VITAMINS IN MILKS[76–78]

Species	Vitamin A ($\mu g\, litre^{-1}$)	Vitamin D ($IU\, litre^{-1}$)	Vitamin E ($mg\, litre^{-1}$)
Cow	410	25	1
Goat	700	23	<1
Human	750	50	3
Rat	1440	5	3
Rabbit	2080	—	—

present or possesses no biological activity.[79,80] Several forms of vitamin E exist in milk. In cows' milk, α-tocopherol is the main component and γ-tocopherol is the only other isomer detectable.[81] However, human milk contains appreciable amounts of β-, γ- and δ-tocopherol and γ-tocotrienol in addition to α-tocopherol.[82,83] Jensen et al.[10] have reviewed the vitamin E status of milk in the light of current attempts to increase the content of polyunsaturated fatty acids in the human diet.

3.3.5. Prostaglandins

Human milk contains prostaglandins E and F, at concentrations 100-fold greater than in adult plasma, and they appear to have a relatively long half-life.[84] In addition, biologically inactive metabolites of thromboxane A_2 and prostacyclin have been detected. The significance of these compounds for the infant is uncertain, but it is possible that they modulate some physiological function, e.g. gut motility. It is known that prostaglandin $F_{2\alpha}$ passes into the milk of cows and goats,[85,86] but it is doubtful whether it could retain its biological activity during storage or processing.

3.3.6. Carnitine and Acylcarnitines

Carnitine is an essential cofactor for the oxidation of fatty acids in animal tissues. It has long been known that it is secreted into the milk of ruminants but more recently it was recognized that a proportion is in an esterified form, i.e. as acetylcarnitine.[87] Human milk also contains appreciable amounts of carnitine[88,89] and some of this is apparently acylated by acetate and long-chain fatty acids.[89] In the human and rat, it has been established that a high proportion of the carnitine requirement of the newborn is met by milk as the enzymes required for synthesis have not developed,[90] but the nutritional significance of acyl-, as opposed to free-, carnitine is not known.

3.3.7. Flavour Compounds

Aliphatic compounds contribute greatly to the flavour and palatability of milk and dairy products and their nature and compositions have been comprehensively reviewed.[1,91,92] Very many different compounds are involved and a high proportion are derived, chemically or biochemically, from milk lipids, e.g. by hydrolysis of milk fat constituents. At the normal trace levels, they impart desirable flavours, but when the proportions are changed or specific components are increased in concentration they can give rise to off-flavours. Other off-flavours can arise through oxidative rancidity or biological degradation during improper storage. For example, it has recently been shown that oxidation of small amounts of (n-3) pentenoic fatty acids in butter fat can give rise to the vinyl ketones, oct-1-en-3-one and octa-1,cis-5-dien-3-one, that impart a metallic flavour.[93,94] The flavour components most studied are lactones and methyl ketones, but short-chain aldehydes and fatty acids are also important. They exist in the free form, or as esterified precursors, and have been detected in the form of conjugates, e.g. as glucuronide or sulphate derivatives.[95,96]

Milk fat contains relatively small amounts of γ- and δ-lactones in the free form. They are mainly C_6 to C_{18} normal saturated compounds, although a small proportion of monoenoic and branched-chain constituents may also be present.[1,91,92] The precursors are 4- and 5-hydroxy acids esterified to a primary position in triacylglycerols (the other positions are occupied by normal fatty acids); these hydroxy-fatty acids form lactones spontaneously when the triacylglycerols are hydrolysed. Similarly, milk fat contains trace amounts of β-ketoacids linked to triacylglycerols, which are readily hydrolysed and the fatty acids decarboxylated to form methyl ketones (alkan-2-ones) with one carbon atom fewer than the parent acid, i.e. saturated C_3 to C_{15} components. The biology of methyl ketones has been reviewed.[97]

3.4. Ceramide and Glycosphingolipids

Free ceramide, the *N*-acyl derivative of a long-chain base sphinogosine, has been isolated from bovine milk and characterized.[98] Like the phospholipids, glycosphingolipids derived from ceramide are located primarily in the milk fat globule membrane.[99,100] Mono- and di-hexosylceramides are the main glycolipid constituents of bovine milk,[101] and these have been shown to be glucosyl- and lactosyl-ceramides.[102,103] On the other hand, galactosylceramide (88%) is the major monohexosylceramide in human milk fat globule membrane, but is also accompanied by lactosylceramide[104] and more complex neutral glycosphingolipids.[105] In addition, a number of gangliosides, i.e. glycolipids containing sialic acid residues, have been isolated and characterized from the milk fat globule membrane of the cow[106,107] and mouse.[108]

4. THE FATTY ACID COMPOSITION OF MILK LIPIDS

4.1. The Main Fatty Acids of the Triacylglycerols and Total Lipids

A greater range of fatty acids has been isolated or identified as components of milk fats than from any other natural source. For example, a compilation of the fatty acids detected in bovine milk up to January 1974 listed 437 distinct constituents,[5] while by February 1980, 183 distinct fatty acids had been recognized in human milk fat.[11] Most of these components could only be detected by using combinations of the most advanced chromatographic and spectroscopic techniques available, and some have been detected only in particular minor lipid classes. In a single analysis of butterfat by highly efficient capillary gas chromatography 80 components were resolved,[109] while in a similar analysis on a packed column 53 components were detected.[110] Fortunately, only a relative few of these are present in appreciable concentrations or are of particular nutritional significance, and need be considered in making interspecies comparisons or nutritional judgements.

The fatty acids in milk are derived from two sources, namely the plasma lipids and synthesis *de novo* in the mammary gland. The former may come from the diet, but also include fatty acids released from body tissues, especially adipose tissue, by lipolysis. As a consequence, the fatty acid composition of the milk of non-ruminants is highly dependent on the fatty acid profile of the diet (c.f. the human, for example[10,11]) and is readily altered, i.e. an increase in the concentration of a particular fatty acid in the diet is likely to soon be reflected by an increase in the concentration of that acid in the milk fat. On the other hand, changes in the level of unsaturated fatty acids in the diet of ruminant animals under normal conditions have comparatively little effect on the composition of the milk fatty acids because extensive biohydrogenation occurs in the rumen. Dietary linoleic acid (9c,12c-18:2), for example, is in part hydrogenated by the rumen micro-organisms to stearic acid (18:0), but vaccenic acid (11t-18:1) and other monoenoic isomers and a conjugated dienoic acid (9c, 11t-18:2) are also found in smaller amounts. All of these components are absorbed from the intestines and are transported via the lymph and then the plasma to the tissues. Within the tissues of the ruminant (and non-ruminant), further modification can take place, e.g. chain elongation, α-oxidation, β-oxidation and especially desaturation, and some of the stearic acid is desaturated to oleic acid (9c-18:1). Therefore, when either stearic acid or linoleic acid is added to ruminant diets, increased levels of stearic and oleic acids are found in milk fat. However, when palmitic acid is used as a supplement this is essentially the only fatty acid to increase in concentration in the milk fat. Biohydrogenation can be circumvented by formulating the diet in a manner such that the fats are protected from the rumen micro-organisms. These and other aspects of manipulating the fatty acid composition of ruminant milk have been comprehensively reviewed by Christie.[111]

Those fatty acids synthesized *de novo* within the mammary gland are generally the short- to medium-chain-length constituents (up to and including some of the C_{16}), the proportions of each being determined by the properties of acylthiol ester hydrolases associated with the fatty acid synthase of each species; this and other facets of lipid biosynthesis in the mammary gland have been reviewed.[5,112,113] The mammary gland is often the only tissue within an animal in which short- and medium-chain-length fatty acids are found in esterified form. The fatty acid composition of the milk of a given species then represents a balance between the contributions of fatty acids of dietary origin and those newly synthesized; it is dependent not only on the nature of each contribution but also on its relative size.

In making comparisons of the fatty acid compositions of milks of different species, especially when the results are simply culled from the literature on the subject, it is important to recognize that they will vary to some extent as the diet is varied, and an average diet for the species cannot always be easily defined. Morrison[1] has made an extensive compilation of the fatty acid compositions of the milks of various species (to mid-1968); Glass and Jenness[114] added data for a further 34 species (their paper should be read in conjunction with an earlier one[115]); and further data have been contributed by other authors. Some representative compositions of the main fatty acids in the milks of a number of species, covering the main groups of mammals, are listed in Table 5. Only those fatty acids that are widely distributed are considered; the list is weighted in favour of the more recent literature, and, as far as is possible or can be judged, the diet for each species was reasonably typical. In some instances, the data are for the triacylglycerol fraction specifically while in others they are for the total lipids, but there is likely to be little difference between the two in the compositions of the main components.

The milk fats of ruminant animals are characterized by the presence of relatively high concentrations of short-chain fatty acids, especially butyric and hexanoic acids, which are rarely found in milks of non-ruminants. Indeed, the method of presenting the results (weight percent of the total fatty acids) tends to undervalue the proportions; if the results are expressed as molar percentages up to 14% of bovine milk fat is butyric acid or up to 40% of the molecules of triacylglycerols contain this component, assuming it is only one of the three moles of fatty acids on average in each mole of triacylglycerol (see Section 5.1). One of the principal biosynthetic precursors of mammary butyric acid, β-hydroxybutyric acid, is available to ruminants in relatively large amounts following uptake of ruminal butyric acid and metabolism in the rumen wall (reviewed by Noble[120]). Appreciable amounts of medium-chain-length fatty acids are also present in ruminant milk fats, but relatively low concentrations of polyunsaturated fatty acids, especially the essential fatty acid linoleic acid, are found because of biohydrogenation in the rumen and the diversion of the main maternal tissue stores of this acid to other functions.[111] Medium-chain-length fatty acids are found in the milks of many quite disparate groups of animals, and this is often indicative of a marked degree of specificity during fatty acid synthesis, usually expressed at the chain-termination step.[5,112,113] In rabbit mammary gland, only C_8 and C_{10} fatty acids are synthesized,[112] for example, and it is possible to surmise that mainly C_{10} and C_{12} fatty acids are synthesized in elephant mammary gland. On the other hand, rat mammary gland is capable of synthesizing the spectrum of fatty acids from C_8 to C_{18}[43] and this may be true of most rodents. Human milk fat contains relatively low concentrations of medium-chain-length fatty acids, but appreciable amounts are found in the milks of most other primates. As in the human,

TABLE 5
THE PRINCIPAL FATTY ACIDS IN MILK TRIACYLGLYCEROLS OR TOTAL LIPIDS FROM VARIOUS SPECIES

Species	4:0	6:0	8:0	10:0	12:0	14:0	16:0	16:1	18:0	18:1	18:2	18:3	C_{20}–C_{22}	Reference
Cow	3·3	1·6	1·3	3·0	3·1	9·5	26·3	2·3	14·6	29·8	2·4	0·8	T[a]	115
Buffalo	3·6	1·6	1·1	1·9	2·0	8·7	30·4	3·4	10·1	28·7	2·5	2·5	T[a]	115
Sheep	4·0	2·8	2·7	9·0	5·4	11·8	25·4	3·4	9·0	20·0	2·1	1·4		115
Goat	2·6	2·9	2·7	8·4	3·3	10·3	24·6	2·2	12·5	28·5	2·2			115
Musk-ox	T[a]	0·9	1·9	4·7	2·3	6·2	19·5	1·7	23·0	27·2	2·7	3·0	0·4	17
Dall-sheep	0·6	0·3	0·2	4·9	1·8	10·6	23·0	2·4	15·5	23·1	4·0	4·1	2·6	18
Moose	0·4	T[a]	8·4	5·5	0·6	2·0	28·4	4·3	4·5	21·2	20·2	3·7		115
Blackbuck antelope	6·7	6·0	2·7	6·5	3·5	11·5	39·3	5·7	5·5	19·2	3·3			20
Elephant	7·4	—	0·3	29·4	18·3	5·3	12·6	3·0	0·5	17·3	3·0			22
Human		T[a]	T[a]	1·3	3·1	5·1	20·2	5·7	5·9	46·4	13·0	0·7	T[a]	115
Squirrel monkey		0·4	4·3	7·9	5·7	4·6	20·0	2·4	3·3	29·3	20·6	1·4		23
Monkey (mean of six species)	0·4	0·6	5·9	11·0	4·4	2·8	21·4	6·7	4·9	26·0	14·5	1·3		24
Baboon		0·4	5·1	7·9	2·3	1·3	16·5	1·2	4·2	22·7	37·6	1·3		116
Lemur macaco			0·2	1·9	10·5	15·0	27·1	9·6	1·0	25·7	6·6	0·6		25
Horse		T[a]	1·8	5·1	6·2	5·7	23·8	7·8	2·3	20·9	14·9	0·5		115

Fatty acid (wt% of the total)

Pig			0.7	0.5	4.0	32.9	11.3	3.5	35.2	11.9	0.7	1.1	115
Rat		1.1	7.0	7.5	8.2	22.6	1.9	6.5	26.7	16.3	0.8	T^a	43
Guinea-pig	T^a				2.6	31.3	2.4	2.9	33.6	18.4	5.7	7.0	115
Marmoset			8.0	8.5	7.7	18.1	5.5	3.4	29.6	10.9	0.9		26
Rabbit	T^a	22.4	20.1	2.9	1.7	14.2	2.0	3.8	13.6	14.0	4.4	T^a	115
Cottontail rabbit		9.6	14.3	3.8	2.0	18.7	1.0	3.0	12.7	24.7	9.8	0.4	27
European hare	T^a	10.9	17.7	5.5	5.3	24.8	5.0	2.9	14.4	10.6	1.7	T^a	117
Snowshoe hare	0.1	3.3	11.1	6.8	5.9	15.5	1.9	13.9	17.8	4.5	9.5	3.0	28
Mink				0.5	3.3	26.1	5.2	10.9	36.1	14.9	1.5		30
Chinchilla				T^a	3.0	30.0			35.2	26.8	2.9		31
Red kangaroo				0.1	2.7	31.2	6.8	6.3	37.2	10.4	2.1	0.1	32
Bottle-nosed dolphin				0.3	3.2	21.1	13.3	3.3	23.1	1.2	0.2	17.3	33
Manatee		0.6	3.5	4.0	6.3	20.2	11.6	0.5	47.0	1.8	2.2	0.4	34
Pygmy sperm whale					3.6	27.6	9.1	7.4	46.6	0.6	0.6	4.5	35
Harp seal					5.3	13.6	17.4	4.9	21.5	1.2	0.9	31.2	36
Hooded seal					3.6	9.5	13.5	2.8	27.2	1.3	0.6	34.9	118
Northern elephant seal					2.6	14.2	5.7	3.6	41.6	1.9		29.3	119
Polar bear	T^a		T^a	0.5	3.9	18.5	16.8	13.9	30.1	1.2	0.4	11.3	37
Grizzly bear	T^a			0.1	2.7	16.4	3.2	20.4	30.2	5.6	2.3	9.5	37
Yeso brown bear	1.1	3.1	0.1	0.5	2.8	20.5	9.2	2.6	41.2	16.9	0.9		38

^{a} T = Trace amount.

the milk of the pig, monotremes, marsupials and insectivores contains mainly C_{16} and C_{18} fatty acids.

Unlike ruminants, most non-ruminants, especially the omnivores, obtain and absorb unchanged, appreciable amounts of polyunsaturated fatty acids from the diet, so generally the linoleic acid and to some extent linolenic acid (18:3(n-3)) contents of their milk fats are reasonably high. The latter is found in elevated concentrations in the milks of non-ruminant herbivores such as the horse and wild rabbit, and reflects the relatively high concentration of this fatty acid in leaf tissue. Similarly, marine animals subsist on a diet of fish and krill which tend to contain high proportions of C_{20} and C_{22} fatty acids and relatively low proportions of the C_{18} polyunsaturated fatty acids; the milks of seals and other species on such a diet therefore differ from those of other species in the content of these constituents. The high concentration of a 16:1 fatty acid in the milk of the pig is due to the activity of palmityl-CoA desaturase in the mammary gland of this species.[121] On the other hand, in marine mammals the presence of this fatty acid in large amounts is a consequence of its relatively high concentration in their diet. It should be recognized that the fatty acid designated 18:1 in Table 5 is not necessarily oleic acid itself; for ruminant animals, it includes a variety of *cis* and *trans* isomers other than *cis*-9, and in marine animals, and to a lesser extent other species, it includes a number of positional isomers (mainly of the *cis* configuration) derived from the diet.

The reason for the high concentration of short- and medium-chain-length fatty acids in the milk fat, but not in other tissues, of many species is not clear. In ruminants, the short-chain fatty acids certainly help to maintain a degree of liquidity in the relatively saturated milk fat at body temperature that may be important for efficient secretion, but this cannot be a reason for the occurrence of medium-chain fatty acids in, for example, the rabbit. Short- and medium-chain-length fatty acids are absorbed directly via the portal blood stream rather than through the lymphatic system,[120] so they can make a more rapid and direct contribution to the energy metabolism (especially to that of the brain) of the newborn. In addition, there is evidence that in some species the capacity to oxidize long-chain fatty acids has not developed at birth.[122] It is possible that this has provided the evolutionary 'push' in such species for the biosynthesis of medium-chain-length fatty acids. Certainly, short- and medium-chain-length fatty acids are entirely catabolized by the newborn and are not laid down in the storage lipids or utilized for membrane lipid synthesis.

At parturition, neonatal ruminants are by some biochemical criteria verging on a state of essential fatty acid deficiency (reviewed by Noble[123]).

In spite of the relatively low content of linoleic acid in ruminant milk, the newborn is able to build up its reserves of this compound and related metabolites very rapidly, partly by means of efficient assimilation and partly because their requirement appears to be lower than that of non-ruminants. On the other hand, there has been some concern that bovine milk may not contain sufficient linoleic acid to meet the needs of the human infant when formulae based on this are used; most manufacturers now supplement commercial infant formulae with vegetable oils to remedy the potential deficiency. This seems a wise precaution, but it is clear from recent reviews of the subject[10,124] that bovine milk with a lower than average linoleic acid concentration would have to be fed to human infants as the sole diet for prolonged periods before clinical signs of essential fatty acid deficiency would be manifest.

Kabara[125] has speculated that some milk fatty acids and their derivatives might function as antimicrobial agents in the intestines of young animals.

4.2. Minor Fatty Acids of Milk Lipids

As described briefly in Section 4.1, an exceedingly large number of minor components have been identified as being present in milk fat. The structure and nature of many of these are of mainly academic interest and can reveal some interesting aspects of the biosynthesis of fatty acids. However, some are of real nutritional significance. It is important to recognize, for example, that part of the component designated '18:2' (or linoleic acid) on analysis by gas chromatography may consist of isomers other than *cis*-9, *cis*-12 and so may lack biological potency as an essential fatty acid. In bovine milk, as little as half the octadecadienoic acid is the essential isomer.[126] Also, a proportion of the monoenoic fatty acids can consist of *trans* isomers—which may have some unwanted biological effects—the nutritional value of which has been the subject of some debate in recent years.[127]

Those fatty acids found in bovine milk by 1974 and listed in a comprehensive compilation[5] include all the odd- and even-numbered normal saturated fatty acids from C_2 to C_{28}, monomethyl-branched fatty acids from C_{11} to C_{28} (including numerous positional isomers), multimethyl-branched fatty acids from C_{16} to C_{26} (115 configurational and positional isomers in total), a number of di- and poly-enoic fatty acids, keto- and hydroxy-fatty acids (many positional isomers, and saturated and unsaturated acyl chains), and cyclohexyl fatty acids. For example, 8–20% of the C_{18} monoene fraction can contain double bonds of the *trans* configuration, and several positional isomers of the *cis* and *trans* forms exist as shown in

Table 6.[128] Most of the *cis* form is the common isomer with the double bond in position 9, and is formed by desaturation of stearic acid within the tissues. The most abundant *trans* isomer has the double bond in position 11 and is a byproduct or intermediate in the biohydrogenation of linoleic acid in the rumen; the origin of the other *trans* isomers is obscure but they are probably also byproducts of biohydrogenation. The *trans*-16 isomer has received some additional attention as it is readily separated from the others by gas chromatography.[129] Since 1974, the branched-chain fatty acids in

TABLE 6
POSITIONAL AND GEOMETRIC ISOMERS OF BOVINE MILK OCTADECENOIC ACID[128]

Position of double bond	Isomers (wt% of the total)	
	cis Isomers	trans Isomers
6		1·0
7		0·8
8	1·7	3·2
9	95·8	10·2
10	T[a]	10·5
11	2·5	35·7
12		4·1
13		10·5
14		9·0
15		6·8
16		7·5

[a] T = Trace amount.

milk fat from goats and cows have been further studied,[130,131] and 9-*cis*,11-*trans*-octadecadienoic acid has been shown to be the main conjugated diene in milk fat (although traces of other isomers are probably present).[132] Non-esterified 3-hydroxy fatty acids (eight homologues from C_8 to C_{16}) have been found in cows' milk.[133] Although their origin is not known, it has been proposed that they are side-products of fatty acid synthesis by the malonyl-CoA pathway rather than degradation products.

Fewer fatty acids have been isolated from human than from bovine milk but if the minor lipid classes in the former were investigated in greater detail many more components would undoubtedly be added to those already found. The list is still extensive and includes 17 normal saturated components from C_4 to C_{23}, 54 branched-chain, 62 monoenoic and 33 polyunsaturated fatty acids.[134-40] The content of *trans* fatty acids in human milk is known to vary markedly with the diet and can range from 0 to 10%.[141] While other milk fats probably contain a similar range of components, few have been examined in detail, but sows' milk contains 37 distinct fatty acids,[142] and the milks of the marine mammals listed in Table 5 are at least as complex as the fish oils that comprise a large part of their diet.

4.3. The Fatty Acid Compositions of the Cholesterol Esters and Glycerophosphatides

Fatty acid compositions of the cholesterol ester and the main glycerophosphatide components of milks from some representative species are listed in Table 7. The cholesterol ester fraction of milk lipids is very small but may be important in the biochemistry of the mammary gland. Unfortunately, few data are available but those for the cow, goat,[143] human[145] and mink,[30] resemble the compositions of the triacylglycerols, except that in the case of ruminants, they do not contain significant amounts of the shorter-chain fatty acids. Also, in ruminants especially, the milk cholesterol esters differ markedly from those of all other tissues and in plasma, for example, the cholesterol esters are relatively rich in polyunsaturated fatty acids.[8] On the other hand, in bovine milk the cholesterol esters of the skim-milk fraction, as opposed to whole milk, that may have originated in the membranous material are very different in composition and contain more than 70% linoleic acid.[149]

Phosphatidylcholine and phosphatidylethanolamine are the main glycerophosphatides of milks where, as described previously, they are located in the membranes, and, as in most tissues, the phosphatidylethanolamines contain somewhat higher concentrations of polyunsaturated fatty acids. In ruminant milks, the concentration of polyunsaturated constituents in these lipid classes is still very low, and indeed is less than in the same lipids from all other tissues of the animal.[8] It appears that the biochemical strategy of the mother is to conserve these essential nutrients and export the minimum that will meet the requirements of the newborn. The same may well be true of non-ruminants, but the effect is not as obvious as they have much larger tissue stores of these components. No short- or medium-chain fatty acids ($<C_{14}$ essentially) have been found in milk phospholipids. In addition to those tabulated, data are available for the compositions of the phosphatidylserine and phosphatidylinositol fractions of the species listed, and further analyses have been recorded for other species.[1,146,147,150]

TABLE 7
THE FATTY ACID COMPOSITIONS OF THE CHOLESTEROL ESTERS, PHOSPHATIDYLCHOLINES AND PHOSPHATIDYLETHANOLAMINES IN THE MILKS (OR MILK FAT GLOBULE MEMBRANES) OF SEVERAL SPECIES

Fatty acid composition (wt% of the total)

Fatty acid	Cow CE[a]	Cow PC[a]	Cow PE[a]	Human CE	Human PC	Human PE	Pig PC	Pig PE	Mink CE	Mink PC	Mink PE	Mouse PC	Mouse PE
12:0	3·4				4·5	1·1	1·8	0·4	0·3				4·5
14:0	11·5	8·4	1·5	3·2	33·7	8·5	39·9	12·4	1·1	1·3	0·8		8·9
16:0	27·6	36·4	11·7	4·8	1·7	2·4	6·3	7·3	25·4	26·4	20·6	20·3	2·7
16:1	6·0	0·6	2·1	23·8	23·1	29·1	10·3	12·3	4·4	1·1	1·2		18·0
18:0	13·6	11·1	10·5	1·5	14·0	15·8	21·8	36·2	14·7	20·8	29·3	30·0	19·8
18:1	28·0	25·7	46·7	8·0	15·6	17·7	15·9	17·8	35·7	31·7	27·8	13·9	17·2
18:2	0·6	5·3	12·4	45·7	1·3	4·1	1·5	1·9	13·5	17·4	19·1	22·8	
18:3		1·1	3·4	12·4	2·1	3·4	0·3	0·7	2·6	2·2	0·5		
20:3		1·0	1·4	T[a]	3·3	12·5	1·3	6·6				8·9	20·0
20:4		0·7	0·9	T[a]	0·4	2·6	0·2	1·6				1·8	6·3
22:6													
Reference	143	144	144	145	146	146	147	147	30	30	30	148	148

[a] Abbreviations: CE = cholesterol esters; PC = phosphatidylcholine; PE = phosphatidylethanolamine; T = trace amount.

4.4. The Fatty Acid and Long-Chain Base Components of the Sphingolipids of Milk

Although glycosphingolipids, chiefly sphingomyelin and glycosphingolipids, are minor components of milk, they have important biochemical roles in the membranes of the mammary gland and in the milk fat globule membrane. They mainly contain very-long-chain-length fatty acids, and the compositions of the principal constituents of bovine and human milk are listed in Table 8.

The simplest natural sphingolipid component is ceramide, which in the cow contains mainly saturated C_{22}, C_{23} and C_{24} fatty acids,[98] in proportions similar to those of the sphingomyelin and ceramide dihexoside fractions,[151] indicating a probable biosynthetic connection. The ceramide monohexoside and ganglioside fractions contain a similar range of fatty acids to the others and are not very different in composition.[107,151] Other authors have obtained comparable results for these lipid classes.[98,100,102,106,152,153] In

TABLE 8
THE COMPOSITION OF THE NON-HYDROXY LONG-CHAIN FATTY ACIDS IN THE SPHINGOLIPIDS OF BOVINE AND HUMAN MILK

	Fatty acid composition (wt% of the total)							
	Cow					Human		
Fatty acid	Ceramide	Sphingo-myelin	CMH[a]	CDH[a]	Ganglio-sides	Sphingo-myelin	CMH	CDH
14:0	0·6	0·4	1·0	0·3	5·7	2·5		
16:0	7·2	7·8	9·3	7·7	25·6	22·5	13·6	16·2
16:1	0·3		1·4			1·0	1·3	0·9
18:0	1·1	1·6	13·7	3·3	10·9	8·1	6·9	8·7
18:1	0·3	0·2	12·2	1·3	10·3	6·2	5·2	7·8
18:2	0·2	0·2	2·0	0·2		0·5		
20:0	0·3	0·6	0·9	1·1	0·9	0·5	3·8	2·8
22:0	17·9	20·7	17·0	24·9	16·7	7·5	13·3	12·5
23:0	38·1	30·4	22·0	29·5	16·0	27·2	3·9	3·4
23:1		5·0	3·4	6·6		1·2	0·4	1·1
24:0	29·5	22·8	9·9	16·5	12·2	17·0	31·9	20·1
24:1	0·2	4·0	2·1	3·7	1·6	2·0	16·8	20·1
25:0	1·8	1·6		0·7			0·3	1·7
25:1		1·6		1·4			0·8	2·6
Reference	98	151	151	151	107	146	104	104

[a] Abbreviations: CMH = ceramide monohexoside; CDH = ceramide dihexoside.

addition, the sphingomyelin, ceramide monohexoside and ceramide dihexoside fractions contain small amounts of 2-hydroxy fatty acids.[151] The compositions tabulated are very different from those of other bovine tissues,[8] and in brain gangliosides, for example, the main component is stearic acid. Although the compositions differ slightly, the same range of fatty acids is present in the sphingolipid classes of human milk; only the ceramide monohexoside contained 2-hydroxy fatty acids.[104] Data for the sphingomyelin composition of milk from many more species are available.[30,42,146-8,150] The glucosyl- and lactosyl-ceramides from the fat globule membrane and skim-milk phase of bovine milk have been isolated and found to contain different relative proportions of the C_{18} and very-long-chain fatty acids.[100]

The sphingolipids of bovine milk contain a complex range of long-chain bases including normal, iso- and anteiso-saturated dihydroxy and trihydroxy compounds, and there are few differences among lipid classes.[151,154,155] Human milk sphingomyelin and monoglycosyl ceramide have relatively simple long-chain base compositions, in which more than 60% is a C_{18}-sphingosine.[156] Similar fatty acid and long-chain base components to these are deposited in brain tissue during myelination, so it is possible that they have some nutritional importance for the newborn.

5. THE STRUCTURES OF MILK LIPIDS

5.1. The Triacylglycerols

Triacylglycerols are synthesized in the mammary gland by enzymic mechanisms that exert some selectivity in the esterification of different fatty acids at each position of the L-(or sn-)glycerol moiety and in various molecular species.[5,112,113] Distinctive structures result that affect the physical properties and digestibility of milk fats. Stereospecific analysis procedures have been devised that have permitted the determination of the positional distributions of fatty acids in the triacylglycerols, and results for animal fats, including milk fat, have been reviewed.[157,158] Although severe technical problems are encountered with milk triacylglycerols because of the presence of short-chain fatty acids, data for a number of species have been obtained in recent years and some representative results are listed in Table 9. Full stereospecific analyses have been performed for several other species,[162] and data are also available for position sn-2 alone.[163,164]

As the overall fatty acid compositions of the triacylglycerols are very

TABLE 9
THE COMPOSITION OF FATTY ACIDS ESTERIFIED TO EACH POSITION OF THE TRIACYL-sn-GLYCEROLS IN THE MILKS OF VARIOUS SPECIES

Fatty acid composition (mol% of the total)

Fatty acid	Cow[159] sn-1	Cow[159] sn-2	Cow[159] sn-3	Human[160] sn-1	Human[160] sn-2	Human[160] sn-3	Rat[a] sn-1	Rat[a] sn-2	Rat[a] sn-3	Pig[161] sn-1	Pig[161] sn-2	Pig[161] sn-3	Guinea-pig[162] sn-1	Guinea-pig[162] sn-2	Guinea-pig[162] sn-3	Seal[162] sn-1	Seal[162] sn-2	Seal[162] sn-3	Echidna[162] sn-1	Echidna[162] sn-2	Echidna[162] sn-3
4:0	—	—	35·4																		
6:0	—	—	0·9																		
8:0	1·4	0·7	3·6	0·2	0·2	1·1															
10:0	1·9	3·0	6·2	1·3	2·1	5·6	7·4	4·0	16·8												
12:0	4·9	6·2	0·6	3·2	7·3	6·9	16·9	17·5	38·5												
14:0	9·7	17·5	6·4	16·1	58·2	5·5	10·9	13·1	14·7	2·4	6·8	3·7	—	0·1	—	0·2	0·3	0·2			
16:0	34·0	32·3	5·4	3·6	4·7	7·6	8·3	15·2	7·0	21·8	57·6	15·4	2·7	4·6	0·2	7·3	23·6	3·8	1·7	0·9	0·4
16:1	2·8	3·6	1·4	15·0	3·3	1·8	19·6	27·4	6·6	6·6	11·2	10·4	24·1	68·1	0·4	13·1	31·0	1·0	31·5	9·0	27·9
18:0	10·3	9·5	1·2	46·1	12·7	50·4	3·0	3·3	0·9	6·9	1·1	5·5	4·1	3·7	1·3	10·2	16·8	14·1	7·1	7·0	8·0
18:1	30·0	18·9	23·1	11·0	7·3	15·0	5·8	2·4	1·4	49·6	13·9	51·7	6·9	0·6	1·0	4·5	0·7	1·0	16·8	2·1	14·3
18:2	1·7	3·5	2·3	0·4	0·6	1·7	13·5	6·2	6·9	11·3	8·4	11·5	40·8	7·3	53·3	53·8	19·4	45·4	33·1	57·6	39·8
18:3							14·5	11·0	8·0	1·4	1·0	1·8	15·6	11·6	33·9	1·3	2·3	2·8	4·1	18·3	4·9
C_{20}–C_{22}													3·4	2·1	9·1	0·3	0·5	0·7	1·0	2·9	2·0
																5·6	0·8	28·7			

[a] Mozes, N. and Smith, S. (1982) Unpublished results.

different for each species, similarities between them are not immediately apparent, but a close perusal can reveal certain common features, especially for the longer-chain fatty acids. For example, for nearly all species a high proportion of the palmitic acid is concentrated in position *sn*-2, the only known exception being the primitive monotreme the echidna. Such a distribution is not common in animal triacylglycerols and in depot fats, the most abundant source of triacylglycerols, unsaturated fatty acids tend to be concentrated in position *sn*-2 and the saturated components are found in positions *sn*-1 and *sn*-3.[158] The main exception is the pig where a high proportion of the fatty acids in position *sn*-2 of the triacylglycerols from many tissues is palmitic acid,[161] due to the presence of an unidentified factor in the cell that modifies the specificities of the acyltransferases involved in triacylglycerol biosynthesis.[165] In addition, metabolically active triacylglycerols such as those of lymph and plasma in some species can have this structure.[166,167] Only a limited amount of information is available about the biosynthetic pathways involved in mammary tissue, but for the cow and rat, at least, the acyltransferases for positions *sn*-1 and *sn*-2 appear to have the intrinsic capacity to confer the required specificity.[5,112,113]

In the milk triacylglycerols of most of the species examined, myristic acid (like palmitic acid) is found in the greatest concentration in position *sn*-2, but stearic acid is concentrated in the primary positions, especially position *sn*-1. Unsaturated fatty acids are also concentrated in the primary positions. Of the species studied, only the seal received appreciable amounts of the very long-chain (C_{20}–C_{22}) fatty acids in the diet and these are most abundant in position *sn*-3,[162] as is also the case with depot fats.[158]

The other distinctive feature of milk fats is the unique distribution of the short-chain fatty acids, which in ruminants are concentrated in position *sn*-3. This was first suggested by Pitas *et al.*[168] for bovine milk, and was confirmed for this species and for other ruminants by means of improved analytical procedures by Breckenridge and co-workers[169-72] and subsequently by other workers.[159,173-6] Indeed, the asymmetry of the triacylglycerols is such that they exhibit a small but significant optical rotation[177] (asymmetry has also been demonstrated by nuclear magnetic resonance techniques[178,179]). It is now evident that butyric and hexanoic acids are esterified virtually entirely to position *sn*-3, octanoic acid is mainly found in position *sn*-3, although a proportion is in the other positions, and as the chain length is further increased, higher proportions tend to be found in position *sn*-2 and then in position *sn*-1. Much remains to be learned about the biochemical mechanisms that introduce these distributions, but it is already apparent that only the acyltransferase responsible for esterifying position sn-3 is capable of utilizing butyric acid.[180]

The monotreme the echidna is exceptional in that unsaturated fatty acids are concentrated in position sn-2 of the milk triacylglycerols, while the compositions of positions sn-1 and sn-3 are almost identical.[162,163] Such a structure is more reminiscent of a vegetable oil than of an animal fat,[158] and may indicate some retardation of biochemical differentiation in the mammary gland of this species.

Although definitive reasons for the distinctive structures of milk fats are awaited, it is possible to speculate on possible advantages to the newborn. During digestion of fats by pancreatic lipase the main products are free fatty acids derived from positions sn-1 and sn-3, and 2-monoacylglycerols. Short-chain fatty acids, such as butyric and hexanoic, which do not form insoluble calcium salts, are therefore released relatively rapidly from milk fats, and can be absorbed immediately via the portal blood system. Also, it has been demonstrated that triacylglycerols containing palmitic acid in position sn-2 are digested more efficiently by both rats and human infants than are similar compounds in which unsaturated fatty acids are in position sn-2 and palmitic acid is in position sn-1 or sn-3.[181,182] The phenomenon is possibly related to the relative ease of incorporation of 2-monopalmitin as opposed to free palmitic acid into micelles,[183] or to the relative insolubility of the calcium soaps of saturated compared to unsaturated fatty acids.[184]

A complete structural analysis of a natural triacylglycerol sample requires that it be fractionated into molecular species containing single specific fatty acid residues in each position. The problem can be extremely complicated as, for example, triacylglycerols with only five different fatty acid constituents may consist of 75 different molecular species (not including enantiomers). As bovine milk contains several hundred different fatty acids, an astronomical number of combinations is possible in the triacylglycerols and it is only feasible technically to undertake some limited fractionation into simpler molecular groups by means of chromatographic procedures. Three techniques are used most often for the purpose, i.e. silver nitrate chromatography which separates fractions according to the total number of double bonds in the molecule, high-temperature gas chromatography which separates fractions according to the sum of the molecular weights of the fatty acid constituents, and reversed-phase liquid chromatography which fractionates according to both molecular weight and degree of unsaturation.[185] It is also possible to separate the triacylglycerols of ruminant milks into high and low molecular weight fractions by means of adsorption chromatography, molecular distillation and low-

TABLE 10
COMPARISON OF OBSERVED AND PREDICTED (ASSUMING A 1-RANDOM, 2-RANDOM, 3-RANDOM ARRANGEMENT) DISTRIBUTIONS OF TRIACYLGLYCEROL ISOMERS FROM PIG MILK[161]

Triacylglycerol type[a]	Isomer distribution (mol% of the total) Observed	Predicted	Triacylglycerol type	Isomer distribution (mol% of the total) Observed	Predicted
SSS	4·7	5·0	SDS	0·5	0·6
SSM	11·3	12·6	DSS	2·1	0·8
SMS	2·2	1·9	SMD + SDM	1·9	2·5
MSS	9·1	9·1	MSD + MDS	5·8	5·4
SMM	5·1	4·8	DSM + DMS	4·2	5·3
MSM	21·8	22·9	M_2D	5·7	6·3
MMS	2·9	3·5	SD_2	1·4	1·4
MMM	9·4	8·8	Remainder	9·1	5·8
SSD	2·8	2·3			

[a] Abbreviations: S, M and D refer to saturated, monoenoic and dienoic acids esterified to positions sn-1 to sn-3 of triacylglycerols in the order shown.

temperature crystallization; while these procedures have practical value in some circumstances, they are of limited analytical worth. As mentioned earlier, molecular species containing acetic acid are completely separable from other triacylglycerols by adsorption chromatography.[53]

Milk triacylglycerols with relatively few fatty acid constituents, such as those of the human[160] or pig,[161] have been fractionated thoroughly by means of combinations of chromatographic techniques coupled with analyses of the positional distributions of the fatty acids in particular fractions. The detailed results of such analyses are only likely to be of interest to the specialist, but it is noteworthy that in the pig[161] the proportions of the various molecular species calculated from the positional distributions of the fatty acids—assuming acylation during biosynthesis with non-correlative specificity, i.e. a 1-random, 2-random, 3-random arrangement—agreed very well with the results obtained experimentally, as shown in Table 10. 'Reasonable to excellent' agreement between experimental data and calculated results have been obtained more recently in comparable experiments for milk triacylglycerols of several further species.[162] In addition, molecular weight distributions have been obtained for milk triacylglycerols of additional species.[186,187]

It is technically much more difficult to separate ruminant milk fats into molecular species, but high-temperature gas chromatography has been used to separate and quantify molecular species in which the combined chain lengths of the fatty acids varied from 26 to 55.[186] By combining liquid–liquid chromatography with analyses of positional distributions of fatty acids, Nutter and Privett[188] were able to identify and quantify 168 molecular species of shorter-chain triacylglycerols from bovine milk. Others have used similar methods in addition to silver nitrate chromatography and, more recently, capillary gas–liquid chromatography to fractionate ruminant milk fats.[189–200] Such studies afford useful fundamental information about the mechanisms of biosynthesis of triacylglycerols, but also have some practical value as they can often be correlated with physical properties such as melting or softening points of butterfats.[201–10]

5.2. The Phospholipids

The specific distributions of fatty acids in positions sn-1 and sn-2 of the glycerophosphatides from milks of a number of species have been determined by means of phospholipase A hydrolysis and some representative results are listed in Table 11. In bovine phosphatidylcholine and phos-

TABLE 11

DISTRIBUTIONS OF FATTY ACIDS IN POSITIONS sn-1 AND sn-2 OF THE PHOSPHATIDYLCHOLINE AND PHOSPHATIDYLETHANOLAMINE FROM MILKS OF THE COW[144] AND HUMAN[146]

| | \multicolumn{4}{c}{Cow} | \multicolumn{4}{c}{Human} |
| | Phosphatidyl-choline | | Phosphatidyl-ethanolamine | | Phosphatidyl-choline | | Phosphatidyl-ethanolamine | |
Fatty acid	sn-1	sn-2	sn-1	sn-2	sn-1	sn-2	sn-1	sn-2
14:0	5·6	10·8	1·9	1·3	3·4	4·9	1·0	1·0
16:0	41·9	30·6	19·7	4·7	34·2	32·3	9·3	8·2
16:1	0·6	1·2	1·2	2·2	1·5	2·2	1·8	3·3
18:0	17·5	2·4	19·0	1·3	43·9	2·1	65·4	1·3
18:1	20·3	27·8	45·8	47·8	14·3	13·7	18·1	15·3
18:2	2·7	9·2	2·9	21·4	2·7	30·9	4·4	30·2
18:3	0·8	1·8	1·1	4·5	—	2·0	—	5·1
20:3	—	1·6	0·2	2·2	—	3·9	—	5·4
20:4	0·2	1·2	0·2	3·0	—	6·6	—	20·9
22:6					—	0·8	—	5·2

phatidylethanolamine,[144,211] for example, saturated fatty acids are found in somewhat greater concentrations in position *sn*-1 (except for myristic acid) and unsaturated (especially those with two or more double bonds) in position *sn*-2, although the asymmetry is not nearly as pronounced as that found in many other tissues.[212,213] Although the fatty acid compositions are somewhat different, a similar general pattern has been found in the corresponding lipids of human milk.[146] Data for other species and other glycerophosphatides have also been obtained, and few differ appreciably from those listed.[146,147,214] A comprehensive fractionation of bovine milk phosphatidylcholine into molecular species has been achieved,[215-17] and the molecular structure of ceramides derived from the sphingomyelin of human milk fat globule membrane has also been determined.[156]

6. PIGEON 'MILK'

Columbiform birds, such as the pigeon, feed their young a white secretion formed in the crop of both the male and female parents. This so-called 'pigeon milk' or 'cropmilk' consists of fat-loaded epithelial cells desquamated from the proliferating lobes of the crop, and secretion is controlled by prolactin. The organ responsible has no functional or morphological relationship to the mammary gland, and the secretory process is very different from that of mammalian milks. Nonetheless, the function of cropmilk is the same, and some similarities are evident in their compositions,[218-20] especially of the lipids. The fat content of cropmilk is 8·1% and most of this is triacylglycerols (81%) with phospholipids making up much of the remainder.[219] The fatty acid compositions of the lipid classes have been determined;[220] only longer-chain (C_{14}–C_{22}) fatty acids were found and oleic acid was the most abundant component. There is also an abundant supply of essential fatty acids.

7. FUTURE RESEARCH TRENDS

The essential features of milk fat composition, especially the fatty acid composition, of a wide range of mammalian species have been obtained, sufficient indeed to give a reasonably clear picture of interspecies differences. The lipids of bovine and human milks, in particular, have been investigated in such detail that it is not at all easy to foresee the direction of the future research effort. However, it seems likely that the minor components of potential physiological importance to the newborn, such as hormones and prostaglandins, will be a focus for further attention. With other species relatively little is known of the composition of the various lipid classes or of their structures, and further work in this area would have both biochemical and nutritional relevance.

REFERENCES

1. MORRISON, W. R., *Topics in Lipid Chem.*, 1970, **1**, 51.
2. JENNESS, R. and SLOAN, R. E., *Dairy Sci. Abstr.*, 1970, **32**, 599.
3. KUZDZAL-SAVOIE, S., *Annls. Nutr. Aliment.*, 1971, **25**, A225.
4. JENNESS, R., In: *Lactation*, Vol. 3, B. L. Larsson and V. R. Smith (eds), 1974, Academic Press, New York, p. 3.
5. PATTON, S. and JENSEN, R. G., *Prog. Chem. Fats*, 1975, **14**, 163.
6. JENSEN, R. G., *J. Am. Oil Chem. Soc.*, 1973, **50**, 186.
7. KURTZ, F. E., In: *Fundamentals of Dairy Chemistry*, 2nd edn., B. H. Webb, A. H. Johnson and J. A. Alford (eds), 1974, Avi Publishing Co., Westport, p. 125.
8. CHRISTIE, W. W., *Progress in Lipid Research*, 1979, **17**, 111.
9. BRACCO, U., In: *Dietary Lipids and Postnatal Development*, C. Galli, G. Jacini and A. Pecile (eds), 1973, Raven Press, New York, p. 23.
10. JENSEN, R. G., HAGERTY, M. M. and MCMAHON, K. E., *Am. J. Clin. Nutr.*, 1978, **31**, 990.
11. JENSEN, R. G., CLARK, R. M. and FERRIS, A. M., *Lipids*, 1980, **15**, 345.
12. BLANC, B., *World Review of Nutrition and Dietetics*, 1981, **36**, 1.
13. PATTON, S. and KEENAN, T. W., *Biochim. Biophys. Acta*, 1975, **415**, 273.
14. PATTON, S. and KEENAN, T. W., *Lipids*, 1971, **6**, 58.
15. MULDER, H. and WALSTRA, P. (eds), *The Milk Fat Globule*, 1974, Centre for Agricultural Publishing and Documentation (Pudoc), Wageningen.
16. RUEGG, M. and BLANC, B., *Biochim. Biophys. Acta*, 1981, **666**, 7.
17. BAKER, B. E., COOK, H. W. and TEAL, J. J., *Canad. J. Zool.*, 1970, **48**, 1345.
18. COOK, H. W., PEARSON, A. M., SIMMONS, N. M. and BAKER, B. E., *Canad. J. Zool.*, 1970, **48**, 629.
19. COOK, H. W., RAUSCH, R. A. and BAKER, B. E., *Canad. J. Zool.*, 1970, **48**, 213.
20. DILL, C. W., TYBOR, P. T., MCGILL, R. and RAMSEY, C. W., *Canad. J. Zool.*, 1972, **50**, 1127.
21. MCCULLAGH, K. G. and WIDDOWSON, E. M., *Brit. J. Nutr.*, 1970, **24**, 109.
22. PETERS, J. M., MAIER, R., HAWTHORN, B. E. and STORVICK, C. A., *J. Mammal.*, 1972, **53**, 717.
23. BUSS, D. H. and COOPER, R. W., *Folia Primatologica*, 1972, **17**, 285.
24. SMITH, L. M. and HARDJO, S., *Lipids*, 1974, **9**, 674.
25. BUSS, D. H., COOPER, R. W. and WALLEN, K., *Folia Primatologica*, 1976, **26**, 301.
26. TURTON, J. A., FORD, D. J., BLEBY, J., HALL, B. M. and WHITING, R., *Folia Primatologica*, 1978, **29**, 64.

27. ANDERSON, R. R., SADLER, K. C., KNAUER, M. W., WIPPLER, J. P. and MARSHALL, R. T., *J. Dairy Sci.*, 1975, **58**, 1449.
28. BAKER, B. E., COOK, H. W., BIDER, J. R. and PEARSON, A. M., *Canad. J. Zool.*, 1970, **48**, 1349.
29. GALANCEV, V. P. and POPOV, S. M., *Vestn. Leningrad. Univ., Ser. Biol.*, 1969, **3**, 99.
30. KINSELLA, J. E., *Internat. J. Biochem.*, 1971, **2**, 6.
31. VOLCANI, R., ZISLING, R., SKLAN, D. and NITZAN, Z., *Br. J. Nutr.*, 1973, **29**, 121.
32. GRIFFITHS, M., MCINTOSH, D. L. and LECKIE, R. M. C., *J. Zool.*, 1972, **166**, 265.
33. ACKMAN, R. G., EATON, C. A. and MITCHELL, E. D., *Canad. J. Biochem.*, 1971, **49**, 1172.
34. BACHMAN, K. C. and IRVINE, A. B., *Comp. Biochem. Physiol.*, 1979, **62A**, 873.
35. JENNESS, R. and ODELL, D. K., *Comp. Biochem. Physiol.*, 1978, **61A**, 383.
36. COOK, H. W. and BAKER, B. E., *Canad. J. Zool.*, 1969, **47**, 1129.
37. COOK, H. W., LENTFER, J. W., PEARSON, A. M. and BAKER, B. E., *Canad. J. Zool.*, 1970, **48**, 217.
38. ANDO, K., MORI, M., KATO, I., YUSA, K. and GODA, K., *J. College of Dairying, Natural Science*, 1979, **8**, 9.
39. JENNESS, R., ERICKSON, A. W. and CRAIGHEAD, J. J., *J. Mammal.*, 1972, **53**, 34.
40. KUMAR, A. and MISRA, U. K., *Ind. J. Anim. Sci.*, 1980, **50**, 273.
41. BRACCO, U., HIDALGO, J. and BOHREN, H., *J. Dairy Sci.*, 1972, **55**, 165.
42. KEENAN, T. W., KING, J.-L. and COLENBRANDER, V. F., *J. Anim. Sci.*, 1970, **30**, 806.
43. KINSELLA, J. E., *Int. J. Biochem.*, 1973, **4**, 549.
44. LOK, C. M., *Receuil des Travaux Chimiques des Pays-Bas*, 1979, **98**, 92.
45. KINSELLA, J. E. and INFANTE, J. P., In: *Lactation—A Comprehensive Treatise*, Vol. IV, B. L. Larson (ed), 1978, Academic Press, New York, p. 475.
46. MORRISON, W. R., *Lipids*, 1968, **3**, 101.
47. KATAOKA, K. and NAKAE, T., *Jap. J. Dairy Sci.*, 1973, **22**, A137.
48. CALBERG-BACQ, C.-M., FRANCOIS, C., GOSELIN, L., OSTERRIETH, P. M. and RENTIER-DELRUE, F., *Biochim. Biophys. Acta*, 1976, **419**, 458.
49. CHEN, C. C. W., ARGOUDELIS, C. J. and TOBIAS, J., *J. Dairy Sci.*, 1978, **61**, 1691.
50. PATTON, S., HOOD, L. F. and PATTON, J. S., *J. Lipid Res.*, 1969, **10**, 260.
51. AHUJA, S. P., SRIVASTAVA, G. and SUKHVINDER, SINGH, *Zentralblatt für Veterinärmedizin*, 1978, **25**, 673.
52. HAY, J. D. and MORRISON, W. R., *Biochim. Biophys. Acta*, 1971, **248**, 71.
53. PARODI, P. W., *J. Chromatogr.*, 1975, **111**, 223.
54. TIMMEN, H. and DIMICK, P. S., *J. Dairy Sci.*, 1972, **55**, 919.
55. PARKS, O. W., KEENEY, M., KATZ, I. and SCHWARTZ, D. P., *J. Lipid Res.*, 1964, **5**, 232.
56. HALLGREN, B., NIKLASSON, A., STÄLLBERG, G. and THORIN, H., *Acta Chemica Scandinavica*, 1974, **28B**, 1029.
57. HALLGREN, B., STÄLLBERG, G. and BOERYD, B., *Progress in the Chemistry of Fats and other Lipids*, 1978, **16**, 45.

58. AHRNE, L., BJÖRCK, L., RAZNIKIEWICZ, T. and CLAESSON, O., *J. Dairy Sci.*, 1980, **63**, 741.
59. SCHOGT, J. C. M., HAVERKAMP BEGEMANN, P. and KOSTER, J., *J. Lipid Res.*, 1960, **1**, 446.
60. SCHOGT, J. C. M., HAVERKAMP BEGEMANN, P. and RECOURT, J. H., *J. Lipid Res.*, 1961, **2**, 142.
61. BREWINGTON, D. R., CARESS, E. A. and SCHWARTZ, D. P., *J. Lipid Res.*, 1970, **11**, 355.
62. FLANAGAN, V. P., FERRETTI, A., SCHWARTZ, D. P. and RUTH, J. M., *J. Lipid Res.*, 1975, **16**, 97.
63. ADACHI, A. and KOBAYASHI, T., *J. Nutrit. Sci. and Vitaminol.*, 1979, **25**, 67.
64. PARODI, P. W., *Australian J. Dairy Technol.*, 1973, **28**, 135.
65. MINCIONE, B., MUSSO, S. S. and FRANCISCIS, G. DE, *Milchwissenschaft*, 1977, **32**, 599.
66. MINCIONE, B., MUSSO, S. S., STINGO, C. and FRANCISCIS, G. DE, *Milchwissenschaft*, 1976, **31**, 282.
67. ADDEO, F., KUZDZAL-SAVOIE, S., CHIANESE, L., MALORNI, A. and SEPE, C., *Lait*, 1981, **61**, 187.
68. MELLIES, M. J. et al., *Am. J. Clin. Nutr.*, 1978, **31**, 1347.
69. MELLIES, M. J., BURTON, K., LARSEN, R., FIXLER, D. and GLUECK, C. J., *Am. J. Clin. Nutr.*, 1979, **32**, 2383.
70. HOFFMANN, B., *Milchwissenschaft*, 1977, **32**, 477.
71. POPE, G. S. and SWINBURNE, J. K., *J. Dairy Res.*, 1980, **47**, 427.
72. PATTON, S., KELLY, J. J. and KEENAN, T. W., *Lipids*, 1980, **15**, 33.
73. MCCARTHY, M. J., KUKSIS, A. and BEVERIDGE, J. M. R., *J. Lipid Res.*, 1964, **5**, 609.
74. AVIGAN, J., MILNE, G. W. A. and HIGHET, R. J., *Biochim. Biophys. Acta*, 1967, **144**, 127.
75. URBACH, G. and STARK, W., *J. Agric. and Food Chem.*, 1975, **23**, 20.
76. HARTMAN, A. M. and DRYDEN, L. P. In: *Fundamentals of Dairy Chemistry*, B. H. Webb, A. H. Johnson and J. A. Alford (eds), 1974, Avi Publishing Co., Westport, p. 325.
77. JENNESS, R., *Semin. Perinatol.*, 1979, **3**, 225.
78. JENNESS, R., *J. Dairy Sci.*, 1980, **63**, 1605.
79. REEVE, L. E., CHESNEY, R. W. and DELUCA, H. F., *Am. J. Clin. Nutr.*, 1982, **36**, 122.
80. HOLLIS, B. W., ROOS, B. A., DRAPER, H. H. and LAMBERT, P. W., *J. Nutr.*, 1981, **111**, 384.
81. HERTING, D. C. and DRURY, E.-J. E., *J. Chromatogr.*, 1967, **30**, 502.
82. KOBAYASHI, H., KANNO, C., YAMAUCHI, K. and TSUGO, T., *Biochim. Biophys. Acta*, 1975, **380**, 282.
83. JANSSON, L., AKESSON, B. and HOLMBERG, L., *Am. J. Clin. Nutr.*, 1981, **34**, 8.
84. LUCAS, A. and MITCHELL, M. D., *Archives of Disease in Childhood*, 1980, **55**, 950.
85. MANNS, J. G., *Prostaglandins*, 1975, **9**, 463.
86. MAULE WALKER, F. M. and PEAKER, M., *J. Physiol. (London)*, 1980, **309**, 65.
87. SNOSWELL, A. M. and LINZELL, J. L., *J. Dairy Res.*, 1975, **42**, 371.
88. SCHMIDT-SOMMERFELD, E. et al., *Pediatr. Res.*, 1978, **12**, 660.
89. DONZELLI, F. et al., *Pediatr. Res.*, 1980, **14**, 179.

90. Borum, P. R., *Nutr. Revs.*, 1981, **39**, 385.
91. Forss, D. A., *J. Am. Oil Chem. Soc.*, 1971, **48**, 702.
92. Forss, D. A., *Progress in the Chemistry of Fats and Other Lipids*, 1972, **13**, 177.
93. Swoboda, P. A. T. and Peers, K. E., *Chemistry and Industry*, 1976, 160.
94. Swoboda, P. A. T. and Peers, K. E., *J. Sci. Food and Agric.*, 1977, **28**, 1019.
95. Brewington, C. R., Parks, O. W. and Schwartz, D. P., *J. Agric. and Food Chem.*, 1973, **21**, 38.
96. Brewington, C. R., Parks, O. W. and Schwartz, D. P., *J. Agric. and Food Chem.*, 1974, **22**, 293.
97. Forney, F. W. and Markovetz, A. J., *J. Lipid Res.*, 1971, **12**, 383.
98. Fujino, Y. and Fujishima, T., *J. Dairy Res.*, 1972, **39**, 11.
99. Hladik, J. and Michalec, C., *Acta Biol. Med. Ger.*, 1966, **16**, 696.
100. Kayser, S. G. and Patton, S., *Biochem. Biophys. Res. Commun.*, 1970, **41**, 1572.
101. Morrison, W. R. and Smith, L. M., *Biochim. Biophys. Acta*, 1964, **84**, 759.
102. Fujino, Y., Saeki, T., Ito, S. and Negishi, T., *Jap. J. Zootech. Sci.*, 1969, **40**, 349.
103. Fujino, Y., Nakano, M. and Saeki, T., *Agric. Biol. Chem.*, 1970, **34**, 442.
104. Bouhours, J.-F. and Bouhours, D., *Biochem. Biophys. Res. Commun.*, 1979, **88**, 1217.
105. Grimmonprez, L. and Montreuil, J., *Biochimie*, 1977, **59**, 899.
106. Huang, R. T. C., *Biochim. Biophys. Acta*, 1973, **306**, 82.
107. Keenan, T. W., *Biochim. Biophys. Acta*, 1974, **337**, 255.
108. Gosselin-Rey, C. et al., *Europ. J. Biochem.*, 1980, **107**, 25.
109. Strocchi, A. and Holman, R. T., *Riv. Ital. Sostanze Grasse*, 1971, **48**, 617.
110. Melcher, F. and Renner, E., *Milchwissenschaft*, 1976, **31**, 70.
111. Christie, W. W., *Prog. Lipid Res.*, 1979, **17**, 245.
112. Dils, R., In: *Lipid Metabolism in Mammals*, Vol. 1, F. Snyder (ed), 1977, Plenum Press, New York, p. 131.
113. Moore, J. H. and Christie, W. W., *Prog. Lipid Res.*, 1979, **17**, 347.
114. Glass, R. L. and Jenness, R., *Comp. Biochem. Physiol.*, 1971, **38B**, 353.
115. Glass, R. L., Troolin, H. A. and Jenness, R., *Comp. Biochem. Physiol.*, 1967, **22**, 415.
116. Buss, D. H., *Lipids*, 1969, **4**, 152.
117. Demarne, Y., Lhuillery, C., Pihet, J., Martinet, L. and Flanzy, J., *Comp. Biochem. Physiol.*, 1978, **61B**, 223.
118. Jangaard, P. M. and Ke, P. J., *J. Fish Res. Bd Can.*, 1968, **25**, 2419.
119. Riedman, M. and Ortiz, C. L., *Physiol. Zool.*, 1979, **52**, 240.
120. Noble, R. C., *Prog. Lipid Res.*, 1978, **17**, 55.
121. Bickerstaffe, R. and Annison, E. F., *Comp. Biochem. Physiol.*, 1970, **35**, 653.
122. Bailey, E., *Biochem. Soc. Trans.*, 1981, **9**, 371.
123. Noble, R. C., *Prog. Lipid Res.*, 1980, **18**, 179.
124. Moore, J. H., *Proc. Nutr. Soc.*, 1978, **37**, 231.
125. Kabara, J. J., *Nutr. Revs.*, 1980, **38**, 65.
126. Kiuru, K., Leppanen, R. and Antila, M., *Fette: Seifen: Anstrichmittel*, 1974, **76**, 401.

127. Kinsella, J. E., Bruckner, G., Mai, J. and Shimp, J., *Am. J. Clin. Nutr.*, 1981, **34**, 2307.
128. Hay, J. D. and Morrison, W. R., *Biochim. Biophys. Acta*, 1970, **202**, 237.
129. Parodi, P. W., *Austral. J. Dairy Technol.*, 1973, **28**, 162.
130. Massart-Leen, A. M., Pooter, H. De, Decloedt, M. and Schamp, N., *Lipids*, 1981, **16**, 286.
131. Lough, A. K., *Lipids*, 1977, **12**, 115.
132. Parodi, P. W., *J. Dairy Sci.*, 1977, **60**, 1550.
133. Parks, O. W., *J. Dairy Sci.*, 1977, **60**, 718.
134. Egge, H., Murawski, U., György, P. and Zilliken, F., *FEBS Letts.*, 1969, **2**, 255.
135. Egge, H., Murawski, U., Ryhage, R., György, P. and Zilliken, F., *FEBS Letts.*, 1970, **11**, 113.
136. Egge, H., Murawski, U., Ryhage, R., Zuiken, F. and György, P., *Ztschr. Anal. Chem.*, 1970, **252**, 123.
137. Egge, H. et al., *Chemistry and Physics of Lipids*, 1972, **8**, 42.
138. Murawski, U., Egge, H., György, P. and Zilliken, F., *FEBS Letts.*, 1971, **18**, 290.
139. Murawski, U., Egge, H., György, P. and Zilliken, F., *Zeitschrift für Naturforschung*, 1974, **29**, 1.
140. Murawski, U. and Egge, H., *J. Chromatogr. Sci.*, 1975, **13**, 497.
141. Aitchison, J. M., Dunkley, W. L., Canolty, N. L. and Smith, L. M., *Am. J. Clin. Nutr.*, 1977, **30**, 2006.
142. Meyer, F. Melcher, F.-W., Senft, B. and Hecklemann, K. H., *Milchwissenschaft*, 1980, **35**, 743.
143. Keenan, T. W. and Patton, S., *Lipids*, 1970, **5**, 42.
144. Morrison, W. R., Jack, E. L. and Smith, L. M., *J. Am. Oil Chem. Soc.*, 1965, **42**, 1142.
145. Clark, R. M., Ferris, A. M., Fey, N., Hundrieser, K. E. and Jensen, R. G., *Lipids*, 1980, **15**, 972.
146. Morrison, W. R. and Smith, L. M., *Lipids*, 1967, **2**, 178.
147. Morrison, W. R., *Lipids*, 1968, **3**, 107.
148. Gosselin, L. et al., *Biochem. Soc. Trans.*, 1977, **5**, 1142.
149. Parks, O. W., *J. Dairy Sci.*, 1980, **63**, 295.
150. Pruthi, T. D., Narayanan, K. M. and Bhalerao, V. R., *Milchwissenschaft*, 1972, **27**, 294.
151. Morrison, W. R. and Hay, J. D., *Biochim. Biophys. Acta*, 1970, **202**, 460.
152. Hladik, J., *Sbornik Vysoke Skoly Chemicko-Technologicke v Praze, E*, 1971, **32**, 123.
153. Fujino, Y., Yamabuki, S., Ito, S. and Negishi, T., *J. Agric. Chem. Soc. Japan*, 1969, **43**, 712.
154. Morrison, W. R., *FEBS Letts.*, 1971, **19**, 63.
155. Morrison, W. R., *Biochem. Biophys. Acta*, 1973, **316**, 98.
156. Bouhours, J. F. and Bouhours, D., *Lipids*, 1981, **16**, 726.
157. Kuksis, A., Marai, L. and Myher, J. J., *J. Am. Oil Chem. Soc.*, 1973, **50**, 193.
158. Breckenridge, W. C., In: *Handbook of Lipid Research*, Vol. 1, A. Kuksis (ed), 1978, Plenum Press, New York, p. 197.

159. CHRISTIE, W. W. and CLAPPERTON, J. L., *J. Soc. Dairy Technol.*, 1982, **35**, 22.
160. BRECKENRIDGE, W. C., MARAI, L. and KUKSIS, A., *Canad. J. Biochem.*, 1969, **47**, 761.
161. CHRISTIE, W. W. and MOORE, J. H., *Biochim. Biophys. Acta*, 1970, **210**, 46.
162. PARODI, P. W., *Lipids*, 1982, **17**, 437.
163. GRIGOR, M. R., *Comp. Biochem. Physiol.*, 1980, **65B**, 427.
164. SMITH, L. M. and HARDJO, S., *Lipids*, 1974, **9**, 713.
165. STOKES, G. B. and TOVE, S. B., *J. Biol. Chem.*, 1975, **250**, 6315.
166. CHRISTIE, W. W. and HUNTER, M. L., *J. Sci. Food Agric.*, 1978, **29**, 442.
167. CHRISTIE, W. W., HUNTER, M. L. and MOORE, J. H., *N.Z. J. Dairy Sci. Technol.*, 1978, **13**, 119.
168. PITAS, R. E., SAMPUGNA, J. and JENSEN, R. G., *J. Dairy Sci.*, 1967, **50**, 1332.
169. BRECKENRIDGE, W. C. and KUKSIS, A., *Lipids*, 1968, **3**, 291.
170. BRECKENRIDGE, W. C. and KUKSIS, A., *J. Lipid Res.*, 1968, **9**, 388.
171. BRECKENRIDGE, W. C. and KUKSIS, A., *Lipids*, 1969, **4**, 197.
172. MARAI, L., BRECKENRIDGE, W. C. and KUKSIS, A., *Lipids*, 1969, **4**, 562.
173. TAYLOR, M. W. and HAWKE, J. C., *N.Z. J. Dairy Sci. Technol.*, 1975, **10**, 49.
174. MILLS, S. C., COOK, L. J., SCOTT, T. W. and NESTEL, P. J., *Lipids*, 1976, **11**, 49.
175. MORRISON, I. M. and HAWKE, J. C., *Lipids*, 1977, **12**, 1005.
176. PARODI, P. W., *J. Dairy Res.*, 1979, **46**, 75.
177. ANDERSON, B. A., SUTTON, C. A. and PALLANSCH, M. J., *J. Am. Oil Chem. Soc.*, 1970, **47**, 15.
178. BUS, J., LOK, C. M. and GROENEWEGEN, A., *Chem. Phys. Lipids*, 1976, **16**, 123.
179. PFEFFER, P. E., SAMPUGNA, J., SCHWARTZ, D. P. and SHOOLERY, J. N., *Lipids*, 1977, **12**, 869.
180. MARSHALL, M. O. and KNUDSEN, J., *Biochim. Biophys. Acta*, 1980, **617**, 393.
181. TOMARELLI, R. M., MEYER, B. J., WEABER, J. R. and BERNHART, F. W., *J. Nutr.*, 1968, **95**, 583.
182. FILER, L. J., MATTSON, F. H. and FOMON, S. J., *J. Nutr.*, 1969, **99**, 293.
183. NORMAN, A., STRANDVIK, B. and OJAMAE, O., *Acta Paediat., Stockh.*, 1972, **61**, 571.
184. AW, T. W. and GRIGOR, M. R., *Biochim. Biophys. Acta*, 1978, **531**, 257.
185. CHRISTIE, W. W., *Lipid Analysis*, 2nd edn., 1982, Pergamon Press, Oxford.
186. BRECKENRIDGE, W. C. and KUKSIS, A., *J. Lipid Res.*, 1967, **8**, 473.
187. BREACH, R. A., DILS, R. and WATTS, R., *J. Dairy Res.*, 1973, **40**, 273.
188. NUTTER, L. J. and PRIVETT, O. S., *J. Dairy Sci.*, 1967, **50**, 1194.
189. ADDEO, F. and KUZDZAL-SAVOIE, S., *Lait*, 1980, **60**, 14.
190. ARUMUGHAN, C. and NARAYANAN, K. M., *Lipids*, 1981, **16**, 155.
191. ARUMUGHAN, C. and NARAYANAN, K. M., *J. Dairy Res.*, 1982, **49**, 81.
192. GROB, K., NEUKOM, H. P. and BATTAGLIA, R., *J. Am. Oil Chem. Soc.*, 1982, **57**, 282.
193. MORRISON, I. M. and HAWKE, J. C., *Lipids*, 1977, **12**, 994.
194. PARODI, P. W., *Austral. J. Dairy Technol.*, 1972, **27**, 140.
195. PARODI, P. W., *Austral. J. Dairy Technol.*, 1980, **35**, 17.
196. PARODI, P. W., *J. Dairy Res.*, 1982, **49**, 73.
197. SHEHATA, A. A. Y., DEMAN, J. M. and ALEXANDER, J. C., *Canad. Inst. Food Sci. and Technol. J.*, 1971, **4**, 61.

198. SHEHATA, A. A. Y., DEMAN, J. M. and ALEXANDER, J. C., *Canad. Inst. Food Sci. and Technol. J.*, 1972, **5**, 13.
199. TAYLOR, M. W. and HAWKE, J. C., *N.Z. J. Dairy Sci. and Technol.*, 1975, **10**, 40.
200. TRAITLER, H. and PREVOT, A., *J. High Res. Chromatogr. and Chromatogr. Comm.*, 1981, **4**, 109.
201. EL-SADEK, G., RIFAAT, I. D., ABD EL-SALAM, M. H. and EL-BAGORY, E., *Ind. J. Dairy Sci.*, 1972, **25**, 167.
202. MORRISON, I. M. and HAWKE, J. C., *Lipids*, 1979, **14**, 391.
203. NORRIS, G. E., GRAY, I. K. and DOLBY, R. M., *J. Dairy Res.*, 1973, **40**, 311.
204. PARODI, P. W., *Austral. J. Dairy Technol.*, 1974, **29**, 20.
205. PARODI, P. W., *J. Dairy Res.*, 1979, **46**, 633.
206. PARODI, P. W., *J. Dairy Res.*, 1981, **48**, 131.
207. SHERBON, J. W., *J. Am. Oil Chem. Soc.*, 1974, **51**, 22.
208. SHERBON, J. W. and DOLBY, R. M., *J. Dairy Sci.*, 1973, **56**, 52.
209. TAYLOR, M. W., NORRIS, G. E. and HAWKE, J. C., *N.Z. J. Dairy Sci. and Technol.*, 1978, **13**, 236.
210. TIMMS, R. E., *Austral. J. Dairy Technol.*, 1980, **35**, 47.
211. HAWKE, J. C., *J. Lipid Res.*, 1963, **4**, 255.
212. KUKSIS, A., *Progress in Chemistry of Fats and Other Lipids*, 1972, **12**, 1.
213. HOLUB, B. J. and KUKSIS, A., *Adv. Lipid Res.*, 1978, **16**, 1.
214. MOORE, G. M., RATTRAY, J. B. M. and IRVINE, D. M., *Canad. J. Biochem.*, 1968, **46**, 205.
215. BLANK, M. L., NUTTER, L. J. and PRIVETT, O. S., *Lipids*, 1966, **1**, 132.
216. PRIVETT, O. S. and NUTTER, L. J., *Lipids*, 1967, **2**, 149.
217. NUTTER, L. J. and PRIVETT, O. S., *J. Dairy Sci.*, 1967, **50**, 298.
218. FERRANDO, R., WOLTER, R., FOURLON, C. and MORICE, M., *Annales de la Nutrition et de l'Alimentation*, 1971, **25**, 241.
219. DESMETH, M. and VANDEPUTTE-POMA, J., *Comp. Biochem. Physiol.*, 1980, **66B**, 129.
220. DESMETH, M., *Comp. Biochem. Physiol.*, 1980, **66B**, 135.

Chapter 2

INFLUENCE OF NUTRITIONAL FACTORS ON THE YIELD, COMPOSITION AND PHYSICAL PROPERTIES OF MILK FAT

J. C. Hawke and M. W. Taylor

Massey University, Palmerston North, New Zealand

SUMMARY

The fatty acid constituents of milk lipids are derived both from blood lipids and synthesis de novo *in the mammary gland. The fatty acids from these two sources are appreciably different and consequently variations in their relative contributions, brought about by nutritional factors, will result in changes in the fatty acid composition of milk lipids. The nature of the fatty acids synthesized by the mammary gland, within a given species, is subject to little variation. On the other hand, the blood lipids have variable fatty acid compositions and in non-ruminants these are highly dietary-dependent. In ruminants the intervention of microbial fermentation in the rumen prior to absorption diminishes the effect of diet on the nature of the fatty acids in the blood lipids. However, the use of feeding techniques which enable the rumen reactions to be bypassed results in ruminant milk lipids being greatly influenced by the fatty acid composition of dietary lipid.*

As a consequence of variations in the fatty acid composition of blood lipids, and in the relative contributions of these lipids and of the fatty acids synthesized in the mammary gland to the total milk fat, the composition of milk triglycerides varies. Therefore factors such as composition of the diet and stage of lactation may influence the chemical and physical properties of milk fat. In bovine milk, changes in the triglyceride composition have been related to the melting characteristics of the fat.

ABBREVIATIONS

CDP	Cytidine diphosphate
FFA	Free fatty acid
G-3-P	Glycerol-3-phosphate
HDL	High density lipoprotein
LDL	Low density lipoprotein
MG	Monoglyceride
NAD	Nicotinamide–adenine dinucleotide
NADP	Nicotinamide–adenine dinucleotide phosphate
TG	Triglyceride (triacylglycerol)
VLDL	Very low density lipoprotein

1. INTRODUCTION

Fat in milk, unlike carbohydrate and protein, is subject to marked variations in both amount and composition due to nutritional factors which are superimposed upon variations due to species, genetic and lactational factors. The nutritional effects specific to fats arise principally because a considerable, but variable, proportion of the fatty acids of milk lipids is derived directly from dietary lipids. Since an appreciation of the factors which alter the composition and amount of milk fat secreted by the mammary gland is dependent upon a working knowledge of the biochemistry of the pathways involved, we have commenced this chapter by considering how dietary constituents are metabolized and made available to the mammary gland and the mechanisms of biosynthesis of fat and its constituent fatty acids in mammary tissue. This information, together with what is known about the regulation of certain key enzymes by hormonal and other factors, provides a basis for understanding many species differences and observed variations in the yield, composition and melting characteristics of bovine milk fat which are discussed later in this chapter.

A. BIOSYNTHESIS OF MILK LIPIDS

2. ORIGINS OF THE FATTY ACIDS IN MILK LIPIDS

The fatty acids of milk lipids arise from two sources: (1) synthesis *de novo* in the mammary gland, and (2) uptake from the circulating blood. The composition of the fatty acids from these two sources is usually very different; in the cow, for example, endogenously produced fatty acids are of carbon chain length C_4 to C_{16} while a proportion of C_{16}, and virtually all the C_{18}, fatty acids arise from blood (Fig. 1).

The relative contributions of the fatty acids taken up by the mammary gland and the fatty acids synthesized within the gland to milk fat appear to be species-dependent. The latter are quantitatively more important in the rabbit and sow than in the goat.[1,2] In rabbit milk fat the contribution of blood lipids to the fatty acids is limited to about 30%, because the medium-chain fatty acids which account for at least 50% of the fatty acids arise from synthesis *de novo*, compared to 45% for the goat.[1] Estimates of the contribution of blood lipids in the cow vary from about 35%[3] to as high as 82%.[4]

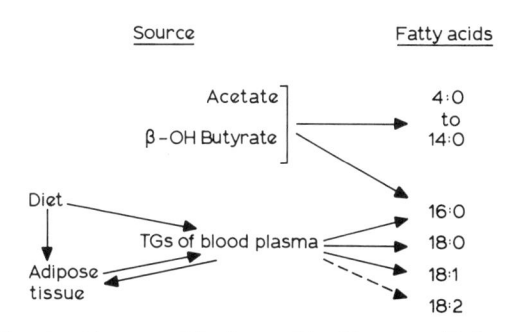

FIG. 1. Sources of the fatty acids of bovine milk fat.

2.1. Contributions from Blood

Under normal dietary conditions only one circulating lipid in blood, the triglyceride of the dextran sulphate precipitable lipoproteins (VLDLs and LDLs),[5] appears to be of quantitative importance in providing fatty acids for bovine milk fat synthesis (Fig. 2). The unesterified fatty acids in blood are in equilibrium with a restricted pool of unesterified fatty acids within the mammary gland[4] but whether or not there is a net uptake of plasma non-esterified fatty acids by the mammary gland, thus making a significant contribution to milk fat, is dietary- and species-dependent. In the ruminant, arteriovenous differences in fatty acids have been observed only in the starved state[6,7] when blood levels are high, but fatty acid uptake occurs in normally fed non-ruminants such as the lactating rabbit,[1] rat[8] and guinea-pig.[9]

The uptake of triglycerides of the VLDLs in the blood stream by mammary tissue involves hydrolysis of the triglycerides by lipoprotein lipase.[10] The lipase appears to be bound to the luminal surface of the endothelium of the blood capillaries and hydrolysis of triglyceride is regarded as a prerequisite to the uptake of lipid by mammary epithelial cells.[11] The role of lipoprotein lipase in triglyceride assimilation by cells has been reviewed recently.[12] After the transfer of apoC_{II}, the lipoprotein lipase activity factor, from HDL particles to VLDLs and chylomicrons the triglyceride cores are hydrolysed by lipoprotein lipase. The mechanism of transport of the resultant fatty acids and monoglycerides through the endothelial layer is not known although the behaviour of fatty acids in model systems suggests that transport across the capillary wall may occur by lateral diffusion in the cell membranes.[13]

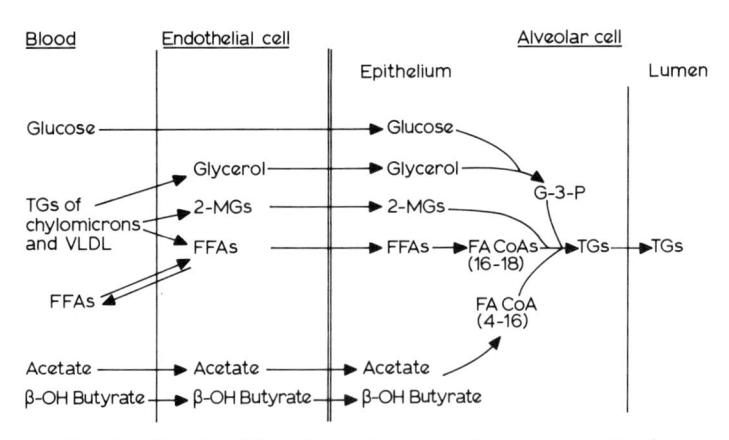

FIG. 2. Uptake of blood constituents by the mammary gland.

A small proportion of the lipoprotein triglycerides may be hydrolysed only to 2-monoglycerides which are then used directly for milk triglyceride synthesis.[14] Prolactin has been implicated in the suppression of lipoprotein lipase activity in adipose tissue and enhancement of mammary gland lipase during lactation[15] resulting in a redirection of circulating triglycerides for lipogenesis in the mammary gland.

Bishop et al.[3] estimated, from the recovery of radioactivity in milk fat and from blood flow rates, that a minimum of 30% of palmitate (16:0) is synthesized *de novo* by the mammary gland. However, intramammary infusions of [^{14}C]-labelled acetate and β-hydroxybutyrate put this figure as high as 60%.[16] In contrast, all the C_{18} fatty acids are derived from plasma sources[16] so that, unlike other animal tissues, mammary tissue appears to be unable to elongate 16:0, synthesized from acetate, to stearate (18:0).[4,17]

There is extensive desaturation of the 18:0 entering the ruminant mammary gland to oleate (18:1)[4,18] by a desaturase located in the microsomes.[19,20] Thus, despite the high 18:0 to 18:1 ratios in triglycerides circulating in the blood of ruminants, arising from extensive biohydrogenation of dietary fat in the rumen, desaturation of 18:0 in mammary tissue results in the production of milk fat not markedly dissimilar in 18:1 content to the milk fat of non-ruminants. Desaturase activity has not been detected in rabbit[1,21,22] and rat[23] mammary tissue but sow mammary tissue is as effective as goat tissue in desaturating 18:0 to 18:1.[20]

In general, the nature of the fatty acids taken up from the blood and utilized for milk fat synthesis reflects the composition of the circulating triglycerides. As mentioned above, biohydrogenation and other microbial reactions concerned with lipid metabolism in the rumen, ensure a measure of consistency in the nature and level of the fatty acids absorbed from the lower gut, irrespective of the dietary lipid. Under normal physiological conditions, 16:0, 18:0 and 18:1 are the principal fatty acids available from ruminant blood for milk fat synthesis. However, in non-ruminants the fatty acids available for absorption by the mammary gland closely reflect the composition of dietary lipids, with consequential effects on the nature of milk fat. Any special provision for dietary lipids to bypass the rumen will of course lead to ruminant milk fat responding to compositional changes in dietary lipids in much the same way as in non-ruminants. There is evidence for some degree of selectivity towards triglycerides containing 18:0 in the assimilation of blood triglycerides by the mammary gland which appears to be related to the distribution of triglycerides of different fatty acid composition between the various serum lipoprotein fractions.[24] The triglycerides of VLDLs and LDLs which are taken up by the mammary gland normally have higher proportions of 18:0 and lower proportions of linoleate (18:2) than the triglycerides of HDLs which are not taken up.[25] The incorporation of 18:2 into the triglycerides of the larger chylomicron and VLDL particles when animals are fed a dietary supplement of 18:2-rich oils protected against biohydrogenation has been advanced as an explanation for the preferential assimilation by the mammary gland of plasma triglyceride fractions containing high concentrations of 18:2.[24]

A unique characteristic of the ruminant mammary gland is its capacity, during lactation, to metabolize large quantities of D(−)β-hydroxybutyrate[16] which is produced in the rumen epithelium during absorption of butyrate from the rumen;[26] Palmquist *et al.*[16] calculated that this substrate contributes a maximum of 8% of the milk fatty acid carbon. β-Hydroxybutyrate is utilized directly during lipogenesis,[27] rather than via acetyl-

CoA, which is consistent with the mitochondrial location of β-hydroxybutyrate dehydrogenase,[28] and contributes about equally with acetate to the initial four carbon atoms at the methyl-terminal end of fatty acids in the range 4:0 to 16:0.[29] Rat mammary cells, which are presumably typical of mammary cells of other non-ruminant species, show much less ability than bovine mammary cells to incorporate β-hydroxybutyrate into fatty acids and there is little evidence for the incorporation of intact C_4 units.[30]

2.2. Fatty Acid Biosynthesis in the Mammary Gland

Essential to fatty acid biosynthesis in the mammary gland is the provision of a carbon source in the form of acetyl-CoA and the availability of large quantities of reducing equivalents as NADPH. Fundamental differences between ruminants and non-ruminants in both of these important aspects

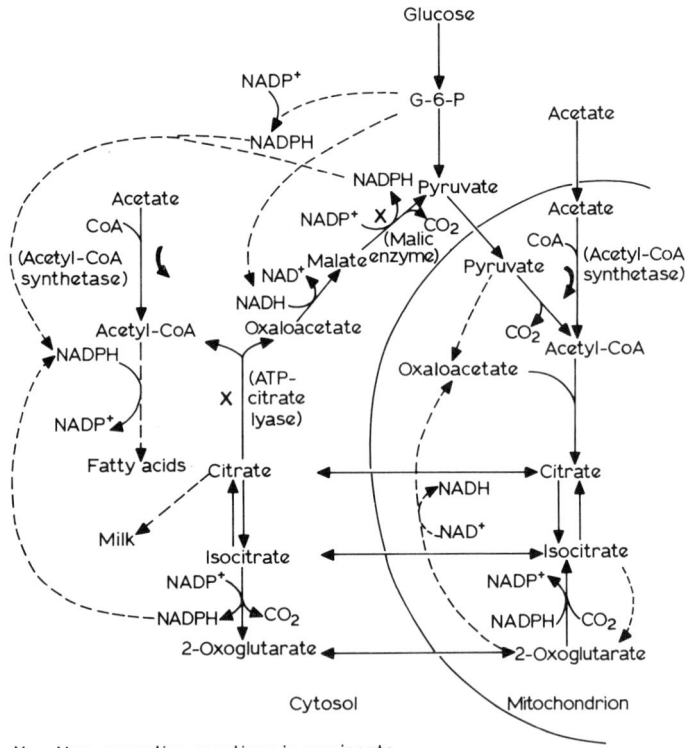

FIG. 3. Acetate and glucose as alternative sources of carbon in fatty acid biosynthesis.

of fatty acid biosynthesis are shown in Fig. 3, and are discussed below. The studies by Folley and co-workers,[31,32] Bauman et al.[28,33] and Hardwick et al.[34] show that ruminant mammary tissue utilizes acetate, and not glucose, for fatty acid synthesis whereas glucose represents the major carbon source for fatty acid synthesis *de novo* in non-ruminant mammary tissue.[32,35,36] The utilization of blood acetate, rather than blood glucose, by ruminants as a carbon source in lipogenesis[29,37,38] represents a unique metabolic adaptation of ruminants to ensure that glucose and glucogenic compounds are conserved for body processes for which these substances are essential. Unlike non-ruminants, ruminants absorb little or no glucose from the small intestine. However, large amounts of acetate, propionate and butyrate, produced by the fermentation of dietary carbohydrates in the rumen, are absorbed through the ruminal wall. Normally, of these compounds, only acetate is available to non-hepatic tissues as an energy or carbon source for biosynthesis although β-hydroxybutyrate formed from butyrate is also available.

2.2.1. Reaction Pathways

The mechanism of synthesis of fatty acids in mammary tissue is similar to that in other tissues. The overall reaction for the synthesis of palmitic acid from acetyl-CoA is

$$8CH_3COSCoA + 7ATP + 14NADPH + 14H^+ \rightarrow CH_3(CH_2)_{14}COOH$$
$$+ 7ADP + 7P_i + 14NADP^+ + 8CoASH + 6H_2O$$

Palmitic acid is the most abundant initial product in all tissues but small amounts of 14:0 and 18:0 are usually also produced. However, mammary tissue of many animals is somewhat unique in also synthesizing lower molecular weight fatty acids by this same basic reaction.[39]

Essential to the synthesis of fatty acids is the carboxylation of acetyl-CoA to malonyl-CoA which is catalysed by acetyl-CoA carboxylase

$$CH_3COSCoA + HCO_3^- + ATP \rightarrow HOOCCH_2COSCoA + ADP + P_i$$

Malonyl CoA is utilized in chain elongation in a multistep reaction (Fig. 4) catalysed by a multifunctional enzyme referred to as fatty acid synthetase. Again this is illustrated for the synthesis of palmitic acid

$$CH_3COSCoA + 7HOOCCH_2COSCoA + 14NADPH + 14H^+$$
$$\rightarrow CH_3(CH_2)_{14}COOH + 7CO_2 + 14NADP^+ + 8CoASH + 6H_2O$$

Bovine mammary tissue was used by Ganguly[40] in one of the earliest experiments to establish this role for malonyl-CoA as an intermediate in the synthesis of fatty acids *de novo*. It follows, therefore, that the two carbons at the methyl-terminal end of the fatty acid chain are derived directly from acetyl-CoA while the remainder are derived from acetyl-CoA via malonyl-CoA.[41]

Functional mammalian and avian fatty acid synthetases have molecular weights of about 500 000, whether isolated from liver, mammary gland or brain, and consist of two subunits. It used to be thought that these subunits

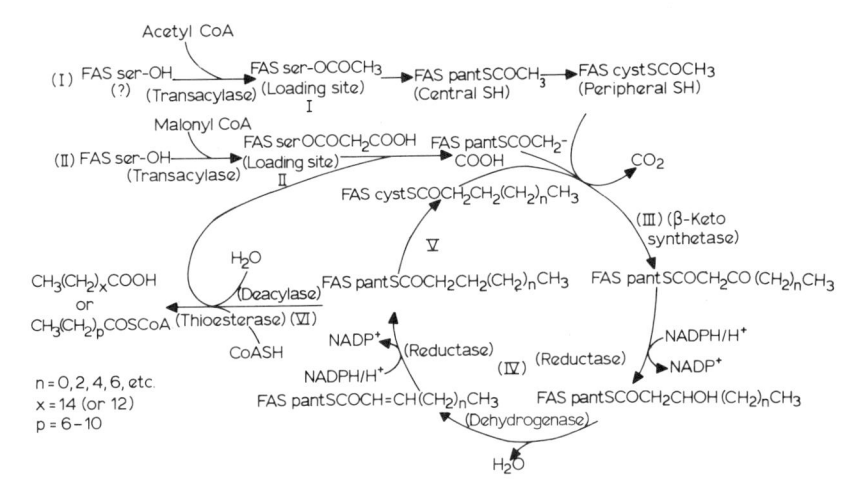

FIG. 4. A schematic representation of fatty acid synthesis involving enzymic components of mammalian fatty acid synthetase (FAS). FAS = Fatty acid synthetase; pant = 4'phosphopanthethiene; cyst = cysteine of β-keto acyl synthetase. 'Central' and 'peripheral' refer to the location of the groups in the molecule.[46] The loading site has been implicated in the release of butyrate and hexanoate from bovine fatty acid synthetase. See text for reference to I, II, etc.

were non-identical polypeptides of slightly different molecular weights,[42] but if proteolysis is avoided during isolation the two subunits appear to be identical.[43-5] A consequence of identical subunits is that the single polypeptide chain contains all seven catalytic centres required for fatty acid biosynthesis and, in addition, contains a 4'-phosphopantetheine binding group which is an essential feature of all fatty acid synthetases studied.[46]

The reaction sequence leading to fatty acid synthesis is initiated by transfer of acetyl and malonyl groups in two transacylations to non-thiol (loading) sites on the polypeptide (Fig. 4 (I and II)). The acetyl group is then transferred from its loading site (–OH group of serine or threonine) via pantetheine–SH, to cysteine–SH of the β-ketosynthetase while the malonyl group is transferred from the loading site (–OH group of serine) to pantetheine–SH. Condensation between the acetyl and malonyl group then follows, catalysed by β-ketoacyl synthetase (Fig. 4 (III)), to form the acetoacetyl moiety. This is the only catalytic transformation missing from the fatty acid synthetase of mutant yeasts lacking 4′-phosphopantetheine; thus it is possible that the ensuing two reductive and dehydratase reactions (Fig. 4 (IV)), also in mammalian synthetases, may not require the substrates to be attached covalently to 4′-phosphopantetheine.[46] The sequence of chain elongation is continued after transfer of the butyryl group to the cysteine–SH of the β-ketosynthetase (V) and transfer of a further malonyl group from the loading site to 4′-phosphopantetheine–SH.

The chain lengths of the products of synthesis *de novo* are believed to be determined by the specificities of two key enzyme components of the fatty acid synthetase, namely β-ketosynthetase which catalyses the condensation (elongation) step and thioesterase (Fig. 4 (VI)) which catalyses the release of acyl groups. Smith and co-workers[47,48] concluded that palmitic acid is the main product of most tissue fatty acid synthetases due to the inability of the β-ketosynthetase to rapidly elongate acyl chains beyond 16 carbon atoms, and to the limited ability of the thioesterase to hydrolyse acyl thioesters shorter than 16 carbon atoms. The thioesterase component, with a molecular weight of about 35 000, has been released from rat tissue synthetases by selective tryptic digestion.[48] It appears that each of the two multifunctional polypeptide components of the fatty acid synthetase contains a thioesterase or deacylase site.

2.2.2. Biosynthesis of Low and Medium Molecular Weight Fatty Acids
Numerous studies have been directed towards understanding the mechanism by which mammary enzymes are able to terminate fatty acid synthesis when the molecule contains from 4 to 14 carbons as well as 16 carbons as in other tissues. The mammary tissue of a number of non-ruminants, e.g. rat, mouse and rabbit, contains an additional soluble thioesterase which is responsible for the termination of fatty acid synthesis at medium-chain length (C_8 to C_{12}). Like the thioesterase component removed from the fatty acid synthetases of other tissues by tryptic digestion,[48] it is a small protein (molecular weight 32 000) with a serine group at the single active site. The soluble thioesterase is most active towards pantetheine esters of decanoate and interaction of the thioesterase with the fatty acid synthetase induces the release of shorter-chain acids

from the multienzyme.[49] The quantity of this soluble thioesterase rises during pregnancy, and rat mammary gland has the capacity to synthesize medium-chain fatty acids in late pregnancy. Thus from the outset of lactation the proportions of medium-chain fatty acids are high even though rates of lipogenesis do not attain a maximum until some days later;[50] this has led to speculation on the importance of medium-chain fatty acids in milk triglycerides to the developing neonate.[48] There is a close relationship between the presence of soluble thioesterase and an ability to synthesize medium-chain fatty acids.[48] Thus the soluble thioesterase has been isolated from rat, mouse and rabbit, but not guinea-pig, mammary gland. The enzyme is not found in other tissues of rodents and rabbit, such as adipose tissue.

In ruminants the synthesis of medium-chain fatty acids in the range C_8 to C_{12} does not appear to be dependent upon a cytosolic thioesterase. Instead, the fatty acid synthetase alone is able to synthesize medium-chain acyl-CoA esters provided that acyl-CoA complexing agents such as albumin and β-lactoglobulin are added. Knudsen and Grunnet[51] suggest that the loading transacylase (Fig. 4 (II)) is involved in the termination of fatty acid synthesis to produce medium-chain compounds. By analogy with yeast fatty acid synthetase, in which the loading transacylase has an affinity for both malonyl- and palmitoyl-CoA, termination in mammary tissue would occur as

$$FAS\text{-Ser-O} \cdot OC(CH_2)_n CH_3 + CoASH \rightleftharpoons$$
$$FAS\text{-Ser-OH} + CH_3(CH_2)_n COSCoA$$

where $n = 6$ to 10 carbon atoms. On the other hand, the synthesis of butyryl-CoA, and to a lesser extent hexanoyl-CoA, is achieved without the addition of acyl-CoA complexing agents or thioesterase. Butyryl- and hexanoyl-CoA, in addition to acetyl-CoA, are effective substrates for mammary gland fatty acid synthetase and the reversal of the loading or transacylase reaction is postulated to be involved (Fig. 4). In this proposal there is a transfer to butyrate from the 4'-phosphopantetheine group of the synthetase to CoA, via the loading site, after one round of the chain elongation reaction.[52] The experiments of Grunnet and Knudsen[53,54] have provided strong evidence for the direct utilization of acyl-CoA by the microsomal enzymes of triglyceride biosynthesis in goat mammary gland.

2.2.3. Sources of Carbon and Reductant
As noted previously, there are striking differences in the utilization of glucose and acetate for lipogenesis by ruminants and non-ruminants.

These are achieved by some fundamental differences in the activity and the cellular location of key enzymes. In the scheme shown in Fig. 3, blood glucose absorbed by the mammary gland is metabolized to pyruvate via the Embden–Meyerhof or the hexose monophosphate pathway in the cytosol, and mitochondrial acetyl-CoA is produced by the oxidative decarboxylation of pyruvate which diffuses into the mitochondria. However, utilization of acetyl-CoA for biosynthesis of fatty acids *de novo* occurs in the cytosol. Mitochondrial membranes have a very low permeability to acetyl-CoA and measured rates of diffusion of the mitochondrial acetyl-CoA derived from glucose to the cytosol are too slow to account for observed rates of fatty acid synthesis from glucose. Of alternative mechanisms for making acetyl-CoA produced in the mitochondria available in the cytosol, there is strong experimental support for the so-called citrate-cleavage pathway.[55,56] This involves citrate synthesis from acetyl-CoA and oxaloacetate by the mitochondrial enzyme citrate synthetase, translocation of citrate to the cytosol and then cleavage to acetyl-CoA and oxaloacetate by ATP-citrate lyase (Fig. 3). The mitochondrial membrane is impermeable to oxaloacetate and the citrate transport system (citrate-cleavage cycle) is completed by reduction of oxaloacetate to malate by NAD-malate dehydrogenase, oxidative decarboxylation of malate to pyruvate and then entry of pyruvate into the mitochondria. During lactation the considerable amount of citrate secreted by the mammary gland is synthesized within the mitochondria and translocated as outlined above.

The non-utilization of glucose for fatty acid biosynthesis in ruminants is a consequence of the low or lack of activity of two key lipogenic enzymes, namely, ATP-citrate lyase and NADPH-malate dehydrogenase (malic enzyme).[38] Much of the acetate absorbed from the blood by the ruminant mammary gland is converted to acetyl-CoA by cytosolic acetyl-CoA synthetase; the remainder enters the mitochondria and after activation contributes to the supply of energy in the form of ATP and to citrate or *iso*citrate translocated to the cytosol.

Three possible mechanisms are available for the provision of the large amount of cytosolic NADPH required for fatty acid biosynthesis.[38] These are the hexose monophosphate, the malate-transhydrogenation and the *iso*citrate cycles. Since the first two of these cyclic pathways require the oxidation of glucose and the third the oxidation of acetate, it would be expected that there are some important quantitative differences in their contributions to NADPH requirements in ruminants and non-ruminants. In rat mammary tissue the hexose monophosphate cycle is, by far, of greatest quantitative importance.[57] Although the transhydrogenation pathway contributes substantially to the pool of NADPH in adipose tissue,[58] in the mammary gland the requirements for milk citrate reduce the proportion of total cytosolic citrate metabolized via this route to pyruvate.

However, ruminant tissues lack NADPH-malate dehydrogenase (Fig. 3) and the hexose monophosphate and *iso*citrate cycles provide the NADPH necessary for fatty acid biosynthesis.[38] An important difference between these two alternative sources of NADPH is that in the former reducing equivalents are generated from glucose metabolism whereas in the latter they may be generated from acetate metabolism. The operative route from acetate begins with the conversion of acetate to intramitochondrial acetyl-CoA (Fig. 3). The citrate which diffuses to the cytosol is oxidized to 2-oxoglutarate by NADP-dependent *iso*citrate dehydrogenase. In the proposed *iso*citrate cycle, glutamate dehydrogenase acts as a transhydrogenase to generate intramitochondrial NADPH which allows for a recycling of 2-oxoglutarate to citrate in mitochondria.[59] Glucose oxidation in ruminant mammary tissue is predominantly via the hexose monophosphate pathway[60,61] and about equal amounts of NADPH are believed to be produced in the ruminant mammary gland by this and the *iso*citrate pathway.[33,59,62] Whereas in the non-ruminant mammary gland the hexose monophosphate pathway contributes 80–100% of the NADPH required for fatty acid biosynthesis, in ruminants there is substantial input from a pathway utilizing acetate.

2.2.4. Activities of Enzymes During Lactogenesis

Measurement of enzyme activities in mammary tissue in the transition from pregnancy to lactation is very useful in the identification of rate-limiting enzymes. There is a high correlation coefficient between acetyl-CoA carboxylase activity and fatty acid synthesis in mammary tissue[36] and the weight of evidence suggests that this enzyme, in the normally fed animal, is the rate-limiting enzyme during fatty acid synthesis.[63] Increased activity of mammary gland acetyl-CoA carboxylase at the onset of lactation is related to 30- to 40-fold increases in its concentration resulting from an increased rate of synthesis relative to degradation[64] and to changes in the proportion of the enzyme in active and inactive forms.[63]

Some marked differences between species are found in the activities of enzymes which are capable of producing acetyl-CoA and NADPH. In ruminants, there are increases in reactions which utilize acetate for acetyl-CoA and NADPH formation due to 25- and 5-fold increases in the activities of acetyl-CoA synthetase and NADP-*iso*citrate dehydrogenase, respectively.[36] In contrast, the enzymes which increase in activity in non-ruminants catalyse reactions which utilize glucose for these purposes. Consequently, substantial increases in the activities of phosphofructokinase (Embden–Meyerhof pathway), ATP-citrate lyase and NADPH-malate dehydrogenase (citrate-cleavage cycle) occur in rat mammary tissue.[38] The activity of NADP-*iso*citrate dehydrogenase remains low throughout.[65] Activities of hexose monophosphate dehydrogenases increase 2- to 3-fold in the cow and sheep from pregnancy to lactation[36,66] but increase 15- to 18-fold in rat mammary tissues,[38] confirming that the oxidation of glucose, via the hexose monophosphate pathway, provides an important source of NADPH in both ruminants and non-ruminants. The greater activity of the dehydrogenases in non-ruminants may be related to competition for substrate being oxidized via the Embden–Meyerhof pathway. It should be noted that NADP-malate dehydrogenase activity is very low or absent in rabbit mammary gland whereas *iso*citrate dehydrogenase activity increases during lactation.[65] Substantial levels of activity of acetyl-CoA synthetase, as well as ATP-citrate lyase, are sustained during lactation which is consistent with both acetate and glucose being important fatty acid precursors in the rabbit[31] and indicative of acetate availability through microbial fermentation in the lower gut.

As discussed above, the activities of a number of the lipogenic enzymes appear to increase from the time of parturition, but others, such as the soluble thioesterase, which induces the release of medium-chain fatty acids from the fatty acid synthetase,[49] and lipoprotein lipase which is concerned with the provision of fatty acid from plasma chylomicrons and VLDLs,[67] show activity before parturition.

2.2.5. Regulation of Fatty Acid Biosynthesis

As discussed previously, the onset of lactation is accompanied by dramatic increases in the capacity of mammary tissue to synthesize fatty acids and, in most species, by an ability of mammary tissue to synthesize milk-specific fatty acids.[68–70] Both changes occur in response to hormonal stimuli. Among the several enzymes which exhibit potential for control, e.g. fatty acid synthetase, ATP-citrate lyase in non-ruminants and *iso*citrate-NADP dehydrogenase, acetyl-CoA carboxylase has been identified as rate-limiting in lipogenesis under most physiological conditions; recently McNeillie and Zammit[63] have shown that its short-term regulation involves activation by insulin whereas prolactin has longer-term stimulatory effects on acetyl-CoA carboxylase concentrations in the mammary gland.

In the cow, only traces of the milk-specific fatty acids, namely those of short- and medium-chain length, are synthesized by mammary tissue 18 days before parturition, whereas 7 days after parturition about 40% of the fatty acids are 4:0 to 12:0.[71] Similarly, in the rabbit between the 18th day of pregnancy and the 2nd day after parturition the fatty acid products of acetate utilization change from being predominantly 4:0, 6:0, 18:0 and 18:1 to being predominantly 8:0 and 10:0.[68] The role of insulin, corticosterone and prolactin in fatty acid biosynthesis is illustrated with reference to the work of Forsyth et al.[72] with rabbit mammary tissue. Explants from 16-day pregnant rabbits utilized acetate for the synthesis of long-chain fatty acids but after culturing for 2 to 4 days with prolactin there was an 8-fold increase in total fatty acid synthesis and 40–50% of the fatty acids synthesized was 10:0. In order to obtain the 30- to 40-fold increases in fatty acid synthesis in pre-lactation tissue which are observed in early lactation insulin and corticosterone, as well as prolactin, were required. It is assumed that prolactin induces medium-chain fatty acid hydrolase in the explants.[39]

A role for prolactin in regulating the relative rates of lipogenesis in mammary, liver and adipose tissues during weaning has been proposed.[73] On removal of pups for 24 h at peak lactation in rats there was a 95% decrease of ^3H_2O incorporation into lipid by mammary gland, a 77% increase by liver and a 330% increase by adipose tissue. These latter increases were prevented by administering prolactin.

3. TRIGLYCERIDE BIOSYNTHESIS

3.1. Reaction Pathways

The so-called glycerol-3-phosphate pathway, the major route to triglyceride biosynthesis in the mammary gland, is outlined in Fig. 5 along with the possible alternative dihydroxyacetone phosphate and 2-monoglyceride pathways. The glycerol-3-phosphate pathway utilizes *sn*-glycerol-3-phosphate (G-3-P) and acyl-CoA's as substrate, and the three transferases involved, namely acyl-CoA:*sn*-glycerol-3-phosphate acyl transferase, acyl-CoA:1-acyl *sn*-glycerol-3-phosphate acyl transferase and acyl-CoA:1,2-diglyceride acyl transferase, appear to be closely associated in the microsomal fraction and are referred to as the triglyceride synthetase. The remaining enzyme required for triglyceride synthesis, phosphatidate phosphatase, appears to be less firmly associated with the endoplasmic reticulum and is partly cytosolic. Each of these individual enzymic steps (Fig. 5) has been studied using mammary gland preparations.[24,74-7]

Fatty acids are utilized as their CoA-esters and, as discussed above, the short- and medium-chain fatty acids seem to be released from the fatty acid

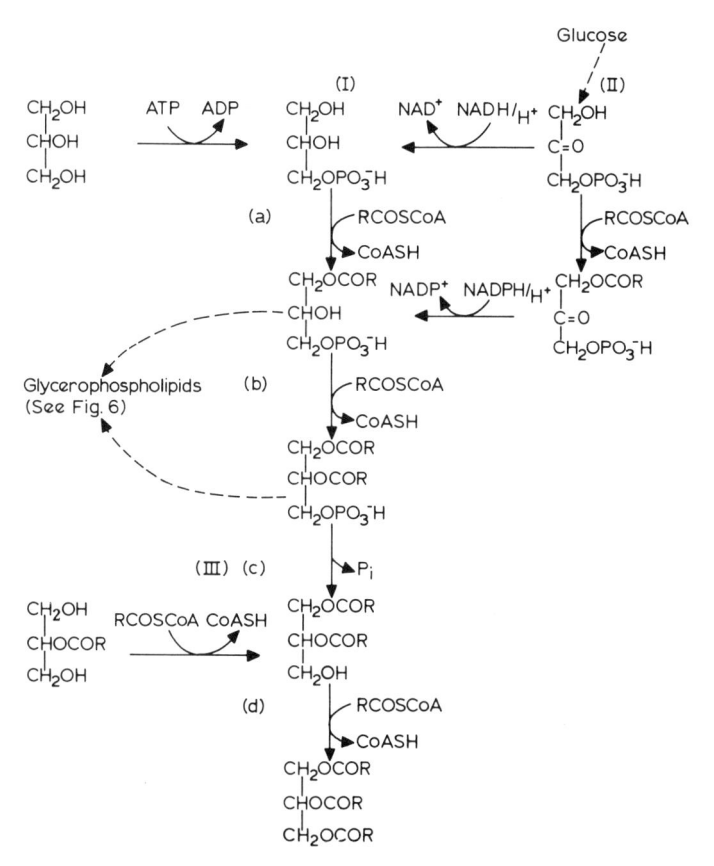

Fig. 5. The biosynthesis of milk glycerolipids by the: (I) glycerol-3-phosphate (II) dihydroxyacetone phosphate, and (III) 2-monoglyceride pathways. (a) Acyl-CoA:sn-glycerol-3-phosphate acyl transferase; (b) Acyl CoA:1-acyl-sn-glycerol-3-phosphate acyl transferase; (c) Phosphatidate phosphatase; (d) Acyl-CoA:1,2-diacylglycerol acyl transferase.

synthetase in this form via transacylase reactions.[51,52] However, the long-chain fatty acids originating either from blood lipids (C_{16}–C_{18}) or from the mammary gland fatty acid synthetase are available as free acids. Therefore before utilization for triglyceride synthesis they must be converted to acyl-CoA's by acyl-CoA synthetases. In mammary gland these synthetases are not well defined but since the major route of fatty acid metabolism is directed to the synthesis of triglycerides rather than oxidation[4] it would be expected that acyl-CoA synthetases are associated with the endoplasmic reticulum and not the mitochondria as in muscle and other tissues concerned with the oxidation of fatty acids.[78,79] In rabbit mammary gland there are co-ordinated increases in the activities of acyl-CoA synthetases and enzymes concerned solely with lipid biosynthesis, i.e. fatty acid synthetase, acyl-CoA:*sn* glycerol-3-phosphate acyl transferase and phosphatidate phosphatase, during pregnancy and lactation.[79] Studies *in vitro* on liver microsomes[80] and mammary gland homogenates[79] suggest that acyl-CoA synthetase is not rate-limiting.

G-3-P is synthesized within the mammary gland from glycerol by cytosolic glycerol kinase and from glucose via the glycolytic or hexose monophosphate pathways, which are dominant in non-ruminants and ruminants respectively. Bauman and Davis[29] reasoned that the proportion of glycerol derived from glucose within the mammary gland should equal the proportion of the total glycerolipid fatty acids synthesized *de novo*, i.e. 50–60% in ruminants and 80% in the rabbit. The remainder arises from the phosphorylation of the glycerol released from blood triglycerides by the action of lipoprotein lipase and taken up by the epithelial cells. There is a substantial increase in the activity of glycerol kinase after parturition.[81] The acylation of dihydroxyacetone phosphate (DHAP) derived from glucose metabolism followed by reduction of 1-acyldihydroxyacetone phosphate to 1-acyl-*sn*-glycerol-3-phosphate provides an alternative route for glycerolipid synthesis in liver[82] and the existence of this pathway is indicated in mouse mammary gland.[83] DHAP and G-3-P were used for glyceride synthesis to a similar degree[83] but mammary tissue contains about 15-times more G-3-P than DHAP.[84]

Although the monoglyceride pathway is operative in intestinal mucosa and adipose tissue,[85] the existence of this route for triglyceride biosynthesis in mammary gland is controversial. Support for the direct utilization of 2-monoglycerides, resulting from lipoprotein lipase hydrolysis of blood triglycerides, arises from the preference of palmitate for position *sn*-2, both in the high molecular weight triglycerides of bovine milk fat and in plasma glycerides, and the incomplete equilibration of *sn*-2-palmitate in the high molecular weight glycerides with free fatty acid pools within the mammary tissue of goats and cows.[86] However, the unequivocal acylation of 2-monoglycerides in mammary tissue has yet to be established.[24,29]

3.2. Specificity of Enzymes

The specificities of the three acylating enzymes which lead to the synthesis of triglycerides via the glycerol-3-phosphate pathway have been examined in attempts to explain the distribution of the individual fatty acids in the constituent triglyceride molecules of milk fat. The complexity of the problem, due to the very large numbers of individual species of triglyceride present in milk fat, will be obvious. Consequently, biochemical explanations of only the most striking features of acyl distribution are being currently sought, in particular the location of butyric and hexanoic acids almost exclusively in position *sn*-3 in the triglycerides of ruminant milk fat and the location of palmitate on positions *sn*-1 and *sn*-2. A further reminder of the complexity of the problem is the preferential esterification of palmitic acid on position *sn*-1 of the triglycerides with short-chain fatty acids on position *sn*-3 (about 40% of the total triglycerides) and a preference of palmitic acid for position *sn*-2 in molecules with long-chain fatty acids on position *sn*-3.[87] Despite this, however, the most widely accepted mechanism of milk fat synthesis predicts that fatty acids are transferred to specific positions of the glycerol moiety but within these positions they are distributed evenly over all species of triglycerides,[88] that is, acylation is non-correlative.[89]

Microsomal preparations synthesize phosphatidic acid and diglyceride when acyl-CoA's and *sn*-glycerol-3-phosphate are used as substrates.[90-2] The two bovine acyl transferases involved (Fig. 5), considered together, show a marked preference for palmitoyl-CoA over myristoyl-, stearoyl- and oleoyl-CoA.[90,91] Acylation of *sn*-glycerol-3-phosphate on position *sn*-1 precedes acylation on position *sn*-2[76] but the partially acylated product does not accumulate. Rat mammary gland microsomes give the highest rates of acylation with palmitoyl-CoA also but the main distinction is between acyl-CoA's in the range C_{14} to C_{18} and C_{10} to C_{12};[92] the latter are very poorly utilized. 1-Palmitoyl- and 1-oleoyl-*sn* glycerol-3-phosphate, the most likely substrates for acyl CoA:1-acyl-*sn*-glycerol-3-phosphate acyl transferase, are acylated by microsomal fractions of bovine mammary gland using acyl-CoA's of chain length C_8–C_{18} but not when butyryl- and hexanoyl-CoA are used.[76] Acyl transferase activity increases with increasing chain length of acyl-CoA from C_8 to C_{16} using both 1-palmitoyl- and 1-oleoylglycerol-phosphate. This is similar to the order of abundance in which the corresponding fatty acids are found at position *sn*-2 of bovine milk triglycerides.[87] Kinsella[93] found that the order of acylation is palmitoyl- >myristoyl->oleoyl->stearoyl->linoleoyl-CoA using saturating levels of monopalmitoyl-*sn* glycerol-3-phosphate but there was no selectivity with less substrate.

Phosphatidate phosphatase, the enzyme that converts phosphatidic acid to 1,2-diglyceride, exists in two forms, one microsomal and the other cytosolic.[94] The microsomal fraction contains all the enzymes required for triglyceride biosynthesis but the relative contributions of the Mg^{++}-dependent microsomal phosphatidate phosphatase and the cytosolic enzyme to overall synthesis are not known.[95] Using microsomes, the major product of sn-glycerol-3-phosphate acylation is phosphatidate but addition of the cytosolic fraction greatly stimulates conversion of phosphatidic acid to di- and tri-glyceride.

The ability of the acyl CoA:1,2-diglyceride acyl transferase of bovine mammary gland to synthesize triglyceride from microsome-bound 1,2-diglyceride and hexanoyl- and butyryl-CoA explains the placement of these short-chain fatty acids at position sn-3.[95] However, two aspects require explanation: first, the mammary gland transferase utilizes palmitoyl-CoA with equal efficiency although considerably less palmitate is esterified in position sn-3 of milk triglycerides;[95] second, the diglyceride acyl transferases of bovine liver and adipose tissue utilize butyryl- and hexanoyl-CoA[96] although butyric and hexanoic acids are not present in the triglycerides of these tissues. Marshall and Knudsen[77] have shown that the incorporation of butyrate from butyryl-CoA by liver and mammary gland into position sn-3 of triglycerides in the presence of palmitoyl-CoA is favoured by a high concentration of butyryl-CoA and by the presence of bovine serum albumin which binds long-chain acyl-CoA. The activities of both mammary gland and liver transferases are independent of the concentration of microsome-bound 1,2-dipalmitoylglycerol. It was interpreted that the intracellular concentration of butyryl- and hexanoyl-CoA relative to non-protein-bound long-chain acyl-CoA, rather than the specificity of the transferase, led to the incorporation of short-chain fatty acids in mammary gland but not in other tissue triglycerides. Recently, Grunnet and Knudsen[53] established that medium-chain fatty acids synthesized *de novo* by goat mammary gland can be directly incorporated into triglycerides. They have speculated on the involvement of a binding factor specific for medium-chain acyl-CoAs and to mammary gland. This could explain the synthesis of these acyl-CoAs only in mammary gland and the synthesis of medium molecular weight triglycerides characteristic of bovine milk fat.[50] Rat mammary gland acyl-CoA:diglyceride acyl transferase shows a broad specificity for C_{10} to C_{18} CoA donors, although palmitoyl-CoA was about twice as effective as decanoyl-CoA.[97] This provides a possible explanation for the preferential positioning of medium-chain fatty acids (C_8–C_{12}) in position sn-3 in rat milk. Tanioka *et al.*[92] also found that the enzymes acylating positions 1 and 2 are highly specific for long-chain acyl-CoA's.

It is interesting to note that approximately 20% of the constituent fatty acids of the vitamin A esters in rat milk have medium-chain (C_8–C_{12}) length.[98] Retinol taken up from plasma, or derived from the hydrolysis of absorbed retinyl esters, is esterified by microsomal acyl-CoA:retinol acyl transferase[98] which, along with the acyl-CoA:diglyceride acyl transferase, has a broad specificity for acyl-CoA donors.

4. BIOSYNTHESIS OF OTHER MILK LIPIDS

4.1. Glycerophospholipids

The small amounts of phospholipid present in milk, mainly as fat globule membranes, appear to be synthesized *de novo* within the mammary gland.[99] Although there may be a small uptake of serum phospholipids by the mammary gland, these are hydrolyzed during absorption and the products, including fatty acids, are utilized by mammary tissue.[100] Phosphatidylcholine, quantitatively the most important phospholipid in milk and mammary tissue, and phosphatidylethanolamine are synthesized in the mammary gland, as in mammalian tissues generally, via phosphatidic acid and *sn*-1,2-diglyceride (Fig. 6). The phosphorylcholine and phosphorylethanolamine moieties are transferred to diglycerides from CDP-choline and CDP-ethanolamine in the synthesis of phosphatidylcholine and phosphatidylethanolamine, respectively. Two soluble kinases which lead to the formation of phosphorylcholine and phosphorylethanolamine in mammary tissue have been detected[101] and, by analogy with other tissues, these compounds are the precursors of the CDP derivatives. Utilization of selected 1,2-diglycerides and these CDP-derivatives in phosphorylcholine and phosphorylethanolamine transferase reactions in phospholipid biosynthesis,[102,103] followed by deacylation–reacylation,[104] appears to be responsible for their characteristic fatty acid composition.[105] The specificities of a re-acylating enzyme, namely acyl-CoA:1-acyl-*sn*-glycerol-3-phosphorylcholine acyl transferase, associated with bovine mammary gland microsomes have been found to vary with the nature of the 1-acyl residue.[104] When 18:1 was a constituent of lysophosphatidylcholine, oleoyl-CoA was preferred over stearoyl- and linoleoyl-CoA while palmitoyl- and myristoyl-CoA were favoured when 16:0 was the lysophospholipid constituent fatty acid. In alternative minor transformations in the mammary gland, phosphatidylserine may be decarboxylated to phosphatidylethanolamine, and phosphatidylethanolamine methylated to phosphatidylcholine.[102] It has been suggested that about 20% of the phosphatidylcholine in rat liver is synthesized via the

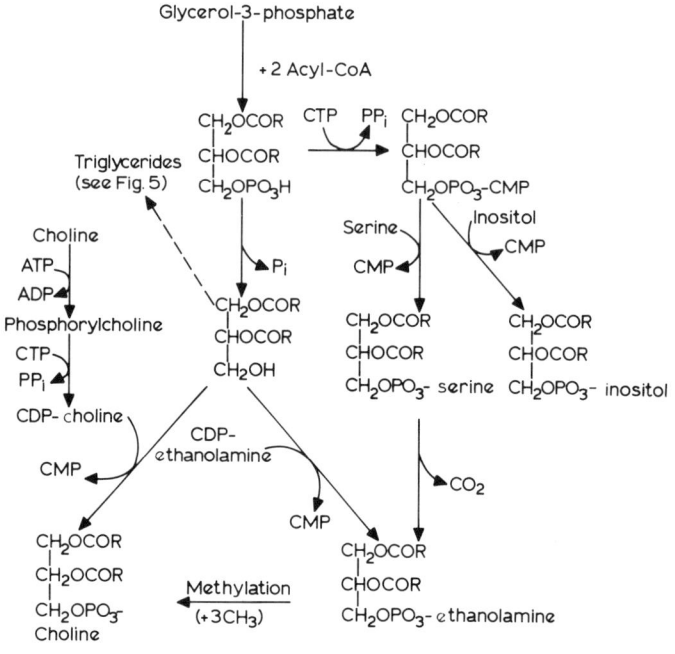

FIG. 6. Pathways in the biosynthesis of milk glycerophospholipids.

methylation pathway.[106] As in other tissues, phosphatidylserine and phosphatidylinositol are synthesized in mammary gland by cytidine-diphosphodiacyl-*sn*-glycerol:serine and :myoinositol transferases, respectively (Fig. 6). Microsomal and soluble phosphatidylinositol phosphohydrolases may lead to the rapid breakdown of the inositide.[107,108]

4.2. Sphingolipids

It seems likely that most, if not all, of the milk lipids containing the sphingosine moiety are synthesized in the mammary gland.[99] Sphingosine is synthesized *de novo* from serine by dispersed bovine mammary cells and, together with fatty acyl groups and the phosphorylcholine group from CDP-choline, is incorporated into sphingomyelins.[104] The biosynthetic pathways outlined in Fig. 7 are based on extensive studies in other mammalian tissues.

4.3. Cholesterol

It is evident that the cholesterol in milk may be of dietary origin, synthesized by other tissues in the animal, or synthesized in the mammary gland itself. The relative contribution of these three sources appears to be subject to considerable interspecies variation and to further variations due to nutritional and other factors.[109]

Dietary [^{14}C]cholesterol[110,111] or [^{14}C]cholesterol placed in the abomasum is secreted in milk, and in the rat[110] and guinea-pig[109] about 11% and 20%, respectively, of the milk cholesterol is of dietary origin. Although the absorptive processes appear to be the same for cholesterol in ruminants, little cholesterol will be ingested by ruminants on normal diets and some of this will be hydrogenated in the rumen.[112]

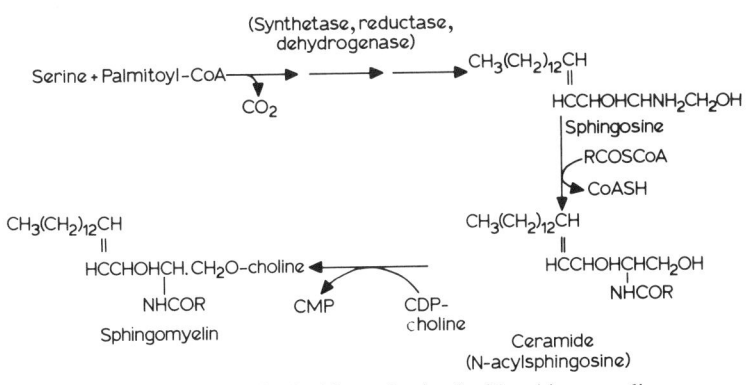

FIG. 7. Pathway in the biosynthesis of milk sphingomyelin.

The relative extents to which the two endogenous cholesterol sources, namely liver and mammary gland, contribute to the balance of the milk cholesterol cannot be determined from dietary labelling experiments until the proportions of dietary cholesterol taken up as chylomicron cholesterol by the liver and the mammary gland are known.[110] There are difficulties in deciding the contributions of various cholesterol-containing components in blood to cholesterol uptake by the mammary gland but it appears that chylomicrons and all three serum lipoproteins (VLDL, LDL and HDL) contribute in the fed animal.[99]

Mammary tissue, like liver and other mammalian tissues, synthesizes cholesterol from acetate by a well-established pathway involving mevalonate and squalene as intermediates.[113] Many experiments *in vivo* and *in vitro*, with both ruminants and non-ruminants, have demonstrated the capacity of the mammary gland to carry out this synthesis *de novo*.[110,114,115] Cholesterol ester synthesis also takes place in the mammary gland.[116]

B. VARIATIONS IN MILK FAT

As discussed above, the composition of the fatty acids synthesized by the mammary gland and those taken up from the blood are very different. Nutritional and other factors may lead to variations in the relative contributions of the fatty acids from these two sources, thus resulting in changes in the composition of milk fat. With non-ruminants in particular, diet may further influence the nature of the fatty acids in blood lipids. However, with ruminants the biochemical events in the rumen considerably reduce this dietary influence leading to a greater constancy of fatty acid composition of their blood lipids. In contrast to the blood lipids, the nature of the fatty acids synthesized by the mammary gland within a given species is subject to little variation due to nutritional and other factors.

5. YIELD AND FATTY ACID COMPOSITION

5.1. Influence of Diet

5.1.1. Fats and Oils
The high calorific content of fats and oils compared with alternative foodstuffs makes them an obvious choice for inclusion in diets if increased energy input is required. In practice, however, the supplementation of ruminant diets with fats to achieve increases in energy intake is limited because fats at concentrations above 5 to 7% of the basal ration impair digestion and metabolism in the rumen and reduce feed intake.[117,118] Despite this limitation, many investigations have been undertaken to determine the effect of fat supplementation on the composition and yield of bovine milk fat and these studies have been reviewed.[119–122]

The inclusion of fats and oils in the diet of ruminants may have the following effects, thereby influencing, either directly or indirectly, lipogenic pathways leading to milk fat secretion.[120,121]

(i) The synthesis *de novo* of fatty acids, such as 16:0 and 18:0, by micro-organisms in the rumen may be reduced, leading to a reduction in blood lipids and mammary gland uptake and to a reduction in the yield of long-chain fatty acids in milk.

(ii) The amounts of long-chain fatty acids released in the rumen by lipolysis may be increased, leading to increases in the blood lipids, mammary gland uptake and the yield of long-chain fatty acids in milk.

(iii) The intramammary synthesis of fatty acids may be reduced thus decreasing the yield of short- and medium-chain fatty acids in milk. Two mechanisms have been suggested for this reduction: firstly, a reduction in the amounts of acetic and butyric acids produced in the rumen, leading to decreased supplies of acetate and β-hydroxybutyrate to the mammary gland; and secondly, an inhibition of mammary gland acetyl-CoA carboxylase.

These effects appear to be influenced by factors such as : (1) stage of lactation; (2) roughage-to-concentrate ratio and fat content of the basal diet; and (3) amount, composition and physical form of fat added to the basal diet.[121]

Possibly the clearest insight into the effect of dietary fat on milk fat synthesis has been gained by examining the effect of the inclusion of specific fatty acids in the diet. Under normal dietary conditions, 16:0, 18:0 and 18:1 are the major fatty acids taken up by the mammary gland from the blood triglycerides. The inclusion of each of these fatty acids at a level of 4% in a high-roughage diet, which increased the fat supplied to the cow from 300 to 800 g day^{-1}, significantly increased the yield of the added acid in milk fat (Table 1).[123,124] These effects reflect the increased blood concentration and uptake of the supplemented fatty acids. The inclusion of 12:0 and 14:0 in high-roughage diets also led to increases of these fatty acids in milk fat[124] and presumably these fatty acids replace a proportion of 16:0, 18:0 and 18:1 normally absorbed from blood.

There is no significant increase in the yield of any fatty acid other than that used to supplement the diet except that the addition of 18:0 leads to an increased secretion of 18:1, presumably as a result of desaturation of 18:0 in the mammary gland.[18] The appreciable increase in the yield of 18:1, when 18:0 is fed, coupled with the lack of significant increases in 14:1 and 16:1 in the milk fat from cows fed the diets containing the fatty acids 14:0 and 16:0, respectively, indicates that mammary gland desaturases have a much higher specificity for 18:0 than for either 14:0 or 16:0. There is no evidence for extensive chain-elongation of dietary fatty acids.

Supplementation of diets with individual fatty acids has a variable effect on the overall yield of fat in milk; 16:0 and 18:0 produce an increase in the yield of milk fat, 12:0 and 18:1 significantly decrease it, while the inclusion of 14:0 has no significant effect. For all diets there is a decrease in one or more of the fatty acids synthesized within the mammary gland,[123,124] thus

TABLE 1
INFLUENCE OF DIETARY SUPPLEMENTS OF PALMITIC, STEARIC AND OLEIC ACIDS ON THE YIELDS OF FATTY ACIDS IN BOVINE MILK FAT

Fatty acid	Yield (g day^{-1})				
	Basal diet[a]	Basal diet[a] + palmitic acid (4% of diet)	Basal diet[a] + stearic acid (4% of diet)	Basal diet[b]	Basal diet[b] + oleic acid (4% of diet)
4:0 to 8:0	26·6	25·9	29·8	32·4	15·6
10:0	1·9	0·5	2·2	1·4	0·6
12:0	8·0	4·2	5·2	6·2	2·4
14:0	43·3	29·1	39·7	45·8	20·6
14:1	1·9	1·9	0·9	—	—
16:0	147·0	280·7	119·5	174·1	90·0
16:1	4·9	16·2	5·2	—	—
18:0	38·4	19·9	80·7	40·4	33·9
18:1	80·9	65·6	129·9	98·8	133·1
18:2 + 18:3	11·0	7·9	6·0	14·6	9·3
Total	364	452	419	414	306

[a] From Steele and Moore.[124]
[b] From Steele and Moore.[123]

the net effect on the total milk fat yield depends on whether the depression of intramammary synthesis, caused by the inclusion of fatty acids in the diet, outweighs the increased uptake of fatty acids by the mammary gland. Where the depression of intramammary synthesis leads to overall decreases in milk fat yield, as occurs with diets containing 12:0 and 18:1, there are appreciable reductions in the total volatile fatty acids in the rumen and in the ratio of acetic to propionic acid.[123,124]

The supplementation of diets with fats and oils, which contain a variety of constituent fatty acids, obviously produces a more complex situation than the inclusion of individual fatty acids. Nevertheless, the numerous studies on the effect of dietary supplements of fat generally show that the yields of fatty acids in bovine milk exhibit the following trends: (1) an increase in the yields of fatty acids 18:0 and 18:1; (2) no significant increase in the yields of polyunsaturated C_{18} acids; and (3) a decrease in the yields of one or more of the fatty acids of chain length C_6 to C_{16}.

These trends appear to hold over wide ranges in the proportion of roughage to concentrate in the basal diets[125,126] and are also maintained whether the added fat or oil contains predominantly saturated or unsaturated fatty acids.[125-8] In addition, the trends are largely unaffected by the level of supplementation of the diets with fat.[126,128,129] Indeed, reference to Table 2 indicates that the magnitude of each trend generally increases as the level of supplementation increases. Coconut oil, which contains about 50% 12:0 and 8% C_{18} acids, is an exception to this. Not surprisingly, its inclusion in diets increases the yield of 12:0 but causes no

TABLE 2
INFLUENCE OF DIETARY SUPPLEMENTS OF TALLOW ON THE YIELDS OF FATTY ACIDS IN BOVINE MILK FAT[a]

Fatty acid	Yield (g day^{-1})				
	Level of supplementation (% of diet)				
	0	1·0	2·0	3·5	5·0
4:0	21·7	21·6	26·6	26·3	30·2
6:0	20·3	22·0	21·4	18·8	19·4
8:0	14·4	17·5	13·4	12·6	8·7
10:0	34·9	36·0	30·0	22·1	17·0
12:0	42·5	41·8	33·8	24·9	18·4
14:0	116·8	122·2	109·0	90·0	76·9
16:0	274·4	271·8	268·0	236·4	247·7
18:0	32·9	53·6	63·6	70·7	84·0
18:1	100·9	137·8	174·4	189·4	236·3
18:2	10·6	15·3	16·9	15·9	11·8
Total	669	740	757	707	750

[a] From Storry et al.[128]

significant increase in 18:0 or 18:1 in milk fat.[130] As might be anticipated, feeding coconut oil produces an effect similar to feeding lauric acid.[124]

The lack of an appreciable increase in the yields of polyunsaturated acids containing 18 carbons when oils such as peanut,[131] cottonseed,[125] and soybean (Table 3)[127] are fed as supplements is consistent with the hydrogenation of these acids by rumen micro-organisms. The increased yields of milk 18:0 and 18:1 acids could either result directly from the increased dietary intake of these acids or, in the case of highly unsaturated oils, indirectly through ruminal hydrogenation of dietary unsaturated fatty acids containing 18 carbons. Additionally, increased 18:1 could arise through desaturation of 18:0 in the mammary gland.[18] Decreased levels of one or more of the fatty acids 6:0 to 14:0 in milk fat are presumably due to a depression of intramammary synthesis of these fatty acids, and in some experiments have been associated with a reduction in the amounts of acetic acid produced in the rumen.[132,133] When no significant reduction in ruminal acetic acid occurs,[128,134] the decreased synthesis *de novo* of fatty acids in the

TABLE 3
INFLUENCE OF DIETARY SUPPLEMENTS OF SOYBEANS AND SOYBEAN OIL ON THE YIELDS OF FATTY ACIDS IN BOVINE MILK FAT[a]

Fatty acid	Yield (g day^{-1})		
	Basal diet	Basal diet + soybeans[b] (25% of diet)	Basal diet + soybean oil (4% of diet)
4:0	20·3	27·6	17·4
6:0	15·2	18·6	9·2
8:0	14·0	11·4	5·8
10:0	47·7	26·4	13·2
12:0	39·8	34·2	15·0
14:0	101·1	90·0	49·8
14:1	20·7	12·0	10·2
16:0	260·9	203·1	121·1
16:1	28·5	21·1	15·3
18:0	19·7	78·3	37·9
18:1	78·3	192·4	202·5
18:2	7·5	19·8	10·4
18:3	0·2	1·1	5·0
Total	654	736	513

[a] From Steele *et al.*[127]
[b] Soybeans contain 16% oil.

mammary gland could be due to inhibition of acetyl-CoA carboxylase by long-chain fatty acids.[120] Milk 16:0 is derived from intra- and extra-mammary sources and therefore its yield in milk fat will depend upon a balance between depression of intramammary synthesis and increased transfer of 16:0 from the blood. The inclusion of fat in the cow's diet generally produces a reduction in overall yield of 16:0 in the milk. However, when fats containing high proportions of 16:0, such as palm oil[129,131] and tallow (Table 2), are added to the diet variations in yield are slight. This contrasts with the increase in the yield of 16:0 in milk fat when 16:0 itself was included in the diet.[124,135] Supplementation of the diet with fat produces a variable effect on the amount of 4:0 in milk, ranging from increases in daily yield to reductions (Tables 2 and 3).[129,133]

The effects of dietary fat supplements on the overall yield of milk fat depend upon the combined effect of the above-mentioned trends in yields of fatty acids which are in turn a reflection of the contributions to milk fat of fatty acids from the different lipogenic pathways. Not surprisingly, this has a variable effect on the daily amount of milk fat secreted. Appreciable increases in milk fat yield have resulted from the inclusion of certain fats and oils in the diet,[125,127,131] but other investigations involving similar supplementations have produced no significant change,[125,126,128–30,134] while the addition of soybean oil[127,136] and coconut oil[130] to diets may reduce yield. The amount of milk fat secreted daily may vary during the course of a feeding trial[125,130] and is affected by the level of fat supplementation.[130,136] Furthermore, the physical form in which fat is added affects yield—soybean oil added as a supplement may decrease yield whereas ground soybean may increase the amount of milk fat (Table 3).

In contrast to ruminant milk fat, the fatty acid composition of human and other non-ruminant milks may be radically altered by changes in total calorific intake as well as by changes in the types of dietary fat.[137] During energy equilibrium or during calorie deficiency the constituent fatty acids of human milk closely resemble those of the dietary or body fats but when excess non-fat is fed, the milk shows a marked increase in 12:0 and 14:0 and a marked decline in all polyenoic fatty acids, e.g. linoleic acid (18:2) falls to about 1·5%. The 18:2 content of infant diets is of particular interest in relation to the provision of essential fatty acids and vitamin E requirements.[138] Variations in the 18:2 content of human milk fat reflect differences in the intake of dietary 18:2 and experimental diets containing corn oil gave in excess of 40·0% 18:2.[137] The mean level of 18:2 in the milk of 12 New Zealand mothers on normal diets for 24 weeks postpartum was 6·3%, which is much higher than the typical mid-lactational level in bovine milk fat (0·9%).[139] North American studies showed average 18:2 levels of 7·1% for 11 mothers[140] but with individuals as high as 16·6%.[141]

5.1.2. Protected Fats and Oils

Australian workers have developed a technique which has enabled large amounts of fats and oils to be fed to ruminants without causing a reduction in dietary intake.[142,143] This has been accomplished by treating fats or oils so that they are not released into the rumen and thus do not affect the metabolic processes in the rumen nor are they subject to ruminal lipolysis and biohydrogenation. Protection is achieved by coating minute droplets of oil with a layer of formaldehyde-treated protein which resists the action of rumen micro-organisms. The oil is released in the acidic conditions prevailing in the abomasum thereby permitting its subsequent digestion and absorption. The technique has generated considerable interest and research and has been the subject of recent reviews.[118,122,144]

The correct preparation of the supplement is necessary to ensure adequate protection and thus efficient transfer of dietary lipid. Maximum encapsulation is achieved by emulsifying the fat in a protein phase prior to treatment with formaldehyde.[144] Direct treatment of oilseeds with formaldehyde does not give adequate protection of the lipid.[145,146] Provided the release of lipid into the rumen is prevented, up to 1500 g day^{-1} of lipid can be effectively digested by the cow.[147] Studies on serum lipids demonstrate the ability of ruminants to absorb, transport and metabolize large quantities of dietary fat without obvious metabolic stress.[148] In one trial, protected oils were fed for two years without harmful effects.[149]

The utilization of long-chain fatty acids, without intervention of rumen reactions, provides an energy source which can be used more efficiently than the major source of digestible energy normally available, namely cellulose and other carbohydrates via the volatile fatty acids. Despite this, the feeding of protected fat supplements does not always result in increased energy intake.[118,150] Lactational effects and the use of inadequately protected supplements may offer explanations for this response. Diets supplemented with protected lipids produce a considerable increase in the daily yield of milk fat and amounts 15 to 30% greater than that from control diets have been recorded.[133,150-3] This is a consequence of the direct absorption of large quantities of long-chain fatty acids, coupled with the lack of depression of intramammary synthesis normally associated with the inclusion of non-protected fat supplements in the diet.[118,133]

One of the most interesting facets of feeding protected lipids is that the fatty acid composition of milk fat is a direct reflection of the component fatty acids of the supplement in a manner similar to that found in non-ruminants. In other words, the technique enables the fatty acid composition, and hence the physical properties of milk fat, to be dramatically altered by inclusion of the appropriate fat or oil. Within 24 to 48 h of commencement of feeding a protected supplement containing an 18:2-rich oil, such as sunflower-seed oil, the 18:2 content of milk fat begins to rise rapidly and may reach levels in excess of 15% (Table 4),[146,151,154] with the transfer of 18:2 to milk in the range of 20 to 40%.[146,147] The increased amount of milk fat secreted means that the yields of most fatty acids do not decrease despite the large percentage rise in 18:2 (Table 4).

The feeding of protected lipids was introduced as a means of producing bovine milk fat with elevated 18:2 levels, which may be regarded as desirable from a nutritional viewpoint and which provides the opportunity

TABLE 4

INFLUENCE OF A DIETARY SUPPLEMENT OF PROTECTED SUNFLOWER-SEED OIL ON THE COMPOSITION AND YIELDS OF FATTY ACIDS IN BOVINE MILK FAT[a,b]

Fatty acid	Fatty acid composition (mol %)		Yield (g day^{-1})	
	Basal diet	Basal diet + sunflower-seed oil	Basal diet	Basal diet + sunflower-seed oil
4:0	11·1	10·4	62·2	71·8
6:0	5·1	5·4	28·6	37·3
8:0	2·2	2·1	12·3	14·5
10:0	3·7	3·2	20·7	22·1
12:0	3·7	2·6	20·7	17·9
14:0	9·6	7·7	53·8	53·1
14:1	0·8	0·5	4·5	3·5
15:0	0·9	0·7	5·0	4·8
16:0	20·5	14·1	114·8	97·3
16:1	1·8	1·1	10·1	7·6
17:0	0·6	0·5	3·4	3·5
18:0	14·0	14·1	78·4	97·3
18:1	22·2	20·6	124·3	142·1
18:2	1·6	15·5	9·0	107·0
18:3	1·0	0·7	5·6	4·8
			553	685

[a] From Morrison and Hawke.[153]
[b] Diet contained 1·25 kg day^{-1} protected sunflower-seed oil.

to produce a butter which is spreadable at low temperatures.[144] However, the use of this technique to feed protected non-polyunsaturated fat supplements as a high-energy source which is easily digestible has general application.[118] Supplements of this type may be used at selected times during the dairying season to maintain milk fat production and reduce body weight loss.

5.1.3. Low-Fat Syndrome

The term 'low-fat syndrome' has been applied to the depression in the fat content of milk and in milk yield observed when diets contain either: (1) high ratios of readily digestible carbohydrate to fibrous material; or (2) supplementary amounts of unsaturated oils or fatty acids. The similar end-effect of these two quite different dietary conditions does not appear to be due to recognizably similar physiological and metabolic changes induced by the diets. As discussed above, the component fatty acids of the fat secreted in milk are either synthesized within the mammary gland or are absorbed as blood lipids which originate in the diet or in other tissues. Reduction in the availability to the mammary gland of fatty acid precursors for synthesis *de novo* or of lipids from the blood supply is likely to be the basic cause of the syndrome.

In the early work on the depressant effect of lipid supplementation of diets on milk fat yield,[155] attention focussed on the effect of polyunsaturated oils, such as cod liver oil, but the effect appears to be a phenomenon of added lipids generally which has been discussed above.[120] In summary, features of the fat-depressant effect of administering oils include: (1) no consistent changes in the ratios of short-chain fatty acids produced in the rumen,[156] (2) the same or greater uptake of long-chain fatty acids from blood by the mammary gland;[120] (3) oil administered in ways that bypass the rumen increasing rather than decreasing milk fat yields;[157] and (4) a failure of fats and oils administered post-ruminally to depress intramammary gland synthesis of fatty acids.[156] In considering these and other effects, Davis and Brown[156] and Storry[120] conclude that the effect of unsaturated oils, and possibly other oils and fatty acids, in lowering milk fat is on rumen fermentation and not directly on the mammary gland.

The decreases in fat content of milk encountered with high-producing cows on high-concentrate and low-roughage diets can be much more dramatic than when diets are supplemented with oils, and in extreme cases the fat content may be less than 2·0%, with 50–60% lower fat yield. In general, the milk fat content falls as the proportion of forage in the diet declines, the decrease being appreciable below about 40% forage.[158]

There appears to be a close relationship between the amount of propionate produced by microbial fermentation in the rumen and lowered milk fat,[159] which provides a link between diet and the low-fat syndrome. Dietary conditions which give increased fermentation rates, such as lowered forage-to-concentrate ratios, reduce physical fibrousness of the forage and increase readily fermentable carbohydrate,[158] are those associated with depressed milk fat. These are the same conditions which lead to greater propionate production and an increased propionate-to-acetate ratio.[160] The milk fat depression observed in grazing cows in early spring[161] is probably diet-related. Perennial ryegrasses with early spring growth characteristics have been identified which are relatively low in crude fibre and high in soluble carbohydrate and which produce high propionate-to-acetate ratios in the rumen and significant reductions in milk fat.[162]

Propionate is the principal glucogenic substance available to ruminants and its greater availability relative to lipogenic compounds such as acetate appears to depress milk fat secretion. It has been proposed that higher rates of glycerol-3-phosphate synthesis in adipose tissue, which arise from increased gluconeogenesis and increased glucose production and elevated blood insulin in cows fed concentrates, increases the capacity of this tissue to synthesize triglycerides.[163] This results in a decreased availability of free fatty acids (FFAs) and lipogenic precursors for milk fat synthesis. Increases in the activities of lipoprotein lipase and triglyceride synthetase have been observed in adipose tissue of cows exhibiting depressed milk fat secretion.[164]

Clearly it is important to maintain an adequate level of fibre in the diet in order to avoid the low-fat syndrome and it is not too difficult to predict, from feeding regimes used, when serious declines in milk fat are likely to occur. However, the more marginal depressions which sometimes occur with more conventional feedstuffs are more difficult to anticipate and rectify, yet may be of considerable financial disadvantage to the producer.

5.2. Influence of Lactation and Season

There is general agreement that, provided the plane of nutrition is maintained at a high level, the amount of milk secreted daily increases slightly during the first four to six weeks of lactation and thereafter declines, while the proportion of fat in milk initially declines and then increases later in lactation.[165] As a consequence of these trends, the yield of milk fat gradually declines as lactation progresses.[165-7]

Decaen and Adda[166] have shown that the yields of fatty acids vary with the stage of lactation in the following manner: (1) 4:0 gradually declines throughout lactation; (2) the acids 6:0 to 14:0 rise from low levels at the beginning of lactation to a maximum six to ten weeks after calving and thereafter decline; (3) 16:0 shows little variation as lactation progresses; and (4) 18:0 and 18:1 decline rapidly for the first six to eight weeks and thereafter at a slower rate. Other workers[167,168] generally support these findings although the increase in 6:0 and 8:0 throughout lactation observed by Stull et al.[167] is an example of a contradictory result.

The seasonal variation in the fatty acid composition of milk fat is affected by a number of factors such as composition of diet, plane of nutrition and stage of lactation. Despite this complexity, studies in different countries have shown a seasonal pattern of fatty acid variation which consistently re-occurs.[169–71] In New Zealand, where the general pattern of dairy husbandry is such that cows are grazed on pasture throughout the year and lactation commences in early spring, the following trends in the proportions of fatty acids of milk fat are observed:[172] (1) a general decline in 4:0 throughout the season; (2) 6:0 to 14:0 increase from low values at the beginning of the season in early spring to a maximum value during early summer and thereafter decline throughout the rest of the season; (3) values for 16:0 rise during early summer and then decrease in autumn; and (4) both 18:0 and 18:1 show a seasonal trend which is the reverse of that exhibited by 16:0. Comparable seasonal variations in fatty acids have been obtained in Australia,[169] where the general pattern of dairy husbandry is similar to New Zealand. The seasonal change in the fatty acids of chain length C_4 to C_{14} appears to be caused by lactational effects since the variation is the same as that mentioned above. In contrast, the seasonal trend in 16:0, 18:0 and 18:1 does not fit the lactational changes shown by these acids. Rather these appear to be diet-related. Hawke[173] has shown that milk fat from cows grazed on mature summer ryegrass, which has less lipid than new spring and autumn growth, contains lower proportions of 18:0 and 18:1 and a greater proportion of 16:0 than milk fat from cows fed immature ryegrass. The finding that the seasonal trends in 16:0, 18:0 and 18:1 are a consequence of dietary effects is not surprising when account is taken of the sources of these fatty acids. Although the short- and medium-chain fatty acids are synthesized in the mammary gland, variations in fermentation reactions in the rumen brought about by seasonal dietary effects may also be important in leading to variations of these fatty acids in milk which may be superimposed upon lactational effects.

These findings are in contrast to results obtained in North America and Europe[170,171,174,175] which generally show the following trends for the proportions of fatty acids: (1) 4:0 to 14:0 show no appreciable variation throughout the season; (2) 16:0 is lower in summer than in winter; and (3) 18:0 and 18:1 are higher in summer than in winter. The lack of seasonal pattern in the fatty acids 4:0 to 14:0 may be attributed to the practice of not confining calving to the early spring while the variations in 16:0, 18:0 and 18:1 may result from a change in feeding conditions, since it has been shown that the change from a winter feed of hay and concentrate to a diet of fresh grass causes an increase in the proportions of fatty acids containing 18 carbon atoms and a decrease in the proportions of 14:0 and 16:0.[176]

It would appear, therefore, that the difference between the seasonal variations in fatty acid composition in northern-hemisphere countries and those in southern-hemisphere countries is a consequence of different dairy husbandry practices.

6. TRIGLYCERIDE COMPOSITION AND MELTING BEHAVIOUR

As noted in Chapter 1, bovine milk fat consists of many different triglycerides which vary considerably in molecular weight (mol. wt.) and degree of unsaturation. Despite this complexity, the general pattern of stereospecific distribution of fatty acids in milk fat triglycerides obtained from cows fed basal diets in different parts of the world is the same.[87,177-9] In bovine milk fat, 4:0 is almost entirely esterified at position sn-3 and about 90% of 6:0 is also esterified at this position. The fatty acids 10:0, 12:0 and 14:0 are preferentially esterified at position sn-2 while 16:0 is incorporated preferentially at positions sn-1 and sn-2. 18:0 is preferentially esterified at position sn-1, and 18:1 shows a preference for positions sn-1 and sn-3. Furthermore, this overall pattern of fatty acid distribution in milk fat applies throughout the dairying season.[87,179] Consequently, the above-mentioned changes in the stage of lactation and in feed conditions which occur with the progress of the dairying season have little effect on the positional specificity of fatty acids.

When a cow is placed on a restricted diet there is a decrease in the proportions of fatty acids of chain length C_4 to C_{16} and an increase in 18:0 and 18:1 as a result of the mobilization of body reserves.[180] This causes a marked increase in the proportion of 18:1 at position sn-3 from 17·1 to 47·6% which is accompanied by reductions in the proportions of 4:0 to 16:0 at this position (Table 5). However, the overall pattern of stereospecific distribution of fatty acids is altered only slightly by these changes. Parodi[179] has proposed that a cow, faced with a reduction in the amounts of short- and medium-chain fatty acids available for biosynthesis, has the ability to increase the amount of 18:1 available for esterification at position sn-3 by desaturation of 18:0 in the mammary gland. This ensures that milk triglycerides remain liquid at physiological temperatures. The proposal is an interesting one in view of the general acceptance that fatty acids at position sn-3 are the last to be incorporated into triglycerides (Fig. 5).

TABLE 5
INFLUENCE OF A RESTRICTED FEEDING REGIMEN ON THE POSITIONAL DISTRIBUTION OF FATTY ACIDS IN BOVINE MILK FAT[a]

Fatty acid	Fatty acid composition (mol %)					
	Basal diet			Restricted diet		
	Position[b]			Position[b]		
	1	2	3	1	2	3
4:0	0	0·4	30·6	0·4	0	26·8
6:0	0	0·7	13·8	0·7	0·4	9·2
8:0	0·3	3·5	4·2	0·7	2·6	1·8
10:0	1·4	8·1	7·5	1·1	5·6	2·4
12:0	3·5	9·5	4·5	2·1	6·7	0
14:0	13·1	25·6	6·9	7·4	17·4	0
16:0	43·8	38·9	9·3	32·6	40·7	2·1
18:0	17·6	4·6	6·0	20·9	5·6	10·1
18:1	19·7	8·4	17·1	34·0	21·1	47·6

[a] Calculated from the data of Parodi.[179]
[b] Relative to sn-glycerol-3-phosphate.

Milk fat with a high 18:2 content, obtained by feeding cows a diet supplemented with a protected 18:2-rich oil, contains considerable proportions of 18:2 in each of the three positions (Table 6).[178] Comparisons between the stereospecific distributions of fatty acids in normal and in 18:2-rich milk fat show that the increases in the proportions of 18:2 are compensated for by: (1) decreases in the proportions of 16:0 in each of the three positions; (2) a decrease in the proportion of 14:0 at position sn-2; and (3) a decrease in the proportion of 4:0 at position sn-3, without significantly altering the general pattern of positional specificity of these acids.

The many different triglycerides of bovine milk fat can be broadly classified into two main structural types—those containing three long-chain fatty acids (high mol. wt. triglycerides) and those containing one short- and two long-chain fatty acids (low mol. wt. triglycerides). In view of the highly specific placement of the short-chain fatty acids 4:0 and 6:0 at position sn-3, it follows that a clearer picture of the positional specificity of fatty acids is provided by a structural analysis of these high and low mol. wt. triglycerides. Investigations of this type show that the overall pattern of stereospecific distribution of fatty acids of chain length C_4 to C_{16} in normal milk fat also applies to the high and low mol. wt. triglycerides of milk fat.

TABLE 6
INFLUENCE OF A DIETARY SUPPLEMENT OF PROTECTED SUNFLOWER-SEED OIL ON THE POSITIONAL DISTRIBUTION OF FATTY ACIDS IN BOVINE MILK FAT[a,b]

Fatty acid	\multicolumn{6}{c}{Fatty acid composition (mol %)}
	Basal diet			Basal diet + sunflower-seed oil		
	Position[c]			Position[c]		
	1	2	3	1	2	3
4:0	0	0	36·3	0	0	31·8
6:0	0	1·8	13·2	0	2·4	12·2
8:0	0·9	3·3	2·2	0·8	4·0	1·3
10:0	2·0	6·2	3·8	1·9	6·8	1·7
12:0	3·3	7·5	1·8	2·9	6·5	0·7
14:0	9·4	23·1	3·9	8·2	15·3	2·6
16:0	32·1	30·9	8·9	21·9	19·1	5·2
18:0	19·5	6·1	9·2	22·1	8·7	11·6
18:1	25·3	12·9	14·9	22·6	16·5	18·4
18:2	1·9	1·2	0·9	13·5	16·4	11·3
18:3	1·1	0·9	0·8	1·1	0·9	1·2

[a] From Morrison and Hawke.[178]
[b] Diet contained 1·25 kg day^{-1} protected sunflower-seed oil.
[c] Relative to *sn*-glycerol-3-phosphate.

However, the pattern of distribution of 18:0 and 18:1 varies according to the mol. wt. of the triglycerides; these fatty acids are preferentially esterified at positions *sn*-1 and *sn*-3 in triglycerides of high mol. wt. and concentrated at position *sn*-1 in low mol. wt. triglycerides.[87,178,181-3]

Similarly, in 18:2-rich milk fat the overall pattern of stereospecific distribution of the fatty acids 4:0 to 16:0 is unaltered by changes in the mol. wt. of the triglycerides, while 18:0, 18:1 and 18:2 have distribution patterns which vary with the mol. wt. of triglycerides.[178] 18:0 and 18:1 are preferentially esterified at positions *sn*-1 and *sn*-3 in triglycerides of high mol. wt. and are concentrated at position *sn*-1 in triglycerides of low mol. wt. On the other hand, 18:2 shows a slight preference for position *sn*-3 in high mol. wt. triglycerides and is preferentially esterified at positions *sn*-1 and *sn*-2 in triglycerides of low mol. wt.

The structural analyses discussed above show that in milk fats, in which fatty acid compositions differ to some extent as a result of dietary and

TABLE 7
FATTY ACID COMPOSITIONS OF BOVINE MILK FATS FROM DIFFERENT SOURCES

Fatty acid	Fatty acid composition (mol %)			
	New Zealand milk fat[a]		Canadian milk fat[b]	18:2-rich milk fat[c]
	Spring	Summer		
4:0	12·0	9·6	9·3	10·4
6:0	4·5	4·5	4·4	5·4
8:0	2·3	2·2	1·9	2·1
10:0	4·2	4·2	3·3	3·2
12:0	4·0	4·1	3·5	2·6
14:0	10·8	11·5	9·9	7·7
14:1	0·8	1·2	0·5	0·5
15:0	1·4	1·7	1·3	0·7
16:0	22·0	27·6	23·7	14·1
16:1	1·4	2·1	2·6	1·1
17:0	0·5	0·8	1·4	0·5
18:0	13·1	10·1	11·8	14·1
18:1	21·5	17·8	23·9	20·6
18:2	0·7	1·4	1·6	15·5
18:3	0·3	0·8	0·4	0·7

[a] From Taylor and Hawke.[185]
[b] From Breckenridge and Kuksis.[182,184]
[c] From Morrison and Hawke.[153]

lactational effects, the pattern of positional specificity of fatty acids varies little. Consideration of this fact, together with the triglyceride and fatty acid compositions of the respective milk fats and their fractions, leads to the conclusion that as the fatty acid composition changes, the composition and structure of the major triglyceride species in these milk fats remain essentially constant but the proportions of the species vary.[87,153,178,181,182,184,185]

Comparison of spring and summer milk fats from New Zealand shows that the greater proportions of 4:0 and 18:1 in spring milk fat result in greater proportions of triglyceride species containing these fatty acids and greater proportions of low mol. wt. unsaturated triglycerides (Tables 7 and 8).[87,185] A similar trend is evident when New Zealand and Canadian milk fats are compared, with the proportion of saturated triglycerides in the high mol. wt. fraction of Canadian milk fat being particularly low (Tables 7 and 8).[182,184]

A study of the contributions of the various milk triglycerides to the

TABLE 8
PROPORTIONS OF TRIGLYCERIDES IN BOVINE MILK FATS FROM DIFFERENT SOURCES

Nature of triglycerides		Proportions of triglycerides (percentage in milk fat)			
Molecular weight	Level of unsaturation	New Zealand milk fat[a] Spring	Summer	Canadian milk fat[b]	18:2-rich milk fat[c]
High	Saturated	11·4	13·2	6·4	8·6
	Monoene	15·6	16·1	14·4	10·7
	Diene	8·7	6·9	10·8	11·8
	Triene	4·2	3·0	5·0	6·7
	Polyene	—	—	2·4	5·2
Medium	Saturated	7·7	9·1	6·6	5·5
	Monoene	6·3	5·5	6·6	6·2
	Diene	1·7	2·0	2·5	4·4
	Triene	1·2	1·5	1·5	1·8
	Polyene	—	—	—	1·7
Low	Saturated	20·1	23·2	19·7	11·7
	Monoene	16·1	13·9	16·7	9·4
	Diene	4·3	3·6	5·1	10·4
	Triene	2·7	2·0	2·3	3·4
	Polyene	—	—	—	2·6
Total	Saturated	39·2	45·5	32·7	25·8
	Monoene	38·0	35·5	37·7	26·3
	Diene	14·7	12·5	18·4	26·6
	Triene	8·1	6·5	8·8	11·9
	Polyene	—	—	2·4	9·5

[a] From Taylor and Hawke.[185]
[b] From Breckenridge and Kuksis.[182,184]
[c] From Morrison and Hawke.[153]

overall melting characteristics of milk fat shows that the saturated triglycerides of high and low mol. wt. melt between 28 and 47°C, and 17 and 32°C, respectively, whereas the unsaturated triglycerides of high and low mol. wt. have melting ranges of −17−+31°C and −50−−3°C.[186] These results, which follow the trend normally found for fats, show that the greater the contribution from the low mol. wt. and unsaturated triglycerides, the greater the proportion of low melting triglycerides and hence the softer the milk fat.

Consideration of melting curves of New Zealand spring and summer

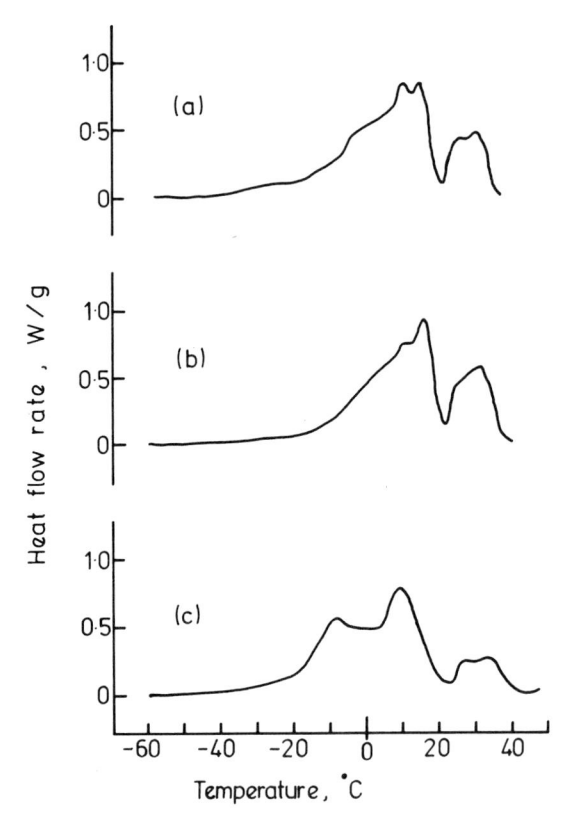

FIG. 8. Influence of different basal diets and of a dietary supplement of protected sunflower-seed oil on the melting thermograms of bovine milk fat: (a) spring pasture; (b) summer pasture; (c) basal diet supplemented with protected sunflower-seed oil.

milk fats shows the expected trend (Fig. 8);[187,188] namely that spring milk fat contains a higher proportion of low melting triglycerides, has less solid fat at any given temperature between -20 and $+20\,°C$, and gives a softer butter. For example, the solid fat contents at $5\,°C$ are 59 and 64% for the spring and summer milk fats, respectively. This trend is due to the greater proportions of triglycerides containing 4:0 and 18:1 in spring milk fat which is, in turn, a consequence of seasonal changes in fatty acid composition. Similarly, the difference between the proportions of unsaturated triglycerides in Canadian and New Zealand milk fats (Table 8) results in Canadian milk fat containing greater proportions of low melting tri-

TABLE 9
INFLUENCE OF A DIETARY SUPPLEMENT OF SOYBEAN OIL ON THE FATTY ACID COMPOSITION OF BOVINE MILK FAT[a]

Type of diet	Fatty acid composition (weight %)									
	4:0	6:0	8:0	10:0	12:0	14:0	16:0	18:0	18:1	18:2
Basal diet	—	2·2	1·3	2·2	2·4	9·2	43·3	8·9	22·8	2·7
Basal diet + soybean oil (4·3% of diet)	—	1·6	0·8	1·1	1·2	5·3	26·0	16·3	39·8	4·1

[a] From Banks et al.[136]

glycerides and presumably explains the observation that Canadian butter is softer than New Zealand butter.[189,190]

The above discussion suggests that changes in the fatty acid composition of milk fat resulting from the feeding of fat supplements will cause variations in the proportions of the major triglyceride species of milk fat but will not change the composition and structure of these species. If this is

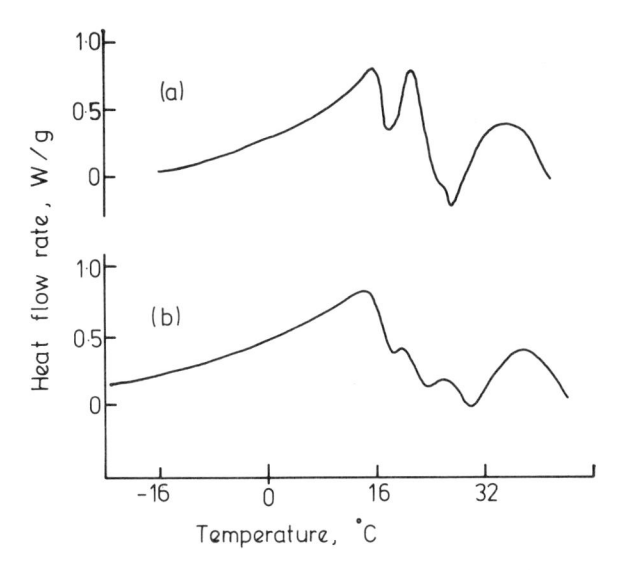

FIG. 9. Influence of a dietary supplement of soybean oil on the melting thermograms of bovine milk fat: (a) basal diet; (b) basal diet supplemented with soybean oil.

the case, then the relationship between the fatty acid composition and melting behaviour of milk fats obtained from cows fed control and fat-supplemented diets can be readily explained. Reference to Table 9 shows that the inclusion of soybean oil in the diet produces the trend discussed earlier, namely a reduction in the proportions of fatty acids 6:0 to 16:0 and an increase in the proportions of 18:0 and 18:1 with 16:0 and 18:1 showing the greatest variation. These changes result in an increase in the proportions of lower melting triglyceride species as shown by the melting curves in Fig. 9.[136]

Feeding cows a diet supplemented with a protected 18:2-rich oil, such as sunflower-seed oil, represents a special case since the resultant milk fat contains appreciable proportions of 18:2 (Table 7) and a new series of major triglyceride species containing 18:2.[153,178] Comparisons between normal and 18:2-rich milk fat show that these triglyceride species result in appreciable increases in the proportions of diene, triene and polyene triglycerides in each of the mol. wt. fractions of milk fat (Table 8) and a much greater proportion of triglycerides melting below 0°C (Fig. 8). As a consequence, butter produced from 18:2-rich milk is significantly softer than normal butter and is considered to be spreadable at domestic refrigerator temperatures.[191,192]

REFERENCES

1. JONES, C. S. and PARKER, D. S., *Biochem. J.*, 1978, **174**, 291.
2. SPINCER, J., ROOK, J. A. F. and TOWERS, K. G., *Biochem. J.*, 1969, **111**, 727.
3. BISHOP, C., DAVIES, T., GLASCOCK, R. F. and WELCH, V. A., *Biochem. J.*, 1969, **113**, 629.
4. ANNISON, E. F., LINZELL, J. L., FAZAKERLEY, S. and NICHOLS, B. W., *Biochem. J.*, 1967, **102**, 637.
5. GLASCOCK, R. F. and WELCH, V. A., *J. Dairy Sci.*, 1974, **57**, 1364.
6. KRONFELD, D. S., *Vet. Rec.*, 1965, **77**, 30.
7. ANNISON, E. F., LINZELL, J. L. and WEST, C. E., *J. Physiol.*, 1968, **197**, 445.
8. HAWKINS, R. A. and WILLIAMSON, D. H., *Biochem. J.*, 1972, **129**, 1171.
9. MCBRIDE, O. W. and KORN, E. D., *J. Lipid Res.*, 1964, **5**, 453.
10. BARRY, J. M., BARTLEY, W., LINZELL, J. L. and ROBINSON, D. S., *Biochem. J.*, 1963, **89**, 6.
11. SCHOEFL, G. I. and FRENCH, J. E., *Proc. Royal Soc. B*, 1968, **169**, 153.
12. NILSSON-EHLE, P., GARFINKEL, A. S. and SCHOTZ, M. C., *Ann. Rev. Biochem.*, 1980, **49**, 667.
13. SCOW, R. O., DESNUELLE, P. and VERGER, R., *J. Biol. Chem.*, 1979, **254**, 6456.
14. DIMICK, P. S., MCCARTHY, R. D. and PATTON, S., In: *Physiology of Digestion and Metabolism in the Ruminant*, Phillipson, A. T. (ed), 1970, Oriel Press, Newcastle-upon-Tyne, p. 529.

15. Zinder, O., Hamosh, M., Fleck, T. R. C. and Scow, R. O., *Am. J. Physiol.*, 1974, **226**, 744.
16. Palmquist, D. L., Davis, C. L., Brown, R. E. and Sachan, D. S., *J. Dairy Sci.*, 1969, **52**, 633.
17. Kinsella, J. E., *Biochim. Biophys. Acta*, 1970, **210**, 28.
18. Lauryssens, M., Verbeke, R. and Peeters, G., *J. Lipid Res.*, 1961, **2**, 383.
19. Kinsella, J. E., *Lipids*, 1972, **7**, 349.
20. Bickerstaffe, R. and Annison, E. F., *Comp. Biochem. Physiol.*, 1970, **35**, 653.
21. Strong, C. R. and Dils, R., *Comp. Biochem. Physiol.*, 1972, **43B**, 643.
22. Bu'Lock, J. D. and Smith, G. N., *Biochem. J.*, 1965, **96**, 495.
23. Lossow, W. J. and Chaikoff, I. L., *J. Biol. Chem.*, 1958, **230**, 149.
24. Moore, J. H. and Christie, W. W., *Lipid Metabolism in Ruminant Animals*, W. W. Christie (ed), 1981, Pergamon Press, Oxford, p. 227.
25. Stead, D. and Welch, V. A., *J. Dairy Sci.*, 1975, **58**, 122.
26. Pennington, R. J., *Biochem. J.*, 1952, **51**, 251.
27. Smith, G. H. and McCarthy, S., *Biochim. Biophys. Acta*, 1969, **176**, 664.
28. Bauman, D. E., Brown, R. E. and Davis, C. L., *Arch. Biochem. Biophys.*, 1970, **140**, 237.
29. Bauman, D. E. and Davis, C. L., In: *Lactation*, Vol. II, Larson, B. L. and Smith, V. R. (eds), 1974, Academic Press, New York and London, p. 31.
30. Kinsella, J. E., *Lipids*, 1970, **5**, 892.
31. Balmain, J. H., Folley, S. J. and Glascock, R. F., *Biochem. J.*, 1952, **52**, 301.
32. Folley, S. J. and French, T. H., *Biochem. J.*, 1950, **46**, 465.
33. Bauman, D. E., Mellenberger, R. W. and Derrig, R. G., *J. Dairy Sci.*, 1973, **56**, 1312.
34. Hardwick, D. C., Linzell, J. L. and Mepham, T. B., *Biochem. J.*, 1963, **88**, 213.
35. Mellenberger, R. W. and Bauman, D. E., *Biochem. J.*, 1974, **138**, 373.
36. Mellenberger, R. W., Bauman, D. E. and Nelson, D. R., *Biochem. J.*, 1973, **136**, 741.
37. Ballard, F. J., Hanson, R. W. and Kronfeld, D. S., *Fed. Proc.*, 1969, **28**, 218.
38. Bauman, D. E. and Davis, C. L., In: *Digestion and Metabolism in the Ruminant*, McDonald, I. W. and Warner, A. C. I. (eds), 1974, University of New England Publishing Unit, Armidale, p. 498.
39. Dils, R. R. and Knudsen, J., *Biochem. Soc. Trans.*, 1980, **8**, 292.
40. Ganguly, J., *Biochim. Biophys. Acta*, 1960, **40**, 110.
41. Becker, M. E. and Kumar, S., *Biochemistry*, 1965, **4**, 1839.
42. Lornitzo, F. A., Qureshi, A. A. and Porter, J. W., *J. Biol. Chem.*, 1975, **250**, 4520.
43. Stoops, J. K. *et al.*, *Proc. Natl. Acad. Sci. USA*, 1975, **72**, 1940.
44. Guy, P., Law, S. and Hardie, G., *FEBS Letts.*, 1978, **94**, 33.
45. Smith, S. and Stern, A., *Arch. Biochem. Biophys.*, 1979, **197**, 379.
46. Bloch, K. and Vance, D., *Ann. Rev. Biochem.*, 1977, **46**, 263.
47. Dileepan, K. N., Lin, C. Y. and Smith, S., *Biochem. J.*, 1978, **175**, 199.
48. Smith, S., *J. Dairy Sci.*, 1980, **63**, 337.
49. Libertini, L. J. and Smith, S., *Arch. Biochem. Biophys.*, 1979, **192**, 47.

50. SMITH, S. and RYAN, P., *J. Biol. Chem.*, 1979, **254**, 8932.
51. KNUDSEN, J. and GRUNNET, I., *Biochem. J.*, 1982, **202**, 139.
52. HANSEN, J. K. and KNUDSEN, J., *Biochem. J.*, 1980, **186**, 287.
53. GRUNNET, I. and KNUDSEN, J., *Eur. J. Biochem.*, 1979, **95**, 503.
54. GRUNNET, I. and KNUDSEN, J., *Biochem. Biophys. Res. Commun.*, 1981, **100**, 629.
55. SPENCER, A. F. and LOWENSTEIN, J. M., *J. Biol. Chem.*, 1962, **237**, 3640.
56. SPENCER, A., CORMAN, L. and LOWENSTEIN, J. M., *Biochem. J.*, 1964, **93**, 378.
57. KATZ, J. and WALS, P. A., *Biochem. J.*, 1972, **128**, 879.
58. ROGNSTAD, R., *Arch. Biochem. Biophys.*, 1969, **129**, 13.
59. GUMAA, K. A., GREENBAUM, A. L. and MCLEAN, P., *Eur. J. Biochem.*, 1973, **34**, 188.
60. WOOD, H. G., PEETERS, G. J., VERBOKE, R., LAURYSSENS, M. and JACOBSON, B., *Biochem. J.*, 1965, **96**, 607.
61. KUHN, N. J., *Biochem. Soc. Trans.*, 1978, **6**, 539.
62. ANNISON, E. F., BICKERSTAFFE, R. and LINZELL, J. L., *J. Agric. Sci.*, 1974, **82**, 87.
63. MCNEILLIE, E. M. and ZAMMIT, V. A., *Biochem. J.*, 1982, **204**, 273.
64. MACKALL, J. C. and LANE, M. D., *Biochem. J.*, 1977, **162**, 635.
65. GUL, B. and DILS, R., *Biochem. J.*, 1969, **112**, 293.
66. BALDWIN, R. L., *J. Dairy Sci.*, 1966, **49**, 1533.
67. OTWAY, S. and ROBINSON, D. S., *Biochem. J.*, 1968, **106**, 677.
68. STRONG, C. R. and DILS, R., *Biochem. J.*, 1972, **128**, 1303.
69. MELLENBERGER, R. W., BAUMAN, D. E. and NELSON, D. R., *Biochem. J.*, 1973, **136**, 741.
70. MELLENBERGER, R. W. and BAUMAN, D. E., *Biochem. J.*, 1974, **138**, 373.
71. STRONG, C. R., FORSYTH, I. and DILS, R., *Biochem. J.*, 1972, **128**, 509.
72. FORSYTH, I. A., STRONG, C. R. and DILS, R., *Biochem. J.*, 1972, **129**, 929.
73. AGIUS, L., ROBINSON, A. M., GIRARD, J. R. and WILLIAMSON, D. H., *Biochem. J.*, 1979, **180**, 689.
74. SMITH, S. and ABRAHAM, S., *Adv. Lipid Res.*, 1975, **13**, 195.
75. CAFFREY, M. and KINSELLA, J. E., *J. Lipid Res.*, 1977, **18**, 44.
76. MARSHALL, M. O. and KNUDSEN, J., *Biochim. Biophys. Acta*, 1977, **489**, 236.
77. MARSHALL, M. O. and KNUDSEN, J., *Biochim. Biophys. Acta*, 1980, **617**, 393.
78. GROOT, P. H. E., SCHOLTE, H. R. and HÜLSMAN, W. C., *Adv. Lipid Res.*, 1976, **14**, 75.
79. SHORT, V. J., BRINDLEY, D. N. and DILS, R., *Biochem. J.*, 1977, **162**, 445.
80. LLOYD-DAVIES, K. A. and BRINDLEY, D. N., *Biochem. J.*, 1975, **152**, 39.
81. BALDWIN, R. L. and MILLIGAN, L. P., *J. Biol. Chem.*, 1966, **241**, 2058.
82. HAJRA, A. K., *Biochem. Soc. Trans.*, 1977, **5**, 34.
83. RAO, G. A. and ABRAHAM, S., *Lipids*, 1978, **13**, 95.
84. RAO, G. A. and ABRAHAM, S., *Lipids*, 1973, **8**, 232.
85. POLHEIM, D., DAVID, J. S. K., SCHULTZ, F. M., WYLIE, M. B. and JOHNSTON, J. M., *J. Lipid Res.*, 1973, **14**, 415.
86. DIMICK, P. S., MCCARTHY, R. D. and PATTON, S., *Biochim. Biophys. Acta*, 1966, **116**, 159.
87. TAYLOR, M. W. and HAWKE, J. C., *N.Z. J. Dairy Sci. Technol.*, 1975, **10**, 49.
88. BREACH, R. A., DILS, R. and WATTS, R., *J. Dairy Res.*, 1973, **40**, 273.

89. Slakey, P. M. and Lands, W. E. M., *Lipids*, 1968, **3**, 30.
90. Kinsella, J. E. and Gross, M., *Biochim. Biophys. Acta*, 1973, **316**, 109.
91. Gross, M. J. and Kinsella, J. E., *Lipids*, 1974, **9**, 905.
92. Tanioka, H., Lin, C. Y., Smith, S. and Abraham, S., *Lipids*, 1974, **9**, 229.
93. Kinsella, J. E., *Lipids*, 1976, **11**, 680.
94. Bickerstaffe, R. and Annison, E. F., *Int. J. Biochem.*, 1971, **2**, 153.
95. Marshall, M. O. and Knudsen, J., *Eur. J. Biochem.*, 1977, **81**, 259.
96. Marshall, M. O. and Knudsen, J., *Eur. J. Biochem.*, 1979, **94**, 93.
97. Lin, C. Y., Smith, S. and Abraham, S., *J. Lipid Res.*, 1976, **17**, 647.
98. Ross, A. C., *J. Lipid Res.*, 1982, **23**, 133.
99. Patton, S. and Jensen, R. G., *Biomedical Aspects of Lactation*, 1976, Pergamon Press, Oxford, p. 63.
100. Easter, D. J., Patton, S. and McCarthy, R. D., *Lipids*, 1971, **6**, 844.
101. Infante, J. P. and Kinsella, J. E., *Lipids*, 1976, **11**, 727.
102. Kinsella, J. E., *Lipids*, 1973, **8**, 393.
103. Patton, S. and Keenan, T. W., *Lipids*, 1971, **6**, 58.
104. Kinsella, J. E. and Infante, J. P., *Lipids*, 1974, **9**, 748.
105. Christie, W. W., In: *Lipid Metabolism in Ruminant Animals*, Christie, W. W. (ed), 1981, Pergamon Press, Oxford, p. 95.
106. Sundler, R. and Akesson, B., *J. Biol. Chem.*, 1975, **250**, 3359.
107. Kinsella, J. E., *Biochim. Biophys. Acta*, 1968, **164**, 540.
108. Patton, S., McCarthy, R. D., Plantz, P. E. and Lee, R. F., *Nature, New Biol.*, 1973, **241**, 241.
109. Connor, W. E. and Lin, D. S., *Am. J. Physiol.*, 1967, **213**, 1353.
110. Clarenburg, R. and Chaikoff, I. L., *J. Lipid Res.*, 1966, **7**, 27.
111. Raphael, B. C., Patton, S. and McCarthy, R. D., *J. Dairy Sci.*, 1975, **58**, 971.
112. Ashes, J. R., Gulati, S. K., Cook, L. J., Mills, S. C. and Scott, T. W., *J. Lipid Res.*, 1978, **19**, 244.
113. Bloch, K., *Science*, 1973, **248**, 1856.
114. Popják, G., French, T. H. and Folley, S. J., *Biochem. J.*, 1951, **48**, 411.
115. Cowie, A. T. et al., *Biochem. J.*, 1951, **49**, 610.
116. Keenan, T. W. and Patton, S., *Lipids*, 1970, **5**, 42.
117. Robertson, J. A. and Hawke, J. C., *J. Sci. Food Agric.*, 1964, **15**, 890.
118. Storry, J. E., Brumby, P. E. and Dunkley, W. L., *Factors Affecting the Yields and Contents of Milk Constituents of Commercial Importance*, Int. Dairy Fed. Bull., Document 125, 1980, Belgium, p. 105.
119. Moore, J. H. and Steele, W., *Proc. Nutr. Soc.*, 1968, **27**, 66.
120. Storry, J. E., *Factors Affecting the Yields and Contents of Milk Constituents of Commercial Importance*, Int. Dairy Fed. Bull., Document 125, 1980, Belgium, p. 88.
121. Storry, J. E., *J. Dairy Res.*, 1970, **37**, 139.
122. Christie, W. W., In: *Lipid Metabolism in Ruminant Animals*, Christie, W. W. (ed), 1981, Pergamon Press, Oxford, p. 193.
123. Steele, W. and Moore, J. H., *J. Dairy Res.*, 1968, **35**, 353.
124. Steele, W. and Moore, J. H., *J. Dairy Res.*, 1968, **35**, 361.
125. Steele, W. and Moore, J. H., *J. Dairy Res.*, 1968, **35**, 223.
126. Steele, W. and Moore, J. H., *J. Dairy Res.*, 1968, **35**, 343.

127. STEELE, W., NOBLE, R. C. and MOORE, J. H., *J. Dairy Res.*, 1971, **38**, 49.
128. STORRY, J. E., HALL, A. J. and JOHNSON, V. W., *J. Dairy Res.*, 1973, **40**, 293.
129. STORRY, J. E., HALL, A. J. and JOHNSON, V. W., *Br. J. Nutr.*, 1968, **22**, 609.
130. STORRY, J. E., HALL, A. J. and JOHNSON, V. W., *J. Dairy Res.*, 1971, **38**, 73.
131. STORRY, J. E., ROOK, J. A. F. and HALL, A. J., *Br. J. Nutr.*, 1967, **21**, 425.
132. STEELE, W., NOBLE, R. C. and MOORE, J. H., *J. Dairy Res.*, 1971, **38**, 43.
133. STORRY, J. E., BRUMBY, P. E., HALL, A. J. and JOHNSON, V. W., *J. Dairy Sci.*, 1974, **57**, 61.
134. MACLEOD, G. K., WOOD, A. S. and YAO, Y. T., *J. Dairy Sci.*, 1972, **55**, 446.
135. BANKS, W., CLAPPERTON, J. L. and FERRIE, M. E., *J. Dairy Res.*, 1976, **43**, 219.
136. BANKS, W., CLAPPERTON, J. L. and KELLY, M. E., *J. Dairy Res.*, 1980, **47**, 277.
137. INSULL, W., JR., HIRSCH, J., JAMES, T. and AHRENS, E. H., JR., *J. Clin. Invest.*, 1959, **38**, 443.
138. HAWLEY, L. E., CANOLTY, N. L. and DUNKLEY, W. L., *J. Am. Dietetic Assn.*, 1978, **72**, 170.
139. HAWKE, J. C. and MICKLESON, K. N. P., *20th Int. Dairy Congr.*, 1978, 1061.
140. INSULL, W., JR. and AHRENS, E. H., JR., *Biochem. J.*, 1959, **72**, 27.
141. BRECKENRIDGE, W. C., MARAI, L. and KUKSIS, A., *Can. J. Biochem.*, 1969, **47**, 761.
142. SCOTT, T. W. et al., *Austral. J. Sci.*, 1970, **32**, 291.
143. SCOTT, T. W., COOK, L. J. and MILLS, S. C., *J. Am. Oil Chem. Soc.*, 1971, **48**, 358.
144. FOGERTY, A. C. and JOHNSON, A. R., *Factors Affecting the Yields and Contents of Milk Constituents of Commercial Importance*, Int. Dairy Fed. Bull., Document 125, 1980, Belgium, p. 96.
145. HUTJENS, M. F. and SCHULTZ, L. H., *J. Dairy Sci.*, 1971, **54**, 1876.
146. BITMAN, J. et al., *J. Am. Oil Chem. Soc.*, 1973, **50**, 93.
147. WRENN, T. R. et al., *J. Dairy Sci.*, 1976, **59**, 627.
148. STORRY, J. E. et al., *J. Agric. Sci.*, 1980, **94**, 503.
149. WRENN, T. R. et al., *J. Dairy Sci.*, 1977, **60**, 521.
150. MACLEOD, G. K., YU, Y. and SCHAEFFER, L. R., *J. Dairy Sci.*, 1977, **60**, 726.
151. PLOWMAN, R. D. et al., *J. Dairy Sci.*, 1972, **55**, 204.
152. PAN, Y. S., COOK, L. J. and SCOTT, T. W., *J. Dairy Res.*, 1972, **39**, 203.
153. MORRISON, I. M. and HAWKE, J. C., *Lipids*, 1977, **12**, 994.
154. COOK, L. J., SCOTT, T. W. and PAN, Y. S., *J. Dairy Res.*, 1972, **39**, 211.
155. HILDITCH, T. P. and WILLIAMS, P. N., *The Chemical Constitution of Natural Fats*, 1964, Chapman and Hall, London.
156. DAVIS, C. L. and BROWN, R. E., In: *Physiology of Digestion and Metabolism in the Ruminant*, Phillipson, A. T. (ed), 1970, Oriel Press, Newcastle-upon-Tyne, p. 545.
157. STORRY, J. E., BRUMBY, P. E., HALL, A. J. and TUCKLEY, B., *J. Dairy Sci.*, 1974, **57**, 1046.
158. SUTTON, J. D., In: *Factors Affecting the Yields and Contents of Milk Constituents of Commercial Importance*, Int. Dairy Fed. Bull., Document 125, 1980, Belgium, p. 126.
159. MCCULLOUGH, M. E., *J. Dairy Sci.*, 1966, **49**, 896.
160. STORRY, J. E. and ROOK, J. A. F., *Br. J. Nutr.*, 1966, **20**, 217.

161. Rook, J. A. F., *Proc. Nutr. Soc.*, 1964, **23**, 71.
162. Wilson, G. F. and Dolby, R. M., *N.Z. J. Agr. Res.*, 1967, **10**, 415.
163. Opstvedt, J., Baldwin, R. L. and Ronning, M., *J. Dairy Sci.*, 1967, **50**, 108.
164. Benson, J. D., Askew, E. W., Emery, R. S. and Thomas, J. W., *Fed. Proc.*, 1969, **28**, 623.
165. Touchberry, R. W., In: *Lactation*, Vol. III, Larson, B. L. and Smith, V. R. (eds), 1974, Academic Press, New York and London, p. 349.
166. Decaen, C. and Adda, J., *17th Int. Dairy Congr.*, 1966, **Vol A**, 161.
167. Stull, J. W., Brown, W. H., Valdez, C. and Tucker, H., *J. Dairy Sci.*, 1966, **49**, 1401.
168. Parodi, P. W., *Austral. J. Dairy Technol.*, 1974, **29**, 145.
169. Parodi, P. W., *Austral. J. Dairy Technol.*, 1970, **25**, 200.
170. Hutton, K., Seeley, R. C. and Armstrong, D. G., *J. Dairy Res.*, 1969, **36**, 103.
171. Hall, A. J., *Dairy Inds.*, 1970, **35**, 20.
172. Gray, I. K., *J. Dairy Res.*, 1973, **40**, 207.
173. Hawke, J. C., *J. Dairy Res.*, 1963, **30**, 67.
174. Huyghebaert, A. and Hendrickx, H., *Milchwissenschaft*, 1971, **26**, 613.
175. Jensen, R. G., Gander, G. W. and Sampugna, J., *J. Dairy Sci.*, 1962, **45**, 329.
176. Reiter, B., Sorokin, Y., Pickering, A. and Hall, A. J., *J. Dairy Res.*, 1969, **36**, 65.
177. Pitas, R. E., Sampugna, J. and Jensen, R. G., *J. Dairy Sci.*, 1967, **50**, 1332.
178. Morrison, I. M. and Hawke, J. C., *Lipids*, 1977, **12**, 1005.
179. Parodi, P. W., *J. Dairy Res.*, 1979, **46**, 75.
180. Luick, J. R. and Smith, L. M., *J. Dairy Sci.*, 1963, **46**, 1251.
181. Breckenridge, W. C. and Kuksis, A., *J. Lipid Res.*, 1968, **9**, 388.
182. Breckenridge, W. C. and Kuksis, A., *Lipids*, 1969, **4**, 197.
183. Parodi, P. W., *J. Dairy Res.*, 1982, **49**, 73.
184. Breckenridge, W. C. and Kuksis, A., *Lipids*, 1968, **3**, 291.
185. Taylor, M. W. and Hawke, J. C., *N.Z. J. Dairy Sci. Technol.*, 1975, **10**, 40.
186. Taylor, M. W., Norris, G. E. and Hawke, J. C., *N.Z. J. Dairy Sci. Technol.*, 1978, **13**, 236.
187. Norris, G. E., Gray, I. K. and Dolby, R. M., *J. Dairy Res.*, 1973, **40**, 311.
188. Taylor, M. W. and Norris, R., *N.Z. J. Dairy Sci. Technol.*, 1977, **12**, 166.
189. de Man, J. M. and Wood, F. W., *J. Dairy Sci.*, 1958, **41**, 360.
190. Wood, F. W. and Dolby, R. M., *J. Dairy Res.*, 1965, **32**, 269.
191. Kieseker, F. G. and Eustace, I. J., *Austral. J. Dairy Technol.*, 1975, **30**, 17.
192. Wood, F. W., Murphy, M. F. and Dunkley, W. L., *J. Dairy Sci.*, 1975, **58**, 839.

Chapter 3

ORIGIN OF MILK FAT GLOBULES AND THE NATURE OF THE MILK FAT GLOBULE MEMBRANE

T. W. Keenan, Daniel P. Dylewski, Terry A. Woodford

and

Rosemary H. Ford

Virginia Polytechnic Institute and State University, Blacksburg, Virginia, USA

SUMMARY

Available evidence suggests that milk fat globules originate from endoplasmic reticulum of mammary epithelial cells. The actual process by which these globules are formed is unknown, but there are indications that triglyceride-containing vesicles which bleb from endoplasmic reticulum may serve as nucleation sites for globules. After formation, lipid droplets grow within the cell and this growth appears to be mediated by fusion of triglyceride-containing vesicles with intracellular lipid droplets. After formation, lipid droplets migrate to apical regions of cells, from where they are secreted into the alveolar lumen. Some evidence that microtubules and microfilaments may be involved in the intracellular migration of lipid droplets has been obtained, but this evidence is equivocal. Lipid droplets are released from the cell by envelopment in specialized regions of apical plasma membrane. Membranes of secretory vesicles may also make a contribution to the membrane which envelops lipid droplets. As reviewed herein, the membrane which surrounds fat globules in milk has been extensively characterized.

1. INTRODUCTION

This chapter reviews the literature on the intracellular origin of milk fat globules, discusses the mechanisms by which they move through and are released from the cell, and summarizes the literature on the nature of the membrane which surrounds fat globules. This membrane, known as the milk fat or milk lipid globule membrane, is acquired by the globules during their release from milk-secreting mammary epithelial cells. The composition of milk lipids is discussed in Chapter 1 of this volume; basically milk fat globules consist of a triglyceride-rich lipid core surrounded by a specialized bilayer membrane composed primarily of proteins, phospholipids, glycolipids, sterols and glycerides. Triglycerides, along with partial glycerides, account for 98 to 99% of the mass of milk fat globules of the cow. More than 95% of the total milk lipid is in the globule fraction with the remainder being present in a heterogeneous membrane fraction found in milk serum.

In this chapter, emphasis will be on the origin, growth and secretion of globules. Related topics, such as the source of precursors and the mechanisms of glyceride biosynthesis, will not be discussed. Where possible, the reader will be referred to reviews covering these and other related areas. Most of the discussion will be from literature of the past two decades with little more than passing mention of earlier studies. Historical aspects of research on milk fat globules have been reviewed by Brunner,[1,2] and several other reviews covering aspects of research on milk lipids are available.[3-12]

2. INTRACELLULAR ORIGIN AND GROWTH OF LIPID DROPLETS

The origin of the fatty acids and glycerol moieties and the nature of enzymes catalyzing reactions leading to the formation of triacylglycerols have been reviewed.[13-17] While acyltransferases which esterify fatty acids into triglycerides are known to be membrane bound (see, for example, Reference 18), the subcellular localization of these enzymes in mammary epithelial cells has not been precisely defined. Since esterification is accomplished by microsomes from the mammary gland, it is generally assumed that the terminal steps in triglyceride synthesis occur in or on the surface of membranes of the endoplasmic reticulum (e.g. References 10 and 17). Electron microscope autoradiographic studies have shown that labelled fatty acids supplied to the mammary gland are rapidly incorporated into lipids and that the label is initially concentrated over the endoplasmic reticulum.[19,20] It has been suggested that triacylglycerols accumulate within endoplasmic reticulum bilayer leaflets and are eventually released coated with one-half of the bilayer.[21,22] When rat mammary tissue was fixed directly with osmium tetroxide, evidence was obtained for association of one-half of the bilayer membrane of the

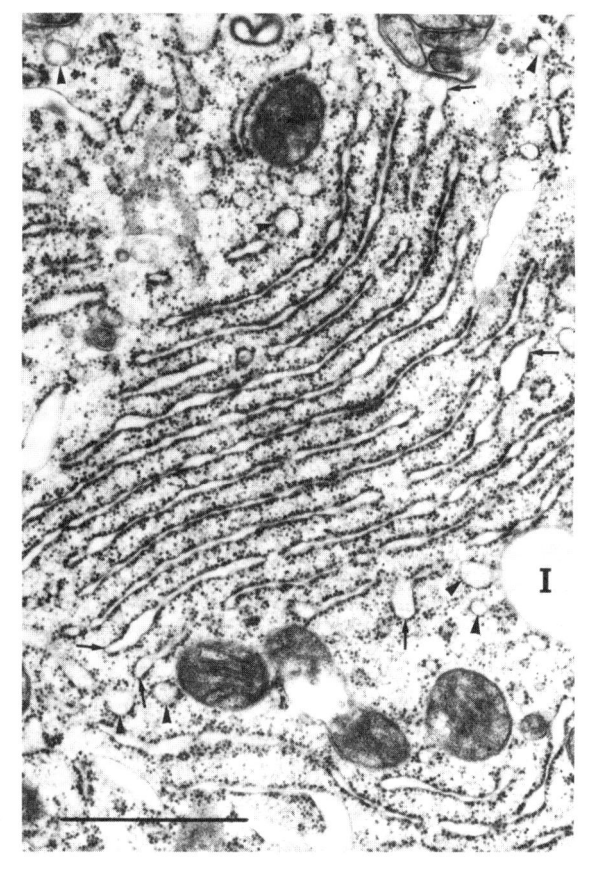

FIG. 1. Stacked endoplasmic reticulum cisternae in an epithelial cell of bovine mammary gland. Distended regions of endoplasmic reticulum are numerous; these appear to be the sites of origin of lipid-containing vesicles (lipovesicles). Certain of these distended regions are denoted by arrows. Arrowheads denote small lipovesicles which are surrounded by a rough or polysome-studded membrane. I, intracellular lipid droplet. Magnification ×36 000. Bar = 1 μm.

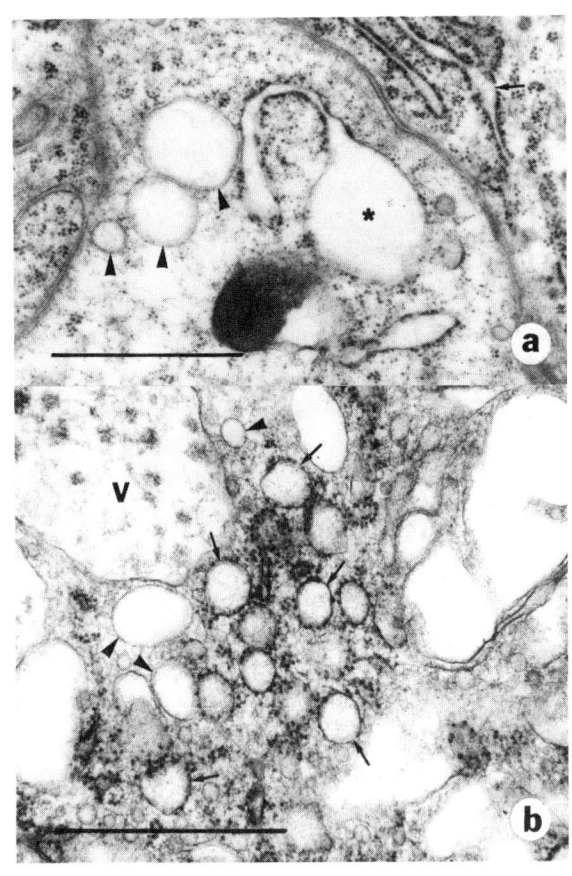

FIG. 2. (a) Section through a mammary epithelial cell of a lactating rat showing a lipovesicle attached to a distended fragment of endoplasmic reticulum (*). Three lipovesicles surrounded by largely agranular membranes are seen in this section (arrowheads); (b) Several lipovesicles are observed in this section through a bovine mammary epithelial cell. Many of these vesicles are surrounded by a polysome-studded membrane (some of which are designated by arrows) while larger lipovesicles (arrowheads) lack polysomes. It appears that, as lipovesicles grow through fusion of smaller vesicles, polysomes are lost from the surface. V, secretory vesicle. Magnifications: (a) ×37 000; (b) ×45 000. Bars = 1 μm.

endoplasmic reticulum with the surface of cytoplasmic lipid droplets.[6,11] Biochemical support for an association of endoplasmic reticulum with intracellular lipid droplets comes from the reported similarity of these preparations in phospholipid composition.[23,24] However, evidence for

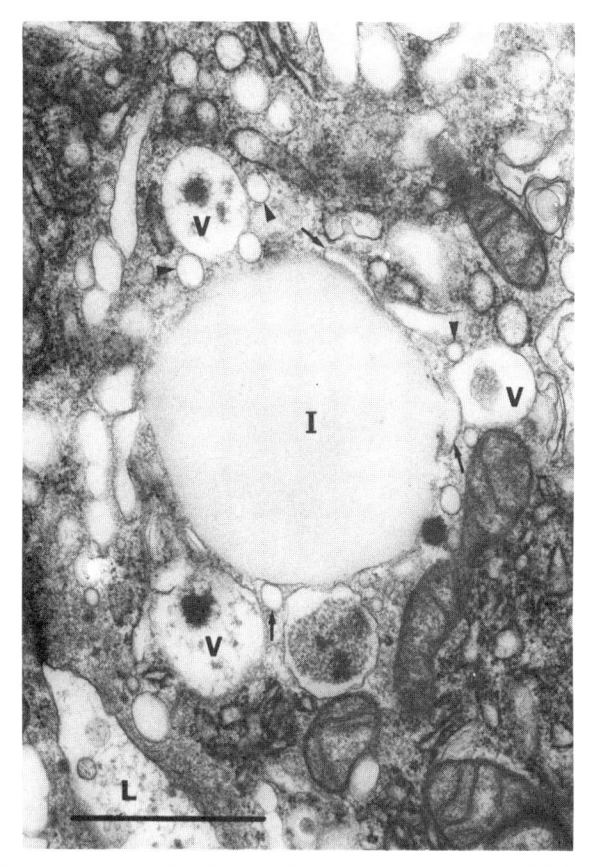

FIG. 3. Numerous agranular lipovesicles are present in the vicinity of a large intracellular lipid droplet (I) in the apical region of a bovine mammary epithelial cell. Sites of apparent fusion of smaller lipovesicles with the intracellular lipid droplet are indicated by arrows. Arrowheads denote some of the many lipovesicles observed. V, secretory vesicle; L, lumen. Magnification: ×38 000. Bar = 1 μm.

association of forming lipid droplets with endoplasmic reticulum is weak and other possibilities for the origin of these droplets are not excluded.

In a morphological study (Keenan *et al.*, 1982, unpublished) of the origin and growth of intracellular lipid droplets, numerous vesicles with a lightly amorphous content which appeared to originate from endoplasmic reticulum (Figs 1 and 2) were observed. These vesicles fused with lipid droplets throughout the cytoplasm (Figs 3 and 4). During and after fusion

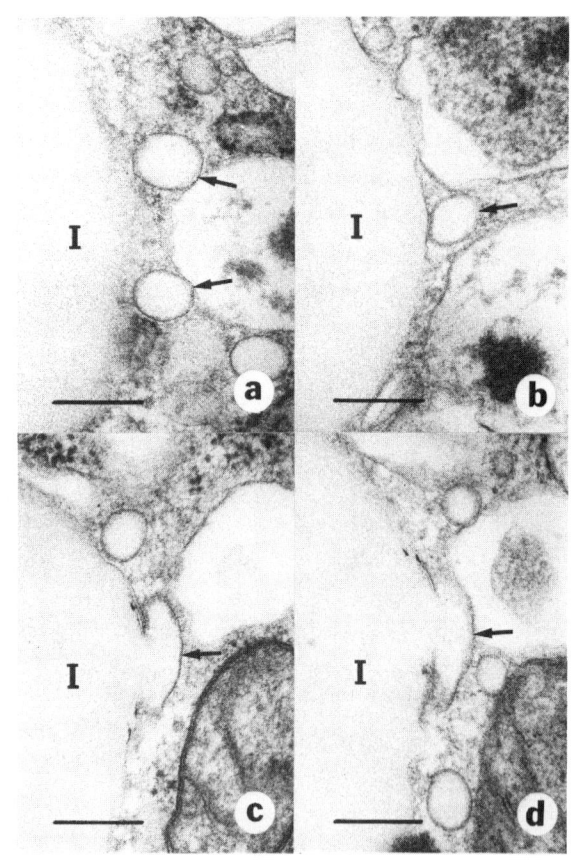

FIG. 4. Apparent stages of fusion of lipovesicles with intracellular lipid droplets (I) are indicated by arrows. In (a) there are apparent sites of contact between the droplet and two lipovesicles. This contact appears to be between the membrane of the lipovesicles and the remnant membrane material on the droplet surface (b). Fusion of the lipovesicles with droplets appears to involve fusion of the vesicle membrane with surface material on the droplet and integration of this membrane onto the droplet surface ((c) and (d)). Magnification: ×70 000. Bars = 0·25 μm.

the membrane of these 'lipovesicles' appeared to be adsorbed onto the surface of lipid droplets and to undergo morphological alteration; in appearance this surface material became less membrane-like and assumed a more granular appearance (Figs 4 and 5(a) and (b)). Granular material

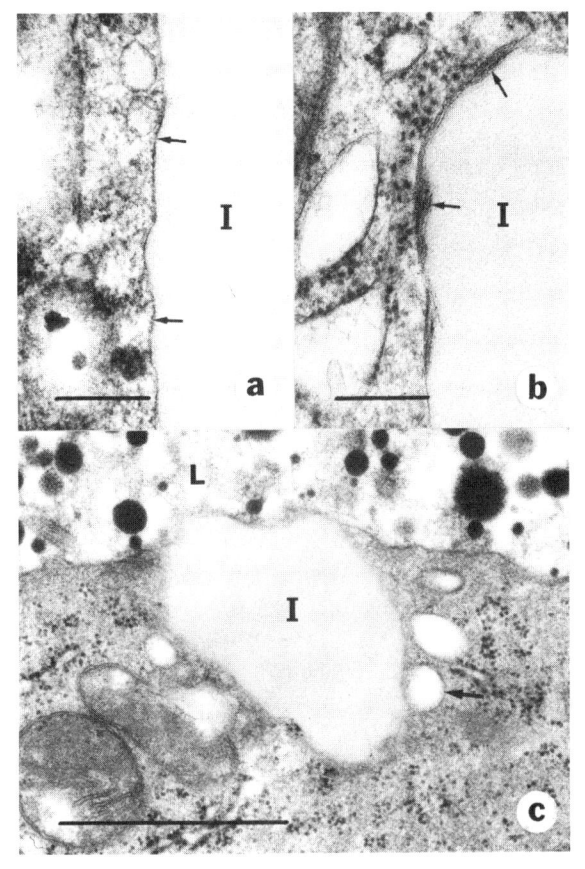

FIG. 5. Material on the surface of intracellular lipid droplets (I) has a finely granular appearance ((a) and (b)). In certain regions a unit-like membrane appearance is evident (arrows in (b)) but in other regions unit-like membrane structure is not observed. It appears that the membrane initially present on lipovesicles undergoes morphological alteration during droplet maturation. Nevertheless, morphologically distinct material, presumably originating from the endoplasmic reticulum membrane, is present on surfaces of intracellular droplets. (c) A lipovesicle (arrow) appears to have been in the process of fusing with a lipid droplet (I) being extruded from the cell. Continued fusion of lipovesicles with large droplets may be the mechanism by which droplets grow during intracellular transit and secretion. (a) and (b) bovine mammary tissue, magnification ×72 000; (c) rat mammary tissue, magnification ×45 000. Bars = 0·25 μm ((a) and (b)), and 1 μm (c).

with a similar appearance was observed over the surface of intracellular lipid droplets of all sizes. These lipovesicles were numerous in mammary epithelial cells of both cows and rats in active lactation. Based on these observations, we suggest that intracellular lipid droplets destined to become milk fat globules originate from triglyceride-containing vesicles which bleb from regions of the endoplasmic reticulum. Growth occurs by continued fusion of these vesicles with each other and with larger lipid droplets. That lipovesicles were observed to fuse with lipid droplets in medial and apical regions, including globules in the process of being budded from the cell (Fig. 5(c)), further suggests that lipovesicles are involved in the growth of globules during transit to apical regions and during their envelopment in apical plasma membrane.[25]

In our study, material associated with the surface of isolated intracellular lipid droplets was found to resemble endoplasmic reticulum in polypeptide composition. However, the polar lipid composition of this material differed from that of endoplasmic reticulum in that it contained appreciable quantities of monohexosyl- and dihexosyl-ceramides. Sphingomyelin accounted for about 14% of the total lipid phosphorus of the droplets compared with about 5% of the lipid phosphorus of mammary endoplasmic reticulum.[26] This implies that the material on the surface of the intracellular droplets may be derived from regions of the endoplasmic reticulum which differ from the bulk reticular network in lipid composition. Taken together, available information clearly favours an endoplasmic reticular origin of precursors of milk lipid globules.

3. MECHANISM OF INTRACELLULAR TRANSPORT OF FAT GLOBULES

Intracellular lipid droplets migrate from their sites of origin in the basal regions of the secretory cell to apical regions from which they are secreted into the alveolar lumen. Mechanisms for this unidirectional transport of droplets are not known with certainty. Evidence that microtubules and microfilaments, elements of the cytoskeletal system, may be involved in this transport has been obtained; however, this evidence is weak and in at least some instances contradictory results have been obtained (review, Reference 14). Cytoplasmic microtubules are numerous in epithelial cells of lactating mammary gland[27] and the content of both total and polymerized tubulin in mammary gland increases in late pregnancy and early lactation.[28] As first shown by Patton,[29] intramammary infusion of colchicine, an inhibitor of microtubule assembly, reversibly suppresses milk secretion. Subsequent observations, both *in vivo* and *in vitro* with a number of species, confirmed that colchicine and other alkaloids which disrupt microtubules inhibit release, but not synthesis, of milk constituents (e.g. References 20, 30–5). Patton and co-workers[32,33] observed inhibition of lipid globule secretion after intramammary infusion of colchicine in rats and goats. Lipid synthesis did not appear to be affected and, in goats, intracellular lipid droplets grew to larger than normal sizes. However, other groups have been unable to duplicate these results. With tissue slices from lactating ewes and rabbits, Daudet *et al.*[20] could not demonstrate inhibition of lipid secretion by colchicine. In contrast to Patton *et al.*,[32] who observed an increase in milk lipid concentration during recovery from colchicine infusion, Henderson and Peaker[31] found no change in the lipid content of milk from a colchicine-treated goat mammary gland. Thus, while colchicine depresses secretion of milk lipid globules when introduced into the mammary gland, this effect may be nonspecific since exocytosis of secretory vesicles is also depressed. The results available are thus equivocal and do not demonstrate direct involvement of microtubules in translocation of intracellular lipid droplets.

Microfilaments are abundant in lactating mammary epithelial cells (e.g. Reference 36), and Amato and Loizzi[37] have provided evidence that these actin-like filaments are concentrated in apical cell regions. Cytochalasin B, a drug which disrupts microfilaments, has been shown to inhibit secretion of milk protein, lactose[38–40] and lipid.[41] However, this drug has other effects, including inhibition of glucose transport. Thus one cannot conclude that microfilaments are involved in intracellular transit or secretion of milk fat globules although this possibility deserves further investigation.

4. SECRETION OF MILK FAT GLOBULES

The manner in which lipid globules are released from mammary epithelial cells into alveolar lumina has been actively investigated over the past 30 years or so. The morphology of the process has been repeatedly described, the biochemistry of milk lipid secretion has been studied, and a biophysical rationale for the secretory process has been developed. While there is still some controversy regarding the mechanism by which fat globules acquire a membrane during release from the cell, knowledge of the secretion process is substantial. Present debate centres around the extent to which secretory

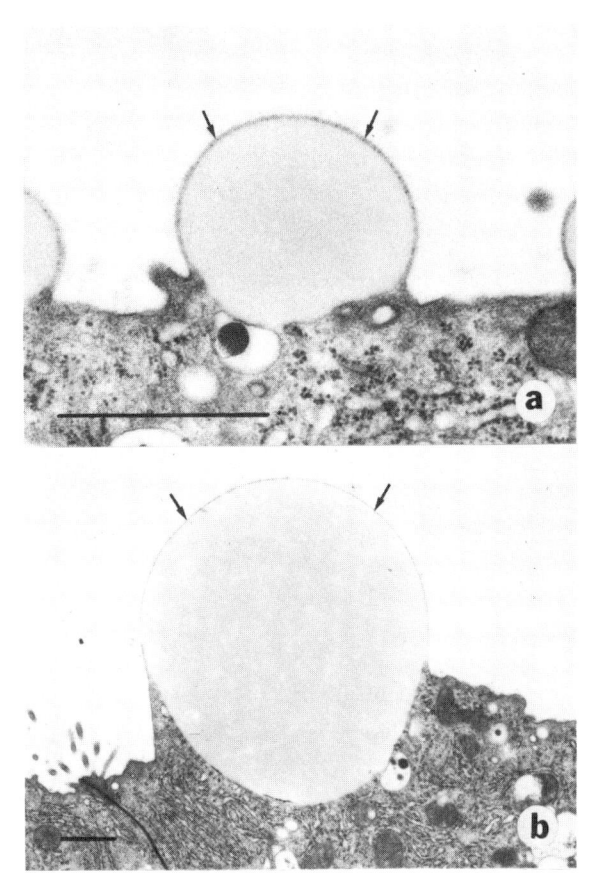

FIG. 6. Fat globules are released from mammary epithelial cells by progressive envelopment in apical plasma membrane (arrows). This same mechanism of secretion, as illustrated here for (a) the rat, and (b) the cow has been observed with a number of species. Magnifications: (a) ×41 000; (b) ×11 000. Bars = 1 μm.

vesicles and the apical plasma membrane contribute membrane to fat globules during their exit from the cell.

During the period from 1959 to 1961, electron micrographs of lactating epithelium from rat,[42,43] cow[44] and mouse[45,46] (see also Reference 47) were published which gave clear evidence that fat globules are released from cells by progressive envelopment in the apical plasma membrane. This process (Fig. 6) was later confirmed for these species and extended to several other species (reviews, References 48 and 49). Globules in alveolar lumina, ducts and in milk would thus be expected to be surrounded by a membrane derived from the apical plasma membrane of the cell from which the globule originated. It is now well established that fat globules in milk are surrounded by a membrane, commonly referred to as the milk fat or milk lipid globule membrane (abbreviated as MFGM hereafter). However, there remains some question as to whether this membrane originates largely or entirely from the apical plasma membrane. Wooding[50,51] has made observations suggesting that lipid droplets become surrounded by secretory vesicles in such a manner that cytoplasmic vacuoles containing lipid droplets partially coated by membrane are formed. Wooding has suggested that release of at least some lipid globules occurs by exocytotic fusion of secretory vesicles with the apical plasma membrane. Since at least some secretory vesicles appear to fuse with the apical plasma membrane and become integrated into this membrane, the resulting MFGM would be expected to be similar or identical irrespective of whether it is derived partially from secretory vesicles or totally from the apical plasma membrane. Available morphological evidence on this matter is equivocal; micrographs showing globules partially enveloped in apical plasma membrane (e.g. References 9, 11, 42, 48, 52 and 53) as well as those showing association of secretory vesicles with intracellular lipid droplets[48,50,51,54,55] have been published. In view of the static nature of electron microscopy, we can learn little about the rate of globule secretion by either of these processes using this method.

Results from biochemical studies can be interpreted as favouring either a plasma membrane or a secretory vesicle membrane origin of the MFGM. Detection of enzymes characteristic of Golgi apparatus in MFGM can be interpreted as favouring a role for Golgi apparatus-derived secretory vesicles in the envelopment of fat globules (e.g. References 56–8). Compositional similarity between MFGM and plasma membrane fractions from mammary gland can be taken as evidence for an apical plasma membrane origin of MFGM.[5,6,11,23,59–62] Both lines of evidence are weak. The presence of Golgi apparatus-associated enzymes may have been due to entrained cytoplasmic material between the globule core and membrane (discussed subsequently). Isolated plasma membrane fractions are enriched in membranes from basolateral surfaces and may not reflect apical plasma membrane composition. Secretory vesicle membranes have been found to be compositionally intermediate between Golgi apparatus and MFGM.[63] However, the fractions isolated contained largely immature secretory vesicles, and it may be that membranes of mature secretory vesicles more closely resemble MFGM in composition. Franke *et al.*[64] have provided the strongest evidence to date that MFGM is derived largely from the apical plasma membrane. By immunomicroscopy they found that antibodies to butyrophilin, a major protein associated with MFGM, stained lipid globules protruding from the cell and globules in alveolar lumina. Staining was observed over localized regions of apical plasma membrane but was not seen over secretory vesicles in apical regions of cells. These observations, taken together with compositional differences between secretory vesicle membranes and MFGM, support the contention that, in cows, apical plasma membrane is the major contributor of MFGM. Even if secretory vesicles do not participate directly in the envelopment of milk fat globules, it is probable that they contribute to replenishment of that portion of the apical plasma membrane expended as MFGM. This replenishment may require an appreciable amount of membrane; it has been calculated that the entire apical surface of milk secreting cells of the cow must be replaced every 8–10 h.[36] Membrane flow and differentiation in relation to milk secretion is beyond the scope of this presentation; this topic is discussed elsewhere;[5,6,14] for a general review of endomembrane differentiation and flow, see Reference 65.

Biochemical and biophysical processes involved in recognition and envelopment of fat globules by plasma membrane remain unknown. Patton and Fowkes[66] proposed that van der Waals forces may be involved in the attraction of the plasma membrane around lipid droplets. Wooding[50,51,67] observed an electron-dense material which separated the fat globule from the plasma membrane by at least 10 nm and concluded that this distance was too great for the generation of effective attractive forces between the plasma membrane and the droplet surface. This dense-staining material, which is also seen along the inner face of the MFGM, appears to be composed of proteins.[68–70] The major constituent proteins of this electron-dense coat material, butyrophilin and xanthine oxidase, appear to contain tightly (perhaps covalently) bound fatty acids.[71] It may be that van der Waals forces of sufficient magnitude to attract plasma membrane around the droplet are generated between these coat proteins and the droplet surface.

The localization and properties of butyrophilin and xanthine oxidase are such as to suggest that they may be involved in recognition and apical plasma membrane envelopment of lipid droplets. As yet, however, we have no specific knowledge of how these proteins may function. Proteins of the apical plasma membrane may be involved in milk fat globule secretion as suggested by the finding of Patton and co-workers[72-4] that introduction of either the lectin Concanavalin A or antibodies to MFGM into the goat mammary gland through the teat canal reversibly suppresses milk secretion. Since secretion of protein and lactose, as well as lipid, is inhibited by these treatments, it cannot be concluded from these observations that apical plasma membrane proteins are specifically involved in fat globule secretion. This possibility does, however, merit further investigation.

This mechanism of lipid secretion, by plasma membrane envelopment of globules, has not been observed in other cell types. Nevertheless, the process of secretion of milk fat globules resembles, in many respects, the budding of enveloped viruses through the plasma membrane of host cells.[12] Thus, it may be instructive to consider mechanisms operative in the budding of enveloped viruses in searching for clues as to mechanisms controlling recognition and envelopment of fat globules.

Secretion of milk fat globules is not always a 'clean' process, resulting in globules surrounded solely by plasma-membrane-derived material. Occasionally cytoplasmic material is entrained between the core lipid and the surrounding membrane (e.g. Fig. 7). Kurosumi et al.[52] and Helminen and Ericsson[75] appear to have been the first to observe this, and these observations were confirmed and extended by Wooding et al.[76] Qualitative electron microscope observations[48] and a limited amount of quantitative morphometric data[77] suggest that the extent of entrainment of cytoplasmic material in fat globules is species-variable. Morphometric and biochemical results show the membrane material associated with milk fat globules from cows in established lactation to be largely of plasma membrane origin; larger amounts of intracellular membrane material appear to be associated with fat globules from milk of rats, humans and cows in the colostral stage of lactation.[56-8,63,78,79] One must take the possibility of cytoplasmic membrane material being associated with fat globules into account when studying MFGM as a model for the apical plasma membrane.

It is probable that some membrane material is lost from the surface of fat globules following release from cells, although estimates of the extent of this loss vary widely.[67,70,77,80,81] In morphological studies, the electron-dense coat material on the inner face of the MFGM was observed to thicken in localized areas, and patches of membrane with associated coat material were lost from the globule surface by vesiculation.[67,70] Since the extent of membrane loss appears to depend to some extent on methods used for fixation,[80] the amount of membrane lost from globules cannot be estimated from published morphological observations. Biochemical results suggest that, under certain storage conditions, there is little loss of membrane from fat globules for periods up to 24 h after withdrawal of milk from cows.[81,82]

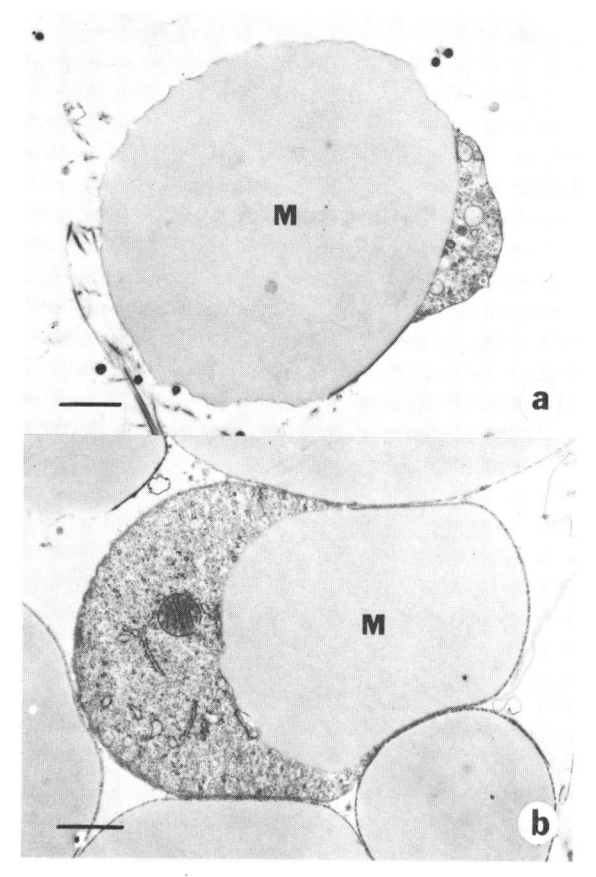

FIG. 7. Cytoplasmic material is sometimes entrained between the core lipid and limiting membrane of milk fat globules (M). In (a) the small cytoplasmic crescent contains several vesicles, and in (b) the large crescent contains a mitochondrian as well as vesicles and strands of endoplasmic reticulum. Both micrographs are of globules in lumina of rat mammary gland. Magnifications: (a) ×12 000; (b) ×13 000. Bars = 1 μm.

5. ISOLATION OF MILK FAT GLOBULE MEMBRANE

Methods for the isolation of MFGM and historical developments which led to present-day methods have been reviewed.[2,3,11,12] Relative to the plasma membrane or many intracellular components, MFGM is easily isolated and can be obtained in large quantities. Before the actual isolation of MFGM, fat globules are usually separated from milk by centrifugal floatation. Globules are then washed by resuspension in and floatation from a suitable wash solution; washing is repeated a sufficient number of times to reduce contamination with milk serum constituents to levels acceptable for the particular study. For certain studies, such as of the distribution of phospholipids or proteins associated with milk fat globules, various constituent classes can be recovered by direct extraction of washed globules. Other types of study require that MFGM be released from globules and collected. Physical methods commonly used to release MFGM from globules include slow freezing and thawing, agitation (rapid mixing or stirring, churning), and exposure to ultrasound. These methods induce a phase inversion, the result of which is a butter phase containing the bulk of the triglycerides and an aqueous phase in which membrane fragments are suspended. These membrane fragments can be collected from the aqueous phase by precipitation with salts or by pH adjustment. More commonly, membrane fragments are collected by ultracentrifugation of the aqueous phase without prior treatment.

It is apparent that, depending on the choice of isolation conditions, the yield of MFGM will vary, as will the composition of the material obtained.[2,12] For certain MFGM constituents, wide ranges in composition have been reported. In particular, there are wide variations in reported protein-to-lipid ratios, relative amounts of triglycerides and specific activities of several membrane-associated enzymes (reviews, References 3 and 11, see also Reference 12). While not necessarily recognized at the time, it is probable that much of the reported variation is due to the use of different isolation methods.

While it is frequently assumed that the membranes isolated from washed fat globules originate from the apical plasma membrane, perhaps with some contribution from mature secretory vesicle membrane as well, this assumption may not always be justified. Any cytoplasmic membranes entrained during globule secretion will be present in MFGM preparations. For certain studies even a small amount of contamination by cytoplasmic membranes can be intolerable. Preparations of MFGM can be fractionated by centrifugation in density gradients, but since MFGM does not appear to be homogeneous in buoyant density, this approach must be used with caution;[83-5] review, Reference 12.

6. GROSS COMPOSITION OF MILK FAT GLOBULE MEMBRANE

Although there has been increased interest in comparative studies of milk composition from different species in recent years, MFGM from bovine

TABLE 1
COMPOSITION OF BOVINE MILK FAT GLOBULE MEMBRANES[a]

Constituent class	Amount
Protein	25–60% of dry weight[a]
Total lipid	0·5–1·2 mg per mg protein[a,b]
Phospholipid	0·13 to 0·34 mg per mg protein[a,b,c]
Phosphatidyl choline	34% of total lipid phosphorus[d]
Phosphatidyl ethanolamine	28% of total lipid phosphorus[a]
Sphingomyelin	22% of total lipid phosphorus[a]
Phosphatidyl inositol	10% of total lipid phosphorus[a]
Phosphatidyl serine	6% of total lipid phosphorus[a]
Neutral lipid	56–80% of total lipid[a]
Hydrocarbons	1·2% of total lipid[a]
Sterols	0·2–5·2% of total lipid[a]
Sterol esters	0·1–0·8% of total lipid[a]
Glycerides	53–74% of total lipid[a]
Free fatty acids	0·6–6·3% of total lipid[a]
Cerebrosides	3·5 nmoles per mg protein[a]
Gangliosides	6 to 7·4 nmoles sialic acid per mg protein[a,b]
Total sialic acids	63 nmoles per mg protein[a]
Hexoses	0·6 μmoles per mg protein[a]
Hexosamines	0·3 μmoles per mg protein[a]
Cytochrome b_5 + P-420	30 pmoles per mg protein[e,f]
Uronic acids	99 ng per mg protein[g]
RNA	20 μg per mg protein[e]

[a] Patton and Keenan.[11]
[b] Keenan et al.[63]
[c] Kitchen.[119]
[d] Compiled from References 23, 26 and 63; for calculation it was assumed that these phospholipids accounted for all of the phospholipid of the membrane.
[e] Jarasch et al.[77]
[f] Bruder et al.[90]
[g] Shimizu et al.;[89] uronic acids were first reported by Lis and Monis[88] but the value calculated from this study, 58 μg per mg protein, appears to be unrealistically high.

milk remains the most highly characterized milk membrane. The gross composition of MFGM from bovine milk is summarized in Table 1. Proteins and lipids account for well over 90% of the dry weight of MFGM preparations, but there is wide variation in reported protein-to-lipid ratios. In addition to differences in methods of preparation of MFGM, some variation may be due to factors such as breed differences, seasonal variation, stage of lactation, age, and treatment of the milk. The neutral lipid content appears to be the largest variable in MFGM composition. Triglycerides, which account for most of the mass of the fat globule, may adhere to the membrane and are probably responsible for much of the variation observed in the neutral lipid content. There is some question as to whether triglycerides are true MFGM constituents, since plasma membranes from tissues such as liver contain only small amounts of triglycerides e.g. References 86 and 87. Phospholipids and glycolipids appear to be more constant in amounts, per unit of protein, in MFGM. Carbohydrates of MFGM appear to be primarily protein- and lipid-bound. Carbohydrates found in glycolipids include sialic acids (primarily N-acetylneuraminic acid), N-acetylgalactosamine, galactose and glucose. Protein-bound carbohydrates include those found in glycolipids and also N-acetylglucosamine, mannose and fucose. Uronic acids have recently been detected in bovine MFGM preparations.[88,89] To date, hyaluronic acid, chondroitin sulphate and heparin sulphate have been identified as constituents of the uronic acid fraction. Shimizu *et al.*[89] found uronic acids in human MFGM at 5–10 times higher levels than in bovine MFGM.

Jarasch and co-workers[77,90] have identified the cytochromes of MFGM as b_5 and P-420. RNA has been found in MFGM preparations from bovine[77,91] and human[57] milks. Whether this RNA, which has not been characterized, is associated with the globule membrane or is contributed by polysomes in entrained cytoplasmic material is not known. DNA appears to be absent from preparations of bovine, human and rat MFGMs.[57,77] Several elements have been found in MFGM preparations from cow's milk; the major elements detected are copper, iron, sulphur, manganese, magnesium, molybdenum, calcium, sodium, potassium and zinc.[92]

7. LIPID COMPOSITION OF MILK FAT GLOBULE MEMBRANE

Neutral lipids account for more than half of the total lipid of MFGM, and triglycerides are by far the most abundant constituent class of the neutral lipid fraction (e.g. Reference 93; Table 1). It is probable that at least a portion of these triglycerides is adsorbed from the globule core onto the inner face of the MFGM. These MFGM triglycerides have been described as high-melting because they contain a higher proportion of longer acyl chains than total milk triglycerides.[59,94] There is evidence that differences in conditions under which MFGM is prepared influence the levels of triglycerides associated with the membrane.[94,95] That these triglycerides are associated with the inner face of the membrane is suggested by results from microelectrophoretic studies which show the outer surface of fat globules to be ionogenic and to contain little neutral lipid.[96] Localization of triglycerides in or on the MFGM and the amount of triglycerides indigenous to the membrane before globule envelopment need further study.

Sterols of bovine MFGM are largely cholesterol and cholesterol esters; lanosterol and dihydrolanosterol are present, in unesterified form, at about 6% of the level of cholesterol.[97,98] Many different free fatty acids and the hydrocarbons β-carotene and squalene have been identified in the MFGM. The fatty acid composition of the neutral lipids of bovine MFGM has been summarized.[2]

The major phospholipids of the MFGM are sphingomyelin and the phosphatides of choline, ethanolamine, inositol and serine (Table 1). Other phospholipids, primarily lyso-derivatives of the major phosphatides, have also been detected in the MFGM. These lyso-derivatives are minor constituents when fresh samples, which are handled so as to minimize lipid degradation, are used. MFGM phospholipids account for about 60% of the total milk phospholipid; the remainder is present in a membrane-containing fraction in milk serum. An identical distribution pattern for the major phospholipids is found for whole milk, globule membrane and milk serum.[93,99,100] Within milk fat globules, the phospholipids are apparently associated entirely with the membrane since no phospholipid is detected in core lipids.[93,99,101] The distributions of phospholipids in the MFGM and plasma membrane from lactating bovine mammary gland are similar.[5,23] However, this pattern is distinct from that of intracellular membranes from mammary gland (nuclear membrane, endoplasmic reticulum, Golgi apparatus, secretory vesicles and mitochondria) in that the sphingomyelin-to-phosphatidyl choline ratio is high (reviews, References 5, 6 and 11; for secretory vesicles, see Reference 63). This distinctive distribution pattern of phospholipids shared by the plasma membrane and the MFGM provides biochemical support for a plasma membrane origin of the MFGM.

Phospholipids and glycolipids are present in preparations of intracellular lipid droplets from bovine mammary gland. Early analyses yielded an endoplasmic reticulum-like distribution pattern (low sphingomyelin-to-phosphatidyl choline ratio) for phospholipids of intracellular lipid droplets.[23,24] However, recent analyses of preparations of intracellular droplets, washed to remove contaminating endoplasmic reticulum, have shown the sphingomyelin-to-phosphatidyl choline ratio of intracellular lipid (0·3) to be much higher than that of endoplasmic reticulum (0·1) (unpublished results; for endoplasmic reticulum, see Reference 26). This ratio for intracellular lipid droplets is intermediate between that of the Golgi apparatus and the MFGM. Intracellular lipid droplets also contain glycolipids with thin-layer chromatographic mobilities corresponding to glucosyl- and lactosyl-ceramides, known constituents of the MFGM (Table 2). This raises the possibility that a portion of the complement of phospholipids and glycolipids found in the MFGM originates from the presecretory droplet and is not contributed by the apical plasma membrane.

The fatty acid composition of bovine milk and MFGM phospholipids has been reviewed elsewhere (see Chapter 1; also Reference 102). The same five major phospholipids of bovine MFGM are present in and distributed similarly in milk or MFGM preparations from mouse,[103] rat,[104] goat,[99] sheep, camel, ass, pig, Indian buffalo and human.[105]

Several different carbohydrate-containing sphingolipids, both neutral (cerebrosides) and acidic (gangliosides) glycosphingolipids, have been identified as constituents of the MFGM. The structures of these glycolipids are given in Table 2. The major cerebrosides of bovine milk are glucosyl- and lactosyl-ceramides, and these occur in nearly equal proportions in the MFGM.[106-9] Cerebrosides more complex than lactosylceramide have not yet been detected in bovine MFGM. Free ceramide (N-acylsphingosine) occurs in milk lipids but has not yet been shown to be a constituent of the

TABLE 2
STRUCTURES OF GLYCOSPHINGOLIPIDS OF BOVINE MILK FAT GLOBULE MEMBRANE[a]

Glycosphingolipid[b]	Structure
Glucosyl ceramide	β-Glucosyl-($1 \rightarrow 1$)-ceramide
Lactosyl ceramide	β-Galactosyl-($1 \rightarrow 4$)-β-glucosyl-($1 \rightarrow 1$)-ceramide
GM_3 (Hematoside)	Neuraminosyl-($2 \rightarrow 3$)-galactosyl-glucosyl-ceramide
GM_2	N-Acetylgalactosaminyl-(neuraminosyl)-galactosyl-glucosyl-ceramide
GM_1	Galactosyl-N-acetylgalactosaminyl-(neuraminosyl)-galactosyl-glucosyl-ceramide
GD_3 (Disialohematoside)	Neuraminosyl-($2 \rightarrow 8$)-neuraminosyl-($2 \rightarrow 3$)-galactosyl-glucosyl-ceramide
GD_2	N-Acetylgalactosaminyl-(neuraminosyl-neuraminosyl)-galactosyl-glucosyl-ceramide
GD_{1b}	Galactosyl-N-acetylgalactosaminyl-(neuraminosyl-neuraminosyl)-galactosyl-glucosyl-ceramide

[a] From Huang,[113] Keenan,[101] and Bushway and Keenan.[114]
[b] Abbreviations for gangliosides are those of Svennerholm.[178]

MFGM.[110] Human MFGM also contains glucosyl- and lactosyl-ceramides, but in this membrane galactosylceramide is the major cerebroside.[111] A total of six different gangliosides have been found in bovine MFGM (Table 2), which on a sialic acid basis total 6–7 nmol/mg protein (Table 1).[101,112-14] The disialoganglioside GD_3 (disialohematoside) is the major constituent of the ganglioside fraction of the MFGM. Glycosphingolipids of the type found in the MFGM are synthesized by stepwise addition of carbohydrates to ceramide (review, Reference 115). Glycosyltransferases which add these carbohydrates are concentrated in the Golgi apparatus from the mammary gland.[116] Sphingomyelin, cerebrosides and gangliosides of milk all contain predominantly long-chain, saturated fatty acids.[101,102,107,108,114,117] Based on similarities in fatty acid composition, it may be that all of these lipids are synthesized from a common pool of ceramides.

8. PROTEINS OF THE MILK FAT GLOBULE MEMBRANE

Knowledge of the protein composition of the MFGM has expanded rapidly in recent years, due primarily to the introduction of sodium dodecyl sulphate (SDS) as an agent for disaggregation of membrane samples. Using sodium dodecyl sulphate–polyacrylamide gel electrophoresis (SDS–PAGE), a number of investigators have characterized MFGM proteins.[70,83,84,118-24] While there are variations in the number and estimated molecular weights of various polypeptides in these reports, electrophoretic patterns obtained by all groups show a striking similarity in the distribution of the major polypeptides and glycoproteins of the membrane. Although calculated M_r values varied, all the above-mentioned groups detected major size classes of polypeptides of M_r of about 155 000 and 67 000. A third major size class, consisting of at least two polypeptides of M_r about 48 000 and 44 000 was observed by some but not all investigators. These polypeptides can be selectively extracted from fat globules by solutions of high ionic strength[123] and variable amounts of these polypeptides remain associated with core lipid or in the supernatant when the MFGM is sedimented. When proteins are extracted directly from washed lipid globules with solutions of SDS, these polypeptides are invariably observed as major components.[83,123]

Typical electrophoretic patterns for the MFGM are shown in Fig. 8. In addition to the major polypeptides, several other polypeptides have also been detected. For example, Mather and Keenan[123] enumerated 21 discrete bands ranging in molecular weight from about 250 000 to 11 000.

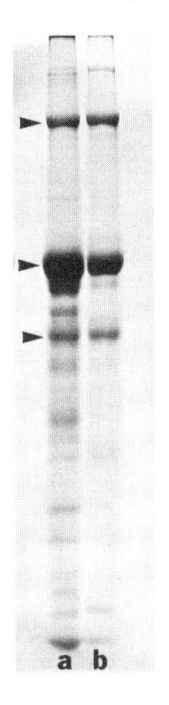

FIG. 8. Electrophoretic patterns (SDS–PAGE) of polypeptides of isolated MFGM (lane a) and the high-salt and Triton-insoluble coat fraction derived from MFGM (lane b). Arrowheads from top to bottom indicate xanthine oxidase (M_r 155 000), butyrophilin (M_r 67 000), and glycoprotein B (M_r 48 000), respectively. The gel was stained with Coomassie blue.

The true protein constituency of the MFGM is not known, but by using a two-dimensional electrophoretic system[125,126] and a sensitive silver stain,[127] a very complex polypeptide pattern was observed (I. H. Mather, personal communication). Since a plasmin-like protease is associated with the MFGM,[128] it is possible that some of the polypeptides observed are proteolytic fragments of larger polypeptides. In this regard it is noteworthy that the major Coomassie-blue-stained polypeptides are immunologically distinct (e.g. References 64, 129–32). In contrast, polypeptides of M_r 49 000 and 46 000 share common immunological determinants.[129]

Five[119,123] or six[83,118,122] glycoproteins have been detected by one-dimensional SDS–PAGE of the MFGM. These glycoproteins give a positive reaction with Schiff reagent after periodate oxidation. It is a common observation that some glycoproteins detected with Schiff reagent stain poorly, if at all, with Coomassie blue (e.g. References 83 and 123). Three MFGM glycoproteins migrate in gel regions where polypeptides are not detected with Coomassie blue. Using radioactive lectins to detect glycoproteins after SDS–PAGE separation, Murray et al.[133] found seven different glycoproteins in bovine MFGM, and at least eight different glycoproteins have been detected by electrofocussing.[134] The true complexity of the glycoprotein composition of the MFGM has not received detailed study.

Species comparisons of MFGM proteins have been limited. Human and goat MFGMs have polypeptide patterns qualitatively similar to that of cow.[57,64,70,133,135] Polypeptides with mobilities similar to the M_r 155 000 and 67 000 size classes in bovine MFGM have been detected in human, pig, rat, goat and sheep globule membrane preparations.[70,133,135] In all these species the M_r 67 000 polypeptide, which has been named butyrophilin,[64] shares the property of being largely insoluble in solutions of high ionic strength and in non-ionic detergent solutions.[64,70,84,135] Butyrophilin and the M_r 155 000 polypeptide, identified as xanthine oxidase (e.g. References 84, 130 and 131), from different species give similar but not identical peptide maps, and antibodies against these proteins from bovine MFGM do not cross-react across with all of the above species.[64,131,133,135]

While we as yet know little about the functional significance of individual proteins of the MFGM, much progress has been made over the past few years in purification and characterization of proteins from the membrane. Polypeptide size class M_r 155 000 consists, at least in part, of xanthine oxidase,[84,130–2,136–8] a complex iron–sulphur–molybdenum flavoprotein with multifunctional enzymatic activities (review, Reference 139). This enzyme oxidizes pyrimidines, purines, pterins and aldehydes and can generate superoxide radicals. The biological function of xanthine oxidase, which occurs in several body tissues and in milk, is unknown. Part of the xanthine oxidase, which may account for up to 20% of the protein of the MFGM, shows a stable association with the membrane.[131,140] When the MFGM is treated with moderately active detergents, xanthine oxidase is one of two polypeptides selectively enriched in the insoluble fraction (coat fraction).[70,84,130] A portion of the xanthine oxidase is released in a soluble form on lysis of fat globules.[84,130,140] In mammary tissue this enzyme is in a predominantly, but not exclusively, soluble form.[131,132] At least four isoelectric variants, focussing in the pH range 7·0–7·5, of MFGM xanthine oxidase have been observed.[129–31,138] While this enzyme promotes lipid peroxidation in systems *in vitro*, no evidence could be found that native xanthine oxidase had a role in the oxidation of MFGM lipids.[131] In bovine mammary gland, xanthine oxidase is concentrated in the apical regions of mammary epithelial cells but is found throughout the epithelial cell cytoplasm.[130] Xanthine oxidase has an unusual distribution; in tissues other than mammary gland it is detected only in capillary endothelial cells.[130,131]

As mentioned earlier, the membrane-bound form of xanthine oxidase in bovine MFGM contains small amounts of tightly, perhaps covalently, bound fatty acids[71] which may promote attraction between the membrane or the membrane-associated coat and the surface of the pre-secretory globule.

Polypeptide size class 67 000, the major SDS–PAGE-resolved constituent of bovine MFGM, consists, largely, of a hydrophobic, difficult-to-solubilize protein. This protein has been named butyrophilin to reflect its association with and affinity for milk lipid.[64] Butyrophilin is insoluble or sparingly soluble in aqueous solution even after treatment with many detergents, chaotrophic agents or solutions of high ionic strength (e.g. References 70 and 84). So far no enzymatic activity has been ascribed to butyrophilin. Following separation of bovine MFGM proteins by electrofocussing under non-denaturing conditions, butyrophilin was not coincident with xanthine oxidase, NADH-iodonitrotetrazolium reductase, 5′-nucleotidase, alkaline phosphatase, phosphodiesterase or γ-glutamyl transpeptidase activities.[129] Butyrophilin, a glycoprotein with at least four isoelectric variants focussing in the pH range 5·2–5·3, tenaciously binds phospholipids.[64,70] Butyrophilin also contains tightly, perhaps covalently, bound fatty acids.[71] On thermal denaturation, this protein forms a disulphide-stabilized aggregate.[141] Within mammary epithelial cells butyrophilin was observed only at the apical cell surface including those portions of the apical plasma membrane over budding milk fat globules.[64] Butyrophilin was neither detected, by immunomicroscopy, elsewhere in mammary secretory cells nor in any of the several other bovine tissues examined. These properties of butyrophilin suggest that it may have a role in membrane envelopment of fat globules.

Polypeptides in the M_r 44 000 to 48 000 size class are not tightly associated with MFGM and can be extracted with salt solution.[123] One of the proteins solubilized by extraction with salt solution may be identical to the glycoprotein that Basch *et al.*[142] purified and characterized. The protein purified by these workers, termed glycoprotein B, was found to have N-terminal serine, C-terminal leucine, and an M_r of 49 500 on SDS–PAGE. The purified protein contained 14% carbohydrate including mannose, galactose, glucose, N-acetylgalactosamine, N-acetylglucosamine and sialic acid.

While a search for actin in the MFGM proved fruitless, a fraction containing two polypeptides of M_r 48 000 and 50 000 was recovered by application of methods used to extract actin from skeletal muscle.[85]

Proteins in this fraction mimicked actin in that they interacted with myosin, aggregated at high ionic strength and were tightly bound by deoxyribonuclease I. These proteins differed from known actins in molecular weights on SDS–PAGE, amino-acid composition, inability to stimulate myosin adenosine triphosphatase activity and in the ultrastructure of their aggregated forms; they displayed a tenacious association with lipids.[85]

Several fractions enriched in glycoproteins, as well as some purified glycoproteins of the MFGM, have been characterized since the last review on this topic.[1-3,11] Kanno and co-workers[143-5] characterized a glycoprotein from bovine MFGM which appeared homogeneous by sedimentation velocity analysis; however, this fraction was subsequently found to contain multiple N-terminal amino acids and several SDS–PAGE-resolvable proteins, at least one of which was not glycosylated.[146] Keenan et al.[147] recovered a fraction containing a major glycoprotein of M_r 182 000 and minor glycoproteins of M_r 135 000 and 86 000 by extraction of the MFGM with aqueous lithium diiodosalicylate. Proteins in this fraction, which contained carbohydrates typical of membrane glycoproteins, avidly bound the lectin Concanavalin A.

As discussed above, a constituent termed glycoprotein B has been purified and characterized. Snow et al.[148] used chloroform–methanol to release a bovine MFGM glycoprotein which they subsequently purified to near homogeneity. By SDS–PAGE this protein had an M_r of 70 000 and was heavily glycosylated, containing nearly 50% by weight carbohydrate.

Two glycoproteins have been purified from human MFGM. One of these has an M_r of 70 000 and contains about 14% carbohydrate.[149] The other is a very large mucin-like glycoprotein containing about 50% carbohydrate linked to the polypeptide through O-glycosidic bonds.[150] This glycoprotein binds wheat germ, castor bean and peanut agglutinins.

Glycoproteins released from fat globules by pronase have both N- and O-linked oligosaccharide chains.[151] The structure of a tetrasaccharide released from bovine MFGM has been established as β-D-galactosyl(1 → 3)-N-acetylgalactosamine substituted by sialic acid at positions C3 of galactose and C6 of galactosamine.[152] This structure has been confirmed in bovine MFGM and also occurs in human MFGM.[153-6]

The amino acid composition of bovine MFGM has been summarized.[11] All investigators have found MFGM to have high levels of leucine, glutamic acid and aspartic acid and to have low levels of sulphur amino acids. Differences in amino-acid composition of MFGM, as determined in different laboratories, appear to reflect differences in preparative methods and in analytical precision.[157]

9. ENZYMES OF THE MILK FAT GLOBULE MEMBRANE

Enzymatic activities which have been detected in bovine MFGM are listed in Table 3. Some enzymes with high specific activity in the MFGM, notably 5'-nucleotidase, phosphodiesterase I and adenosine triphosphatase, serve as marker enzymes for plasma membranes from some tissues. Marker enzymes for the plasma membrane of the mammary gland have not been established, but 5'-nucleotidase is enriched in plasma-membrane-rich fractions from lactating bovine and rat mammary glands.[61,62,158] Xanthine oxidase, an abundant protein of the MFGM, can catalyze the oxidation of

TABLE 3
ENZYME ACTIVITIES DETECTED IN BOVINE MILK FAT GLOBULE MEMBRANE

Enzyme	EC number	Reference[a]
Alkaline phosphatase	3.1.3.1	3, 11
Acid phosphatase	3.1.3.2	3, 11
5'-Nucleotidase	3.1.3.5	3, 11
Phosphodiesterase I	3.1.4.1	3, 11, 179
Inorganic pyrophosphatase	3.6.1.1	3, 11
Nucleotide pyrophosphatase	3.6.1.9	3, 11
Phosphatidic acid phosphatase	3.1.3.4	165
Adenosine triphosphatase	3.6.1.15	3, 11
Lipoamide dehydrogenase	1.6.4.3	11
Cholinesterase	3.1.1.8	3, 11
Aldolase	4.1.2.13	3, 11
Xanthine oxidase	1.2.3.2	11
Thiol oxidase	1.8.3.2	3, 11
γ-Glutamyl transpeptidase	2.3.2.1	3, 11, 180
UDP-glucose hydrolase	3.2.1.–	3, 11, 79
UDP-galactose hydrolase	3.2.1.–	3, 11, 79
NADH-cytochrome c reductase	1.6.99.3	3, 77, 90, 131
NADH-ferricyanide reductase	1.6.99.3	77, 90, 131
NADPH-cytochrome c reductase	1.6.99.1	77, 90, 131
Glucose-6-phosphatase	3.1.3.9	3, 165
Galactosyl transferase	2.4.1.–	58
Plasmin	3.4.21.7	128
β-Glucosidase	3.2.1.21	3
β-Galactosidase	3.2.1.23	3
Catalase	1.11.1.6	3
Ribonuclease I	3.1.4.22	3

[a] Where possible, reference is made to a review where the primary references have been cited.

NADH, and there is evidence that part of the NADH- and NADPH-cytochrome c reductase activities of the MFGM are due to xanthine oxidase.[77,90,131] However, the NADH-ferricyanide reductase activity of the membrane appears to be distinct from xanthine oxidase.[131] A cytochrome-linked redox system is present in the MFGM, and the cytochromes in this system are b_5 and P-420.[77,90] Cytochrome P-450 was not detected in MFGM or endoplasmic reticulum from mammary gland, suggesting that P-420 is a native membrane constituent and not a degradation product of P-450.[90]

Aldolase, β-glycosidases and acid phosphatase, enzymes usually associated with lysosomes, have been reported in the MFGM. Specific activities of aldolase and β-glycosidases are low and variable, and may have been due to entrainment of leucocytes in lipid globule preparations.[159] However, acid phosphatase is present in high specific activities in the MFGM, and this enzyme may well be a true constituent of the MFGM (review, Reference 3). Galactosyltransferase, normally considered to be a marker for the Golgi apparatus, has been detected in preparations of bovine[58] and human[56,160] MFGM, but others failed to detect this activity in bovine MFGM.[79] Whether galactosyltransferase and other enzymes not ordinarily found in plasma membranes occur in the primary MFGM or in the cytoplasmic membranes entrained during globule secretion remains to be determined. Another possibility which must be considered is that certain enzymes may be present on the pre-secretory globule surface and subsequently adhere to the enveloping membrane. Plasmin, an alkaline protease of milk, is associated with the MFGM.[128] Under appropriate incubation conditions the MFGM proteins undergo autoproteolysis, presumably catalyzed by plasmin.

It is, perhaps, surprising that few enzymes of bovine MFGM have been purified or extensively characterized. The notable exception is xanthine oxidase, which has been purified and extensively characterized (review, Reference 139; for recently developed purification methods see References 136–8 and 161). 5'-Nucleotidase has been partially purified, and evidence for two forms of this enzyme, both of which were tenaciously associated with phospholipid, was obtained.[162] On electrofocussing, MFGM 5'-nucleotidase was resolved into several distinct isoenzymes.[129] Both membrane-bound and solubilized forms of 5'-nucleotidase were inhibited by Concanavalin A, confirming that this enzyme is a glycoprotein.[163,164] Adenosine triphosphatase activity of bovine MFGM is stimulated by K^+ and Mg^{2+} but, in contrast to an earlier report,[165] not by Na^+.[60,166] Acid and alkaline phosphatases have been extensively purified from milk (reviews, References 3 and 11) but not specifically from the

MFGM. Some properties of the membrane-bound form of alkaline phosphatase have recently been described; interestingly, this enzyme is inhibited by sucrose.[167] Remarkably little is yet known about the function(s) of MFGM enzymes in globule secretion or in catalyzing post-secretion alterations in milk.

10. MOLECULAR ORGANIZATION OF THE MILK FAT GLOBULE MEMBRANE

The pioneering biochemical studies of Morton and co-worker[168,169] and the ultrastructural studies of Bargmann and Knoop[42] led to the realization that the MFGM is a true biological membrane originating from the secretory cell. Prior to this, the popular view was that milk fat globules were stabilized by interfacial material comprised of various layers ranging from apolar constituents on the inner surface to highly polar constituents on the outer face. Brunner[1,2] has thoroughly reviewed the history of this subject.

Available morphological evidence shows that the MFGM has a typical bilayer membrane structure with specializations including a layer of dense-staining, proteinaceous coat material along the inner face (the face closest to the globule core) of the membrane. Examination of MFGM negatively stained with phosphotungstate or fixed with glutaraldehyde and osmium tetroxide (e.g. Fig. 9) has revealed the isolated membrane to exist partially in the form of plate-like structures; this is in contrast to isolated plasma membrane which is largely vesiculated.[23,59,64,67,68,70,77] Resistance to vesiculation may be due to the presence of coat material along the inner face of the membrane. Freeze-fractured replicas of the MFGM have shown that the intramembranous particles are aggregated, leaving large areas of the membrane faces depleted of these particles.[170-2] There is a stark contrast between the apparently reduced particle density on fat globule membrane faces and the high particle density on apical plasma membrane faces. This suggests that membrane constituents which form intramembranous particles are cleared from or rearranged in plasma membrane regions which envelop fat globules.

That carbohydrates are asymmetrically distributed with respect to the bilayer of the MFGM has been shown by a number of investigators. Horisberger et al.[173] used lectins coupled to gold granules to show that carbohydrates are uniformly distributed over the outer face of the membrane on bovine and human milk fat globules. Sasaki and Keenan[174]

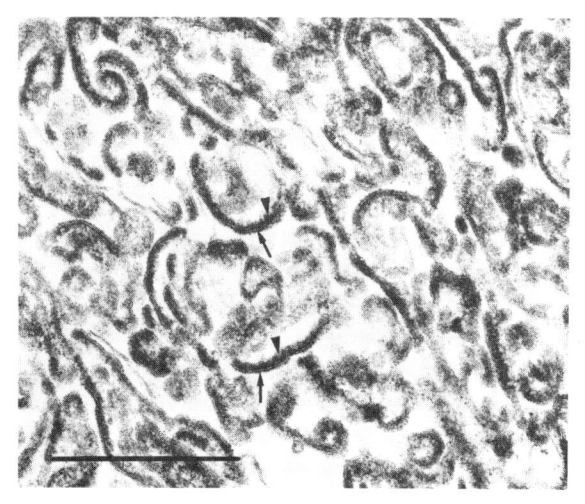

FIG. 9. Isolated MFGM fixed with glutaraldehyde and osmium tetroxide exists as plate-like structures and vesicles. A typical trilaminar or unit-like membrane appearance is evident. A thick layer of dense-staining coat material is present along the inner face of the membrane (arrowheads) while the more fibrillar glycocalyx is observed on the outer face of the membrane (arrows). Magnification: ×74 000. Bar = 0·5 μm.

used a number of carbohydrate-selective staining procedures to show a similar distribution of carbohydrates over the outer face of rat MFGM. Biochemical confirmation for an outer surface localization for much of the carbohydrate of the MFGM has also been obtained. The same amounts of sialic acid are released, at nearly the same rates, when intact globules or isolated MFGM are incubated with neuraminidase.[123] Radiolabelled Concanavalin A is bound to nearly the same extent by intact globules and isolated MFGM.[175] There is also evidence that carbohydrates of MFGM gangliosides are oriented along the outer face of the lipid bilayer.[176]

Proteins of the MFGM also display an asymmetric orientation. Through comparison of specific activities of enzymes in washed lipid globules and released MFGM, Patton and Trams[60] obtained results suggesting that the active site of Mg^{2+}-adenosine triphosphatase was accessible to substrates on the inner face of the membrane, and that of 5'-nucleotidase on the outer face of the membrane. An outer surface localization of the active site of 5'-nucleotidase was also suggested by results of studies on inhibition of this enzyme by Concanavalin A.[163,164] Mather and Keenan[123] observed major differences in rates of hydrolysis of proteins when isolated membranes or intact globules were incubated with trypsin. That other workers obtained a different result appears to be due to the use of inadequate methods for trypsin inactivation.[118] Mather and Keenan[123] also found that more membrane proteins were accessible to lactoperoxidase-catalyzed iodination in isolated membranes than in intact globules. Based on their results, Mather and Keenan[123] concluded that, of the major protein constituents, polypeptides of M_r 67 000 (butyrophilin), 48 000 and 44 000 were exposed on the globule surface and that a polypeptide of M_r 155 000 (subsequently identified as xanthine oxidase) was accessible only when the membrane was released from globules. Using immunological methods, Nielsen and Bjerrum[177] identified four major protein complexes in the MFGM; of these, xanthine oxidase was found on the internal face of the membrane and the other three complexes, as well as 5'-nucleotidase and Mg^{2+}-adenosine triphosphatase, were accessible on the outer surface of the membrane. It has not yet been directly demonstrated that any of the MFGM proteins span the lipid bilayer. However, it is possible that butyrophilin does span the bilayer. Abundant evidence suggests that butyrophilin is present in the coat on the inner membrane face, yet this protein is accessible on the outer surface of milk fat globules. All of the carbohydrate of butyrophilin is present in an acidic polypeptide released by controlled proteolysis of the parent molecule.[135] It is probable that the carbohydrate-containing segment of the molecule is exposed on the outer membrane face and that other regions of the polypeptide, together with xanthine oxidase, form the structure identified in micrographs as the coat on the inner face of the MFGM.

11. CONCLUSIONS

Knowledge of the nature and origin of the milk fat globule membrane has greatly expanded over the past 15 to 20 years due to the efforts of an impressively large number of scientists. The increasing popularity of this research area suggests that knowledge will continue to accumulate rapidly in the coming years. Much remains to be learned about fat globules and their membranes, which will make this an exciting research area well into the future. Our knowledge of the intracellular origin of the globule and the processes which control its growth and movement through the cell is, at best, rudimentary. How intracellular fat globules recognize the apical cell surface and the mechanisms involved in their envelopment in membrane remain open questions. The contribution made by the apical plasma membrane, the secretory vesicle membrane and the membrane material on the surface of intracellular globules to the material known as the milk fat globule membrane remains to be quantified. One can envision that loss of cytoplasmic materials, entrained during globule secretion, may in some manner be related to the rate of milk formation or the length of productive lactation.

While we have learned a great deal about the composition of the milk fat globule membrane, we are a long way from knowing how various constituents are arranged within or on the membrane. The roles played by membrane-associated constituents in globule secretion and in controlling or influencing globule stability in milk are largely unknown.

This research area also presents challenges to those interested in evolutionary biology. The lipids of milk are secreted by a mechanism which appears to be unique to the mammary gland; no other tissue or organ has yet been observed to secrete lipids by this mechanism. All of the species which have been studied (about 20) utilize the same basic mechanism for milk lipid secretion, although the lipid content of the milk produced by these species ranges from a few percent to nearly 50%. The evolutionary advantage of this unique lipid secretory mechanism remains a mystery.

In our view, one of the major advantages of this research area is that it offers a wide variety of fundamental questions, yet important practical benefits may well arise from studies which focus on these fundamental questions. Virtually all basic studies on the origin, secretion and nature of milk lipids impinge on production, processing, nutrition or health considerations.

ACKNOWLEDGEMENTS

We are grateful to Dr Ian H. Mather for many helpful suggestions and for information on the complexity of proteins of the milk fat globule membrane. Research conducted in our laboratory was supported by grants from the National Science Foundation (PCM8241913 and predecessors) and the National Institute of General Medical Science (GM 31244 and predecessors).

REFERENCES

1. BRUNNER, J. R., In: *Structural and Functional Aspects of Lipoproteins in Living Systems*, Tria, E. and Scanu, A. M. (eds), 1969, Academic Press, New York, p. 545.

2. Brunner, J. R., In: *Fundamentals of Dairy Chemistry*, 2nd edn., Webb, B. H., Johnson, A. H. and Alford, J. A., (eds), 1974, Avi, Westport, Connecticut, p. 474.
3. Anderson, M. and Cawston, T. E., *J. Dairy Res.*, 1975, **42**, 459.
4. Blanc, B., *Wld. Rev. Nutr. Diet.*, 1981, **36**, 1.
5. Keenan, T. W., Morré, D. J. and Huang, C. M., In: *Lactation*, Vol. 2, Larson, B. L. and Smith, V. R. (eds), 1974, Academic Press, New York, p. 191.
6. Keenan, T. W., Franke, W. W., Mather, I. H. and Morré, D. J., In: *Lactation*, Vol. 4, Larson, B. L. (ed.), 1978, Academic Press, New York, p. 405.
7. Jenness, R. G., In: *Lactation*, Vol. 3, Larson, B. L. and Smith, V. R. (eds), 1974, Academic Press, New York, p. 3.
8. Jensen, R. G., Clark, R. M. and Ferris, A. M., *Lipids*, 1980, **15**, 345.
9. Patton, S., *J. Am. Oil Chem. Soc.*, 1973, **50**, 178.
10. Patton, S. and Jensen, R. G., *Biomedical Aspects of Lactation*, 1976, Pergamon Press, Oxford.
11. Patton, S. and Keenan, T. W., *Biochim. Biophys. Acta*, 1975, **415**, 273.
12. Keenan, T. W., Mather, I. H. and Dylewski, D. P., In: *Fundamentals of Dairy Chemistry*, 3rd edn., Wong, N. P. (ed.), Avi, Westport, Connecticut. (In press.)
13. Linzell, J. L., In: *Lactation*, Vol. 1, Larson, B. L. and Smith, V. R. (eds), 1974, Academic Press, New York, p. 143.
14. Mather, I. H. and Keenan, T. W., In: *Biochemistry of Lactation*, Mepham, T. B. (ed.), Elsevier, Amsterdam. (In press.)
15. Smith, S. and Abraham, S., *Adv. Lipid Res.*, 1975, **13**, 195.
16. Davis, C. L. and Bauman, D. E., In: *Lactation*, Vol. 2, Larson, B. L. and Smith, V. R. (eds), 1974, Academic Press, New York, p. 3.
17. Bauman, D. E. and Davis, C. L., In: *Lactation*, Vol. 2, Larson, B. L. and Smith, V. R. (eds), 1974, Academic Press, New York, p. 31.
18. Cooper, S. M. and Grigor, M. R., *Biochem. J.*, 1980, **187**, 289.
19. Stein, O. and Stein, Y., *J. Cell Biol.*, 1967, **34**, 251.
20. Daudet, F., Augeron, C. and Ollivier-Bousquet, M., *Eur. J. Cell Biol.*, 1981, **24**, 197.
21. Long, C. A. and Patton, S., *J. Dairy Sci.*, 1978, **61**, 1392.
22. Scow, R. O., Blanchette-Mackie, E. J. and Smith, L. C., *Fed. Proc.*, 1980, **39**, 2610.
23. Keenan, T. W., Morré, D. J., Olson, D. E., Yunghans, W. N. and Patton, S., *J. Cell Biol.*, 1970, **44**, 80.
24. Hood, L. F. and Patton, S., *J. Dairy Sci.*, 1973, **56**, 858.
25. Stemberger, B. H. and Patton, S., *J. Dairy Sci.*, 1981, **64**, 422.
26. Keenan, T. W. and Huang, C. M., *J. Dairy Sci.*, 1972, **55**, 1586.
27. Nickerson, S. C. and Keenan, T. W., *Cell Tiss. Res.*, 1979, **202**, 303.
28. Guerin, M. A. and Loizzi, R. F., *Proc. Soc. Exp. Biol. Med.*, 1980, **165**, 50.
29. Patton, S., *FEBS Letts.*, 1974, **48**, 85.
30. Guerin, M. A. and Loizzi, R. F., *Am. J. Physiol.*, 1978, **234**, C177.
31. Henderson, A. J. and Peaker, M., *Quart. J. Exp. Physiol.*, 1980, **65**, 367.
32. Patton, S., Stemberger, B. H. and Knudson, C. M., *Biochim. Biophys. Acta*, 1977, **499**, 404.

33. KNUDSON, C. M., STEMBERGER, B. H. and PATTON, S., *Cell Tiss. Res.*, 1978, **195**, 169.
34. NICKERSON, S. C., SMITH, J. J. and KEENAN, T. W., *Cell Tiss. Res.*, 1980, **207**, 361.
35. NICKERSON, S. C., SMITH, J. J. and KEENAN, T. W., *Eur. J. Cell Biol.*, 1980, **23**, 115.
36. FRANKE, W. W., LÜDER, M. R., KARTENBECK, J., ZERBAN, H. and KEENAN, T. W., *J. Cell Biol.*, 1976, **69**, 173.
37. AMATO, P. A. and LOIZZI, R. F., *Cell Motility*, 1981, **1**, 329.
38. AMATO, P. A. and LOIZZI, R. F., *Eur. J. Cell Biol.*, 1979, **20**, 150.
39. SMITH, J. J., NICKERSON, S. C. and KEENAN, T. W., *Int. J. Biochem.*, 1982, **14**, 87.
40. SMITH, J. J., NICKERSON, S. C. and KEENAN, T. W., *Int. J. Biochem.*, 1982, **14**, 99.
41. ZULAK, I. M., Ph.D Thesis, Purdue University, 1982.
42. BARGMANN, W. and KNOOP, A., *Z. Zellforsch.*, 1959, **49**, 344.
43. BARGMANN, W., FLEISCHHAUER, K. and KNOOP, A., *Z. Zellforsch.*, 1961, **53**, 545.
44. FELDMAN, J. D., *Lab. Invest.*, 1961, **10**, 238.
45. WELLINGS, S. R., DEOME, K. B. and PITELKA, D. R., *J. Nat. Cancer Inst.*, 1960, **25**, 393.
46. WELLINGS, S. R., GRUNBAUM, B. W. and DEOME, K. D., *J. Nat. Cancer Inst.*, 1960, **25**, 423.
47. HOLLMANN, K. H., In: *Lactation*, Vol. 1, Larson, B. L. and Smith, V. R. (eds), 1974, Academic Press, New York, p. 3.
48. WOODING, F. B. P., In: *Comparative Aspects of Lactation*, Peaker, M. (ed.), 1977, Academic Press, London, 1.
49. PITELKA, D. R. and HAMAMOTO, S. T., *J. Dairy Sci.*, 1977, **60**, 643.
50. WOODING, F. B. P., *J. Cell Sci.*, 1971, **9**, 805.
51. WOODING, F. B. P., *J. Cell Sci.*, 1973, **13**, 221.
52. KUROSUMI, K., KOBAYASHI, Y. and BABA, N., *Exp. Cell Res.*, 1968, **50**, 177.
53. BARGMANN, W. and WELSCH, U., In: *Lactogenesis: The Initiation of Milk Secretion at Parturition*, Reynolds, M. and Folley, S. J. (eds), 1969, University of Pennsylvania Press, Philadelphia, p. 43.
54. MORRÉ, D. J., In: *Cell Surface Reviews*, Vol. 4, Poste, G. and Nicholson, G. L. (eds), 1977, Elsevier/North-Holland, Amsterdam, p. 1.
55. FRANKE, W. W. and KEENAN, T. W., *J. Dairy Sci.*, 1979, **62**, 1322.
56. MARTEL-PRADAL, M. B. and GOT, R., *FEBS Letts.*, 1972, **21**, 220.
57. MARTEL, M. B., DUBOIS, P. and GOT, R., *Biochim. Biophys. Acta*, 1973, **311**, 565.
58. POWELL, J. T., JÄRLFORS, U. and BREW, K., *J. Cell Biol.*, 1977, **72**, 617.
59. KEENAN, T. W., OLSON, D. E. and MOLLENHAUER, H. H., *J. Dairy Sci.*, 1971, **54**, 295.
60. PATTON, S. and TRAMS, E. G., *FEBS Letts.*, 1971, **14**, 230.
61. HUGGINS, J. W. and CARRAWAY, K. L., *J. Supramol. Struct.*, 1976, **5**, 59.
62. HUGGINS, J. W., TRENBEATH, T. P., CHESNUT, R. W., CARRAWAY, C. A. C. and CARRAWAY, K. L., *Exp. Cell Res.*, 1980, **126**, 279.
63. KEENAN, T. W. *et al.*, *Exp. Cell Res.*, 1979, **124**, 47.

64. FRANKE, W. W. et al., *J. Cell Biol.*, 1981, **89**, 485.
65. MORRÉ, D. J., KARTENBECK, J. and FRANKE, W. W., *Biochim. Biophys. Acta*, 1979, **559**, 71.
66. PATTON, S. and FOWKES, F. M., *J. Theor. Biol.*, 1967, **15**, 274.
67. WOODING, F. B. P., *J. Ultrastruct. Res.*, 1971, **37**, 388.
68. WOODING, F. B. P. and KEMP, P., *Cell Tiss. Res.*, 1975, **165**, 113.
69. WOODING, F. B. P. and KEMP, P., *J. Dairy Res.*, 1975, **42**, 419.
70. FREUDENSTEIN, C. et al., *Exp. Cell Res.*, 1979, **118**, 277.
71. KEENAN, T. W., HEID, H. W., STADLER, J., JARASCH, E. D. and FRANKE, W. W., *Eur. J. Cell Biol.*, 1982, **26**, 270.
72. PATTON, S., *FEBS Letts.*, 1976, **71**, 154.
73. PATTON, S., STEMBERGER, B. H., HORTON, A. and McCARL, R. L., *Biochim. Biophys. Acta*, 1980, **630**, 530.
74. PATTON, S., BOGUS, E. R., STEMBERGER, B. H. and TRAMS, E. G., *Biochim. Biophys. Acta*, 1980, **597**, 216.
75. HELMINEN, H. J. and ERICSSON, J. L. E., *J. Ultrastruct. Res.*, 1968, **25**, 193.
76. WOODING, F. B. P., PEAKER, M. and LINZELL, J. L., *Nature*, 1970, **226**, 762.
77. JARASCH, E. D., BRUDER, G., KEENAN, T. W. and FRANKE, W. W., *J. Cell Biol.*, 1977, **73**, 223.
78. PATTON, S., HOOD, L. F. and PATTON, J. S., *J. Lipid Res.*, 1969, **10**, 260.
79. KEENAN, T. W. and HUANG, C. M., *J. Dairy Sci.*, 1972, **55**, 1013.
80. BAUER, H., *J. Dairy Sci.*, 1972, **55**, 1375.
81. BAUMRUCKER, C. R. and KEENAN, T. W., *J. Dairy Sci.*, 1973, **56**, 1092.
82. PATTON, S., LONG, C. and SOKKA, T., *J. Dairy Sci.*, 1980, **63**, 697.
83. KOBLYKA, D. and CARRAWAY, K. L., *Biochim. Biophys. Acta*, 1972, **288**, 282.
84. MATHER, I. H., WEBER, K. and KEENAN, T. W., *J. Dairy Sci.*, 1977, **60**, 394.
85. KEENAN, T. W., FREUDENSTEIN, C. and FRANKE, W. W., *Cytobiologie*, 1977, **14**, 259.
86. RAY, T. K., SKIPSKI, V. P., BARCLAY, M., ESSNER, E. and ARCHIBALD, F. M., *J. Biol. Chem.*, 1969, **244**, 5528.
87. KEENAN, T. W. and MORRÉ, D. J., *Biochemistry*, 1970, **9**, 19.
88. LIS, D. and MONIS, B., *J. Supramol. Struct.*, 1978, **8**, 173.
89. SHIMIZU, M., URYU, N. and YAMAUCHI, K., *Agric. Biol. Chem. (Japan)*, 1981, **45**, 741.
90. BRUDER, G., FINK, A. and JARASCH, E. D., *Exp. Cell Res.*, 1978, **117**, 207.
91. SWOPE, F. C. and BRUNNER, J. R., *J. Dairy Sci.*, 1965, **48**, 1705.
92. SWOPE, F. C. and BRUNNER, J. R., *Milchwissenschaft*, 1968, **23**, 470.
93. HUANG, T. C. and KUKSIS, A., *Lipids*, 1967, **2**, 453.
94. VASIC, J. and DEMAN, J. M., *Proc. 17th Int. Dairy Congr.*, 1966, **C**, 167.
95. WALSTRA, P., *Neth. Milk Dairy J.*, 1974, **28**, 3.
96. NEWMAN, R. A. and HARRISON, R., *Biochim. Biophys. Acta*, 1973, **298**, 798.
97. SCHWARTZ, D. P., BURGWALD, L. H. and BREWINGTON, C. R., *J. Am. Oil Chem. Soc.*, 1966, **43**, 472.
98. SCHWARTZ, D. P., BURGWALD, L. H., SHAMEY, J. and BREWINGTON, C. R., *J. Dairy Sci.*, 1968, **51**, 929.
99. PATTON, S. and KEENAN, T. W., *Lipids*, 1971, **6**, 58.
100. PATTON, S., DURDAN, A. and McCARTHY, R. D., *J. Dairy Sci.*, 1964, **47**, 489.
101. KEENAN, T. W., *Biochim. Biophys. Acta*, 1974, **337**, 255.

102. MORRISON, W. R., In: *Topics in Lipid Chemistry*, Gunstone, F. D. (ed.), 1970, Wiley-Interscience, New York, p. 51.
103. CALBERG-BACQ, C. M., FRANCOIS, C., GOSSELIN, L., OSTERRIETH, P. M. and RENTIER-DELRUE, F., *Biochim. Biophys. Acta*, 1976, **419**, 458.
104. KEENAN, T. W., HUANG, C. M. and MORRÉ, D. J., *J. Dairy Sci.*, 1972, **55**, 51.
105. MORRISON, W. R., *Lipids*, 1968, **3**, 101.
106. MORRISON, W. R. and SMITH, L. M., *Biochim. Biophys. Acta*, 1964, **84**, 759.
107. HLADIK, J. and MICHALEC, C., *Acta Biol. Med. Germania*, 1966, **16**, 696.
108. KAYSER, S. G. and PATTON, S., *Biochem. Biophys. Res. Comm.*, 1970, **41**, 1572.
109. FUJINO, Y., NAKANO, M. and SAEKI, T., *Agr. Biol. Chem. (Japan)*, 1970, **34**, 442.
110. FUJINO, Y. and FUJISHIMA, T., *J. Dairy Res.*, 1972, **39**, 11.
111. BOUHOURS, J. F. and BOUHOURS, D., *Biochem. Biophys. Res. Comm.*, 1979, **88**, 1217.
112. KEENAN, T. W., HUANG, C. M. and MORRÉ, D. J., *Biochem. Biophys. Res. Comm.*, 1972, **47**, 1277.
113. HUANG, R. T. C., *Biochim. Biophys. Acta*, 1973, **306**, 82.
114. BUSHWAY, A. A. and KEENAN, T. W., *Lipids*, 1978, **13**, 59.
115. BASU, S., BASU, M., CHIEN, J. L. and PRESPER, K. A., In: *Structure and Function of Gangliosides*, Svennerholm, L., Dreyfus, H. and Urban, P. F. (eds), 1980, Plenum Press, New York, p. 213.
116. KEENAN, T. W., *J. Dairy Sci.*, 1974, **57**, 187.
117. MORRISON, W. R., JACK, E. L. and SMITH, L. M., *J. Am. Oil Chem. Soc.*, 1965, **42**, 1142.
118. KOBYLKA, D. and CARRAWAY, K. L., *Biochim. Biophys. Acta*, 1973, **307**, 133.
119. KITCHEN, B. J., *Biochim. Biophys. Acta*, 1974, **356**, 257.
120. KITCHEN, B. J., *J. Dairy Res.*, 1977, **44**, 469.
121. ANDERSON, M., *J. Dairy Sci.*, 1974, **57**, 399.
122. ANDERSON, M., CAWSTON, T. and CHEESEMAN, G. C., *Biochem. J.*, 1974, **139**, 653.
123. MATHER, I. H. and KEENAN, T. W., *J. Membrane Biol.*, 1975, **21**, 65.
124. MANGINO, M. E. and BRUNNER, J. R., *J. Dairy Sci.*, 1975, **58**, 313.
125. O'FARRELL, P. H., *J. Biol. Chem.*, 1975, **250**, 4007.
126. O'FARRELL, P. Z., GOODMAN, H. M. and O'FARRELL, P. H., *Cell*, 1977, **12**, 1133.
127. SWITZER, R. C., MERRIL, C. R. and SHIFRIN, S., *Anal. Biochem.*, 1979, **98**, 231.
128. HOFMANN, C. J., KEENAN, T. W. and EIGEL, W. N., *Int. J. Biochem.*, 1979, **10**, 909.
129. MATHER, I. H., TAMPLIN, C. B. and IRVING, M. G., *Eur. J. Biochem.*, 1980, **110**, 327.
130. JARASCH, E. D. *et al.*, *Cell*, 1981, **25**, 67.
131. BRUDER, G., HEID, H., JARASCH, E. D., KEENAN, T. W. and MATHER, I. H., *Biochim. Biophys. Acta*, 1982, **701**, 357.
132. MATHER, I. H., SULLIVAN, C. H. and MADARA, P. J., *Biochem. J.*, 1982, **202**, 317.
133. MURRAY, L. R., POWELL, K. M., SASAKI, M., EIGEL, W. N. and KEENAN, T. W., *Comp. Biochem. Physiol.*, 1979, **63B**, 137.

134. Mather, I. H., *Biochim. Biophys. Acta*, 1978, **514**, 25.
135. Heid, H. W., Winter, S., Bruder, G., Keenan, T. W. and Jarasch, E. D., *Biochim. Biophys. Acta*, 1983, **728**, 228.
136. Waud, W. R., Brady, F. O., Wiley, R. D. and Rajagopalan, K. V., *Arch. Biochem. Biophys.*, 1975, **169**, 695.
137. Mangino, M. E. and Brunner, J. R., *J. Dairy Sci.*, 1977, **60**, 841.
138. Sullivan, C. H., Mather, I. H., Greenwalt, D. E. and Madara, P. J., *Mol. Cell. Biochem.*, 1982, **44**, 13.
139. Bray, R. C., In: *The Enzymes*, Vol. 12, 3rd edn., Boyer, P. D. (ed.), 1975, Academic Press, New York, p. 299.
140. Briley, M. S. and Eisenthal, R., *Biochem. J.*, 1974, **143**, 149.
141. Appell, K. C., Keenan, T. W. and Low, P. S., *Biochim. Biophys. Acta*, 1982, **690**, 243.
142. Basch, J. J., Farrell, H. M. and Greenberg, R., *Biochim. Biophys. Acta*, 1976, **448**, 589.
143. Kanno, C., Shimizu, M. and Yamauchi, K., *Agr. Biol. Chem. (Japan)*, 1975, **39**, 1835.
144. Kanno, C., Shimizu, M. and Yamauchi, K., *Agr. Biol. Chem. (Japan)*, 1977, **41**, 83.
145. Kanno, C. and Yamauchi, K., *Agr. Biol. Chem. (Japan)*, 1978, **42**, 1697.
146. Shimizu, M., Kanno, C. and Yamauchi, K., *Agr. Biol. Chem. (Japan)*, 1976, **40**, 1711.
147. Keenan, T. W., Powell, K. M., Sasaki, M., Eigel, W. N. and Franke, W. W., *Cytobiologie*, 1977, **15**, 96.
148. Snow, L. D., Colton, D. G. and Carraway, K. L., *Arch. Biochem. Biophys.*, 1977, **179**, 690.
149. Imam, A., Laurence, D. J. R. and Neville, A. M., *Biochem. J.*, 1981, **193**, 47.
150. Shimizu, M. and Yamauchi, K., *J. Biochem. (Tokyo)*, 1982, **91**, 515.
151. Harrison, R., Higginbotham, J. D. and Newman, R., *Biochim. Biophys. Acta*, 1975, **389**, 449.
152. Newman, R. A., Harrison, R. and Uhlenbruck, G., *Biochim. Biophys. Acta*, 1976, **433**, 344.
153. Farrar, G. H. and Harrison, R., *Biochem. J.*, 1978, **171**, 549.
154. Glöckner, W. M., Newman, R. A., Dahr, W. and Uhlenbruck, G., *Biochim. Biophys. Acta*, 1976, **443**, 402.
155. Newman, R. A. and Uhlenbruck, G. G., *Eur. J. Biochem.*, 1977, **76**, 149.
156. Farrar, G. H., Harrison, R. and Mohanna, N. A., *Comp. Biochem. Physiol.*, 1980, **67B**, 265.
157. Mangino, M. E. and Brunner, J. R., *J. Dairy Sci.*, 1977, **60**, 1208.
158. Huang, C. M. and Keenan, T. W., *J. Dairy Sci.*, 1972, **55**, 862.
159. Anderson, M., *J. Dairy Sci.*, 1977, **60**, 1217.
160. Martel, M. B. and Got, R., *Biochim. Biophys. Acta*, 1976, **436**, 789.
161. Nathans, G. R. and Hade, E. P. K., *Biochim. Biophys. Acta*, 1978, **526**, 328.
162. Huang, C. M. and Keenan, T. W., *Biochim. Biophys. Acta*, 1972, **274**, 246.
163. Carraway, C. A. and Carraway, K. L., *J. Supramol. Struct.*, 1976, **4**, 121.
164. Snow, L. D., Doss, R. C. and Carraway, K. L., *Biochim. Biophys. Acta*, 1980, **611**, 333.

165. Dowben, R. M., Brunner, J. R. and Philpott, D. E., *Biochim. Biophys. Acta*, 1967, **135**, 1.
166. Huang, C. M. and Keenan, T. W., *Comp. Biochem. Physiol.*, 1972, **43B**, 277.
167. Diaz-Maurino, T. and Nieto, M., *Biochim. Biophys. Acta*, 1976, **448**, 234.
168. Morton, R. K., *Biochem. J.*, 1954, **57**, 231.
169. Bailie, M. J. and Morton, R. K., *Biochem. J.*, 1958, **69**, 44.
170. Zerban, H. and Franke, W. W., *Cell Biol. Intern. Reports*, 1978, **2**, 87.
171. Peixoto de Menezes, A. and Pinto da Silva, P., *J. Cell Biol.*, 1978, **76**, 767.
172. Pinto da Silva, P., Peixoto de Menezes, A. and Mather, I. H., *Exp. Cell Res.*, 1980, **125**, 127.
173. Horisberger, M., Rosset, J. and Vonlanthen, M., *Exp. Cell Res.*, 1977, **109**, 361.
174. Sasaki, M. and Keenan, T. W., *Cell Biol. Intern. Reports*, 1979, **3**, 67.
175. Keenan, T. W., Franke, W. W. and Kartenbeck, J., *FEBS Letts.*, 1974, **44**, 274.
176. Tomich, J. M., Mather, I. H. and Keenan, T. W., *Biochim. Biophys. Acta*, 1976, **433**, 357.
177. Nielsen, C. S. and Bjerrum, O. J., *Biochim. Biophys. Acta*, 1977, **466**, 496.
178. Svennerholm, L., *J. Neurochem.*, 1963, **10**, 613.
179. Diaz-Maurino, T. and Nieto, M., *J. Dairy Res.*, 1977, **44**, 483.
180. Baumrucker, C. R., *J. Dairy Sci.*, 1979, **62**, 253.

Chapter 4

PHYSICAL CHEMISTRY OF MILK FAT GLOBULES

P. WALSTRA

Agricultural University, Wageningen, The Netherlands

SUMMARY

The physical and colloidal properties of milk fat globules depend on their size, surface layer (membrane), and presence and orientation of fat crystals; and on external conditions such as agitation and incorporation of air. Stability of the globules to creaming, flocculation, coalescence and disruption are related to the (presumed) interaction energy between approaching globules and to properties of the fat globule membrane. The latter is considerably affected by homogenization, as is globule size. Consequently, natural, homogenized and recombined fat globules are considered. The flocculation of fat globules in raw milk at low temperatures (cold agglutination) is discussed in some detail, including new results; this intricate phenomenon is still insufficiently understood. Effects of fat content, cold agglutination, (partial) coalescence and clustering of fat globules on rheological properties are briefly reviewed.

1. INTRODUCTION

In this chapter, physical and colloidal aspects of the milk fat globules of cows' milk are discussed. They are of importance for the stability of milk and milk products as emulsions; this concerns creaming, flocculation, coalescence and churning, and also some aspects of lipolysis. Foaming and whipping are strongly affected by properties of the fat globules, as are such product properties as viscosity and colour; the properties of high-fat products may especially be dominated by those of the fat globules.

Most aspects considered here have been treated extensively in a monograph,[1] to which we will often refer.

2. GLOBULE SIZE DISTRIBUTION

Milk fat is predominantly present in spherical droplets which range in diameter from less than 0·2 to about 15 μm. The bulk of the fat is in globules 1–8 μm diameter.[2] The size distributions found may depend greatly on the measurement method employed[3] and many older data are unreliable; most newer results are essentially identical (e.g. References 2, 4 and 5). Figure 1 gives some examples of number frequency distributions, i.e. the number of globules per unit volume, N, in a certain size class divided by the width of the size class, Δd, as a function of size, d. $N/\Delta d$ is plotted on a log scale because of the wide range.

Figure 1 suggests that the frequency distribution is composed of three sub-distributions:

(i) 'Small particle', essentially an exponential distribution, comprising about 80% of the number of globules but only a few percent of the fat.
(ii) 'Main', comprising about 95% of the fat, following a mathematical expression that can be derived from a log-normal distribution by introducing a minimum (at 0·1 d_{vs}) and a maximum (at 2·5 d_{vs}) possible diameter.[6]
(iii) 'Large globule', comprising about 2% of the fat.

It is tempting to explain the latter sub-distribution as arising from a limited coalescence of globules after their expulsion from the secretory cell.[6] The biological significance of the 'small globule' sub-distribution can only be guessed.[2] The size distribution of the fat globules in human milk can also be separated into three sub-distributions.[7]

Average globule size can be expressed in many ways; it has become more and more common to employ the volume–surface average (or surface-weighted mean) diameter, $d_{vs} = \Sigma Nd^3/\Sigma Nd^2$. For most milks, $d_{vs} = 2\cdot 5$ to 5 μm.[1,2] The specific surface area of the fat globules is given by $A = 6\phi/d_{vs}$, where ϕ is the volume fraction of fat. For most milks, $A = 5$ to 11 $m^2/100$ g milk. The width of the size distribution can, for instance, be expressed as

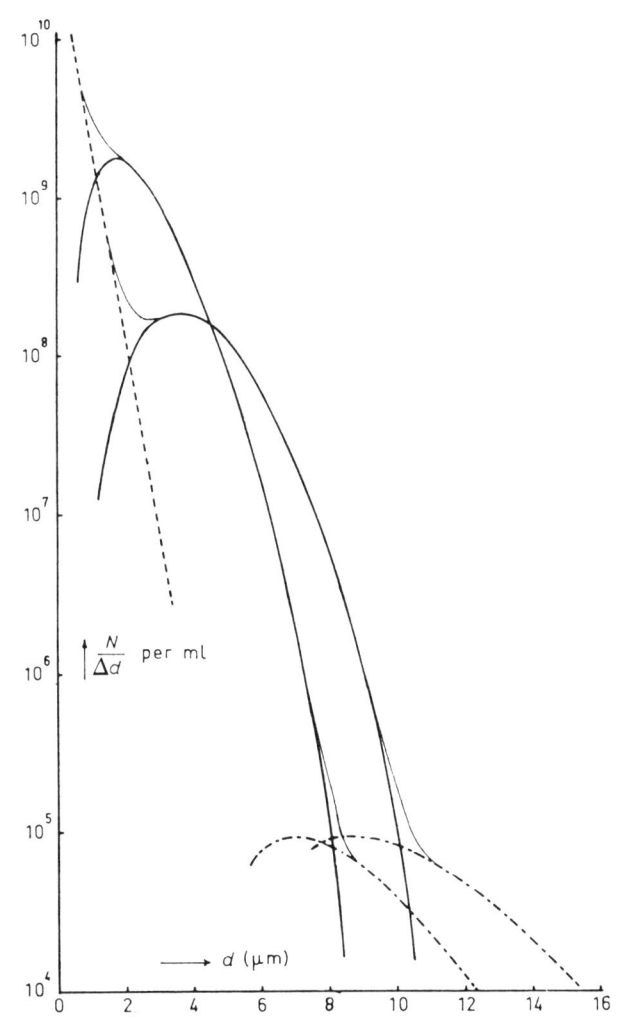

FIG. 1. Two idealized examples of the number frequency distribution of the diameter of fat globules in milk from individual cows. The three sub-distributions are depicted by ----, ——— and –·–·–·–. See text for explanation.[2]

the relative standard deviation of the surface-weighted distribution, c_s. Milks of individual cows show a remarkably constant c_s of about 0·44;[2] in mixed milk, c_s is slightly higher, e.g. 0·5.

In fact, the shape of the size distribution is almost constant between samples of individual milkings,[2,5] as is illustrated in Fig. 2. Undoubtedly, it is the 'main' sub-distribution which is essentially constant. The only variables are d_{vs} and total fat content. These vary between cows; those of Channel Island breeds give milk with larger fat globules than most other breeds, the average d_{vs} being, for instance, 4·5 and 3·4 μm, respectively.[1] There are considerable differences between cows of the same breed, and there is a fair correlation between average globule size and fat content. A ration that causes an increase in milk fat content usually causes an increase

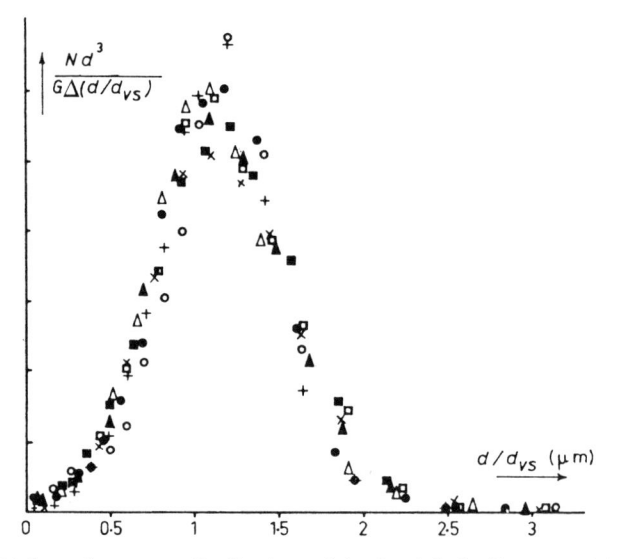

FIG. 2. Volume frequency distributions of the fat globule diameter of the milk of eight cows. The results are reduced to equal fat content, G (ordinate) and to equal d_{vs} (abscissa). Further explanation in text.[2]

in fat globule size rather than in number of globules.[5] Average globule size decreases with advancing lactation, for instance from d_{vs} = 4·4 to d_{vs} = 2·9 μm.[1]

Globule size can be considerably altered by various treatments, particularly homogenization. Results depend on the type of machine and conditions during homogenization, and primarily upon homogenizing pressure.[8] Figure 3 gives some examples of volume frequency distributions. In most homogenized products, d_{vs} = 0·3 to 0·8 μm and c_s = 0·8 to 1·2.[1] Changes in globule size by other treatments are discussed elsewhere in this chapter (Section 9).

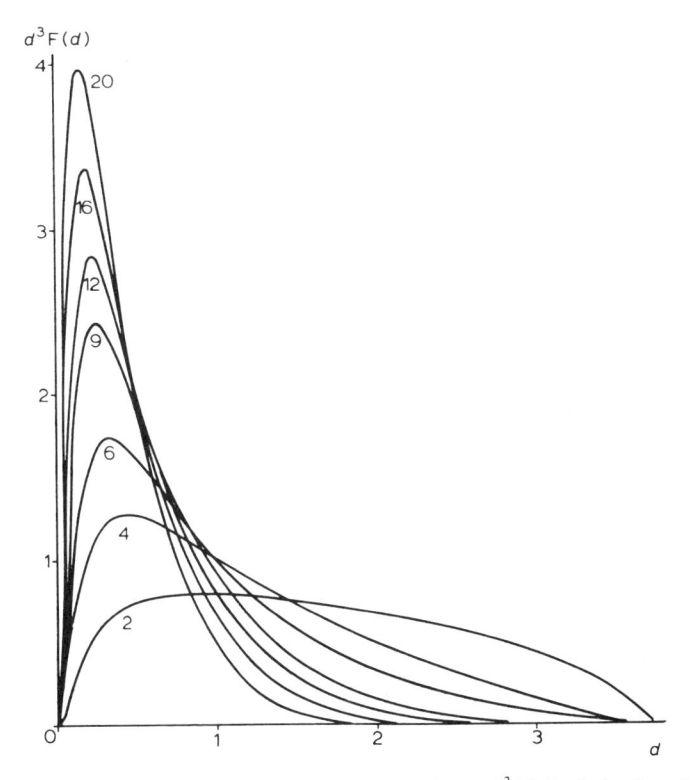

FIG. 3. Examples of volume frequency distributions, $d^3 F(d)$ of the fat globule diameter, d, of homogenized milk. Parameters on curves are homogenization pressures in MPa.[8]

3. COMPOSITIONAL DIFFERENCES

The fat globules of milk differ in composition. In samples of one milking from a single cow, globules of equal diameter (in the range 3·5–6·5 μm) show[9] a standard deviation in refractive index (at 589 nm) of 7×10^{-4}, which corresponds to considerable differences in composition.

The composition of globules may also vary with size. It has been established that this is because the quantity of membrane lipids (predominantly phospholipids) per unit mass of fat is higher for smaller globules.[10,11] Other variations of fat composition with globule size have not been clearly established.[1]

4. FAT CRYSTALS IN GLOBULES

Whether a quantity of fat is present in numerous small globules or as a continuous mass profoundly influences its crystallization behaviour. The most striking effect is on supercooling, but the extent of compound crystal formation and the abundance of the various polymorphs are also affected.[12] These aspects are discussed in Chapter 5. The compositional variation among globules mentioned above is a further cause of differences.

Crystals in fat globules generally cannot grow larger than the globule diameter. Most crystals are much smaller than globule size and they flocculate into a network, giving the globule a certain rigidity.[1] Crystals in fat globules may show up under the polarizing microscope because of their birefringence, but they must be large enough or have a common orientation to be visible. Some globules show a birefringent outer layer. From a series of experiments and observations, the author[13] concluded that such a layer must consist of small fat crystals tangentially oriented at or near the boundary of the globule. Nucleation of crystals at the boundary was never observed at temperatures $>5\,°C$, and was considered most unlikely on the basis of the relationship between globule size and nucleation kinetics.[12,14] Fat globules with birefringent outer layers are rarely observed in cold milk, but they often occur after a globule has somehow been deformed.[13,15]

A different view is held by Buchheim[16] and several others. Electron micrographs of freeze-etched or freeze-fractured specimens of milk or cream often show concentric layers of apparently crystalline fat at the boundary of some fat globules[16-18] and it is assumed that such a crystalline shell represents the birefringent outer layer observed under the polarizing microscope. From similar studies on cream during churning and on butter[19,20] it was inferred that the globules with a crystalline shell are particularly stable to coalescence and remain as such in butter. Butter indeed shows many fat globules with a birefringent outer layer (e.g. References 1 and 13). However, in extensive studies of coalescence of milk fat globules and several other emulsions with fat crystals in the droplets, it was observed that instability was invariably coupled with the occurrence of fat droplets with a birefringent outer layer;[21] i.e. such globules are the unstable ones.

Other electron microscopic techniques applied to fat globules or to preparations of fat globule membrane material never show a crystalline shell, but distinct, platelet-shaped crystals.[13,18,22,23] Consequently, the appearance of a crystalline shell on freeze-fracture electron micrographs of milk fat globules may be an artefact.

5. TYPES OF INSTABILITY

Emulsions like milk and cream can show various kinds of physical change, as illustrated in Fig. 4. We will follow here the discussions in Reference 1.

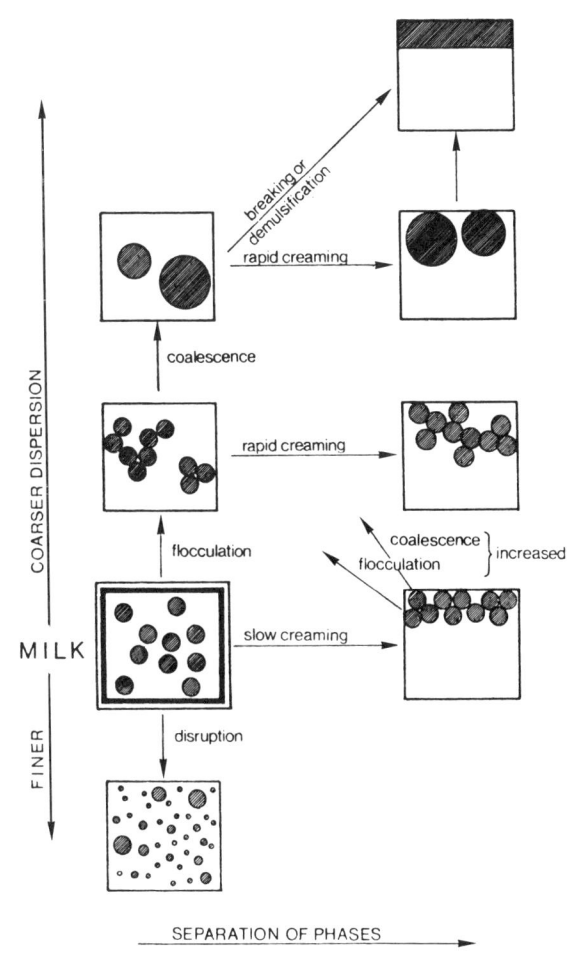

FIG. 4. Different types of emulsion instability, highly schematic. Fat is grey.[1]

Some changes occur spontaneously, for instance creaming under most conditions. For other changes, such as coalescence and flocculation, an activation energy must be overcome in most cases, and this causes retardation. An emulsion is thermodynamically unstable, since segregation into two layers (Fig. 4, upper right-hand corner) leads to a lower interfacial free energy. But the activation energy for coalescence of any pair of globules is generally so large as to retard segregation almost indefinitely. If no activation energy existed, any encounter between two globules would cause their coalescence. Encounters caused by Brownian motion alone would then lead to visible coalescence (fat lenses) within a few minutes, whereas milk may remain stable to the eye for months. To achieve rapid coalescence, energy must be applied, e.g. by agitation. (Some types of flocculation may occur spontaneously; see Section 13.) Disruption of globules can only be achieved by applying much energy.

The changes depicted in Fig. 4 may occur simultaneously. Coalescence can only occur if two globules are very close to each other, either because they are in a flocculated state or in a fleeting encounter. Agitation enhances encounter frequency and thus possibly flocculation and coalescence; it also disturbs creaming and it may disrupt floccules and possibly even globules (Section 9). The higher the fat content, the higher the probability of coalescence while the probability of disruption is hardly affected.

6. INTERACTION ENERGY

Emulsions may be considered as lyophobic colloids. The stability of such colloids is commonly explained by the interaction energy between two particles, i.e. the energy needed to bring two particles from an infinite distance apart closer to each other; the energy is thus a function of the interparticle distance (see, for example, Fig. 5). If the interaction energy is negative at all distances, the particles will flocculate when they meet each other. If the energy at some distance is positive and much larger than the average difference in kinetic energy of two particles in Brownian motion (i.e. kT), the particles will not flocculate. More precisely, the probability of two colliding particles flocculating is $\exp(-E_a/kT)$, where E_a is the maximum interaction energy; it is thus an activation energy for flocculation. This statement, however, needs modification. The existence of a large positive interaction energy at close distances (often 1–5 nm) does not exclude the possibility of a negative interaction energy at larger distances.

Such a minimum is often found, e.g. at 5–20 nm, and if it is deep enough (a few times kT), particles may flocculate 'in the minimum', i.e. without touching each other. These aspects are treated in most texts on colloid or emulsion science (e.g. References 24–7).

The difficulty is to determine the interaction energy. The classical DLVO theory[24] considers electrostatic repulsion and van der Waals attraction. Repulsion can presumably be calculated with fair accuracy since the parameters needed are known. The zeta potential (which is commonly taken to be equal to the electrostatic potential at the surface) of fat globules in milk ≈ -13 mV,[28–30] the globule size is known and the

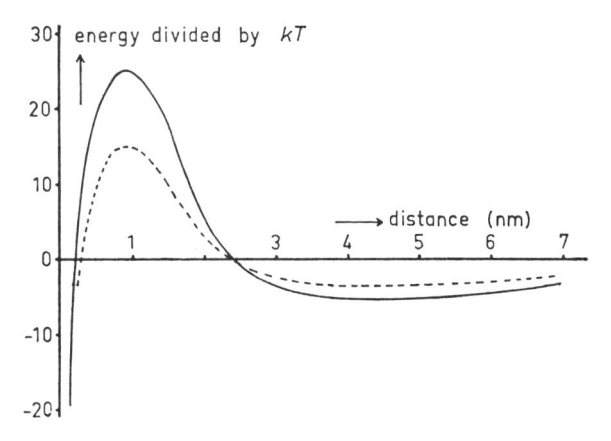

FIG. 5. Interaction energy between two fat globules, of 5 (———) or 3 μm diameter (------), according to the DLVO theory and assuming a very small Hamaker constant ($\sim 2 \times 10^{-21}$ J). From Reference 1, after data in Reference 29.

thickness of the electric double layer in milk at room temperature is $\sim 1\cdot 1$ nm, as follows from its total ionic strength of $\sim 0\cdot 08$ M. To calculate the van der Waals attraction, the Hamaker constant must be known; the compilations available[31] point strongly to a value of 1 to 2×10^{-20} J for milk fat in water. Applying these data in the DLVO theory results in a negative interaction energy at all distances, so that milk fat globules should immediately flocculate.[21] However, this does not occur except at low temperatures under some conditions (Section 13). Attempts have been made[29,30] to explain the stability of milk fat globules by assuming a very low Hamaker constant, based on the consideration that the outer layer of the globule predominantly determines the van der Waals attraction and that this layer (essentially the membrane) may have a much lower Hamaker constant. The approximate result is shown in Fig. 5. Even if this were true, it cannot explain the simultaneous stability of small and large globules. Since the interaction energy between two equally sized globules is, at any distance, proportional to their diameter, it follows that either the maximum in the interaction energy is too low to prevent flocculation of small globules or that the energy minimum is so deep as to lead to flocculation of large globules. An even stronger argument is that milk fat globules have an isoelectric pH of $\sim 3 \cdot 7$[32] and that they do not flocculate at this pH,[21] even though electrostatic repulsion must be zero. Likewise, washed fat globules are particularly stable to churning at the isoelectric pH.[33]

Consequently, another repulsive force must be acting. This is probably steric repulsion, caused by flexible, hydrophilic molecular chains protruding from the globule surface into the milk plasma. In such a case, the interpenetration of the layers of chains of two approaching globules may cause repulsion, primarily by osmotic effects.[34] The number (per unit area), length and solubility characteristics of such chains must be known to permit even a rough calculation of the ensuing repulsive energy, and none of these is known. Careful interpretation of the coalescence stability of milk fat globules in laminar flow of varying velocity gradient shows the results to be consistent with a moderately strong steric repulsion, keeping the globules some 20 nm apart.[21] It should, however, be taken into account that such a layer of protruding chains may cause the absolute value of the electrokinetic or zeta potential (which is the electrostatic potential at the slipping plane of a moving globule) to be much lower than the surface potential (since the slipping plane is at some distance from the surface). Hence electrostatic repulsion may be stronger than previously assumed. Calculation of the interaction energy will thus be very difficult.

It is interesting to note that the fat globule membrane contains several glycoproteins, mostly derived from the plasmalemma of the lactating cell. It is known that such membrane proteins have very hydrophilic parts (containing polysaccharide chains) that can stick out from the surface over a considerable distance;[35] they may thus cause steric repulsion. It has been shown[36] that treatment of milk with papain, which hydrolyses glycoproteins, results in flocculation of the fat globules. It has also been reported that a glycocalyx (a layer of polysaccharide chains) can be seen around fat globules on micrographs, using special staining techniques.[37]

7. COALESCENCE

Coalescence of fat globules may have several consequences:

(i) It may lead to rapid creaming and to the formation of fat lenses or butter granules in milk products.
(ii) It may cause cream to become more viscous, or a fairly solid cream plug to be formed in a bottle of milk.
(iii) It permits the isolation of fat from milk products, as in churning.
(iv) It is desirable, when controlled, during the whipping of cream and the freezing of ice-cream for the acquirement of desirable texture and stability.[1]
(v) It leads to transfer of fat globule membrane substances (e.g. phospholipids, glycoproteins, xanthine oxidase) to the milk plasma because of the decrease in total surface area of the globules.[1]

Milk fat globules should thus be stable to coalescence during handling and storage, but unstable under other conditions. These requirements can usually be met.

Coalescence is the flowing together of two emulsion droplets into one. Its ultimate cause is the rupture of the film of the continuous phase (in this case milk plasma) between two droplets that are close to each other; the thickness of a film that can rupture is generally of the order of 10 nm. Coalescence stability should thus be treated in terms of the stability of a thin film. General theory[38,39] explains the stability of such a film on the basis of the interfacial tension, γ, between the oil and water phases and of the interaction energy between the droplets as a function of film thickness. These factors determine whether any disturbance of the film (i.e. a local fluctuation in thickness) will be damped-out or enhanced. In the latter case, the film ruptures. It turns out that the probability of rupture is higher as the size of the film is larger (hence globule diameter larger), γ is smaller and the droplets can come closer together (hence net repulsion is lower). The theory is, however, insufficient. Generally, if it predicts instability the emulsion is unstable, but if it predicts stability the emulsion may still show coalescence. It should be taken into account that, even in an unstable emulsion, only a very small proportion (say 10^{-5}) of the encounters between droplets leads to coalescence.[1,21] Coalescence is thus a chance effect and is not easy to predict.

The presence of crystals (solid fat) in the globules strongly affects coalescence.[1] If all (or by far, most) of the fat is solid, coalescence is definitely impossible; solid particles can flocculate but not coalesce. If part of the fat is solid, the globules may show partial coalescence; irregular clumps or granules are formed that keep their shape because of the crystals (that form a network), while liquid fat acts as a sticking agent—subsequent melting of the crystals causes complete coalescence. The presence of crystals actually promotes coalescence if some crystals are located in the oil–water interface; presumably such crystals may pierce the film between globules.[21,40] Streaming in the liquid (e.g. caused by agitation) may promote (partial) coalescence, but can also cause disruption of globules, particularly of large globules. Consequently, coalescence during agitation may never lead to visible oil separation, since the larger droplets formed are disrupted again. But if part of the fat is solid, partial coalescence is possible and disruption of the granules is mostly impossible; consequently, if the fat is partly solid, rapid granule formation may be observed in an agitated cream, while no visible change occurs when the fat is either fully solid or fully liquid.[1,41]

The following are the main factors affecting (partial) coalescence:

(i) *Streaming or agitation.* The main variable is the velocity gradient or shear rate, S, although the type of flow is also important.[21] For liquid milk fat globules (at 40°C), laminar flow at $S = 10^2$ to $10^3 \, \text{s}^{-1}$ causes coalescence at such a rate that $\sim 10^{-7}$ of the calculated number of encounters between globules lead to coalescence;[42] the initial coalescence rate shows first-order kinetics with time, but soon levels off, presumably because of simultaneous disruption. Liquid fat globules show no significant coalescence in milk or cream at rest.[21]

If part of the fat is solid, coalescence during agitation is more pronounced.[21,40,41] Examples are shown in Fig. 6. Turbulence causes even faster coalescence.[21] The presence of birefringent outer layers in the globules appears to be necessary for rapid coalescence, and the deformation of droplets, even if very slight, may enhance the genesis of such globules.[21]

(ii) *Temperature.* Temperature primarily affects the proportion of solid fat and thereby partial coalescence (Fig. 7). Pre-cooling may be needed to obtain solid fat in the globules and thus may affect coalescence but it also promotes coalescence by some other mechanism, perhaps via changes induced in the globule membrane.[1] Temperature history is thus of importance. A special case is the 'rebodying' of cream;[43] cream is cooled to $\sim 4\,°\text{C}$, kept for ~ 2 h, warmed to $\sim 30\,°\text{C}$ and cooled again. This causes the (apparent)

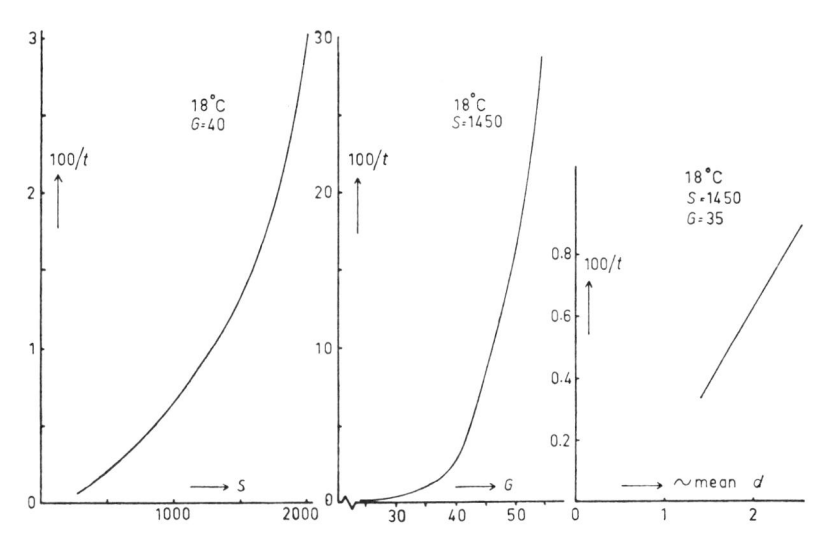

FIG. 6. Influence of apparent shear rate, S (s^{-1}) fat content, G (% m m^{-1}) and globule diameter, d (μm) on the effective rate of partial coalescence in cream in a shear field (t = time(s) needed to form visible granules). From Reference 1, after data in Reference 41.

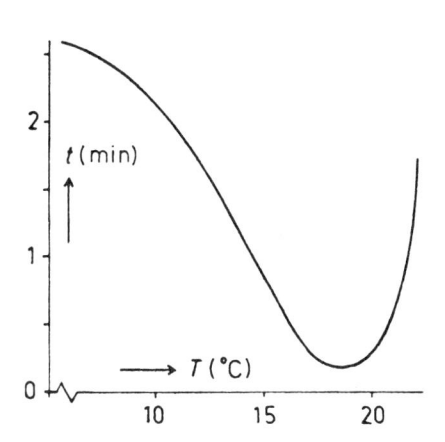

FIG. 7. Time needed, t, for visible granules to form in cream of 40% fat in a shear field (S = 1900 s^{-1}) at various temperatures, T; the cream was pre-cooled for 18 h at 5 °C. From Reference 1, after data in Reference 41.

viscosity of the cream to increase greatly; the main cause is a limited partial coalescence.[44]

If the fat is liquid, coalescence rate (during agitation), is faster at higher temperatures.[45,46] In some types of heat exchanger, considerable coalescence may occur in cream.[45]

(iii) *Fat content*. The higher the fat content, the higher the rate of coalescence during agitation[21,41] (Fig. 6). The dependence of the coalescence rate on fat content is very strong, much stronger than the quadratic relation that would be expected from the effect on encounter frequency. Presumably, the fat globules are pressed closer together during agitation in a high-fat cream.[1]

(iv) *Creaming*. Creaming, as such, causes little or no coalescence of liquid globules.[21] However, if the fat globules are closely packed in the cream layer (thus in the absence of agglutination; see Section 14), changes in crystallization may cause partial coalescence[1] and thus the formation of a cream plug.[47]

(v) *Freezing*. Freezing of cream causes considerable partial coalescence, presumably because the ice crystals damage the fat globules.[1]

(vi) *Surface layer or membrane of the globules*. Undoubtedly, the properties of the surface layer considerably affect the coalescence stability, but unequivocal explanations or correlations have not been established. The composition of the surface layer affects interfacial tension; for natural milk fat globules it is low, mostly 1 to 2 mN m^{-1},[55] and a low value has an adverse effect on coalescence stability. On the other hand, the composition of the surface layer determines the interaction energy, and thereby affects stability. Moreover, clear correlations between surface-rheological properties and coalescence stability of emulsion droplets with widely varying surface layers have been established (see Reference 1 for a discussion), but not explained. Modification of the natural membrane of the fat globules probably causes them to be less stable (assuming other conditions to be unaltered). Partial degradation of membrane phospholipids by added[36] or bacterial enzymes[48] leads to coalescence. *Bacillus cereus* may in this way produce 'bitty cream'.

(vii) *Globule size*. The smaller the globules, the better their stability to coalescence[1,21,41,49] (Fig. 6). Part of the explanation that a milk or cream with small globules is more stable than systems with large globules follows from the greater number of coalescence events that must take place for a visible change to occur, but the rate constant for coalescence is also smaller, often much smaller for smaller globules (e.g. Reference 21). Consequently, homogenized milk products are usually very stable to coalescence, despite their modified surface layers.

To conclude, milk fat globules are fairly stable, but partial coalescence may occur especially during agitation of cream at temperatures (say 15–30°C) where part of the fat is solid. [1,21,40,41,50,51]

8. INTERACTION WITH AIR BUBBLES

Very little has been published recently on this subject, and a discussion given previously will be largely followed.[1] Possible interactions are illustrated in Fig. 8. An oil globule (O) coming into contact with a water–air interface (WA) will spread over that interface if the spreading pressure

$$P_s = \gamma_{WA} - (\gamma_{OW} + \gamma_{OA}) > 0$$

If the interface is clean, spreading will certainly occur since γ_{WA} then is high (72 mN m^{-1}), while γ_{OA} is much lower (e.g. 34 mN m^{-1}) and γ_{OW} is very low for a milk fat globule (e.g. 2 mN m^{-1}, though it will increase rapidly as membrane material spreads over the WA interface). Still, fat globules rarely become attached to the air–plasma interface,[52,53] presumably because the interface rapidly acquires an adsorption layer of plasma proteins, which causes the fat globule and the WA interface to repel each other (moreover, the spreading pressure may be negative in this case). If air bubbles are beaten into the milk (or formed in another way) it will take some 10–100 ms before the protein is adsorbed, and there is thus some time for fat globules to become attached to a bubble. Consequently, when bubbling air through a column of milk or cream, fat globules are collected in the foam layer forming on top; such a process is called floatation. In model experiments on a paraffin oil-in-water (OW) emulsion with a suitable polymer as surfactant, it is found that the rate of floatation decreases with increasing polymer concentration, undoubtedly because of the increased rate of polymer adsorption to the WA interface, which decreases the time available for globules to become attached to this interface.[54]

Further events depend on conditions. Some material from the fat globule membrane will spread over the WA interface; consequently, the milk plasma will become enriched in membrane materials such as phospholipids. If the fat is predominantly solid it will not spread and the globules just become attached to air bubbles. If the fat is fully liquid it will spread, particularly if the ratio of fat to protein is high, as in cream. The thin layer of spread oil is easily disrupted if the air bubble is disrupted, as occurs during agitation. Consequently, vigorously shaking milk or cream or agitating it while air is beaten in can cause considerable disruption of fat globules. A similar effect occurs when milk is boiled (as in an evaporator)

$$P_s = \gamma_{WA} - (\gamma_{OW} + \gamma_{OA}) < 0$$

$$P_s > 0$$

$$P_s > 0, \text{ interfacial layers}$$

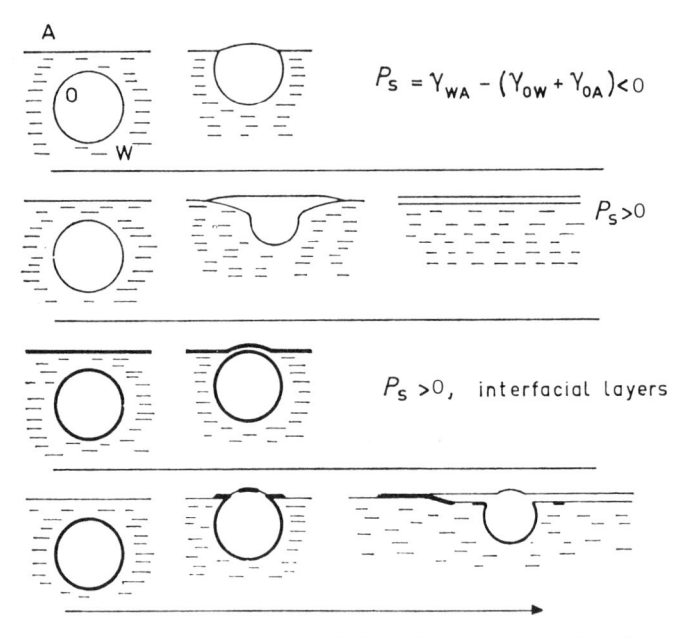

FIG. 8. Interaction between an oil globule and an air–water interface under various conditions. A, air; O, oil phase; W, water phase; interfacial adsorption layer or membrane is indicated by a heavy line. The lower row refers to a milk fat globule approaching a clean air–plasma interface.[1]

because of spreading over vapour bubbles. If part of the fat in the globules is solid, shaking leads to churning. Fat globules adhere to air bubbles, some liquid fat leaves the globules by spreading, and subsequent coalescence of air bubbles causes their surface area to diminish, thereby driving the adhering globules closer together; the liquid fat acts as a sticking agent and granules of globules are formed. This is an efficient method for achieving partial coalescence. Vigorous shaking of milk at 30°C may lead to visible granules within 30 s.

9. DISRUPTION

Fat globules can be disrupted by surface forces (as discussed in Section 8) and by forces transmitted by the liquid. These can be viscous forces (originating from shearing action exerted by the streaming liquid) or inertial forces (as in turbulent flow and cavitation). A general discussion is given, for instance, in References 1, 25 and 27.

A globule resists deformation, and thus disruption, by its Laplace pressure; the pressure at the concave side of a curved interface is higher than that at the convex side by an amount

$$\Delta P = \gamma \left(\frac{1}{R_1} + \frac{1}{R_2} \right)$$

where γ is the interfacial tension and R_1 and R_2 are the principal radii of curvature. For a spherical droplet, $R_1 = R_2 = d/2$. Any deformation causes a decrease in the radius of curvature and thus a rise in ΔP, which must be overcome. In the absence of outside forces, a droplet will thus assume a spherical shape which is, incidentally, the cause of complete coalescence as soon as two droplets have made contact.

A globule can be deformed by the shearing action of the liquid, and the viscous stress acting on a particle is of the order of ηS, where η is the viscosity of the continuous phase (milk plasma) and S is the velocity gradient. If $\eta S \geq 4\gamma/d$, the globule may be disrupted. For natural milk fat globules, $\gamma \approx 10^{-3}$ N m^{-1},[55] and is thus very small, although it will increase as the globule is deformed. Since $\eta \approx 10^{-3}$ Pa s, disruption of a globule of $d = 4$ μm will need $S \approx 10^6$ s^{-1}, an extremely high velocity gradient. Even then, simple shear flow (i.e. flow with parallel streamlines, where velocity gradient = shear rate) does not cause disruption because of the unfavourable viscosity ratio of fat to plasma;[27] simple shear with $S = 16 \times 10^6$ s^{-1} fails to cause any disruption of fat globules in milk.[56] Larger droplets, as may result from coalescence, can be disrupted in elongational flow (i.e. flow with a velocity gradient in the direction of flow), if S is high enough. However, such conditions only exist under very exceptional circumstances.

Disruption can occur in turbulent flow. The relevant variable is the energy density, ϵ (quantity of turbulent energy dissipated per unit time and unit mass of liquid). Broadly speaking, globules are disrupted if

$$d^5 \gg \gamma^3 \rho^{-3} \epsilon^{-2}$$

where ρ is the mass density of the liquid ($\sim 10^3$ kg m^{-3}). This implies that milk fat globules of 4 μm diameter can be disrupted if ϵ is at least 100 kW kg^{-1}, which is a very high energy density (it would cause the temperature of the milk to rise at a rate of some 25 °C per second). Such energy densities can occur only for very short times (as in the slit of a homogenizer valve) or at local sites (e.g. near the top of a fast-moving stirrer blade or where a rapid stream impinges upon an obstacle). Disruption of milk fat globules (decrease of d_{vs} by a few percent up to ~20%) has been observed under conditions that favour intense turbulence, particularly in heat exchangers.[1,57-61] Disruption tends to increase with increasing temperature,[1,60] and under extreme conditions d_{vs} may be halved; γ probably decreases with increasing temperature, but reliable measurements have not been published.

Fat globules can be disrupted by cavitation (i.e. the formation and sudden collapse of small vapour bubbles, initiated by local negative pressures), for instance as induced by ultrasonic waves.[62,63] It is, however, questionable whether this occurs during normal milk processing operations.

Homogenization is, of course, applied to disrupt fat globules into much smaller ones; it is usually achieved in high-pressure homogenizers, although some other types of machinery are occasionally used.[1] The mechanism of disruption is, in all probability, pressure fluctuation in turbulent flow;[1,27,63] ϵ is typically 10^9 W kg^{-1}. Under certain conditions, cavitation may materially add to disruption in homogenizers.[27,64] The extent of fat globule disruption in homogenizers depends on various conditions,[1,8,65] notably homogenizing pressure (see Fig. 3).

10. CHANGES IN SURFACE LAYERS

The membranes of newly formed fat globules seem to alter spontaneously, although part of the changes may be induced by cooling. This is discussed in Chapter 3 and in References 1, 23 and 66. It appears that part of the original membrane material is released into the plasma. At low temperatures, some lipoproteins present in milk plasma tend to adsorb onto the fat globules.[1,67]

Processing treatments of milk may alter the globule membranes. Heat treatment certainly causes changes;[1] several membrane proteins are susceptible to heat denaturation, although there is no direct experimental evidence. The fat globules release hydrogen sulphide (H_2S) on heating.[68] Some serum protein and calcium phosphate are probably deposited onto the fat globules during heating, although, again, there is no direct evidence. Heat treatment (e.g. 15 s at 80°C) causes a distinct transfer of copper from the plasma to the globules[69] and causes a change in the zeta potential.[70] Heat treatment probably causes a slight decrease in coalescence stability.[1]

The most conspicuous change in the surface layer of the fat globules is the adsorption of new material on the OW interface if (part of) the original material is lost by contact with air bubbles or if the fat surface area is increased (by disruption of globules). The material adsorbed primarily consists of plasma proteins; it takes 10–100 ms for these to diffuse to the OW interface and form an adsorption layer.[1] If milk is homogenized, the greater part of the surface layers will be new, because fat surface area increases, usually by a factor of 5 to 10; but part of the surface still has a more-or-less natural membrane since most of it is retained.

In recombined milk (made by homogenizing a mixture of skim milk and butterfat) the surface layers are completely new. Consequently this product offers a good opportunity to study those layers and numerous publications have appeared (e.g. References 71–80). The results may be summarized as follows. The proteins are adsorbed irreversibly, at least for the time-scales considered (1 h to a few days). Some slow surface denaturation presumably occurs, since at least part of the proteins are more tenaciously held at the globule surface after the recombined milk is kept for, say, 1 day.[77] Casein micelles are adsorbed, but they are partly spread into a thinner layer consisting of fragments of micelles and possibly submicelles.[1,75,81-3] Serum proteins, but apparently no proteose–peptone, are also adsorbed. Figure 9 gives a schematic picture, derived mainly from electron microscopy.

The composition of the adsorbed surface layer depends on conditions during adsorption. Some important variables are:

(i) *Composition of the continuous phase.* In first approximation, proteins are adsorbed in proportion to their concentration. So milk plasma gives a layer of casein and serum proteins (total protein load, Γ, ≈ 8 mg of protein/m^2 fat globule surface), but whey gives only serum proteins ($\Gamma \approx 2\cdot5$ mg m^{-2}).[73,79] Sweet-cream buttermilk contains much phospholipid (presumably in the form of small lipoprotein particles) and this can also be adsorbed.

(ii) *Intensity of agitation.* If agitation is slight, protein will arrive at the interface by diffusion; the smallest molecules or particles diffuse fastest and are thus preferentially adsorbed. If agitation is very

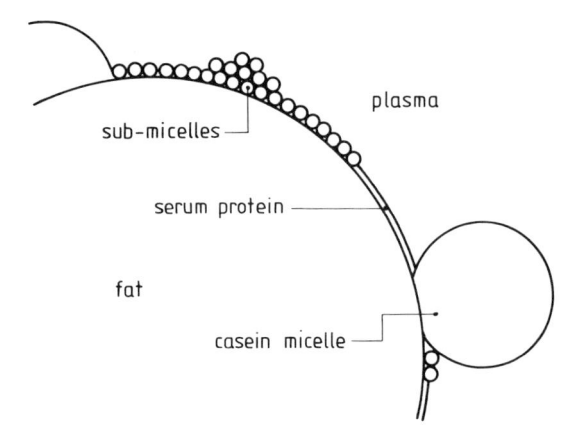

FIG. 9. Schematic representation of the surface layer of synthetic fat globule (milk fat emulsified into skim milk).[66]

vigorous, as in a homogenizer, the protein arrives by convection and larger particles or, more precisely, particles that are closer in size to the fat globules are preferentially adsorbed.[80] In a recombined milk (butterfat homogenized with low-pasteurized skim milk) serum proteins will thus make up only about 5% of the adsorbed protein, covering about 25% of the surface area.[80] Moreover, the preponderance of adsorption of large casein micelles is strongest for small globules. Consequently, small globules acquire much more protein per unit surface area if casein micelles and serum proteins are both present; an example of recombined milk gave $\Gamma \approx 15$ mg m^{-2} for globules of $d \approx 0.4$ μm, and $\Gamma \approx 3$ mg m^{-2} for $d \approx 1.6$ μm.[80] More intense agitation (higher energy density) thus gives thicker adsorbed layers, mainly because it leads to smaller globules.

(iii) *Temperature.* In general, a higher temperature (up to about 70°C) during adsorption (e.g. during homogenization) causes a thinner protein layer. For instance, 40°C gave $\Gamma = 10.7 \pm 0.7$ mg m^{-2}, and 60°C gave $\Gamma = 6.0 \pm 0.8$ mg m^{-2}.[79] The explanation is presumably that casein can spread faster over an OW interface at a higher temperature.

(iv) *Heat treatment.* A heat treatment of the (skim) milk prior to adsorption causes the adsorbed layer to be thicker. For instance, a heat treatment of 10 min at 65°C gave $\Gamma \approx 10$ mg m^{-2}, while 10 min at 95 °C gave $\Gamma \approx 15$ mg m^{-2}.[79] Part of the explanation may be the association of serum proteins with the casein micelles caused by heating, but some other effect must also play a part.[79] When the plasma is concentrated skim milk, which has comparatively large micelles, Γ may be very high—up to 25 mg m^{-2}.[1]

(v) The ratio of fat to protein (or rather fat surface area to be covered to plasma protein available) has little effect on Γ if the continuous phase is milk plasma. If it is whey, Γ may vary considerably, for example from 0·4 mg m^{-2} if the ratio is high, to 3 mg m^{-2} if protein is in excess.[79] A high ratio causes considerable depletion of protein from the continuous phase; for instance, when homogenizing a 20% fat cream to $d_{vs} = 0·5$ μm approximately all the plasma protein is needed to cover the globules ($\Gamma = 10$ mg m^{-2}).

The presence of surfactants may considerably affect the surface layer. If a (small-molecule) surfactant gives a lower γ_{OW} than the existing surface layer does, the latter is (partly) desorbed and replaced by the surfactant. Since γ is very low for natural fat globules (1 to 2 mN m^{-1}), few surfactants can cause desorption of the natural membrane. Excess of sodium deoxycholate[84] or Triton X-100[85] can release part of the membrane material, apparently without gross coalescence. Some surfactants, when added in excess to cream while stirring at high temperature (e.g. 15 min at 75 °C), may cause almost complete coalescence of the globules;[86,87] This implies extensive removal of natural membranes. The interfacial tension imparted by plasma proteins is much higher, perhaps as high as 15 mN m^{-1} or more,[88] which implies that adsorbed plasma proteins can be desorbed more easily. Small quantities of Tween 20, for instance, can cause a marked decrease in Γ (mg protein/m^2), whether added before or after formation of the surface layer.[79] Free fatty acids and monoglycerides, for instance formed by lipolysis, may also affect the surface layers (e.g. lowering of Γ), but higher concentrations are needed and they must be present in the fat phase.[79]

11. PROPERTIES OF MODIFIED FAT GLOBULES

Homogenization and recombination give rise to fat globules that differ markedly in properties from the natural ones but these properties depend on conditions during treatment. Most conspicuous is the change in size, the globules usually being smaller. Smaller fat globules are more stable in almost every respect, particularly to creaming and coalescence. In recombined milk all fat globules have an identical fat composition, while there is a spread in composition between globules in untreated and homogenized milks. This may have some consequences for fat crystallization, although the effect is small.[12] Of course, the fat in recombined milk may have been modified, for instance by fractionation based on partial crystallization, and this may considerably affect crystallization behaviour.

The surface layer is different, and we will primarily consider recombined milk since here the change is complete. Natural membrane material evidently contributes to the creamy flavour of milk products; the lack of such a flavour note in recombined products can be remedied by the addition of 5–10% sweet-cream buttermilk (e.g. Reference 89). Because of the different surface layers, the fat in recombined (and homogenized) milk is far less prone to auto-oxidation but far more prone to lipolysis compared to natural milk.[1]

Recombined fat globules are very stable against coalescence, more stable than natural ones even if of the same size.[1] This implies, for instance, that churning and whipping[90] are difficult to achieve. Adding phospholipids or sweet-cream buttermilk before homogenization may considerably enhance churning rate,[44] whipping properties of cream,[91] and textural properties of ice-cream;[92] in all these cases, decreased coalescence stability is responsible.[1] Likewise, addition of surfactants, e.g. Tweens, diminishes the freeze–thaw stability of recombined cream,[93] improves textural properties of ice-cream,[92,94,95] and enhances whippability of homogenized cream.[96] Pre-heating of the skim milk (denaturing the serum proteins) before recombination gives an even better coalescence stability of the fat globules in recombined milk.[97] Emulsifying milk fat into whey protein solutions gives less stable emulsions[1,44,98] and spontaneous coalescence occurs at high temperatures, particularly if Γ is low. Sodium caseinate gives very stable emulsions. Protein-stabilized emulsions have been reviewed.[99]

Fat globules that are covered mainly with casein will take part in any casein reaction.[1] If the pH is lowered to near the isoelectric point of casein, if rennet is added, or if the product is heated to the extent that the casein micelles become unstable, the globules join in the ensuing coagulation. The fat globules act to some extent as if they were large casein particles and the effective casein concentration thus is higher; hence coagulation proceeds faster. Consequently, homogenized cream is far less stable to heat coagulation than is homogenized milk; in fact homogenization conditions primarily determine the stability.[100] A similar relationship is seen in the feathering stability of cream in hot coffee.[76] The stability can be improved by replacing part of the surface layers with other material, e.g. phospholipids.[100] The heat stability of evaporated milk, and particularly of recombined evaporated milk,[101] is similarly affected by the amount and composition of protein in the surface layers; intense heat treatment of the (skim) milk before homogenization, which gives more casein in the surface coat, diminishes heat stability.[100,102] The forces holding the heat-coagulated globules together are strong, and the floccules cannot be redispersed by stirring.

In cultured milk, and particularly in sour cream, the fat globules participate in the casein network formed if they have casein in their surface coat. The same happens in rennet curd. This has a considerable effect on rheological properties. Homogenization or recombination is essential in obtaining a sour cream of the desired viscosity.[1] In acid and rennet milk gels, firmness is much higher if the fat globules contain casein as compared to natural globules.[103,104]

12. HOMOGENIZATION CLUSTERS

Homogenization of cream may result in a considerable increase in its viscosity due to the formation of extensive clusters of fat globules which can be dispersed into separate globules by adding reagents that cause disintegration of casein micelles. The subject has been reviewed[1] and some further work has been published since.[105,106] It is clear that the globules in a cluster are held together because one casein micelle is simultaneously held in the OW surface of two globules. This occurs, broadly speaking, if the amount of protein present per unit volume of product is less than the quantity needed to cover the newly formed interface, $\Gamma \Delta A$. The specific surface area of the globules is $A = 6\, \phi/d_{vs}$, where ϕ is the fat volume fraction. Consequently, high ϕ (high fat content), small d_{vs} (i.e. high homogenization pressure) and high Γ (depending on several conditions, for instance homogenization temperature) favour cluster formation. These trends are illustrated in Fig. 10. Intense heat treatment before homogenization causes a high Γ, and hence an increased tendency to form clusters.

Homogenization clusters can only be formed in high-pressure homogenizers; here the homogenized globules can leave the valve slit only partly covered with protein (partly because of the very short passage time) and form clusters immediately afterwards in a zone where turbulence is too weak to disrupt these clusters again. In other machines (such as high-speed

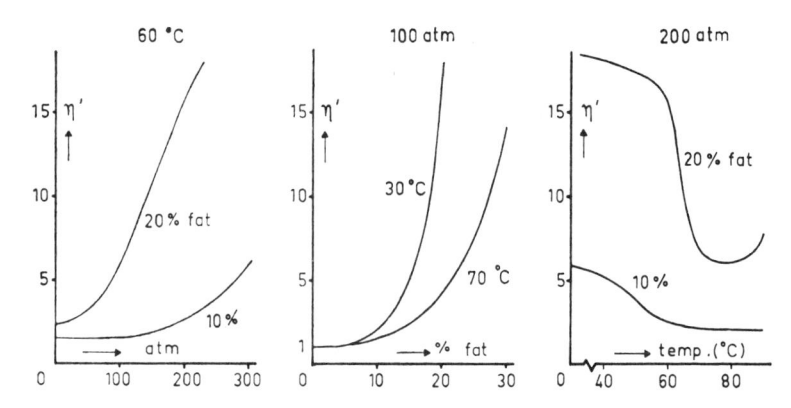

FIG. 10. Approximate relationship between apparent viscosity, η' (mPa s) of pasteurized cream and conditions during homogenization.[1]

mixers) most of the clusters formed are disrupted in zones of somewhat lower turbulent intensity, whereby some globules coalesce. This can to some extent be achieved in a homogenizer by two-stage homogenization if the pressure drop over the second valve is small compared to (e.g. 20% of) the total homogenizing pressure. In this way, the extent of clustering can be diminished considerably; for instance, the apparent viscosity of a 20% cream homogenized at 20 MPa may be reduced five-fold by a second-stage pressure drop of 5 MPa.

The consequence of homogenization clusters is thus an increase in viscosity; high-fat (e.g. 30%) cream becomes a paste upon homogenization. Sour cream (20%) may attain an almost solid consistency if it contains sufficient homogenization clusters. Sometimes partial homogenization of milk is applied to reduce costs; milk is separated, the cream homogenized, and mixed again with the skim milk. If homogenization clusters have been formed, these lead to rapid creaming, and conditions should be carefully optimized.[107] Cream with less than 9% fat usually gives no clusters upon homogenization, whatever the conditions, and cream of over 18% fat always does.

13. COLD AGGLUTINATION

Cows' milk shows rapid creaming in the cold (Fig. 11) due to flocculation of the fat globules into large aggregates, up to 1 mm size, containing up to 10^6 globules and 10–60% (v/v) fat. The flocculs can easily be distinguished from homogenization clusters; they can be disrupted by fairly gentle stirring, and they spontaneously and completely disperse on warming to 37°C or higher. The large flocculcs rise to the surface far quicker than do separate globules. The phenomenon is by no means universal; the milks of buffalo, sheep and goat do not exhibit it[108] and the milks of some cows may exhibit little or none[1]—a genetic trait.[109] The review by Mulder and Walstra,[1] supplemented with newer results and considerations, will be followed.

The flocculation of fat globules bears considerable resemblance to the agglutination caused by the specific action of immunoglobulins of the class IgM (macroglobulins). Dependence on pH, concentration and valency of

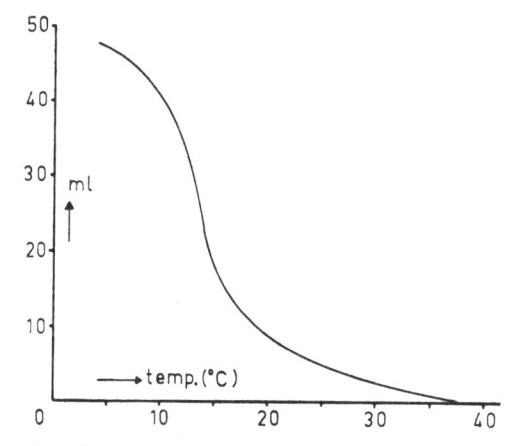

FIG. 11. Example of creaming of fresh milk, expressed as volume of cream obtained after 2 h at various temperatures.[1]

cations, and presence of several reagents is roughly the same as for the agglutination of, say, erythrocytes or bacteria. Milk contains immunoglobulins, predominantly IgG, but also some IgA and IgM (e.g. Reference 110). IgM is a large molecule (mol. wt. ≈900 000), being a pentamer of IgG-like molecules, and thus having 10 active sites for adhesion to foreign particles. Because of its large diameter (~30 nm) the molecule can act as a bridge between particles, despite any (strong) interparticle repulsion at close distance. The adhesion and thereby the agglutination is, however, a specific, antigen–antibody reaction. The IgM antibody is synthesized by leucocytes in response to the presence of a specific antigen (e.g. bacteria) in the blood. In the blood of some humans, cryoprecipitation can occur; on cooling below physiological temperatures, the erythrocytes are agglutinated. This is a non-specific reaction; the particular blood contains a cryoglobulin, a protein complex in which IgM is predominant, which becomes insoluble in the cold and then precipitates on particles present, thereby flocculating them.

The cold agglutination of milk fat globules appears to follow the same mechanism. Milk fat globules dispersed in cold skim milk, whey or solutions containing immunoglobulins from milk or colostrum exhibit flocculation. The fraction designated euglobulin[117] is particularly active; it adsorbs onto washed fat globules and causes them to flocculate.[111] If milk is heated to such an extent (e.g. 20 s at 76 °C) that the immunoglobulins are denatured (as judged by their becoming insoluble at pH 4·6), cold agglutination no longer occurs, but it can be restored by adding euglobulin isolated from milk.[1] If milk is separated by centrifugation at 40 °C the material causing cold agglutination is found in the skim milk, while if separation is at 5 °C the material is in the cream; it can be released from the fat globules by warming to 40 °C.[1] All types of fat globule appear to be agglutinated, including recombined globules (recombined in skim milk or in whey) and oil droplets covered with polyvinylalcohol.[1,112]

Flocculation proceeds mostly as shown in Fig. 12 (which also gives information about the effect of ionic strength). A time lag of 20–30 min is observed before flocculation attains its maximum rate, which is somewhat surprising. By application of Smoluchovski's theory on perikinetic flocculation,[24,26] and assuming the IgM concentration in milk to be about 5×10^{-8} M, it turns out that the time needed for the IgM to reach the fat globules would be about 10 s, i.e. $< 10^{-2}$-times the time lag observed. The maximum slope of the curve is, however, in approximate agreement with the predicted flocculation rate of fat globules sticking together as soon as they meet.[1] The time lag can be explained by assuming that a cryoglobulin molecule will only adhere to a fat globule if the contact involves a reactive site of the molecule; these sites occupy $\sim 10^{-2}$ of the (apparent) surface area of the molecule.[113] The mutual flocculation of cryoglobulin molecules, in the absence of fat globules, should then proceed even more slowly, $\sim (10^{-2})^2$-times the predicted rate. Flocculation of immunoglobulins in cold whey was indeed found to proceed as a bimolecular reaction, the rate constant being $\sim 10^{-5}$-times the predicted value for unhindered flocculation.[113]

It thus seems as if a consistent model of the mechanism of cold agglutination can be given (except for the forces responsible for adhesion at low temperature). However, there are many observations, often of a bewildering complexity (e.g. References 112, 114 and 115), that cannot be reconciled with the model outlined above. Some of these complications will be discussed briefly:

(i) Gentle stirring (shear rate ~400 s^{-1}) suffices to completely disrupt the floccules.[1] Such stirring cannot affect the rate of mutual flocculation of cryoglobulin molecules, as follows from the theory of orthokinetic flocculation,[24,26] and will somewhat enhance the encounter frequency of cryoglobulin with fat globules. Consequently, stirring cold milk for 20–30 min at 400 s^{-1} should bring all cryoglobulin to the fat globules and subsequently stopping the

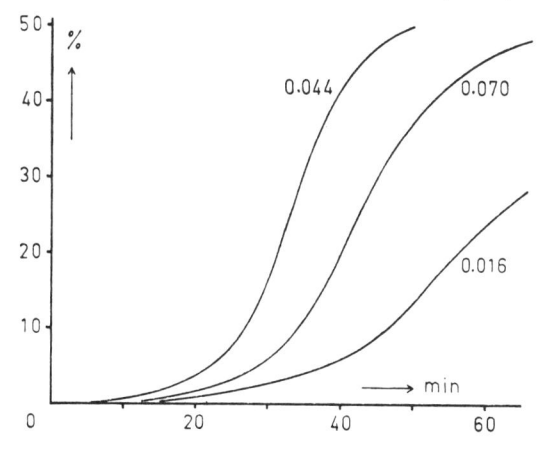

FIG. 12. Flocculation of fat globules in milk diluted (1:5) with NaCl solution as a function of time since cooling to 5 °C. Flocculation is given as the decrease in turbidity at wavelength 1·2 μm. Parameter is ionic strength (M).[1]

stirring should lead to rapid flocculation of the fat globules without any time-lag. However, it turns out that the fat globules do not flocculate at all, or only slightly after a considerable time lag.[112] The agglutinating ability could be restored by warming the milk to 40 °C and re-cooling it. This is in agreement with the long-established fact that milk which has been kept cold and stirred or shaken exhibits poor creaming, which can be remedied by warming and re-cooling.[1]

(ii) Different lots of milk may give different flocculation curves and, although an S-shape is common, the distinct time-lag shown in Fig. 12 is not always observed.[112] Moreover, numerous factors affect flocculation (e.g. temperature, concentration of several substances, pre-treatment) and it is found that almost invariably the apparent time-lag and the reciprocal of the maximum slope are almost proportional; in other words, the curves could be made to coincide (at least up to a certain degree of flocculation) by altering the time-scale.[112] These observations do not fit the simple model.

(iii) It has been pointed out[115] that the quantity of euglobulin in milk is far smaller than is required in the model experiments mentioned.[111] Such small quantities do not cause flocculation of washed fat globules in, for instance, milk ultrafiltrate. However, if skim milk is ultracentrifuged an opalescent layer is found just above the sedimented casein micelles; this layer contains a rather ill-defined lipoprotein, apparent mol. wt. $\gg 10^6$. This lipoprotein does not cause fat globule agglutination in the cold, but lipoprotein plus a little euglobulin (quantity as present in milk) does.[115] Material isolated from whey by cooling and sedimentation causes copious cold agglutination when added to model systems in quantities as present in milk.[112,114] The material presumably contains lipoproteins as well as immunoglobulins, as is also suggested by other studies.[116]

(iv) Heat-treated milk, e.g. 70°C for 15 min, shows no cold agglutination, but this can be restored by adding euglobulin. Even euglobulin fractions containing no IgM can, to some extent, cause flocculation in the cold, if added to heat-treated milk.[112,118] However, if the heat treatment is 2 min at 95°C cold agglutination cannot be restored.[112,114,119]

(v) Homogenization of milk inactivates cold agglutination. This has nothing to do with altering the fat globules, since homogenizing skim milk (if obtained by separation at, say, 40°C) has the same effect.[103,120] Homogenization at very low pressures (e.g. 1 Pa) suffices.[103,121] Even stirring (3 h at 400 s^{-1} and 40°C) causes almost complete inactivation of cold agglutination;[103] it is remarkable that such mild treatments can affect such very small particles, but similar stirring (without air) may even cause inactivation of some enzymes under some conditions (e.g. Reference 122). In one study,[121] it was shown that homogenization of a euglobulin solution does not cause its inactivation, but that homogenization in the presence of κ-casein (either as such, in whole caseinate or in casein micelles) does; it was presumed that the cryoglobulin and κ-casein somehow react during homogenization. The euglobulin thus complexed still adsorbs onto fat globules in the cold, but does not cause their flocculation.[121] However, homogenization of rennet whey or neutralized acid whey also causes inactivation of cold agglutination.[112]

(vi) Pasteurized skim milk, homogenized skim milk, or skim milk that has been pasteurized and homogenized are ineffective in inducing cold agglutination, but a mixture of pasteurized unhomogenized and homogenized unpasteurized skim milk may induce copious cold agglutination.[112,114,123] This is designated the Samuelsson effect, and has been interpreted as cows' milk having two factors needed for cold agglutination, one of which is heat sensitive, and the other homogenization sensitive. This seems to be borne out by similar experiments with mixtures of cows' and buffaloes' skim milk treated in various ways, which could be interpreted as buffaloes' (carabaos') milk missing the homogenization-sensitive but not the heat-sensitive factor.[108] However, experiments with mixtures of cows' skim milk, half of which was homogenized but unheated and the other half unhomogenized but heated for 5 min at temperatures of 60, 70, 80 or 90°C, consistently gave positive, negative, positive, and negative results, respectively, for cold agglutination.[112] This implies three different reactions; the first (between 60 and 70°C) coincides with the normal heat inactivation, and the third (between 80 and 90°C) coincides with the inactivation that cannot be undone by adding euglobulin (see above). All three reactions had a temperature dependency ($Q_{10} \approx 14$, apparent activation energy $\approx$270 kJ mol^{-1}) typical for heat denaturation of globular proteins. Mixtures of pasteurized skim milk and homogenized whey caused cold agglutination, while mixtures of pasteurized whey and homogenized skim milk did not.[112]

It can only be concluded that the mechanism of cold agglutination is still unclear. It appears that a complex of substances is involved, and that this complex is formed in the cold *in situ* (i.e. at the fat globule surface). Under some conditions it is active, but not under others. The complex contains immunoglobulins, and (part of the) IgM is probably the most active, but lipoproteins (possibly also lipoproteins in the fat globule membrane) are also needed. It remains a question why other particles, notably the numerous casein micelles, do not interfere by taking part in the agglutination. It has been found, however, that casein micelles and denatured serum proteins do somewhat inhibit cold agglutination.[112] These and several other disturbing factors may confuse the interpretation of experimental results.

There is considerable variation in the agglutination activity among lots of milk.[1,109,116] There is a poor correlation with euglobulin content[124] (which need not be surprising in view of the role of the lipoprotein), although the activity decreases with advancing lactation, as does immunoglobulin content. There is a distinct effect of globule size,[116] which may be due to the fact that smaller globules have a larger specific surface area, and thus need more 'agglutinin' than larger globules. Homogenized milk can still exhibit cold agglutination if sufficient 'agglutinin' (e.g. obtained from cold whey) is added. It has been observed that agglutination is inversely related to the 5'-nucleotidase activity of the fat globules, which in turn is related to their phospholipid content.[125]

It is well known that several bacteria can be agglutinated in milk, but this is a specific, antigen–antibody reaction which also (and predominantly) occurs at physiological temperatures. Still, several bacteria are enriched in the cream when cold agglutination occurs; it appears as if the presence of both the specific antibody and the cryoglobulin is needed.[126] It may well be that aggregates of bacteria resulting from 'normal' agglutination subsequently participate in the cold agglutination of the fat globules.

14. CREAMING

Since milk fat has a lower density than plasma, the fat globules rise under the influence of a gravitational field force. As the resultant of buoyancy and friction the globules attain a constant rising speed, v, which, for perfect spheres, is given by the Stokes equation:

$$v_s = a(\rho_p - \rho_f)d^2/18\eta_p$$

where a is the acceleration defining the field, ρ is the mass density, d is the globule diameter and η is the viscosity; subscripts f and p refer to fat and plasma, respectively. For gravity creaming, $a = g$ ($\approx 9\cdot 8$ m s^{-2}); for creaming in a centrifugal field $a = R\omega^2$, where R is the effective centrifugal radius and ω is the angular velocity (in radians/second; thus, $\omega = 2\pi n/60$, where n is the revolutions/minute).[1,127] This equation was shown to accurately predict the rising speed of separate fat globules ($d > 2$ μm) under ideal conditions (very low fat content, no convection currents) (e.g. Reference 128). Several prerequisites must, however, be met for v_s to be correct;[127] the most relevant of these being:

(i) Disturbance by Brownian motion or streaming (e.g. convection currents). The smaller the globules, the larger the deviation. For gravity creaming, a diminishing rising speed (as compared to v_s)

was observed for $d < 3$ μm, and for $d = 1$ μm rising was virtually absent; for centrifugal creaming ($a \geq 200$ g), the disturbance is probably negligible.[127]

(ii) Thick surface layers, as formed by homogenization. These may considerably retard the rising rate of small globules.[127] In vigorously homogenized evaporated milk, many globules may even sediment instead of cream because of their high protein load.[129]

(iii) Fat content. This may considerably affect creaming rate, as illustrated in Fig. 13. The decrease with increasing fat content observed in gravity creaming is mainly caused by counterflow of liquid and by mutual hindrance of the fat globules; the relationship is $v \approx v_s(1 - \phi)^{8.6}$, where ϕ is the fat volume fraction. The initial increase in rate observed in centrifugal creaming is presumably caused by the rising of fat globules in groups formed by chance due to their Brownian movement; this mechanism can only act if the time needed for a globule to move over a distance comparable to its own diameter by Brownian movement is large compared to the time needed for its rising over the same distance.[127]

The disturbances mentioned imply that creaming rate may be much less than predicted by the Stokes equation, particularly for gravity creaming of homogenized products. If the equation is obeyed, the creaming rate of a globule is proportional to d^2; this implies that the creaming rate in the emulsion as a whole is proportional to $H = \Sigma Nd^5/\Sigma Nd^3$.[1] This indicates that the large globules, in particular, affect H and thus creaming rate; for

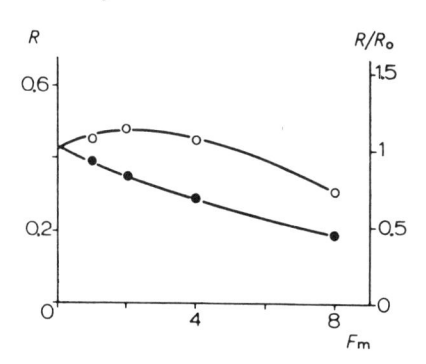

FIG. 13. Example of the effect of fat content (F_m, % w/w) of slightly homogenized milk on the proportions of fat creamed, R. Centrifugal creaming (○), and gravity creaming (●) of the same samples; the product of a and duration of creaming was kept constant.[127]

the same average globule size, d_{vs}, the sample with the widest spread in size, c_s, will cream fastest. Creaming rate (defined as the proportion of the fat arriving in the cream layer per unit time) is proportional to H only as long as the milk is not depleted of the largest globules. In practice, the proportionality often holds if less than ~40% of the fat is creamed.[127]

In the centrifugal separation of milk, the aim is generally to obtain the lowest possible fat content in the skim milk. This depends greatly on temperature, since the factor $(\rho_p - \rho_f)/\eta_p$ varies; it increases almost linearly from 2×10^{-4} s m^{-2} at 0°C to 24×10^{-4} s m^{-2} at 80°C (including a correction for change in globule size).[1] The fat content of the skim milk also depends on the proportion of fat in very small globules (say <1 μm), which varies among cows[2] and can increase due to disruption of fat globules. Good separators can produce a skim milk with 0·05% fat[130] and half of this is, in fact, non-globular fat.

As soon as fat globules are clustered, the creaming rate may be much higher than predicted by the Stokes equation. This happens if homogenization clusters are formed (Section 12). This phenomenon has been used to produce 'viscolized milk' which shows very copious creaming; milk is separated into skim milk and cream, the latter is homogenized (which leads to extensive clustering) and added to the skim milk.[131] During sterilization of evaporated milk, small clusters of fat globules may be formed at the onset of heat coagulation;[132] the extent of such clustering is probably an important variable in the creaming (sometimes wrongly called fat separation) exhibited by this product.

Cold agglutination considerably enhances creaming as discussed in Section 13 (see particularly Fig. 11). There is an induction period of 30–60 min, followed suddenly by fast creaming. The rising of floccules causes them to increase in size as the larger floccules overtake the smaller ones (and individual globules). Even in a tank several metres high, cream rising is nearly as fast as in a much thinner layer.[1] Under optimum conditions (milk warmed to >40°C and then quickly cooled to 4°C) creaming can be fairly exhaustive, and a fat content of the skim milk as low as 0·1% can, sometimes, be obtained; however, in most lots of milk, higher fat contents remain.[1]

An important difference is observed between a cream layer obtained by natural creaming (i.e. with cold agglutination) and one obtained by rising of separate globules (as, for example, in milk pasteurized for 15 s at ≥75°C). The former has a low fat content (say 20%), and the cream can easily be redispersed throughout the milk by stirring.[1] In the latter, the fat globules are much more closely packed (fat content, for example, 50%)[133]

and the cream is much more difficult to redisperse, partly because the fat globules may show some (partial) coalescence, especially when crystallization occurs in the globules. In extreme cases this leads to a solid cream plug.[107]

The cream layer is often enriched in bacteria, particularly in natural creaming (Section 13). The cream may also contain many of the leucocytes in milk, on average about 40%.[134] This is probably caused by leucocytes swallowing fat globules (phagocytosis),[135] whereby they become buoyant.

15. RHEOLOGICAL PROPERTIES

The fat globules have a distinct effect on some physical properties of the product, the more so at high fat contents. One aspect is physical stability itself, another is optical properties since fat globules scatter light.[1,136] With increasing fat content, mass density decreases slightly, electrical conductivity decreases considerably (by half for 40% fat),[137] and specific heat and thermal conductivity also decrease.[138] The most striking effect of the fat globules may be on rheological properties. Such properties of emulsions[25] and milk products as emulsions[1] have been reviewed.

If the fat globules are present as separate particles, if shear rate is not too low (say >10 s^{-1}) and fat content $<40\%$, milk and cream behave as Newtonian liquids (e.g. Reference 139). In such cases the Eilers equation[140] is usually obeyed; this relates the viscosity, η, to that of the continuous phase, η_0, and the volume fraction, ϕ, of the (spherical) particles present by

$$\eta = \eta_0 \left[1 + \frac{1 \cdot 25 \phi}{1 - \phi/\phi_{max}} \right]^2$$

where ϕ_{max} is the hypothetical volume fraction when the particles are in the closest packing possible. The Eilers equation gives accurate predictions[141] if ϕ is taken as the total volume fraction of fat globules, casein micelles, protein molecules and lactose molecules; then $\eta_0 \approx 1 \cdot 02 \ \eta_{water}$ and $\phi_{max} \approx 0 \cdot 9$. This is shown in Fig. 14. For skim milk, $\phi \approx 0 \cdot 16$ is found experimentally. The Eilers equation will give results that are too low if $\phi = \phi_{fat}$ and $\eta_0 = \eta_{plasma}$ are taken. Viscosity of milk and cream decreases considerably with increasing temperature because η_0 decreases ($\eta_0 \approx 1$ mPa s at room temperature); moreover, ϕ will decrease somewhat since the casein micelles decrease in voluminosity.[142]

Deviations from Newtonian behaviour usually occur if $\eta/\eta_0 > 10$,

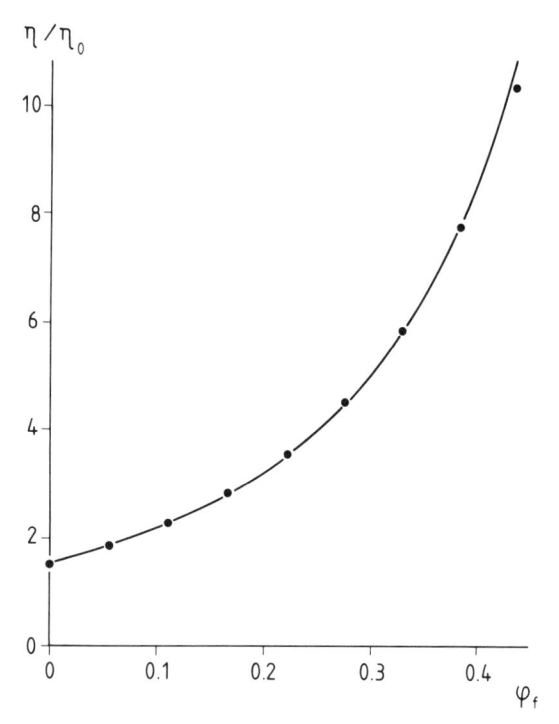

FIG. 14. Example of the relative viscosity, η/η_0, of milk and cream as a function of the volume fraction of fat (ϕ_f). The points give experimental values (from Reference 139), the line is according to the Eilers equation. After Reference 141.

particularly at low shear rates. One then speaks of the apparent viscosity, η', since it depends on shear rate. Cream of 50% fat may give $\eta' \approx 100$ mPa s at, say, $S = 2\,\text{s}^{-1}$, and 30 mPa s at $S = 100\,\text{s}^{-1}$.[143] In high-fat cream, η' may increase during measurement since the shear promotes partial coalescence.[144] Thereby, networks are formed throughout the cream which may in extreme cases cause a fairly solid consistency. This may also occur during the so-called 'rebodying' of cream.[43,44] Note that full coalescence will have a negligible effect on viscosity since the Eilers equation does not contain a factor for globule size. Partial coalescence may increase η, even in a dilute system (milk) since the effective ϕ increases as the granules formed have irregular shapes.[1]

Homogenization clusters considerably increase η since they contain interstitial liquid, thus increasing ϕ; examples are given in Fig. 10. The liquid is distinctly non-Newtonian and η' decreases markedly with

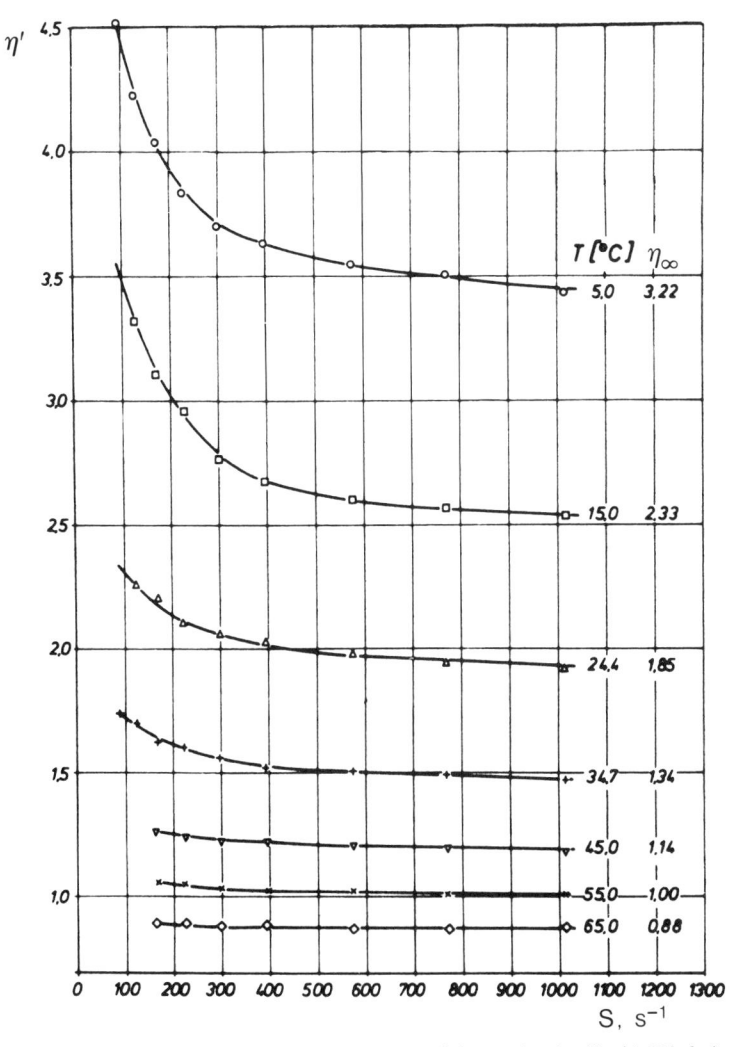

FIG. 15. Example of the apparent viscosity, η' (mPa s) of milk (4·5% fat) as a function of shear rate, S (s^{-1}) at various temperatures.[146]

increasing shear rate; moreover the shearing forces disrupt some of the clusters thereby causing a permanent decrease in η'.[105,106] Plots of η' versus S may thus show hysteresis, η' being higher when measurements start at low and subsequently at higher S than when S is decreased again.

Cold agglutination causes a considerable increase in ϕ since the floccules contain much interstitial liquid.[1] It is found that η' of raw milk increases with decreasing temperature and decreasing shear rate.[145,146] Examples are shown in Fig. 15; undoubtedly η' will increase further for still lower S. It is seen that fairly low shear rates suffice to disrupt most of the floccules. They re-form only partly after the shearing is stopped. The slight increase of η' with decreasing shear rate at temperatures over 40°C, as shown in Fig. 15, has not been observed by others (e.g. Reference 139).

REFERENCES

1. MULDER, H. and WALSTRA, P., *The Milk Fat Globule. Emulsion Science as Applied to Milk Products and Comparable Foods*, Commonwealth Agricultural Bureaux, Farnham Royal and Centre for Agricultural Publishing and Documentation, Wageningen, 1974.
2. WALSTRA, P., *Neth. Milk Dairy J.*, 1969, **23**, 99.
3. WALSTRA, P., OORTWIJN, H. and DE GRAAF, J. J., *Neth. Milk Dairy J.*, 1969, **23**, 12.
4. CORNELL, D. G. and PALLANSCH, M. J., *J. Dairy Sci.*, 1966, **49**, 1371.
5. HOOD, R. L., *J. Dairy Sci.*, 1981, **64**, 19.
6. WALSTRA, P., *Neth. Milk Dairy J.*, 1969, **23**, 111.
7. RÜEGG, M. and BLANC, B., *Biochim. Biophys. Acta*, 1981, **666**, 7.
8. WALSTRA, P., *Neth. Milk Dairy J.*, 1975, **29**, 279.
9. WALSTRA, P. and BORGGREVE, G. J., *Neth. Milk Dairy J.*, 1966, **20**, 140.
10. KERNOHAN, E. A., WADSWORTH, J. C. and LASCELLES, A. K., *J. Dairy Res.*, 1971, **38**, 65.
11. KUCHROO, T. K. and NARAYANAN, K. M., *Ind. J. Dairy Sci.*, 1977, **30**, 99 and 225.
12. WALSTRA, P. and VAN BERESTEYN, E. C. H., *Neth. Milk Dairy J.*, 1975, **29**, 35.
13. WALSTRA, P., *Neth. Milk Dairy J.*, 1967, **21**, 166.
14. WALSTRA, P., *Neth. Milk Dairy J.*, 1974, **28**, 3.
15. KING, N., *Neth. Milk Dairy J.*, 1950, **4**, 30.
16. BUCHHEIM, W., *Proc. XVIII Int. Dairy Congr.*, 1970, **1E**, 73.
17. HENSON, A. F., HOLDSWORTH, G. and CHANDAN, R. C., *J. Dairy Sci.*, 1971, **54**, 1752.
18. BAUER, H., *J. Dairy Sci.*, 1972, **55**, 1375.
19. BUCHHEIM, W. and PRECHT, D., *Milchwissenschaft*, 1979, **34**, 657.
20. PRECHT, D. and BUCHHEIM, W., *Milchwissenschaft*, 1979, **34**, 745.
21. VAN BOEKEL, M. A. J. S., *Agr. Res. Rep. (Wageningen)*, 1980, **901**.
22. KEENAN, T. W., MORÉ, D. J., OLSON, D. E., YUNGHANS, W. N. and PATTON, S., *J. Cell Biol.*, 1970, **44**, 80.
23. PINTO DA SILVA, P., PEIXOTO DE MENEZES, A. and MATHER, I. H., *Exp. Cell. Res.*, 1980, **125**, 127.
24. OVERBEEK, J. T. G., In: *Colloid Science*, Vol. I, Kruyt, H. R. (ed.), 1952, Elsevier, Amsterdam, p. 278.
25. SHERMAN, P. (ed.), *Emulsion Science*, 1968, Academic Press, London.

26. IVES, K. J. (ed.), *The Scientific Basis of Flocculation*, 1978, Sijthoff and Noordhoff, Alphen aan de Rijn.
27. BECHER, P. (ed.), *Encyclopedia of Emulsion Technology*, Vol. 1, 1983, Dekker, New York.
28. SIRKS, H. A., *Versl. Landbouwk. Onderz.*, 1924, **29**, 137.
29. PAYENS, T. A. J., *Neth. Milk Dairy J.*, 1963, **17**, 150.
30. PAYENS, T. A. J., *Kieler Milchw. Forsch Ber.*, 1964, **16**, 457.
31. VISSER, J., *Adv. Colloid Interface Sci.*, 1972, **3**, 331.
32. MOYER, L. S., *J. Biol. Chem.*, 1940, **133**, 29.
33. KOOPS, J., PhD Thesis, University of Wageningen, 1963.
34. VINCENT, B., *Adv. Coll. Interf. Sci.*, 1975, **4**, 193.
35. COLIN HUGHES, R., *Membrane Glycoproteins*, 1976, Butterworths, London.
36. SHIMIZU, M., YAMAUCHI, K. and KANNO, C., *Milchwissenschaft*, 1980, **35**, 9.
37. MORALES, C. R., *Gaceta Veterinaria*, 1978, **40**, 483.
38. VRIJ, A., *Disc. Faraday Soc.*, 1966, **42**, 23.
39. VRIJ, A., HESSELINK, F. T., LUCASSEN, J. and VAN DEN TEMPEL, M., *Proc. K. Ned. Akad. Wet.*, 1970, **B73**, 124.
40. VAN BOEKEL, M. A. J. S. and WALSTRA, P., *Colloids Surfaces*, 1981, **3**, 109.
41. LABUSCHAGNE, J. H., PhD Thesis, University of Wageningen, 1963.
42. VAN BOEKEL, M. A. J. S. and WALSTRA, P., *Colloids Surfaces*, 1981, **3**, 99.
43. SOMMER, H. H., *Market Milk and Related Products*, 3rd edn., 1952, published by the author, Madison.
44. OORTWIJN, H. and WALSTRA, P., *Neth. Milk Dairy J.*, 1982, **36**, 279.
45. DOLBY, R. M., *J. Dairy Res.*, 1953, **20**, 201.
46. MELSEN, J. M. and WALSTRA, P., Unpublished results, 1982.
47. THOMÉ, K. E., SAMUELSSON, E. G., FRENNBORN, T. and GYNNING, K., *Milchwissenschaft*, 1958, **13**, 115.
48. STONE, M. J. and ROWLANDS, A., *J. Dairy Res.*, 1952, **19**, 51.
49. POULSEN, P. R., MADSEN, H. and MOGENSEN, G., *Beretn. Statens Forsøgsmejeri*, 1981, **242**.
50. REUTER, H., *Milchwissenschaft*, 1978, **33**, 97.
51. KAMMERLEHNER, J. and KESSLER, H. G., *Deutsche Milchwirtsch.*, 1980, **31**, 1746.
52. KING, N., *The Milk Fat Globule Membrane*, 1955, Commonwealth Bureau Dairy Sci., Reading.
53. VAN KREVELD, A., *Neth. Milk Dairy J.*, 1959, **13**, 141.
54. VAN DEN BOOGAARD, C., OORTWIJN, H. and WALSTRA, P., Unpublished results, 1976.
55. PHIPPS, L. W. and TEMPLE, D. M., *J. Dairy Res.*, 1982, **49**, 61.
56. REES, L. H., *J. Soc. Dairy Technol.*, 1968, **21**, 172.
57. DOLBY, R. M., *J. Dairy Res.*, 1953, **20**, 201.
58. DOLBY, R. M., *J. Dairy Res.*, 1957, **24**, 77.
59. SURKOV, V. and FOFANOV, J., *Proc. XVII Int. Dairy Congr.*, 1966, **A**, 525.
60. ZADOW, J. G., *Aust. J. Dairy Technol.*, 1969, **27**, 44.
61. OBERMAIER, O. and FORMAN, L., *Prumysl Potravin*, 1976, **27**, 326.
62. SURKOV, V., BARKAN, S. and GARLINSKAYA, E., *Moloch. Prom*, 1960, **21**(4), 24; also in *Techq. Lait.*, 1960, **15**(307), 12.
63. WALSTRA, P., *Neth. Milk Dairy J.*, 1969, **23**, 290.

64. KURZHALS, H. A., PhD Thesis, University of Hannover, 1977.
65. GOULDEN, J. D. S. and PHIPPS, L. W., *J. Dairy Res.*, 1964, **31**, 195.
66. WALSTRA, P., *The Milk Fat Globule: Natural and Synthetic*, Review paper **75 ST** of the XX Int. Dairy Congr., Paris, 1978.
67. JELLEMA, A., *Neth. Milk Dairy J.*, 1980, **34**, 133.
68. BADINGS, H. T., *Officieel Org. K. Ned. Zuivelbond*, 1969, **61**, 958.
69. VAN DUIN, H. and BRONS, C., *Officieel Org. K. Ned. Zuivelbond*, 1967, **59**, 136.
70. DAHLE, C. D. and JACK, E. L., *J. Dairy Sci.*, 1937, **20**, 605.
71. BRUNNER, J. R., DUNCAN, C. W. and TROUT, G. M., *Food Res.*, 1953, **18**, 454, 463 and 469.
72. SASAKI, R., TSUGO, T. and MIYUZAWA, K., *Proc. XIV Int. Dairy Congr.*, 1956, **1–2**, 223.
73. WALSTRA, P. and MULDER, H., *Proc. XIX Int. Dairy Congr.*, 1974, **1E**, 217.
74. TODT, K., *Milchwissenschaft*, 1976, **31**, 83.
75. OORTWIJN, H., WALSTRA, P. and MULDER, H., *Neth. Milk Dairy J.*, 1977, **31**, 134.
76. ANDERSON, M., BROOKER, B. E., CAWSTON, T. E. and CHEESEMAN, G. C., *J. Dairy Res.*, 1977, **44**, 111.
77. DARLING, D. F. and BUTCHER, D. W., *J. Dairy Res.*, 1978, **45**, 197.
78. VAITKUS, V. V. and ZIBERKAITE, R. B., *Proc. XX Int. Dairy Congr.*, 1978, **E**, 279.
79. OORTWIJN, H. and WALSTRA, P., *Neth. Milk Dairy J.*, 1979, **33**, 134.
80. WALSTRA, P. and OORTWIJN, H., *Neth. Milk Dairy J.*, 1982, **36**, 103.
81. EGGMAN, H., *Milchwissenschaft*, 1969, **24**, 479.
82. BUCHHEIM, W. and KNOOP, E., *Kieler Milchw. ForschBer.*, 1970, **22**, 323.
83. HENSTRA, S. and SCHMIDT, D. G., *Neth. Milk Dairy J.*, 1970, **24**, 45.
84. HAYASHI, S. and SMITH, L. M., *Biochem. (Easton)*, 1965, **4**, 2550.
85. PATTON, S., *J. Dairy Sci.*, 1980, **63** (Suppl. 1), 48.
86. PATTON, S., *J. Dairy Sci.*, 1952, **35**, 324.
87. STINE, C. M. and PATTON, S., *J. Dairy Sci.*, 1952, **35**, 655.
88. JACKSON, R. H. and PALLANSCH, M. J., *J. Agr. Food Chem.*, 1961, **9**, 424.
89. SANDERSON, W. B., *N.Z. J. Dairy Sci. Technol.*, 1970, **5**, 139.
90. GRAF, E. and MÜLLER, H. R., *Milchwissenschaft*, 1965, **20**, 302.
91. BUCHANAN, R. E. and SMITH, D. R., *Proc. XVII Int. Dairy Congr.*, 1966, **E/F**, 363.
92. KLOSER, J. J. and KEENEY, P. G., *Ice Cream Rev.*, 1959, **42**(10), 36.
93. DESAI, R. J. and HARPER, W. J., *J. Dairy Sci.*, 1967, **52**, 578.
94. JOHN, M. G. and SHERMAN, P., *Proc. XVI Int. Dairy Congr.*, 1962, **C**, 61.
95. STISTRUP, K. and ANDREASEN, J., *Proc. XVI Int. Dairy Congr.*, 1962, **C**, 29.
96. MIN, B. S. and THOMAS, E. L., *J. Food Sci.*, 1977, **42**, 221.
97. KAWANARI, M., TAKAHASHI, K., AHIKO, K. and HAYASHI, H., *Rep. Res. Lab. SnowBrand Milk Prod. Co.*, 1977, **75**, 1.
98. YAMAUCHI, K., SHIMIZU, M. and KAMIYA, T., *J. Food Sci.*, 1980, **45**, 1237.
99. HALLING, P., *Crit. Rev. Food Sci. Nutr.*, 1981, **15**, 155.
100. KOOPS, J., *Neth. Milk Dairy J.*, 1967, **21**, 29 and 50.
101. KIESEKER, F. G., *Proc. XVIII Int. Dairy Congr.*, 1970, **1E**, 265.
102. KIESEKER, F. G. and PEARCE, R. J., *Proc. XX Int. Dairy Congr.*, 1978, **1E**, 975.

103. WALSTRA, P., *Neth. Milk Dairy J.*, 1980, **34**, 181.
104. VAN VLIET, T. and DENTENER-KIKKERT, A., *Neth. Milk Dairy J.*, 1982, **36**, 261.
105. OGDEN, L. V., PhD Thesis, University of Minnesota, 1973.
106. OGDEN, L. V., WALSTRA, P. and MORRIS, H. A., *J. Dairy Sci.*, 1976, **59**, 1727.
107. THOMÉ, K. E., ÅNÄS, K. E. and ELOVSON, G., *Rep. Milk and Dairy Res., Alnarp*, 1963, **67**.
108. GONZALES-JANOLINO, V. T., *Milchwissenschaft*, 1968, **23**, 204.
109. BOTTAZZI, V. and ZACCONI, C., *Proc. XXI Int. Dairy Congr.*, 1982, **1**(2), 159.
110. BUTLER, J. E., In: *Lactation: a Comprehensive Treatise*, LARSON, B. R. and SMITH, V. R. (eds.), Part III, 1974, Ch. 5.
111. PAYENS, T. A. J., KOOPS, J., and KERKHOF MOGOT, M. F., *Biochim. Biophys. Acta*, 1965, **94**, 576.
112. WALSTRA, P., DE JONG, D., HOSSAIN, M. A. and DE KLERK, S., Unpublished results, Wageningen, 1970–78.
113. PAYENS, T. A. J., *Milchwissenschaft*, 1968, **23**, 325.
114. SAMUELSSON, E., BENGTSSON, G., NILSSON, S. and MATTSSON, N., *Svenska Mejeritidn.*, 1954, **46**, 163 and 193.
115. GAMMACK, D. B. and GUPTA, B. B., *Proc. XVIII Int. Dairy Congr.*, 1970, **1E**, 20.
116. PAYENS, T. A. J., *Kieler Milchw. ForschBer.*, 1964, **16**, 457.
117. SMITH, E. L., *J. Biol. Chem.*, 1946, **165**, 665.
118. HLADIK, J., DOLEZÁLEK, J. and SMRTOVÁ, S., *Sbornik Vysoké Školy Chemicko-Technologické v Praze*. E (1976) **47**, 117; vide *Dairy Sci. Astr.*, 1978, **40** (3686).
119. ORLA-JENSEN, S., LE DOUS, C., JACOBSEN, J. and OTTE, N. C., *Lait*, 1929, **9**, 622, 724, 816, 914 and 1032.
120. MERTENS, E., *Milchw. Forsch.*, 1933, **14**, 1.
121. KOOPS, J., PAYENS, T. A. J. and KERKHOF MOGOT, M. F., *Neth. Milk Dairy J.*, 1966, **20**, 296.
122. GUILBOT, A., MULTON, J. L. and DRAPRON, R., *Dechema Monogr.*, 1972, **10**, 279.
123. KENYON, A. and JENNESS, R., *J. Dairy Sci.*, 1958, **41**, 716.
124. BOTTAZZI, V., CORRADINI, C. and MONTESCANI, G., *Scienza Tec. Latt. Casear.*, 1971, **22**, 321.
125. BOTTAZZI, V. and PREMI, L., *Scienza Tec. Latt. Casear.*, 1977, **28**, 7.
126. STADHOUDERS, J. and HUP, G., *Neth. Milk Dairy J.*, 1970, **24**, 79.
127. WALSTRA, P. and OORTWIJN, H., *Neth. Milk Dairy J.*, 1975, **29**, 263.
128. TROY, H. C. and SHARP, P. F., *J. Dairy Sci.*, 1928, **11**, 189.
129. FOX, K. K., HOLSINGER, V. H., CAHA, J. and PALLANSCH, M. J., *J. Dairy Sci.*, 1960, **43**, 1396.
130. FJAERVOLL, A., *J. Soc. Dairy Techn.*, 1968, **21**, 180.
131. DOAN, F. J., *J. Dairy Sci.*, 1927, **10**, 501.
132. SCHMIDT, D. G., BUCHHEIM, W. and KOOPS, J., *Neth. Milk Dairy J.*, 1971, **25**, 200.
133. DAHLE, C. D. and JACK, E. L., *J. Dairy Sci.*, 1937, **20**, 605.
134. PHIPPS, L. W. and NEWBOULD, F. H. S., *J. Dairy Res.*, 1966, **33**, 51.
135. PAAPE, M. J. and WERGIN, W. P., *J. Am. Vet. Med. Ass.*, 1977, **170**, 1214.
136. WALSTRA, P., *Neth. Milk Dairy J.*, 1965, **19**, 93.

137. PRENTICE, J. H., *J. Dairy Res.*, 1962, **29**, 131.
138. PEEPLES, M. L., *J. Dairy Sci.*, 1962, **45**, 297.
139. PHIPPS, L. W., *J. Dairy Res.*, 1969, **36**, 417.
140. EILERS, H., *Kolloid Z.*, 1941, **97**, 313.
141. VAN VLIET, T. and WALSTRA, P., *J. Texture Stud.*, 1980, **11**, 65.
142. WALSTRA, P., *J. Dairy Res.*, 1979, **46**, 317.
143. PRENTICE, J. H., *S.C.I. Monogr.*, 1968, **27**, 265.
144. ROTHWELL, J., *J. Dairy Res.*, 1966, **33**, 245.
145. RANDHAHN, H. and REUTER, H., *Dechema Monogr.*, 1974, **77**, 233.
146. RANDHAHN, H., PhD Thesis, University of Hannover, 1976.

Chapter 5

PHYSICAL PROPERTIES AND MODIFICATION OF MILK FAT

B. K. MORTENSEN

*Danish Government Research Institute for Dairy Industry,
Hillerød, Denmark*

SUMMARY

The fundamental aspects of crystallization of milk fat and the rheology of milk fat products are discussed in this chapter.

The various aspects of formation and growth of fat crystals and the complicated crystallization patterns exhibited by milk fat, including polymorphism and formation of mixed crystals, are explained as well as the differences in crystallization of bulk fat and crystallization of globular fat.

Special attention is drawn to the proportion of solid fat in a fat mixture, because the rheological properties of many dairy products depend more on this than on size and form of the crystals. The content of solid or liquid fat can be determined by dilatometry, differential scanning calorimetry (DSC) and nuclear magnetic resonance (NMR), and the principles of the different methods are discussed.

The rheological properties of milk fat products such as milk, cream and butter are explained, and different measuring methods (sectility, extrusion and penetration) are described.

In Section 4, different ways of modifying the rheological properties of cream and butter are discussed. These include homogenization and rebodying of cream, and where butter is concerned, temperature treatment of cream prior to churning, work softening and alteration of the composition of the fat either by fractionation of fat into portions with different melting points or by direct admixture of other fats, usually vegetable fats with low melting points.

1. INTRODUCTION

Milk fat makes up an important component of most dairy products. The physical properties of milk fat therefore greatly influence the rheological properties of these products and are thus of great technical importance.

The following discussion will be concerned mainly with physical properties related to the crystallization of milk fat. With regard to other physical properties of milk fat the reader is referred to a comprehensive review by Mulder and Walstra.[1] The rheology of milk fat products, in particular the spreadability of butter, will be discussed in detail; the possibilities of modifying the rheological properties by means of the processing technology, fractionation of milk fat and blending with other fats will also be mentioned.

2. CRYSTALLIZATION OF MILK FAT

2.1. Formation of Crystals

The main part of milk fat is triglycerides with different chemical compositions and different physical properties. When the triglyceride molecules are in a molten state, i.e. when they have high kinetic energy, the individual molecules have a rather free mobility since the intermolecular forces tending to hold the molecules together are not strong enough to counteract the thermal motions. However, if molten fat is cooled the thermal motions of the molecules decrease, and the intermolecular forces, i.e. hydrogen bonds and van der Waals' forces, draw the triglyceride molecules closer together simultaneously with an incipient parallel-ordering of the fatty acid chains.

Crystallization starts with the formation of crystal nuclei (centres of crystallization) in the molten fat as a few molecules gather in molecular aggregates where the potential energy is reduced to a minimum. To make these aggregates, in which molecules are continuously replaced, grow into real crystals it is necessary that the probability of a molecule being adsorbed is greater than the probability of a molecule being liberated. The smaller the aggregates, the higher the potential energy, and thereby the probability of not retaining the molecules. This implies a difficulty in getting the crystallization process started unless the melt is inoculated with crystallization centres in the form of pre-formed crystals or the energy level of the molecules is lowered by a strong supercooling of the melt. Lowering the temperature strongly influences the rate of nucleus formation.

According to Tammann's[2] now classic investigation of crystal formation, the nucleation rate is increased by falling temperature until a maximum is reached. The reason why further cooling results in a reduced nucleation rate is due to an increase in the viscosity of the melt and thereby a reduction in the rate of diffusion.

Walstra and van Beresteyn[3] found that for milk fat in globules in the temperature range 5–25 °C the nucleation rate approximately doubles for every 1 to 1·5 °C decrease in temperature. They distinguish between homogeneous and heterogeneous nucleation. Homogeneous nucleation occurs when crystal nuclei are formed spontaneously from a pure melt. The rate of nucleation is negligible at temperatures slightly below the melting point and, usually, a supercooling of 30–50 °C is needed to start the nucleation, but then a further temperature decrease of only a few degrees leads to extremely rapid nucleation. However, in practice, nucleation is mostly heterogeneous since the melt always contains impurities that permit the formation of nuclei at their surface at far less supercooling. It has been suggested[4] that monoglycerides in the fat form micelles which can act as catalytic impurities.

2.2. Growth of Crystals

The growth of the crystal nuclei formed occurs by successive single layers of molecules being deposited on an already ordered crystal surface. First and foremost, the rate of growth depends on the probability of the incorporation of these molecules into the crystal lattice as well as on the material density and on the temperature, which influences the rate of diffusion.

Crystal growth is relatively slow in natural fats and the kinetics of crystallization seem to be similar to those of a chemical reaction, i.e. there is a free energy barrier opposing the aggregation which can be surmounted only by molecules in a relatively high energy state.

Grishchenko[5] studied the kinetics of milk fat crystallization and found the activation energy to be 9·2 kcal mol^{-1}, while deMan,[6] assuming that the process corresponds to a first-order reaction, found 11·0 kcal mol^{-1}. Bryzgin and Eres'ko,[7] who studied the rate of crystallization in the temperature range 0·5–17 °C, confirmed the latter value. A corresponding activation energy can be calculated for bulk fat from the data of Mortensen and Danmark,[8] who furthermore proved that the constants of the crystallization process are related to the iodine value, i.e. they depend on the composition of the fat.

There can be no doubt that during the crystallization process kinetic phenomena occur which cannot be explained from the theory mentioned.

On the other hand, the approximation is fairly good, and for practical applications the crystallization process can therefore be considered as a first-order reaction. This makes it possible, with relatively good approximation, to calculate the length of the crystallization period on the basis of the constants for the process.

Crystallization of fat is a complex phenomenon. DeMan[9] mentioned that one of the complications is the fact that during crystallization there is no distinct difference between solute and solvent. Lowering the temperature will cause some of the solvent to change to the role of solute. Thus the solubility of a given solute fraction is decreased at the same time as the amount of available solvent is diminished. But this is not the only complication and the information available about the crystallization process is far from being complete. Further experimental work, especially concerning crystallization kinetics, would be useful.

2.3. Polymorphism

Fats containing mainly long-chain fatty acids may exhibit polymorphism, i.e. the existence of more than one crystal form due to different patterns of molecular packing in the crystal. The different polymorphic forms have different crystal lattices and different melting points. Only one of the forms is stable, the others are metastable and will gradually transform into the stable form.

Several more or less complicated crystallization patterns have been discussed, based mainly on the pioneering work of Malkin[10] and Lutton.[11] The main opinion nowadays seems to be that pure triglycerides show the following polymorphic pattern[12,13]

$$\text{Fluid} \rightleftharpoons \alpha \rightarrow \beta' \rightarrow \beta \quad (1)$$

$$\text{Fluid} \longleftarrow \beta'$$

$$\text{Fluid} \longleftarrow \beta$$

On rapid cooling, a metastable α-form is produced reversibly from the liquid phase. The α-form may then be transformed irreversibly into the more stable β'-form and further into the most stable β-form. It should be mentioned that not all triglycerides are known to form all three crystal forms. During the rapid cooling a certain structuring takes place from the initially chaotically arranged carbon chains in the melt. In the α-form, however, the chains are still highly disarranged and able to oscillate or rotate; the chain packing is hexagonal. In the β'-form a certain arrangement of the chains has taken place as every second chain plan is orthogonal to the rest; the chain packing here is orthorhombic. A further increase in stability is achieved when all chain plans become parallel, which happens during the rearrangement from β' to β, where the chain packing is triclinic. The rate of polymorphic transformation is strongly dependent on the chain length of the compound, being greatest for short chains.

To illustrate the influence of crystal structure on the melting point, Table 1 shows melting points for some simple triglycerides.[14]

TABLE 1
MELTING POINT OF SOME PURE TRIGLYCERIDES

Triglyceride	Melting point (°C)		
	α-form	β'-form	β-form
Tricaprylin	−51·0	−18·0	10·0
Tricaprin	−10·5	17·0	32·0
Trilaurin	15·0	34·5	46·5
Trimyristin	33·0	46·0	58·0
Tripalmitin	45·0	56·5	66·0
Tristearin	54·7	64·0	73·3

Polymorphism is very often exhibited by fats of a relatively uniform composition which are probably also the only fats showing the simple pattern stated in reaction (1). Thus, Hoerr[15] stated that it is observed only for highly purified individual triglycerides containing only one specific fatty acid group. Less pure samples and triglycerides with a more complex composition may exhibit intermediate forms which are difficult to identify. Larsson[13] concluded that many complex triglyceride mixtures, e.g. margarine, exhibit four crystal forms, α, β', β_2 and β_1, in the order of increasing stability. Milk fat probably exhibits a similar pattern.

One of the first to mention the occurrence of polymorphism in milk fat was Mulder[16] who found that milk fat can show a double melting point—the classic demonstration of polymorphism. The first extensive study of polymorphism in milk fat was carried out by Thomas,[17] who investigated milk fat and milk fat fractions. Three melting points were observed in the high-melting fractions and two in the low-melting fractions. DeMan,[18] using x-ray diffraction, identified α-, β'- and β-forms in milk fat, and Tverdokhleb[19] and Belousov and Vergelesov[20] also detected signs of polymorphism in the high-melting fraction of milk fat. Van Beresteyn[21] confirmed that rapid cooling of milk fat leads to crystals in the α-form; on holding, transition into more stable crystal forms soon occurs but even after long storage α and β' crystals can still be detected. Precht,[22] combining the x-ray diffraction techniques with the electron microscopic examination of butter, found that the β-crystal form dominates in the outer shell of the fat globules while the predominant part of the β' modification is found in the lower-melting crystal layers in the interior of the globules and in the free interglobular fat phase where the content of unsaturated fatty acids is particularly high.

Concerning the effect of polymorphism on the consistency of butter, deMan[23] concluded that there is no direct evidence indicating any effect of polymorphism on the rheological properties of butter. Although this was stated several years ago, new information altering this conclusion is not available when speaking of traditionally manufactured butter or butter manufactured according to the continuous Fritz method. However, the results of Tverdokhleb[19] and Belousov and Vergelesov[20] indicate that the situation may be different for butter made according to the emulsification and concentration methods.

Whether polymorphism influences the consistency of products made of blends of milk fat and vegetable oils has not yet been reported. This might quite well be the case because the crystal form of the fat is a critical quality factor in other products containing vegetable fats, e.g. in margarine and chocolate, where changes in consistency may occur during storage due to crystal transitions.

2.4. Mixed Crystals

It is well known that multicomponent systems exhibit very complicated crystallization patterns. This is also the case with fat composed of triglycerides containing a large number of fatty acids; although somewhat different in chain length the size of the molecules is so uniform that a significant interaction during solidification is possible.

Rossell[24] examined many phase diagrams of triglyceride systems ranging from binary mixtures of simple triglycerides to multicomponent mixtures of more complex glycerides. He stated that, 'Mixtures of triglycerides similar in melting points form solid solutions over extensive, but not usually complete, ranges of composition. This leads to several types of phase behaviour, eutectic formation being the most common, although peritectics and monotectics are also formed.'

According to Bailey[25] a solid solution, often referred to as mixed crystals, is exactly analogous to a liquid solution. He stated that, 'It consists simply of a lattice in which the component atoms or molecules have been partially replaced with dissimilar atoms or molecules. As in a liquid, the foreign molecules are distributed through the structure at random.' The matter is even more complicated because polymorphism must also be considered when the formation of mixed crystals is discussed. According to Rossell,[24] the metastable crystal modifications (α, β') form mixed crystals more easily than the more stable β-modification. This is in accordance with the results of Frede and Precht[26] which also show that in heterogeneous triglyceride systems the incorporation of different molecules in the same crystal lattice implies that the life of metastable crystal forms is prolonged.

Mulder[27] was the first to mention the formation of mixed crystals in milk fat when, based on the assumption that milk fat forms a series of mixed crystals, he tried to explain the fact that the melting point of milk fat is highly dependent on the rate and temperature of crystallization. If cooling occurs very rapidly, a considerable number of the low-melting triglycerides is built into a lattice formed by high-melting glycerides. This involves the formation of relatively uniform mixed crystals with nearly the same melting point. Such crystals adsorb a considerable amount of the low-melting triglycerides, and therefore milk fat that has been cooled rapidly contains less liquid fat at a given temperature than milk fat that has been cooled slowly or stepwise with suitable holding times. In the last case, the solid phase consists of a heterogeneous blend of mixed crystals, characterized by different melting points depending on the temperature treatment employed. On slow or stepwise cooling a considerable part of the crystallization process probably takes place by glyceride molecules being deposited on pre-formed crystal surfaces built up of other triglycerides. The so-called 'overlaid crystals' formed in this way have a sort of laminated structure in which high-melting glycerides often form the nuclei of the crystals with the lower-melting components located in the outer layers.

The ability of milk fat to form mixed crystals has been confirmed by several investigators. For example, Cantabrana and deMan[28] have shown by differential thermal analysis that during rapid cooling of milk fat a rather uniform crystal formation takes place, while the thermograms clearly show that on slow or stepwise cooling a certain fractionation of the fat occurs. Walstra and van Beresteyn[3] have presented additional evidence for the presence of mixed crystals in milk fat based on thermodynamic considerations.

The theory of mixed crystallization was initially based on the assumption that milk fat could form a uniform solid solution of mixed crystals. Belousov and Vergelesov,[29] however, found that triglycerides whose chain lengths differ by more than six carbon atoms do not form uniform solid solutions but crystallize in separate groups. The number of groups and the melting range of the different groups depend on the method of cooling. Similar results have been obtained by Krautwurst.[30]

As already mentioned, the formation of mixed crystals influences the content of liquid fat in the fat mixture. Therefore it is obvious that mixed crystallization has a great influence on the rheological properties of butter. Among others, deMan and Wood[31] and Mortensen and Samuelsson[32] showed that at least part of the effect of temperature treatment of butterfat and cream is due to its influence on mixed crystal formation. Similar considerations have been made by Sherbon and Coulter[33] who furthermore indicate that the formation of mixed crystals influences the rheological properties of products made from mixtures of different fat fractions. Addition of liquid fat to solidified fat results in a greater reduction in firmness of the product than addition to the melt.

2.5. Crystallization of Bulk Fat

According to Mulder and Walstra,[1] bulk fat contains sufficient catalytic impurities to initiate heterogeneous nucleation with little supercooling. Tverdokhleb *et al.*[34] stated that a liquid-crystal phase is formed in the melt during cooling after which a monocrystalline nucleus emerges from the high-melting glycerides. The nucleus has the shape of a spherolite consisting of crystals radiating outward from a common centre. During the growth of the spherolite the rather elongated needle-shaped crystals thicken and assume a feather-like structure characteristic of a typical spherolite.

The formation of spherolites seems to be a rhythmical process. The crystals are arranged along the outer surface giving the spherolite the shape of concentric waves. This is due to the fact that during crystallization the concentration of glycerides crystallizable under the given conditions decreases at the same time as the thermal energy emitted during crystallization accumulates. The growth of the spherolite will therefore temporarily stop until a state of equilibrium between glycerides is established through diffusion, and the heat emitted is dissipated (this might take some time because of the low thermal conductivity of bulk fat). The crystals formed may belong to the hexagonal, orthorhombic or triclinic system depending on whether they crystallize in the unstable or more-stable polymorphic form.

The external form and the size of the crystals grown from the nucleus depend not only on the internal structure but also on the external treatment. The size of milk fat crystals may vary considerably depending on the rate of crystallization; if milk fat is cooled rapidly, numerous very small crystals with a maximum diameter of 1–2 μm are formed while slow cooling results in the formation of a few large crystals with diameters up to 40 μm.[35] Recrystallization of the fat affects the size of the crystals but a much greater effect was found when the fat was cooled stepwise.[36] In this latter case, large spherolitic crystal aggregates were formed.

Kankare[37] showed that not only single crystals but also spherolites can agglomerate; such spherolite agglomerates can vary considerably in size, i.e. from 100 to 1000 μm, and their shape frequently diverges from spherical. Spherolite agglomerates can be disrupted very easily by mechanical treatment.[38] Stirring during the cooling process causes crystallization of fat into smaller spherolites which only agglomerate loosely. Very low and very high agitation speeds result in the formation of very fine crystals and a high agitation speed seems to prevent flocculation of the crystals.[39] In the absence of stirring, the solidified fat tends to flocculate into a network held together by van der Waals' attraction forces.[40] When crystallization is so far advanced that almost all of the remaining liquid phase is bound into the network, the mass appears as a complete solid.

2.6. Crystallization of Globular Fat

Crystallization of globular fat differs considerably from the crystallization of bulk fat. This was shown many years ago by van Dam[41] and since then this subject has been examined in a number of studies in the Netherlands.

Mulder and Walstra[1] enumerate several differences between the crystallization of fat in globules and the crystallization of bulk fat, but indicate that from a rheological point of view the main difference is that crystals in the emulsified state cannot grow larger than the globules; thus a solid network of crystals can form only within the globules unless the globules are clumped. Concerning nucleation, Mulder[27] and Phipps[42] showed that deeper supercooling is needed to initiate crystallization when the fat is in the emulsified state; this was confirmed by Sherbon and Dolby.[43] Also Walstra and van Beresteyn[3] found that, in addition to the requirement for deeper supercooling, a slower crystallization rate is obtained in a more finely dispersed fat which they attributed to differences in nucleation. At least one nucleus must be formed in every globule to achieve full crystallization and the time needed to obtain the first nucleus is proportional to the volume of the globule. This implies that a lower temperature is needed for a finer dispersion.

It should be emphasized that there is not a sharp temperature at which crystallization suddenly starts in all globules. The formation of nuclei is a stochastic process and the probability of the presence of a catalytic impurity that will start nucleation depends on the size of globules—this varies throughout the emulsion. The surface layer of a fat globule probably acts as a catalytic impurity. King[44] observed tiny tangentially orientated fat crystal needles in the outer layer of the globules, and Buchheim[45] later indicated that crystallization might start at the globule boundary. Walstra,[46] using a polarizing microscope, investigated fat globules at temperatures where part of the fat is solidified. He found four types of globule: in one type nothing could be seen except possibly a reflection at the globule boundary; in a second type there were tiny needle-shaped crystals throughout the globule; a third type had a birefringent outer layer which was thought to be formed by the rearrangement of the needle crystals into a tangential orientation along the globule boundary; and a fourth type showed small crystals throughout the globule as well as in the bright outer layer. A similar classification of fat globules has been made by Precht and Peters.[47]

The main conclusion from these investigations seems to be that very small, more-or-less needle-shaped, crystals are initially formed and flocculate into a random network giving the globule a certain firmness. On holding, growth, transformation and rearrangement of the crystals into a tangential orientation along the globule boundary take place.

2.7. Proportion of Solid Fat

The above-mentioned crystallization behaviour implies that milk fat has no sharp, well-defined melting point but melts over a wide temperature range. It is normally anticipated that milk fat is liquid above 40°C and completely solidified below −40°C. At intermediate temperatures it is a mixture of solid and liquid fats. The content of solid fat in the mixture is very important because the rheological properties of many dairy products depend more on this than on the size and form of the crystals. Information about the ratio of solid to liquid fat at a given temperature is, therefore, essential in many aspects of cream and butter manufacture. Furthermore, the ratio determined directly in butterfat, after a standardized pre-treatment, can be used to characterize the physical properties of the butterfat and is highly correlated to the iodine value of the fat.[48,49] The ratio measured corresponds to the solid fat index (SFI) measured by the official American Oil Chemists' Society (AOCS) method.[50]

The content of solid or liquid fat can be determined by different methods which are usually divided into three categories:[51]

(i) Dilatometry.
(ii) Differential scanning calorimetry.
(iii) Nuclear magnetic resonance.

2.7.1. Dilatometry

The determination of the content of solid fat by dilatometry is based on the specific volume change that occurs when fat goes from the solid to the liquid state. This change in specific volume can be observed when fat is heated in a co-called 'dilatometer'. Based on the expansion of the fat sample during heating, the specific volume can be recorded as a function of temperature, and from the graph obtained the ratio of the solid and liquid fractions can be calculated.

Furthermore, the phase distribution in a fat sample can be deduced from knowledge of the thermal expansions of completely solid and completely liquid fat. Hannewijk and Haighton[52] found these values for butterfat to be 0·57 and 0·85 mm^3 g^{-1} °C^{-1}. Similar values have been found by deMan and Wood[31] and by van Beresteyn.[21]

The results obtained are highly dependent on how the crystallization of the fat sample takes place and, in order to obtain comparable results, a standard procedure is needed. In the official method of the AOCS[50] for the determination of the solid fat index, a standard temperature conditioning procedure is laid down. The procedure is that the dilatometer, after the melting of the fat, is cooled to 0 °C for 15 min, heated to 26·7 °C for 30 min, and cooled to 0 °C for 15 min before heating to the measurement temperature.

Compared with the time for crystallization of butterfat,[6] the above-mentioned crystallization times are relatively short and therefore it is questionable whether the AOCS method, without modifications, can be used for butterfat. Kapsalis *et al.*[53] stated that longer holding times have no significant importance for measurements at temperatures above 10 °C but, nevertheless, considerable deviations were found when the holding time was varied.

Most dilatation measurements have been made on pure fats, e.g. butterfat, but the method can also be used on fat emulsions, e.g. cream. Naturally, in such cases the thermal expansion of the water phase must also be considered in the calculations. The method cannot be used directly on butter.

Dilatometric examinations are rather time-consuming and the calculation of solid or liquid fat is based on the assumption that the dilatation of solid and liquid fats, respectively, is constant over the complete melting range. However, this is not the case, because different triglycerides and different crystal modifications have different melting dilatations. On the other hand, the equipment is cheap and the analysis is easy to perform.

2.7.2. Differential Scanning Calorimetry

Thermal examinations of milk fat have previously been carried out using calorimetry and differential thermal analysis, but it was not until the development of differential scanning calorimetry (DSC) that it became possible to make measurements with sufficient precision to allow calculation of the content of solid fat within the complete melting range.

The method is based on the thermal transitions that occur in milk fat during heating or cooling. In a thermogram of a fat sample the energy transfer to or from the sample necessary to raise or lower the temperature, respectively, is recorded as a function of the temperature of the sample. Such a graph gives an illustration of the phase transitions which occur within the complete melting range.

In order to obtain comparable and stable results, which are not affected by the thermal history of the sample, it is necessary to carry out a complete melting of the fat followed by a well-defined cooling before recording the melting thermogram. A frequently used procedure is that suggested by Norris et al.,[54] which employs melting the fat at 60°C, transfer to the calorimeter and holding at 60°C for 30 min to erase the previous thermal history of the sample, cooling to $-60\,°C$ at a rate of $8\,°C\,min^{-1}$ and holding at this temperature for 5 min prior to recording a melting thermogram at $8\,°C\,min^{-1}$.

Integrated DSC curves can be used to give a relative measure of the liquid fat content of a sample at a given temperature. Automated data collection systems, which determine the base-line of the melting curve, are used in New Zealand.[54,55] The method, which is based on the equations of Heuvel and Lind,[56] corrects contemporary measurement data for the temperature delay which is caused by the dynamic character of the method and the finite time required for heat transfer.

The method can be used for the examination of pure milk fat, cream or butter. The water content in cream and butter complicates interpretation of the melting curve because the aqueous phase transitions mask a significant amount of lipid melting below 0°C; furthermore, the position of the base-line in the melting thermogram is rather uncertain.

The analysis is time-consuming and the calculation of solid or liquid fat is based on the assumption that the heat of fusion is constant over the complete melting range, which is not the case.[57] On the other hand, the analysis gives a good picture of the phase transitions that occur over the whole melting range.

2.7.3. Nuclear Magnetic Resonance

Determination of the content of solid or liquid fat in a fat sample by nuclear magnetic resonance (NMR) spectroscopy has been widely used in recent years, after the introduction of the method by Chapman et al.[58] Protons placed in a strong magnetic field can, under certain conditions, absorb energy from electromagnetic waves. This absorption, called nuclear magnetic resonance, depends on the physical state of the protons. By wide-line or continuous-wave NMR only protons in the liquid phase are registered. The content of liquid phase in a sample can be calculated from registration of the energy absorbed by the sample and knowledge of the corresponding absorption by a sample of totally liquid fat at the same temperature.[59] It is also possible to determine the liquid signal directly at the required temperature in a standard sample, e.g. triolein or olive oil, which is liquid within the complete temperature range normally used for the investigation of milk fat.[60,61]

By pulsed NMR, both liquid- and solid-phase protons are registered. Pulsed NMR analysers emit a short intense pulse of electromagnetic radiation at the resonance frequency into the fat sample and the free induction decay of the signal following the pulse is observed. The relaxation time is strongly related to the mobility of the protons and hence the physical state of the sample. Based on registration of the signal a suitable time after the pulse, the content of the solid phase can be calculated.[51]

The use of NMR methods on pure fat samples raises few problems. NMR methods can also be used to determine solid fat in fat emulsions but the contribution of water protons to the signal represents a complication. The relaxation time of water is somewhat higher than that of oil, but not enough to distinguish the oil and water signal on that basis. However, it is possible to make the relaxation time of water so small, by adding paramagnetic ions (Cu^{2+}, Mn^{2+}), that the oil and water signals can be distinguished. This method has been used on cream by Samuelsson and Vikelsøe[59] who added manganese chloride ($MnCl_2$) to damp the water-phase signal, while Walstra and van Beresteyn[3] used copper sulphate ($CuSO_4$) for this purpose. The calculation of solid or liquid fat is based on the assumption that the signals from the fat and water phases can be added directly, which was confirmed by van Boekel[62] who found no interaction between the signals. Therefore the signals do not seem to depend on the interfacial area between the phases, which might have been expected.

Determination of the content of liquid or solid fat in butter by NMR methods has been investigated. Thus, Mortensen[63] tried to separate the signals of the water and fat phases by comparing wide-line NMR signals in butter samples with different water contents made from the same cream. The method is uncertain because, among other things, the content of non-fat dry matter in the butter is of importance. Meriläinen and Antila[64] found that elimination of the water interference by adding $MnCl_2$ to the water fraction of butter enabled the results obtained by NMR directly on butter to be compared with those from waterless fats. In spite of these promising investigations, it must be concluded that the question of separating the water and fat signals has not yet been solved satisfactorily and more fundamental research work in this field is needed.

The performance of the actual NMR measurements is easy and very rapid, but one of the disadvantages of the method is that the equipment is rather expensive. The results obtained seem to be quite similar to those obtained by dilatometry and DSC measurements. A direct comparison between results obtained by the different methods, however, is hardly advisable as it must be remembered that dilatometric and NMR results are obtained under static conditions while DSC measurements are made under dynamic conditions.

3. RHEOLOGY OF MILK FAT PRODUCTS

The rheological properties of dairy products are strongly influenced by composition and vary considerably from products where protein has a major influence on rheology, e.g. cheese and cultured products like yoghurt, to products like cream and butter where the fat phase is especially important.

The rheological properties of dairy products are discussed below, mainly in relation to the physical properties of milk fat.

3.1. Milk and Cream

Milk and cream are oil-in-water emulsions, in which the fat phase is divided into separate fat globules surrounded by a membrane which prevents coalescence. However, the fat emulsion is not entirely stable and, of course, the state of dispersion increasingly influences the rheological properties as the fat content is increased.

Milk behaves as a fluid with almost Newtonian flow behaviour, i.e. its viscosity is independent of the shear rate. This means that the flow characteristics of milk can be described almost completely by the viscosity of the product. This will change when the fat content is increased since the rheological behaviour of cream is far more complicated even though its emulsion structure is basically unchanged. Unlike milk, cream shows clearly non-Newtonian flow characteristics. This means that the viscosity measured is strongly dependent on the shear rate. Viscosity decreases at increasing shear rate, and increases very markedly when the shear rate falls and approaches zero.

According to Prentice,[65] the dependence of viscosity on shear rate for normal market cream can be expressed by a power equation

$$\eta_{app} = \eta_1(\dot{\gamma})^{-\beta} \qquad (2)$$

where η_{app} is the measured apparent viscosity, η_1 is the value of η_{app} at unit shear rate, $\dot{\gamma}$ is the shear rate, and β is a 'coefficient of abnormality' which is zero for Newtonian liquids and of finite value for non-Newtonian liquids. According to Prentice, this equation is valid only within certain limits and 'should not be extrapolated beyond the region for which it can be verified experimentally'.

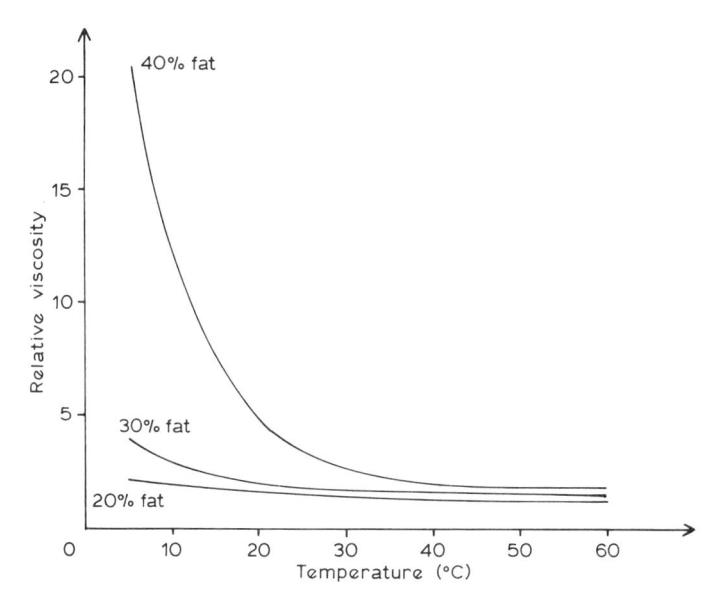

FIG. 1. Apparent viscosity relative to water. Data of Babcock.[68]

The viscosity of cream is highly affected by fat content and by temperature. The results obtained by Phipps[66] indicate that if the temperature is so high that all the fat is liquid, i.e. temperatures above 40°C, cream shows Newtonian behaviour. He found that at 40°C and fat contents up to 40%, the following equation is valid

$$\eta = \eta_0 \exp[3 \cdot 07(\phi + \phi^{5/3})] \qquad (3)$$

where η_0 is the viscosity of the continuous phase, and ϕ is the concentration (w/w) of fat. The maximum shear rate used was $100\,\text{s}^{-1}$.

The relationship between fat content and the viscosity of cream is far more complicated if the fat is partly solidified. Scott Blair *et al.*[67] found considerable deviations from Newtonian behaviour, especially at lower temperatures and higher fat contents. The interaction between temperature and fat content was clearly demonstrated by Babcock[68] who found a very marked increase in the viscosity of cream with a fat content of 40% within the range of temperature where fat starts to solidify (Fig. 1).

3.2. Butter

3.2.1. Structure of Butter

As already mentioned, milk and cream are emulsions where water is the continuous and fat the disperse phase. During the churning process this emulsion is transformed into a water-in-oil emulsion. According to the floatation theory of churning, one of the essential points in the churning process is the creation of a foam formed by the increased volume of air beaten into the cream. During the process, the fat globule membrane is damaged and liquid fat is squeezed out of the globules causing them to clump so that the emulsion is broken and phase inversion occurs. After the churning process, fat is the continuous and water the disperse phase.

The fat phase in butter made by the traditional churning process or continuously by the so-called 'Fritz method' is, however, not a homogeneous phase. Storch[69] was the first to observe more-or-less intact fat globules in butter and, later, King[70] based his churning theory on the presence of two different fat phases in butter, namely a continuous phase of free fat and, dispersed in this, a globular phase consisting of relatively intact fat globules.

The weakening of the fat globule membrane in cream starts at the very beginning of processing. Thus, Mohr and Baur[71] showed that during cooling of the cream prior to churning an incipient breakdown of the globule membrane takes place. Later, Wortmann,[72] using electron microscopy, visualized the destruction of membranes occurring during temperature treatment and ripening of cream, and Buchheim and Precht[73] showed that clustering of fat globules also occurs during the ripening process.

Precht and Buchheim[74] demonstrated by electron microscopic studies of butter that the only fat globules in cream having sufficient stability to withstand the shear forces of the churning process are those with a 0·1–0·5-μm thick shell of solid fat and with small crystalline aggregates of varying size and shape in the liquid fat in the interior; consequently the amount of globular fat present in butter strongly depends upon the mechanical treatment during the manufacturing process. This was confirmed by King[75] who showed that the amount of globular fat falls rapidly with increasing working intensity. On the other hand, Fisker[76] found that between 26 and 35% of the total fat content is present in globular form, more-or-less independently of the treatment of the cream.

Based on studies of numerous electron microscopic pictures of the structure of butter, Knoop[77] distinguished between two typical structures, namely a globular or grain structure and an almost completely homogeneous structure caused by intensive mechanical working at high temperatures. Between these two extremes a number of intermediate states can be shown, where the grain structure is more or less pronounced. Knoop[77] found a strong correlation between the type of structure and the consistency of butter. With increasing homogeneity, i.e. a larger free fat phase, increasing firmness is observed at low temperatures and an increasing oiling-off tendency at higher temperatures.

Precht and Buchheim[78] found many crystalline aggregates of very variable shape in the continuous interglobular fat phase. The number of crystal platelets, which generally have a parallel orientation, increases during storage of the butter. This phenomenon is, no doubt, responsible for the increased firmness—the so-called 'setting' of butter—which occurs during storage.

During the post-manufacture cooling of butter, solidification of liquid fat occurs both inside and outside the globules. The internal crystallization increases the rigidity of the globules, which scarcely influences the consistency, but external crystallization in the continuous fat phase, which results in the formation of a solid network of fat crystals grown together or held together by van der Waals forces, strongly influences the consistency and results in a considerable increase in firmness of the butter. The crystal network can be disrupted by mechanical working of the butter but part of the resulting softening is unstable and the butter later regains some of its firmness due to the fact that the disrupted crystals tend to reflocculate. This phenomenon, which is called thixotropy, has been described by Mulder,[79] and by deMan and Wood.[80] Thixotropy is well known in colloid dispersions and is defined as a reversible, isothermic formation of the colloid particles in such a way that a coherent structure is formed. A certain freedom of movement of the dispersed fat crystals is necessary for the formation of such a structure. The importance of this is demonstrated by Mulder[79] who showed that it is possible to stop the thixotropic process if the temperature is low enough.

3.2.2. Rheology of Butter

Most measuring systems commonly used to characterize the rheological behaviour of butter register either the stress necessary to obtain a given deformation or the deformation obtained at a given stress. The relationship between stress and the rate of deformation of butter shows clearly that butter does not exhibit Newtonian behaviour. The form of the flow curve deviates considerably from a straight line and the shape of the curve, especially at low shear stress, has been a much-debated question.

It is well known that the flow behaviour of many dispersed systems with very high apparent viscosity[81] can be described by a power law of the form

$$D = kP^n \tag{4}$$

where D is the shear rate, P is the stress applied, and k is a constant related to viscosity. The power n, which is related to the interior structure of the product, is often referred to as the 'flow behaviour index'.

This consistency model was related to butter by Scott Blair[82] and later applied by Dolby[83] who, based on this model, explained the results he had obtained for firmness measurements on butter. A similar relationship was applied by Foley *et al.*,[84] and by Dixon and Williams[85] who found that both the viscosity and the structure parameters of eqn (4) have a close correlation to the results of a subjective valuation of spreadability.

Under increasing stress, butter initially shows a reversible deformation while it is deformed irreversibly under heavier stress as the structure in the product is broken. At high deformation speed butter flows almost like a liquid. The relationship between deformation speed and stress for three butter samples is shown in Fig. 2.[86] The flow behaviour shown is characteristic of a pseudoplastic material with a yield stress or yield value. Although unequivocal proof of the existence of a yield point in butter has never been made, several studies of the consistency of butter are based on the presence of a yield stress. Knoop[87] described the relationship between deformation

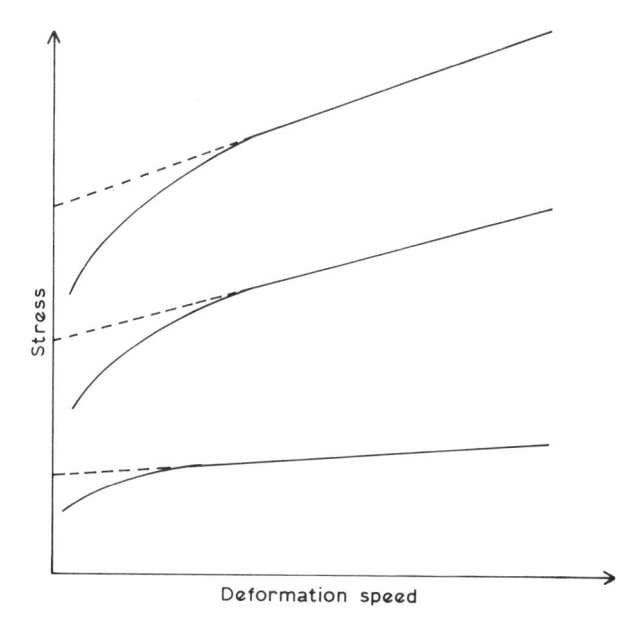

FIG. 2. Relationship between deformation speed and stress for three butter samples.

speed or shear rate and stress by the following equation

$$\sigma = f + \eta(dh/dt) \quad (5)$$

where σ is the stress applied, f is the yield stress, η is the apparent viscosity, and dh/dt is the shear rate.

The investigations of Kruisheer and den Herder[88] and Tanaka et al.[89] were based on a similar model.

Whether butter shows a yield stress or not at low shear rate has, as mentioned earlier, never been established. Prentice,[65] however, stated that the question of whether the yield stress is more apparent than real is purely academic. From a practical point of view there can hardly be any doubt that eqn (5) gives adequate possibilities of calculating relevant consistency parameters even though the equation has not necessarily the same validity all over the consistency range. Thus, Mortensen and Danmark[86] found that deviations from the straight line given by eqn (5) increase with increasing firmness. This is possibly due to the fact that the straight-line relationship only appears at a much higher shear rate than normally used for this type of measurement.

Results obtained by Kruisheer and den Herder[88] indicate that there is a close relationship between the apparent viscosity and the yield stress, indicating that one consistency parameter is sufficient for a complete description of the stress–deformation relationship. Also, Mulder[16] and Knoop[87] found the ratio between apparent viscosity and yield stress to be constant. On the other hand, Mortensen and Danmark (results cited in Reference 86), who examined a large number of butter samples ranging from very soft to very firm, did not confirm the presence of a constant ratio between these two parameters. They even found that stress–deformation curves might cross each other.[90] Nevertheless, the conclusion of the investigation was that yield stress in most cases would be a sufficient measure of the spreadability of a butter sample for practical application.

3.2.3. Measuring Methods

Several methods have been developed to evaluate the consistency of butter. The most frequently applied methods can be divided into the following groups: sectility; extrusion; penetration.

3.2.3.1. Sectility. The sectility or cutting methods, which were developed mainly in Germany,[91] are based on measuring the force required for a wire to cut through a piece of butter at a given constant rate. Nowadays the method is used as the official standard method[92] in the German Federal Republic. The wire used has an effective length of 25 mm and a diameter of 0·3 mm. The cutting speed is 0·1 mm s^{-1} and the measurement temperature is 15 °C.

Dolby,[93] and later Dixon,[94] applied similar methods, but with a somewhat heavier wire. Several authors[90] have shown that there is a high correlation between the results of sectility measurements and spreadability evaluated subjectively. The sectility method has excellent reproducibility and is particularly suitable for measurements in butter with a firm consistency.[86] The measurements are easily performed and can be highly automated. However, Dixon[94] reported a low operator preference for the method because of the rather elaborate sample preparation required.

3.2.3.2. Extrusion. In the extrusion method, a butter sample, placed in a cylinder, is extruded by a plunger through a nozzle. The pressure necessary to maintain a constant extrusion speed is registered and used as an expression of the consistency of the butter.

The so-called 'FIRA–NIRD extruder' was developed by Prentice,[95] who found an excellent correlation between extruder thrusts and spreadability scores obtained by subjective assessment. Similar results have been obtained by Hoffer and Sobeck-Skal[96] and by Johansson and Joost.[97] The excellent reproducibility reported[90] must surely be ascribed to the fact that the method is less sensitive to local inhomogeneities in the butter than other methods since a larger sample volume is used.[65]

Pre-treatment and mounting of the butter sample in the extruder are rather time-consuming and control of the measurement temperature is difficult.[94] The method, therefore, is not very suitable for large-scale routine examinations.

3.2.3.3. Penetration. In the disc penetrometer method developed by Kruisheer and den Herder,[88] the stress necessary to push a plunger with a cross-sectional area of 4 cm^2 into butter at a constant speed of 0·33 mm s^{-1} is measured. The authors claim that the results obtained are equivalent to approximately 1·4-times the yield stress of the sample tested. Alternatively, using a cone penetrometer the distance a cone can penetrate into butter by its own weight when released directly above the surface of the butter is measured.

The latter method was used by Mohr and Wellm[98] to estimate the yield stress of butter—defined as the load per unit area of the cone cross-section. For a cone with a cone angle of $\alpha°$, yield stress is calculated from

$$f = \frac{G}{\pi h^2 \tan^2 \frac{\alpha}{2}} \qquad (6)$$

where f is the yield stress; G is the weight of the cone; h is the penetration depth; and α is the cone angle.

Haighton[99] also used a cone penetrometer and showed that for margarine the exponent of penetration depth is not 2 but 1·6. Using a constant-speed penetrometer, Tanaka *et al.*[89] showed that the value of the exponent can vary to some extent; for a number of foodstuffs, such as margarine, processed cheese and butter, values found ranged from 1·40 to 1·99. The fact that the exponent is always smaller than 2 is probably the result of the deformations caused by the cone not being purely viscoplastic.[89] This possibly also influences to some extent the other terms in eqn (6). The yield stress does not depend directly on $\tan^2(\alpha/2)$, the relationship between the cone angle (α) and the yield stress is far more complicated.[100,101]

Because of the absence of an unequivocal interpretation of the calculation of yield stress, it was suggested by an International Dairy Federation working group[90] that an apparent yield stress be calculated based on eqn (6). Mortensen and Danmark[102] obtained maximum accuracy using a cone with a cone angle of 40° and a weight of 200 g. The apparent yield stress, expressed in kPa, using such a cone can then be calculated to be $470\,000/p^2$, where p is the penetration depth measured in tenths of a millimetre.

The reproducibility obtained by both penetration methods is not quite as good as that obtained with the sectility and extrusion methods, especially with firm butter samples. However, in general, the methods give results which correlate well with the results of a subjective assessment of spreadability.[90] Concerning the practical use of the penetration methods, it appears that they are easy to employ and are most suitable for routine examinations.

The results obtained by sectility, extrusion and penetration methods are highly correlated; the small differences between results obtained by the three methods are probably not very important.

4. MODIFICATION OF RHEOLOGICAL PROPERTIES

The rheological properties of milk fat in products such as cream and butter can, to some extent, be modified during processing. If, however, more extensive modifications are required, e.g. in the rheological properties of butter, it is necessary to alter the composition of the fat either by fractionation of the fat into portions with different melting points, or by direct admixture of some other fats, usually vegetable fats with low melting points.

4.1. Processing

4.1.1. Cream
Two very effective ways of modifying the rheological properties, mainly the viscosity, of cream are homogenization and the so-called 'rebodying process'.

4.1.1.1. Homogenization. The main reason for homogenizing cream is to increase viscosity and to prevent creaming-off and phase separation during storage. The effect achieved by homogenization primarily depends on the physical state of the fat. To achieve a full effect the fat must be completely melted, but the effect will also increase with increasing temperature above the melting point due to decreasing viscosity of the fat.

Homogenization causes disruption of the fat globules into much smaller ones with a large surface area. What happens immediately after the disruption is determined by the collision rate between these newly formed globules without a protective membrane and the adsorption rate of surface-active material from the serum phase. If the adsorption rate of new surface material is much greater than the collision rate, the result of homogenization will be a micronization of the fat emulsion.

Increasing fat content increases the collision rate while at the same time decreases the adsorption rate since the amount of available surface material, relative to the fat, is lowered. At a certain fat content, which depends on homogenization pressure and temperature, the amount of surface material is no longer sufficient to ensure complete coverage of the newly formed fat globules, and homogenization causes a so-called 'viscolization' of the cream, characterized by markedly increased viscosity. This effect is due to the formation of clusters of newly formed globules sharing the same casein micelles in thin surface layers causing the formation of a structural matrix throughout the cream and immobilizing a large part of the milk plasma.

If the fat content is further increased, the collision rate will dominate completely and many of the small fat globules will coalesce into bigger ones before they are emulsified. Also, these big globules will flocculate into clusters and form a structural matrix. Besides high viscosity, such cream will show considerable liability to age thickening.

It should be mentioned that homogenization clusters are quite different from the clusters formed by natural creaming of raw milk or cream, where the cluster formation is caused by the agglutinins of the milk. These latter clusters are rather weak and are easily broken by stirring or heating. Homogenization clusters are very strong and are not broken by gentle stirring or heating but, of course, they can be disrupted by vigorous agitation or by another homogenization.

More details about the homogenization process and its influence on the physical properties of fat emulsions are given in comprehensive reviews by Trout,[103] and Mulder and Walstra.[1]

4.1.1.2. Rebodying. The viscosity of pasteurized cream can be increased substantially by temperature treatment of the cream, the so-called 'rebodying process'. According to Sommer,[104] the rebodying process comprises cooling of the pasteurized cream to 4–5 °C, holding the cream at this temperature for 1·5–2 h, re-warming the cream to 30 °C and re-cooling to 4–5 °C. The apparent viscosity of cream containing 18–20% fat may be increased from 50 to 100% by this treatment. If the fat content is higher, the increase in viscosity is far greater, and in extreme cases the cream may actually solidify.[105]

In earlier literature,[106] the effect of rebodying cream was primarily ascribed to components in the serum phase, the cryoglobulins and lipoproteins which were believed to concentrate on the surface of the fat globules and, due to their swelling, to contribute to a higher viscosity of the cream. Swelling of the fat globule membrane was also considered to be of importance.[107] However, more recent investigations[62,105,108] strongly indicate that the effect of the rebodying process is due to the formation of clumps of globules adhering to each other by patches of liquid fat on the globule surfaces. The liquid fat is squeezed out from the globules during the crystallization process because the formation of a shell of solid fat in the periphery of the globule builds up a pressure in the interior of the globule where the liquid fat is located. The larger the fat globules, the greater the pressure created inside the globules and the greater the effect achieved by the rebodying process.

If rebodied cream is heated above the melting point of the fat the effect is lost and some of the adhering fat globules will coalesce into larger ones. Due to the increase in globule size, the viscosity of the cream will increase further with repeated rebodying.

Age thickening of cream is caused by the same phenomenon, and therefore cream with large fat globules and rebodied cream are susceptible to age thickening.

Rebodying of cream is not used very much in the dairy industry of today because the effect of the process may vary considerably from cream to cream. Furthermore, as mentioned earlier, the increase in viscosity and plasticity is lost if the temperature of the cream rises too much.

4.1.2. Butter

The possibilities of influencing the consistency of butter by processing are based mainly on temperature treatment of the cream and on working of the butter after 'setting'—the so-called 'work softening'.

4.1.2.1 Temperature treatment of cream. Cream is normally cooled immediately after pasteurization. As early as the 1920s, van Dam[109] observed that the temperature at which the cream is stored before churning influences the physical properties of butter and the loss of fat in buttermilk. Samuelsson and Petterson[110] later showed that the firmness of butter can be reduced if the cream is cooled immediately after pasteurization to a low temperature and kept at this temperature for a suitable time until it is heated to a temperature above the churning temperature. The cream is finally cooled to the churning temperature. The reduction in firmness obtained by this treatment, often referred to as the Alnarp method, was attributed to the formation of fewer and bigger fat crystals which adsorb less liquid fat on the crystal surface than the many small fat crystals formed if the cream is cooled directly to the churning temperature.[110]

Mulder[27] later explained the effect of the temperature treatment by the theory of formation of mixed crystals. The treatment reduces the amount of supercooled fat and hence the rate of crystallization and the formation of mixed crystals; the final result of the temperature treatment is, therefore, a higher content of liquid fat in the cream. Such an effect has been shown by Samuelsson and Vikelsøe,[111] but its magnitude is small, and the even smaller variations which can be shown in the content of liquid fat in butter cannot explain the great influence of the Alnarp method on the consistency of butter.

An important factor, no doubt, is the amount of supercooled fat in the cream at the beginning of the churning process. During the churning process many fat globules are disrupted and liquid fat is squeezed out into the continuous phase. Thus a high content of supercooled fat will result in a large amount of high-melting fat being led to the continuous phase where it crystallizes and increases the size of the solid network, which again increases the firmness of butter.

When the content of low-melting fat is relatively high, i.e. iodine value above ~35, excellent spreadability of the butter can be obtained by stepwise cooling of the cream, e.g. cooling the cream after pasteurization to 20°C, holding at 20°C for a few hours prior to cooling to 14°C, and holding for a further 2–3 h prior to cooling to the churning temperature (10°C), i.e. the 20–14–10 treatment.

However, if the iodine value is below ~35 the butter will become too firm, and in this case it would be advantageous to use the Alnarp method. A commonly used procedure is to cool the cream to 8°C after pasteurization, and after 2-h holding time at this temperature to heat it to 20°C. After at least 3-h holding time at this temperature the cream is cooled to the churning temperature, e.g. 10°C.

Other combinations of temperatures and holding times have been examined by Fisker and Jansen[112] who found that cooling the cream after pasteurization to temperatures below 8°C does not further reduce the firmness of butter, but it is important that the cream is held at the cooling temperature for at least 2 h. Further extension of the holding time has some effect on the consistency; holding for 24 and 48 h reduces firmness of the butter by 10 and 20%, respectively.[113]

The effect of different temperature treatments of cream on the firmness of butter is shown in Fig. 3. When measured at 7°C, the firmness of butter made from cream treated by an 8–20–10 sequence was decreased by ~25% compared with butter made from cream which had only been cooled to 8°C. There is no interaction between iodine value and temperature

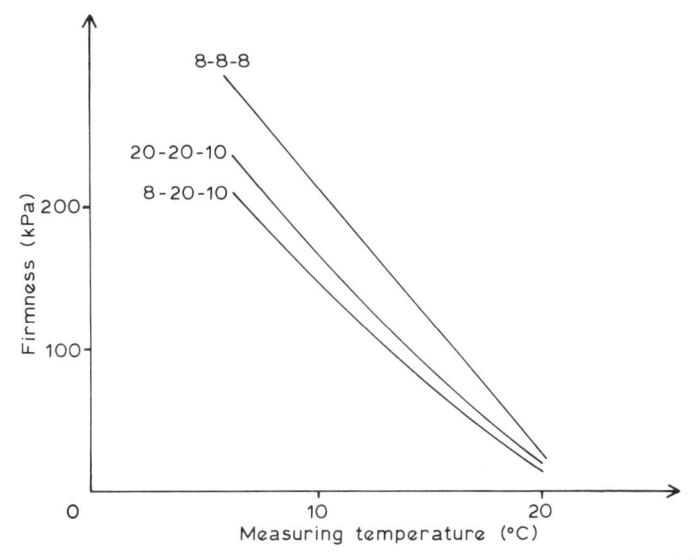

FIG. 3. Influence of temperature treatment of cream on firmness of butter measured at different temperatures. (Iodine value ~36.)[114]

treatment; altering the iodine value therefore only results in parallel displacements of the curves.

The effect of the temperature treatment of the cream can be further increased if the treatment is extended by more steps. Thus, Mortensen and Samuelsson[32] showed that the firmness of butter can be reduced by a further 15% by employing an 8–25–8–21–12 cooling sequence. Certainly, such a temperature treatment of the cream can be automated, but nevertheless it is complicated and rather energy-consuming and so is used only under special conditions where the manufacture of butter with acceptable consistency is not possible in any other way.

In New Zealand a certain interest has been shown in shortening the treatment of cream by heating immediately after the first cooling, i.e. a very short holding time at the cooling temperature in order to accomplish the whole treatment in a plate heat exchanger. A certain reduction of the firmness of butter is obtained with a holding time of 8 s but not as much as the reduction obtained with 2-h holding time.[115,116]

Usually the temperature treatment of cream is planned according to the iodine value of the butterfat but Precht and Peters[117] indicate that it is an advantage to also consider melting and solidification diagrams for the fat. Under normal production conditions, however, detailed knowledge of the properties of butterfat is usually not available prior to manufacture, and it should also be remembered that there are considerable differences between the crystallization behaviour of butterfat in the emulsified state and in bulk fat.

Independently of the treatment of the cream, the consistency of butter is influenced if the storage temperature of the product is temporarily increased. Since a temporary increase in temperature does not influence either the size or the composition of the continuous fat phase, the increase in firmness is due to a rearrangement of the solid fat network. Theoretically, some effects of the temperature pre-treatment should also be detectable after a temporary increase in temperature. However, Taylor and Jebson,[118] using a reduced Alnarp treatment (holding time 8 s), reported that raising the temperature of the butter above 20°C and re-cooling eliminate all effects of the cream treatment, while Mortensen and Danmark,[114] using a full Alnarp treatment (holding time 2 h), showed that, compared to samples manufactured from cream cooled to 8°C, Alnarp-treated samples would still have the softest consistency after a temporary increase in temperature of the butter to 20°C.

4.1.2.2. Work softening. As mentioned earlier a considerable increase in firmness takes place due to crystallization in the continuous fat phase when butter is cooled after manufacturing. If butter is worked mechanically after setting it loses much of its firmness due to disruption of the crystal network. This work softening was studied by Haighton[119] who expressed the effect as the decrease in firmness caused by the treatment as a percentage of the initial firmness

$$W = \frac{C_u - C_w}{C_u} \times 100 \qquad (7)$$

where W is the work softening; C_u is the firmness of the unworked sample; and C_w is the firmness of the worked sample. Haighton stated that butter generally undergoes lower work softening (50–55%) than margarine (70–75%) which means that the firmness of margarine decreases to a much lower value during spreading than butter, even though the initial firmness of both products is the same; this is one of the main reasons why butter is considered less spreadable than margarine.[119] DeMan[120] demonstrated that the greater part of the softening is almost instantaneous and that continued working has only very little additional effect.

Norris[121] showed that working of the butter during cooling in a scraped surface cooler immediately after manufacture has a substantial influence on consistency but, in general, the effect of working depends on the

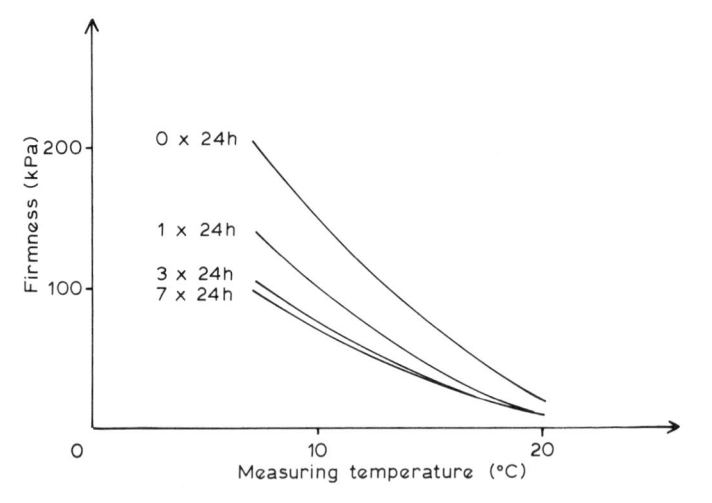

Fig. 4. The effect of working after storage at 5°C for 0, 1, 3 and 7 days. Firmness measured 2 weeks after manufacture.[114]

original firmness of butter. Naturally, the effect is greatest if a strong crystal network has been formed, i.e. the effect of work softening is greatest if the setting of butter is in an advanced stage. The results of Mortensen and Danmark,[114] who treated the butter in a continuous butter mixer, confirmed this (Fig. 4). Working of the butter following storage for 1 day at 5°C reduced the firmness measured at 7°C by ~30%, while working after storage for 3 days reduced the firmness by ~50%. The effect of a further extension of the storage period before working was negligible. If the firmness was measured at 13°C the effect was somewhat smaller, and if the measurements were made at 20°C the effect of the working was scarcely detectable.

The effect of working is partly eliminated by a temporary heating of the butter samples to 20°C, indicating that the crystal network is partly re-created.[114] Similar results have been obtained by Taylor and Jebson.[118]

When butter is stored after mechanical treatment it will regain some but not all of its original firmness. A good portion of the work softening is permanent, so only part of the crystal network seems to regenerate; besides, the recovery of firmness is quite slow and can still be observed to be taking place after several months.[122]

4.2. Fractionation

The fact that milk fat consists of many different triglycerides with different physical properties makes possible a division of the fat into various fractions with different compositions and melting points. Fractionation can be made either by crystallization of fat dissolved in an organic solvent or by crystallization directly from the molten fat. Fractionation of fat from organic solvents such as ethanol or acetone is commonly used in the laboratory. Separation of the fat crystals is easy and the fractions obtained can be re-crystallized easily and purified. However, the application of organic solvents is not suitable for foodstuffs and the method is therefore not used in spite of the obvious technical advantages.

Fractionation of milk fat without the use of solvents is more complicated. Separation of the crystals from the melt is especially difficult and the efficiency of the fractionation process is considerably lower than when solvents are used. The principle is that suitable crystallization is promoted by programmed cooling after which the crystals are separated from the melt by filtration or centrifugation, or by a combination of these methods. The cooling temperatures and the rate of crystallization applied strongly influence the composition of the fat crystals and the cooling rate influences the quantity of the crystallized fat and the size of the crystals.

Several investigations have shown that there is a direct correlation between the size of the crystals and the efficiency of separation. To some extent, optimum crystal size depends on the separation method applied. Black,[123] who used a vacuum leaf filter, preferred a crystal size of 200–350 μm, while Antila,[124] using a filtering centrifuge, preferred a crystal size of 150–250 μm.

A number of methods for separating butterfat crystals from the liquid fat have been studied. An uncomplicated method was used by Wilson;[125] the melted fat is cooled slowly for 2 days in a tank without stirring; a thick cake of crystals is formed on the wall of the tank, while the liquid fat remains in the centre of the tank.

Black[123] compared several methods for separating the crystals. He preferred vacuum filtration and filtration combined with centrifugation. Similar methods have been used in several other investigations.[126-8] The production capacity using these filtration processes is rather limited but fractionation of milk fat is also possible on an industrial scale by application of the 'Lipofrac' method where partially crystallized fat is centrifuged after the addition of water containing surface-active substances. The fat crystals concentrate in the water phase and accompany this during centrifugation, after which the fat crystals can be melted and re-emulsified. The emulsion formed is centrifuged once more and the fat phase is washed with water and finally dried. The substances added are sodium laurylsulphate and magnesium chloride and even though most of these substances are removed by the washing, the use of surface-active agents is forbidden in a number of countries.

The composition and physical properties of the fat fractions obtained depend on the cooling procedure as well as on the efficiency of the separation of the fat crystals from the liquid fat. Fjaervoll,[129] using the industrialized Lipofrac method, found differences in softening points between the liquid and the solid fractions of 15–20°C, while the differences in iodine value were 5–6 units. Similar results were obtained by Norris et al.[130] using the same method. Using filtration systems with lower capacity, Baker[131] found differences of up to 16 units between the iodine values of the liquid and the hard fractions.

Possible uses of the fat fractions obtained have been studied by Dolby,[132] who tried to improve the spreadability of butter by recombining the fractions in different ratios. The fat fractions were emulsified with skim milk to form a cream with 40% fat which was processed into butter in the normal way. Kankare and Antila[133] simply added the liquid fat fraction to cream before churning and showed that the spreadability of the resulting butter at low temperature was markedly improved by adding the liquid fat fraction in amounts of up to 33%; the temperature stability of the butter was good.

Butter with an increased content of liquid fat usually lacks the 'stand-up' characteristics of normal butter so that increasing temperature results in oiling-off. To avoid this, a method was developed in New Zealand in which different fractions of milk fat were combined in order to produce butter with physical properties similar to margarine but with a butter flavour.[134] This is obtained by removing the intermediate fractions to produce a butter with a 'melting gap' of 25°C. This means that little melting occurs in the butter when warmed from refrigerator to ambient temperatures.

From a theoretical point of view, fractionation of butterfat is very interesting, but it is obvious that the technology presently available is not optimal. It is still not possible, without the use of solvents, to obtain sufficiently large differences in the physical properties between the hard and the soft fractions.

Under all circumstances fractionation of butterfat is an expensive process and it does not seem very likely that consumers would be willing to pay the greatly increased production costs while obtaining only a minor improvement in spreadability of the butter.

4.3. Blending

Modification of the physical properties of milk fat during processing and fractionation is, as already mentioned, an expensive procedure. It is much cheaper to alter the composition of fat by adding vegetable oils, e.g. cottonseed, soybean, rapeseed or sunflower-seed oils. This kind of product is marketed in several countries, but the best known is, no doubt, the Swedish product Bregott®,* the manufacture of which has been described by Zillén.[135] It contains 80% fat, of which 80% is milk fat and 20% is soybean oil.

Few technical problems arise in the manufacture of such products. The vegetable oil is usually added to the cream before churning but can also be added to the butter grains. Dixon *et al.*[136] compared several methods of production of blends and concluded that the time of admixture is not very important provided sufficient blending takes place.

By addition of soybean oil, for example, the average melting point of the fat blend is reduced and the content of liquid fat at a given temperature is increased. The influence of 5, 10 and 20% added soybean oil on the spreadability of butter has been examined by Mortensen and Danmark[114] (Fig. 5). The firmness of the samples measured at low temperature is reduced by ~35% if the addition of soybean oil is increased from 5 to 20%; at higher temperatures the firmness is decreased so much that the product becomes extremely soft and shows considerable oiling-off. Special oil-proof packaging material is therefore necessary for such products.

Concerning the use of milk fat in the food industry, it might be a question of making the fat more firm than normal butterfat. This could be done, for example, by blending with beef tallow. The effect of such a procedure has been examined in detail by Timms;[137] blends of milk fat with tallow or its fractions have satisfactory physical properties for use as shortenings.

---

* Bregott® is a registered trade mark of the Swedish Dairies' Association.

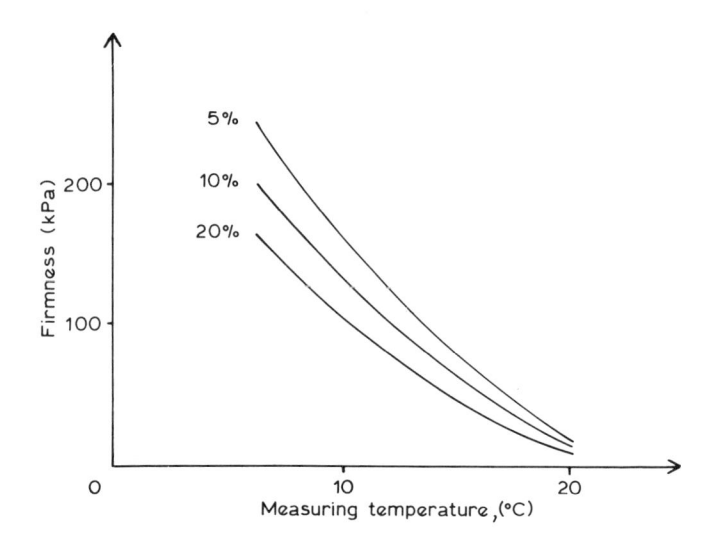

FIG. 5. The influence of the addition of soybean oil on firmness of butter at different measuring temperatures.

Thus the manufacture of tailor-made blends of milk fat and other fats for a wide-ranging field of applications, where special physical properties are required, is possible.

5. CONCLUSIONS

The physical properties of milk fat greatly influence the rheological properties of many dairy products and are thus of great importance.

Much information is available regarding crystallization of milk fat but further investigations of the kinetics of crystallization would be useful because the length of the crystallization period is essential in connection with temperature treatment of cream and fractionation of milk fat.

The content of solid fat in dairy products is very important because the rheological properties depend more on this than on the size and form of the fat crystals. The content of solid or liquid fat can be determined by different methods but the most promising seems to be the pulsed nuclear magnetic resonance method. The performance of the measurements is easy and very rapid, and raises few problems when used on pure fat samples or on fat emulsions. The question of determination of the content of liquid or solid fat in butter has not yet been solved satisfactorily and more fundamental research work is needed in this field.

The conclusion of the numerous investigations of the rheological properties of dairy products seems to be that in most cases the apparent viscosity is sufficient to describe the flow behaviour of milk and cream, while yield stress is a sufficient measure of the spreadability of butter. Several measuring methods have been developed to evaluate spreadability, among them the use of a cone penetrometer. The instrument is easy to employ and is most suitable for routine examinations.

The rheological properties of dairy products can, to some extent, be modified during processing, e.g. by homogenization, temperature treatment of cream or work softening of butter. If, however, more extensive modifications are required in the rheological properties of butter, it is necessary to alter the composition of the fat either by fractionation or by direct admixture with some other fats.

Fractionation is an expensive procedure and the technology available is not optimal. It is still not possible to obtain sufficiently large differences in the physical properties between the hard and soft fractions. It is much cheaper and more effective to alter the composition of fat by adding vegetable oils. Few technical problems arise in the manufacture of such products but it should be noted that the procedure is not permitted in many countries.

REFERENCES

1. Mulder, H. and Walstra, P., *The Milk Fat Globule*, CAB, Farnham Royal and Pudoc, Wageningen, 1974.
2. Tammann, G., *Kristallisieren und Schmelzen*, Barth, Leipzig, 1903.
3. Walstra, P. and van Beresteyn, E. C. H., *Neth. Milk Dairy J.*, 1975, **29**, 35.
4. Skoda, W. and van den Tempel, M., *J. Crystal Growth*, 1967, **1**, 207.
5. Grishchenko, A. D., *XV Int. Dairy Congr.*, 1959, **2**, 1030.
6. deMan, J. M., *Milchwissenschaft*, 1963, **18**, 67.
7. Bryzgin, M. I. and Eres'ko, G. A., *Mol. Prom.*, 1973, **34**, 14.
8. Mortensen, B. K. and Danmark, H., *XIX Int. Dairy Congr.*, 1974, **1E**, 226.
9. deMan, J. M., *J. Dairy Sci.*, 1964, **47**, 1194.
10. Malkin, T., *Progress in Chemistry of Fats and Other Lipids*, Vol. II, Pergamon Press, London, 1954.
11. Lutton, E. S., *J. Am. Oil Chem. Soc.*, 1950, **27**, 276.
12. Chapman, D., *Chem. Reviews*, 1962, **62**, 433.
13. Larsson, K., *Acta Chem. Scand.*, 1966, **20**, 2255.
14. Lutton, E. S. and Fehl, A. J., *Lipids*, 1970, **5**, 90.
15. Hoerr, C. W., *J. Am. Oil Chem. Soc.*, 1960, **37**, 539.

16. MULDER, H., *Versl. Landbouw. Onderz.*, 1940, **46**, 21.
17. THOMAS, A., MSc Thesis, University of Minnesota, USA, 1950.
18. DEMAN, J. M., *J. Dairy Res.*, 1961, **28**, 117.
19. TVERDOKHLEB, G. V., *XVI Int. Dairy Congr.*, 1962, **II**, 155.
20. BELOUSOV, A. P. and VERGELESOV, V. M., *XVI Int. Dairy Congr.*, 1962, **II**, 122.
21. VAN BERESTEYN, C. H., *Neth. Milk Dairy J.*, 1972, **26**, 117.
22. PRECHT, D., *Fette-Seifen-Anstrichmittel.*, 1980, **82**, 142.
23. DEMAN, J. M., *Dairy Sci. Abstr.*, 1963, **25**, 219.
24. ROSSELL, J. B., *Adv. Lipid Res.*, 1967, **5**, 353.
25. BAILEY, A. E., *Melting and Solidification of Fats*, Interscience Publishers, Inc., New York, 1958.
26. FREDE, E. and PRECHT, D., *Deutsche Molkerei-Zeitung.*, 1978, **99**, 1514.
27. MULDER, H., *Neth. Milk Dairy J.*, 1953, **7**, 149.
28. CANTABRANA, F. and DEMAN, J. M., *J. Dairy Sci.*, 1964, **47**, 32.
29. BELOUSOV, A. P. and VERGELESOV., V. M., *Mol. Prom.*, 1963, **24**, 5.
30. KRAUTWURST, J., *Kieler Milchwirtschaftliche Forschungsberichte*, 1970, **22**, 255.
31. DEMAN, J. M. and WOOD, F. W., *J. Dairy Res.*, 1959, **26**, 17.
32. MORTENSEN, B. K. and SAMUELSSON, E.-G., *Mælkeritidende*, 1972, **85**, 1183.
33. SHERBON, J. W. and COULTER, S. T., *J. Dairy Sci.*, 1966, **49**, 1126.
34. TVERDOKHLEB, G. V., STEPANENKO, T. A. and NESTEROV, V. N., *XIX Int. Dairy Congr.*, 1974, **1E**, 214.
35. DEMAN, J. M. and WOOD, F. W., *XV Int. Dairy Congr.*, 1959, **2**, 1010.
36. DEMAN, J. M., *Dairy Industries*, 1964, **29**, 244.
37. KANKARE, V., *Valtion Maitotalouskoelaitos, Jokioinen*, 1974, Publication No. 28.
38. VOSS, E., BEYERLEIN, V. and SCHMANKE, E., *Milchwissenschaft*, 1971, **26**, 605.
39. SCHAAP, J. E. and RUTTEN, G. A. M., *Neth. Milk Dairy J.*, 1976, **30**, 197.
40. VAN DEN TEMPEL, M., *J. Colloid Sci.*, 1961, **16**, 284.
41. VAN DAM, W., *Versl. Landbouwk. Onderz.*, 1915, **16**, 1.
42. PHIPPS, L. W., *J. Dairy Res.*, 1957, **24**, 51.
43. SHERBON, J. W. and DOLBY, R. M., *N.Z. J. Dairy Sci. Technol.*, 1971, **6**, 118.
44. KING, N., *Neth. Milk Dairy J.*, 1950, **4**, 30.
45. BUCHHEIM, W., *XVIII Int. Dairy Congr.*, 1970, **1E**, 73.
46. WALSTRA, P., *Neth. Milk Dairy J.*, 1967, **21**, 166.
47. PRECHT, D. and PETERS, K.-H., *Milchwissenschaft*, 1981, **36**, 673.
48. MORTENSEN, B. K. and DANMARK, H., *XIX Int. Dairy Congr.*, 1974, **1E**, 228.
49. ANDERSSON, K., JÖNSSON, H. and JOOST, K., Meddelande no. 95, 1977, Svenska Mejeriernas Riksförening.
50. AMERICAN OIL CHEMISTS' SOCIETY, *Official and Tentative Methods*, CD 10-57, 1973, Illinois.
51. MORTENSEN, B. K., International Dairy Federation, F-Doc 84, 1981.
52. HANNEWIJK, J. and HAIGHTON, A. J., *Neth. Milk Dairy J.*, 1957, **11**, 304.
53. KAPSALIS, J. G., BETSCHER, J. J., KRISTOFFERSEN, T. and GOULD, I. A., *J. Dairy Sci.*, 1961, **44**, 358.
54. NORRIS, G. E., GRAY, I. K. and DOLBY, R. M., *J. Dairy Res.*, 1973, **40**, 311.
55. NORRIS, R. and MUNIRO, D. S., *XIX Int. Dairy Congr.*, 1974, **1E**, 210.

56. HEUVEL, H. M. and LIND, K. C. J. B., *Analytical Chemistry*, 1970, **42**, 1044.
57. TIMMS, R. E., *Aust. J. Dairy Technol.*, 1978, **33**, 130.
58. CHAPMAN, D., RICHARDS, R. E. and YORKE, R. W., *J. Am. Oil Chem. Soc.*, 1960, **37**, 243.
59. SAMUELSSON, E.-G. and VIKELSØE, J., *Milchwissenschaft*, 1971, **26**, 621.
60. SWINDELLS, C. E. and FERGUSON, P. A., *Can. Inst. Food Sci. Technol.*, 1972, **5**, 82.
61. MERTENS, W. G. and DEMAN, J. M., *Can. Inst. Food Sci. Technol.*, 1972, **5**, 77.
62. VAN BOEKEL, M. A. J. S., *Versl. Landbouwk. Onderz.*, 1980, No. 901.
63. MORTENSEN, B. K., Statens Forsøgsmejeri, Hillerød, 1973, report No. 203.
64. MERILÄINEN, V. and ANTILA, V., *Meijerit. Aikakauskirja*, 1976, **34**, 117.
65. PRENTICE, J. H., *J. Texture Studies*, 1972, **3**, 415.
66. PHIPPS, L. W., *J. Dairy Res.*, 1969, **36**, 417.
67. SCOTT BLAIR, G. W., HENING, J. C. and WAGSTAFF, A., *J. Phys. Chem.*, 1939, **43**, 853.
68. BABCOCK, C. J., US Dep. Agric. Tech. Bull., 1931, No. 249.
69. STORCH, V., *Milchztg.*, 1897, **26**, 257.
70. KING, N., *Milchw. Forsch.*, 1929, **8**, 95.
71. MOHR, W. and BAUR, K., *Milchwissenschaft*, 1949, **4**, 100.
72. WORTMANN, A., *Nord. Mejeritidsskr.*, 1964, **30**, 172.
73. BUCHHEIM, W. and PRECHT, D., *Milchwissenschaft*, 1979, **34**, 657.
74. PRECHT, D. and BUCHHEIM, W., *Milchwissenschaft*, 1979, **34**, 745.
75. KING, N., *Neth. Milk Dairy J.*, 1947, **1**, 19.
76. FISKER, A. N., *Nord. Mejeritidsskr.*, 1954, **20**, 55.
77. KNOOP, E., *Kieler Milchwirtschaftliche Forschungsberichte*, 1965, **17**, 73.
78. PRECHT, D. and BUCHHEIM, W., *Milchwissenschaft*, 1980, **35**, 393.
79. MULDER, H., *XII Int. Dairy Congr.*, 1949, **2**, 81.
80. DEMAN, J. M. and WOOD, F. W., *Dairy Ind.*, 1958, **23**, 265.
81. SHERMAN, P., *Industrial Rheology*, Academic Press, London, 1970.
82. SCOTT BLAIR, G. W., *J. Dairy Res.*, 1938, **9**, 208.
83. DOLBY, R. M., *J. Dairy Res.*, 1941, **12**, 337.
84. FOLEY, J., Ó. SÉ, M. L., FAHY, E. F., *XVI Int. Dairy Congr.*, 1962, **B**, 42.
85. DIXON, B. D. and WILLIAMS, T., *Aust. J. Dairy Tech.*, 1977, **32**, 177.
86. MORTENSEN, B. K. and DANMARK, H., *Milchwissenschaft*, 1982, **37**, 530.
87. KNOOP, E., International Dairy Federation, F-Doc 14, 1972.
88. KRUISHEER, C. I., DEN HERDER, P. C., KROL, B. M. and MULDERS, E. M., *J. Chem. Weekbl.*, 1938, **35**, 719.
89. TANAKA, M., DEMAN, J. M. and VOISEY, P. W., *J. Texture Studies*, 1971, **2**, 306.
90. INTERNATIONAL DAIRY FEDERATION, DOC. 135, 1981.
91. MOHR, W. and OLDENBURG, F., *Milchwirtschaftliche Zeitung*, 1933, **38**, 907.
92. DEUTSCHE INDUSTRIENORM 10331, 1978, Beuth Verlag GmbH, Berlin.
93. DOLBY, R. M., *J. Dairy Res.*, 1941, **12**, 329.
94. DIXON, B. D., *Austr. J. Dairy Technol.*, 1974, **29**, 15.
95. PRENTICE, J. H., *Lab. Practice*, 1943, **3**, 186.
96. HOFFER, H. and SOBECK-SKAL, E., *Milchwirtschaftliche Berichte*, 1971, **26**, 47.
97. JOHANSSON, S. and JOOST, K., *Svenska Mejeritidn.*, 1972, **64**, 15.

98. Mohr, W. and Wellm, J., *Milchwissenschaft*, 1948, **3**, 234.
99. Haighton, A. J., *J. Am. Oil Chem. Soc.*, 1959, **36**, 345.
100. Rehbinder, P. A. and Semenko, N. N., *Dokl. Akad. Nauk. SSSR*, 1949, **64**, 835.
101. Agranat, N. N. and Volarovich, M. P., *Kolloid Zhur.*, 1957, **19**, 1.
102. Mortensen, B. K. and Danmark, H., *Milchwissenschaft*, 1981, **36**, 393.
103. Trout, M. G., *Homogenized Milk*, Michigan State College Press, East Lansing, 1950.
104. Sommer, H. H., *Market Milk and Related Products*, 3rd edn., Published by the author, Madison, 1952.
105. Te Whaiti, I. E. and Fryer, T. F., *N.Z. J. Dairy Sci. Technol.*, 1975, **10**, 2.
106. Skelton, F. M. and Herried, E. O., *J. Dairy Sci.*, 1941, **24**, 289.
107. Thomé, K. E. and Eriksson, G., *Milchwissenschaft*, 1973, **28**, 502.
108. Norlund, J. and Heikonen, M., *XIX Int. Dairy Congr.*, 1974, **1E**, 176.
109. van Dam, W., *Versl. Landbouwk. Onderz.*, 1927, **32**, 234.
110. Samuelsson, E. and Petterson, K. I., Årsskrift för Alnarps lantbruks-mejeri- og trädgårdsinstitut, 1937.
111. Samuelsson, E.-G. and Vikelsøe, J., *Milchwissenschaft*, 1971, **26**, 621.
112. Fisker, A. N. and Jansen, K., *XVI Int. Dairy Congr.*, 1962, **B**, 65.
113. Precht, D. and Peters, K.-H., *XXI Int. Dairy Congr.*, 1982, **1**, Book 1, 335.
114. Mortensen, B. K. and Danmark, H., *Milchwissenschaft*, 1982, **37**, 193.
115. Dolby, R. M., *Aust. J. Dairy Technol.*, 1959, **14**, 103.
116. Wood, F. W. and Dolby, R. M., *J. Dairy Res.*, 1965, **32**, 269.
117. Precht, D. and Peters, K.-H., *Milchwissenschaft*, 1981, **36**, 727.
118. Taylor, M. W. and Jebson, R. S., *XIX Int. Dairy Congr.*, 1974, **1E**, 673.
119. Haighton, A. J., *J. Am. Oil Chem. Soc.*, 1965, **42**, 27.
120. deMan, J. M., *J. Texture Studies*, 1969, **1**, 109.
121. Norris, R., *XXI Int. Dairy Congr.*, 1982, **1**, Book 1, 358.
122. Prentice, J. H., *XIII Int. Dairy Congr.*, 1953, **2**, 723.
123. Black, R. G., *Aust. J. Dairy Tech.*, 1975, **30**, 153.
124. Antila, V., *Milk Industry*, 1979, **81**(8) 17.
125. Wilson, B. W., *Aust. J. Dairy Tech.*, 1975, **30**, 10.
126. Kankare, V., *Meijerit. Aikakauskirja*, 1974, **33**, 1.
127. Schaap, J. E. and van Beresteyn, E. C. H., *NIZO-nieuws*, 1970, **9**, 3.
128. Voss, E., Beyerlein, U. and Schmanke, E., *Milchwissenschaft*, 1971, **26**, 605.
129. Fjaervoll, A., *XVIII Int. Dairy Congr.*, 1970, **1E**, 239.
130. Norris, R., Gray, I. K., McDowell, A. K. R. and Dolby, R. M., *J. Dairy Res.*, 1971, **38**, 179.
131. Baker, B. C., *XVIII Int. Dairy Congr.*, 1970, **1E**, 241.
132. Dolby, R. M., *XVIII Int. Dairy Congr.*, 1970, **1E**, 243.
133. Kankare, V. and Antila, V., *XIX Int. Dairy Congr.*, 1974, **1E**, 671.
134. Norris, R., New Zealand Patent, 172101, 1976.
135. Zillén, M., International Dairy Federation, Doc. 107, 1978.
136. Dixon, B. D., Cracknell, R. H. and Tomlinson, N., *Austr. J. Dairy Technol.*, 1980, **35**, 43.
137. Timms, R. E., *Austr. J. Dairy Technol.*, 1979, **34**, 60.

Chapter 6

LIPOLYTIC ENZYMES AND HYDROLYTIC RANCIDITY IN MILK AND MILK PRODUCTS

H. C. DEETH and C. H. FITZ-GERALD

*Queensland Department of Primary Industries,
Hamilton, Queensland, Australia*

SUMMARY

Enzymic hydrolysis of milk lipids to free fatty acids and partial glycerides has both beneficial and detrimental effects. Free fatty acids contribute to the desirable flavour of milk and milk products but, when present in high concentrations as a result of excessive lipolysis, can impart rancid off-flavours to the product.

The enzymes responsible are of two main types: those endogenous to milk, and those of microbial origin. The major endogenous milk enzyme is a lipoprotein lipase. It is active against milk fat in natural globules only after their disruption by physical treatments or if certain blood serum lipoproteins are present. The major microbial lipases are produced by psychrotrophic bacteria. Many of these enzymes are heat stable and are particularly significant in stored products.

Human milk contains two lipases, a lipoprotein lipase and a bile-salt-stimulated lipase. The ability of the latter to cause considerable hydrolysis of ingested milk lipids has important nutritional implications.

1. INTRODUCTION

Hydrolytic rancidity in milk and milk products has been a problem in the dairy industries of most countries.[1] Although it is not considered a serious problem in many countries today,[2] the potential for problems exists at all times and a constant vigilance is necessary to ensure effective control.

Hydrolytic rancidity results from the hydrolytic degradation of milk lipids. The hydrolysis is catalysed by lipases and produces free fatty acids (FFAs), some of which have low flavour thresholds and are responsible for the unpleasant flavours associated with the problem. These flavours are variously described as rancid, butyric, bitter, unclean, soapy and astringent.

The lipases involved are of two types: the endogenous milk enzyme(s), and those enzymes of microbial origin. That milk contains an enzyme capable of hydrolyzing triglycerides and producing rancidity was recognized early this century (for reviews of this early work see References 3–6). During the first half of the century considerable research was carried out into the causes and effects of the action of the lipase in milk. 'Bitter milk of late lactation'[7] was recognized as a consequence of milk lipase action, as was the increase in fat acidity following shaking,[8] homogenization,[9] and certain temperature manipulations of raw milk.[10] It was also found that some developments in milking and processing methods (e.g. cold storage, mechanization) could exacerbate the problem.

During the 1950s and 1960s, studies focussed on the milk lipase system, the mechanism of its activation and the physico-chemical properties of the enzyme(s) involved. From these studies it was concluded that there was more than one lipase present[11–13] and several attempts to purify a milk lipase were reported.[14–16] Sjöström,[17] Jensen[18] and Schwartz[19] have reviewed much of this work.

Since the early 1970s, there have been several new developments. Of particular importance has been the widespread acceptance of the postulate that bovine milk contains one lipase, a lipoprotein lipase.[20-2] Its purification and characterization have been major advances.[23-6] Similarly, the elucidation of both the lipase system in human milk and the possible role of one of the enzymes in infant nutrition have been noteworthy.[27-9]

Microbial lipolysis has assumed considerable significance with the widespread use of low-temperature storage of raw milk. Heat-stable lipases produced by psychrotrophic bacteria are considered to be a major cause of flavour problems in stored dairy products.[30]

This chapter discusses the lipolytic enzymes in milk and milk products and the causes, consequences and assessment of their action. The significance of lipases in human nutrition and in the production of characteristic flavours in certain dairy products is also covered.

2. THE ENZYMES

2.1. Bovine Milk Lipase

Until the 1970s, bovine milk was thought to contain more than one lipolytic enzyme. Up to six different molecular species with lipase activity had been separated by gel filtration[13,31] and, from substrate specificity studies, it appeared that at least two lipases were present, a 'plasma lipase' in the skim milk and a 'membrane lipase' associated with the milk fat globule membrane.[11]

Several attempts were made to purify a lipase from milk. Gaffney *et al.*[15] purified a lipase a thousand-fold by ion-exchange chromatography of a rennet curd extract, while Fox and Tarassuk[16] used a combination of solubilization and chromatographic steps to obtain a homogeneous lipase preparation with a specific activity 500 times that of skim milk. A B-esterase which hydrolyzed tributyrin and which was almost certainly a lipase[18] was partially purified by Montgomery and Forster.[32] In addition, a low molecular weight (*c.* 7000) lipase was isolated from separator slime and purified to homogeneity.[14] It was thought to be a representative milk enzyme, but subsequent work has shown that it probably originated from somatic cells.[33]

In 1962, Korn[34] showed that milk contained a lipoprotein lipase (EC 3.1.1.34) (LPL) with properties very similar to those of post-heparin plasma, adipose tissue and heart LPLs. The properties used to characterize this enzyme as an LPL were the enhancement of its activity against emulsified triglycerides by blood serum lipoproteins, its inhibition by polylysine and by high ionic strength, e.g. 1 M NaCl, and its requirement for a fatty acid acceptor such as albumin.[34,35] The distinction between a lipase and a lipoprotein lipase is not clear-cut[21] but it is generally accepted that the milk enzyme can be classified as an LPL.

LPL can be readily purified from milk by affinity chromatography on immobilized heparin[23,24,26] and has been widely used as a model for human post-heparin LPL. Antisera to the purified milk enzyme cross-react with the LPL in human post-heparin plasma and human milk[36] and inhibit essentially all of the tributyrinase activity in milk.[20] From this latter fact and from the similarity in response of the lipase (tributyrinase) and lipoprotein lipase activities to various agents such as NaCl, heat and light it has been concluded that both activities can be attributed to the same enzyme molecule and that this is the only major lipolytic enzyme in bovine milk.[20–2,37] Flynn and Fox[38] compared milk lipase purified by the method of Fox and Tarassuk[16] with the LPL purified on heparin–Sepharose and concluded that both procedures yielded the same enzyme. Some workers have however expressed the view that there are two different milk lipases.[39,40] It is assumed here that the LPL accounts for most, if not all, of the triglyceride hydrolase activity in milk.

LPL is synthesized in the mammary gland secretory cells and most is transported to the capillary endothelium where it hydrolyzes triglycerides in circulating lipoproteins to FFAs and 2-monoglycerides. These products are absorbed by the mammary gland and used for milk-fat synthesis. The LPL in milk is thought to be identical with the enzyme in the mammary gland[41] and to be the result of a spill-over. It has no known function in milk and is unlikely to aid in milk fat digestion in the suckling calf because of its instability to the acid conditions of the stomach.[11,34,42] Its level in milk is low at parturition but rises rapidly during the first few days of lactation.[43,44] This rise parallels the requirements of the mammary gland for milk fat precursors. Prolactin stimulates its production.[45]

Under normal circumstances, most of the LPL in milk is in the skim-milk fraction and the major part of this is associated with the casein micelles.[46] Some is in soluble form[47] and a small amount is associated with the milk fat globule membrane.[11,34,48] The enzyme is thought to be bound to the caseins principally by electrostatic interactions since 0·75–1 M NaCl releases most of the lipase activity into a soluble form where it is associated with casein in aggregates of molecular weight of $c.$ $0·5 \times 10^6$.[49] The electrostatic binding of lipase in the micelle appears to be via positive charges on the enzyme to negative charges on the caseins, possibly on κ-casein.[50] LPL binds strongly to the negatively charged heparin, enabling it to be dissociated from the casein micelle by low concentrations of sodium heparin (5 μg ml^{-1}).[49] Hydrophobic association is probably also involved in the lipase–casein interaction since the lipase can be dissociated from the complex by dimethylformamide.[46] Kinnunen[26] made use of the amphipathic nature of milk lipase binding in an efficient purification procedure in which the enzyme was removed from heparin–Sepharose by a combination of NaCl and the non-ionic detergent Triton X-100.

The enzyme has been highly purified and characterized.[24–6] Reviews of its properties, particularly in relation to lipolysis in milk,[42] have been published.[39,51] LPL is a glycoprotein with a native molecular weight of around 100 000 and a monomer subunit of about 50 000.[24,25] It is a relatively unstable enzyme, being inactivated by ultra violet light, heat, acid and oxidizing agents.[19,52,53] Several substances such as blood serum albumin, glycine, glycerol, heparin and sodium oleate[20,24,54] have been used to stabilize the purified enzyme. In milk it is believed to be stabilized by some factor in the skim fraction,[55] possibly a heparin-like glycosaminoglycan.[56,57] The casein also stabilizes it with respect to heat inactivation.[58] It is generally accepted that high-temperature short-time (HTST) treatment (72 °C/15 s) of milk almost completely inactivates the enzyme[59] so that little if any lipolysis caused by milk lipase occurs in pasteurized milk.[60] Somewhat higher temperatures are required for cream pasteurization because of the protective effect of the fat.[61,62] However, some workers have observed considerable lipolysis in HTST-pasteurized milks and have found that more severe heat treatments, e.g. 79 °C/20 s,[63] are required to completely destroy the lipase activity. Furthermore, it has been suggested that the enzyme in milk may be reactivated after being inactivated by heating at 80 °C.[61] Whether this can happen or whether heat-stable bacterial enzymes have been responsible for the reported post-pasteurization lipolysis has not been determined.

The normal substrates for LPLs are long-chain triglycerides in blood chylomicrons and lipoproteins. These particles contain the apolipoproteins (apo-LPs) which activate the enzyme. In human blood the apo-LP which is activatory for milk LPL is apo-LP CII (apo-LP-glu),[64] a polypeptide of molecular weight 9110,[65] while bovine blood high density lipoprotein (HDL) contains two activatory polypeptides with molecular weights 8000–10 000 which have been designated D_1 and D_3,[66] and CI and CII.[67] Both human and bovine blood also contain inhibitory apo-LPs, e.g. in human, CI and CIII.[67,68] However, blood serum, either bovine or human, has an overall activatory effect in assays of LPL using emulsified long-chain triglyceride substrates (e.g. 10–20 fold using Ediol[34] and Intralipid[69]). In such assays, a fatty acid acceptor such as bovine serum albumin (BSA) is required because LPL is susceptible to product inhibition by FFAs which accumulate at the lipid–water interface.[70] LPL is also active against tributyrin, but in this case requires neither serum cofactors nor fatty acid acceptors,[54] and a catalytic rate of about 50% of its 'lipoprotein lipase activity' (measured against serum-activated long-chain triglycerides) is observed.[69] p-Nitrophenyl esters,[71] Tween 20 and monoglycerides are also hydrolyzed in the absence of serum cofactors.[69]

In milk, LPL is not normally active against milk fat because of the protection afforded the fat by the milk fat globule membrane. However, addition of blood serum facilitates the interaction between the enzyme and the fat, and lipolysis ensues.[72,73] The mechanism of this serum-mediated induction of lipolysis is not known. Conversely, the reason why LPL does not normally attack intact milk fat globules is not clear. The discovery of lipolysis-inhibiting glycoproteins in skim milk[74] and in the milk fat globule membrane[75] supports an earlier observation[48] that milk contains inhibiting factors. A glycoprotein from human aorta, lipolipin, has been shown to be a potent inhibitor of milk LPL.[76]

The interaction of LPL and heparin appears to be a characteristic feature of the enzyme.[77] The binding is quite specific, with some heparin fractions binding more strongly than others.[78] Heparin enhances lipolysis by LPL in milk and model systems[43,57,72] but it is not considered to be an activating cofactor for the enzyme. Rather it has a stabilizing effect and reverses the inhibition by agents such as NaCl.[57,79,80] From studies involving LPL associated with immobilized heparin, Bengtsson and Olivecrona[81] concluded that the LPL molecule has at least three functional regions other than the active site, one of which is a heparin–polyanion interaction region. The other two regions are the apo-LP interacting region and the lipid-binding region or interface recognition site. Thus heparin binds to the enzyme independently of apo-LP CII.[82]

Phosphatidyl choline is also hydrolyzed by milk LPL in the presence of serum cofactors[83] to yield a free fatty acid and lysophosphatidyl choline.[84] The importance of this function of LPL appears to be in facilitating its access to the triglyceride core of particles having a phospholipid-containing membrane. The lysophospholipids have a high affinity for both LPL and lipoproteins[85] and are powerful membrane-perturbing agents[86] which may aid lipolysis of milk fat in its natural globular form.[87]

Phospholipids also have a role in the LPL-catalyzed hydrolysis of triglycerides. The activator apo-LPs exhibit enhanced activation in the presence of phospholipids such as phosphatidyl choline[68,88,89] and in milk there is evidence that apo-LPs in the absence of phospholipids are unable to initiate lipolysis of intact milk fat globules by the endogenous LPL.[90,91] The phospholipids are believed to be involved in the reaction through their interaction with the substrate rather than with the enzyme.[89]

LPL exhibits no fatty acid specificity during hydrolysis of mixed triglycerides but does have strong positional specificity. It attacks the primary ester bonds with some preference for position sn-1 over sn-3,[92,93] and can hydrolyze sn-2-monoglycerides only after their conversion to the sn-1 or sn-3 isomer.[94] Against phosphatidyl choline it shows phospholipase A_1 activity, i.e. it hydrolyzes a fatty acid from position sn-1 only.

2.2. Human Milk Lipase

Human milk contains two distinct lipases: a lipoprotein lipase (LPL) similar to that found in bovine milk, and a bile-salt-stimulated lipase (BSSL).

The LPL, which is associated mainly with the cream phase, is activated by blood serum but inhibited by bile salts. It resembles most other serum-stimulated lipases, being inhibited by 1 M NaCl and by protamine sulphate. While it is capable of hydrolyzing triglycerides in the absence of exogenous serum factors, its activity is increased several-fold by blood serum.[27]

LPL activity is very low in colostrum, and in mature milk varies considerably between donors and from day to day. In general, human milk contains less than half the LPL activity found in bovine milk.[27,95] Human milk can become rancid on cold storage,[96] with the release of up to 3 meq. FFA/litre of milk in 24 h. In contrast to the case of bovine milk, this 'spontaneous lipolysis' is strongly correlated ($r = +0.90$) with the LPL activity of the milk.[97]

Rabbit antiserum to bovine milk LPL cross-reacts immunologically with human milk and human post-heparin plasma LPLs.[27,98] Human milk LPL has been purified from acetone–ether extracts of cream using heparin–Sepharose.[27,98] Its molecular weight has been estimated to be 63 000[27] and 95 000.[98] The enzyme exhibits a preference for hydrolysis of medium-chain saturated and long-chain unsaturated triglycerides. Its stereospecificity with triglyceride substrates has been shown to be in the order $sn\text{-}1 > sn\text{-}3 > sn\text{-}2$.[99]

Human milk is unusual in containing a lipase which is activated by bile salts. This enzyme has been described as an 'evolutionary newcomer' to milk.[100] To date, the gorilla is the only other animal in whose milk it has been found;[101] it is not present in the milk of the rhesus monkey or various lower animals.[100] Human BSSL shows immunological identity with the carboxyl ester hydrolase from pancreatic juice, and the two enzymes are very similar in molecular and kinetic properties.[102]

BSSL has been purified from human whey by chromatography on heparin–Sepharose and Affi-Gel blue[103] and on cholic acid-coupled Sepharose.[104] The purified enzyme was characterized as a single-chain glycoprotein with molecular weight 90 000[103] (or 125 000[104]), a high content of acidic amino-acid residues, and 11–13% proline. The isoelectric point is about 4.[103,105]

BSSL is located mainly in the milk serum and is optimally active at pH 8–9 and 30–40 °C in the presence of 8–14 mM sodium taurocholate.[106,107] In the absence of bile salts it can hydrolyze soluble esters (e.g. p-nitrophenyl acetate) and tributyrin, but it is inactive against high molecular weight triglycerides. Bile salts promote the lipolysis of triolein and milk fat, and also enhance activity against tributyrin and soluble esters.[106,107] Unlike LPL or pancreatic lipase, BSSL is reported to have no positional specificity, hydrolyzing triolein to mainly FFA and glycerol.[29,107]

The activation of BSSL is specific to primary bile salts (cholate and chenodeoxycholate) and their taurine and glycine conjugates.[108] Bile salts, secondary as well as primary, protect BSSL against inactivation by intestinal proteases. BSSL is inactivated by heating at 50°C for 1 h, but sodium taurocholate prevents activity loss.[108] The enzyme is stable in buffer for 1 h at 37°C between pH 3·5 and 9. It is inhibited by blood serum, 1 M NaCl, protamine sulphate, eserine and diisopropylfluorophosphate.[27]

Milk BSSL is present in high amounts, even in colostrum, and varies little throughout lactation.[95] In mature milk, its activity is some 200 and 100 times greater than the LPL in human and bovine milks, respectively.[27,95,109] It has been postulated to have a role in digestion in the newborn.[28,29]

2.3. Milk Lipases of Other Species

Lipase activity has been found in the milk of a number of other species. Chandan et al.[110] compared the activities of the milks of human, cow, goat, sheep and sow and found the milk of cow and sow to have the highest levels. Lipoprotein lipase has been found in various milks (cow,[34] human,[27] guinea-pig,[111] goat,[112] rabbit,[39] rat[113]), and it is probable that, with few exceptions (see above), LPL is the only significant natural milk lipase.

Goat's milk contains an LPL, about half of which is in the cream fraction. Its properties, which are typical of LPLs, include activation by blood serum, inhibition by protamine sulphate and 1 M NaCl, and a pH optimum of 8–9. LPL activity varies between individual goats.[112,114] One study[115] noted a significant relationship between stage of lactation and LPL activity, while a second found no consistent trends.[114] As with cow's milk, homogenization, agitation and temperature manipulation (cooling–warming–cooling) can initiate lipolysis in goat's milk. 'Spontaneous lipolysis' also occurs in some milks and a correlation between LPL activity and lipolysis has been found ($r = +0·68, +0·81$). Lipolysis is associated with the production of a 'goaty' flavour in the milk.[114]

Lipase in buffalo's milk has been investigated, but not in terms of LPL activity. Rifaat and co-workers[116] isolated two lipases, of MW 17 000 and 42 000, active against milk fat, with pH optima of 5 and 7, respectively.[117] Buffalo's milk lipase activity showed a positive correlation with stage of lactation[118,119] and was increased by feeding dry rations.[120] Shaking and homogenization induced lipolysis.[121]

Guinea-pig milk contains high levels of LPL (20–50 times that of bovine milk).[113,122,123] Over 90% of the activity is in the skim milk. This LPL has been purified and found to be very similar to the bovine milk enzyme in molecular weight and composition as well as in catalytic properties. The milk LPL is closely related immunologically to guinea-pig tissue LPLs.[123]

In contrast to guinea-pig milk, rat milk contains very low levels of LPL. Both enzymes hydrolyze chylomicron and serum-activated milk triglycerides at pH 8–9 and are inhibited by 0·5 M NaCl. It is considered that the low LPL activity in rat's milk reflects a difference in this animal's milk secretory process compared to that of the guinea-pig or the cow.[113]

Few detailed comparisons of the LPLs from the milk of different species have been reported, but it appears that considerable similarities exist between the different enzymes;[27,36,123,124] some also closely resemble tissue LPLs.[36,123]

2.4. Bovine Milk Esterases

In addition to the now well-documented lipase system, bovine milk contains several other carboxyl ester hydrolases collectively referred to as esterases. These are distinguished from lipases by their ability to act on ester substrates in solution rather than in an emulsified form[125] and/or by their preference for hydrolyzing esters of short- rather than long-chain acids.[126]

Although several reports concerning esterases in milk have appeared in the literature, little detailed information on the individual enzymes is presently available. The designation of the esterase types is difficult because of the lack of a universally accepted classification system. Probably the most commonly used system is that proposed by Holmes and Masters,[127] which primarily differentiates the enzymes on the basis of their specific inhibition by various agents rather than on their substrate specificity. Four types, aryl- or A-esterase (EC 3.1.1.2),[128–32] carboxylesterase (EC 3.1.1.1),[132,133] cholin- or C-esterase (EC 3.1.1.7; EC 3.1.1.8),[128,129,131,132] and acetylesterase (EC 3.1.1.6),[134] have been identified. A fifth type, B-esterase, which hydrolyzes tributyrin[128,129] is almost certainly milk (lipoprotein) lipase.[18]

Arylesterase has received considerable attention because of its elevated levels in colostrum and mastitic milk.[129,135] Since its level in mastitic milk correlates well with other indices of mastitis,[136] it has been suggested as a sensitive indicator of the disease.[60,129] The enzyme is believed to originate from the blood, where its activity is up to 2000-times that in milk.[135] The blood serum arylesterase has been partially purified and characterized. In blood it is associated with a lipoprotein of high molecular weight, but its molecular weight can be decreased to 140 000 by delipidation.[137] In milk it has a molecular weight >500 000.[130]

Carboxylesterase has been demonstrated in milk using β-naphthylacetate[132] and 4-methylumbelliferyl heptanoate[133] as substrates in combination with specific inhibitors. Like arylesterase, the activity of this enzyme is elevated in colostrum and mastitic milk.[133,138] The activity in these two abnormal milks may correspond to the previously reported 'lipases' from somatic cells[33] and colostrum,[139] respectively.

Although some esterase activity in milk has been attributed to a cholinesterase on the basis of inhibitor studies using non-choline ester substrates,[128,131,132] the presence of a true cholinesterase in milk has been questioned.[140]

Purr et al.[134] reported the presence of esterase activity against indoxyl acetate which had considerable thermal stability (inactivation temperature in milk, 88°C). They concluded that this was due to an acetylesterase. No other reports of this enzyme have been published.

The total esterase activity in normal milk can be estimated to be of the order of 0.05 μmol ml^{-1} min^{-1}. Hence, compared with the total LPL activity (0.25–2.5 μmol ml^{-1} min^{-1} [21]), the esterase activity is quite low. This may not be so for some abnormal milks where esterase levels are markedly elevated (10–12-times,[141] and up to 37-times[133]). The significance of these esterases in milk and their relationship to each other, to LPL, and to esterases of other tissues remain to be determined.

2.5. Lipases of Psychrotrophic Bacteria

Extracellular lipases produced by psychrotrophic bacteria have considerable potential for causing hydrolytic rancidity in milk and milk products. Two reviews concerning these enzymes have been published.[30,142] The bacteria principally responsible for these lipases are pseudomonads, particularly *Pseudomonas fluorescens*, Enterobacteriaceae such as *Serratia*, and *Acinetobacter* spp. Other significant organisms include *Achromobacter*, *Aeromonas*, *Alcaligenes*, *Bacillus*, *Flavobacterium*, *Micrococcus* and *Moraxella*.[143-7]

Despite the upsurge of interest in the lipases of psychrotrophic bacteria, the enzymes themselves have not been extensively studied. *Pseudomonas* species usually constitute the largest percentage of lipolytic psychrotrophs in raw milk and cream and hence their lipases have attracted most attention. Purification of only a few has been reported, e.g. *P. fluorescens*[148-50] and *P. fragi*.[151,152] One of the *P. fluorescens* lipases has been obtained in a crystalline state.[148] However, most studies of the properties of the lipases have been carried out on crude cell-free preparations. The discussion below is based mainly on data for pseudomonads, but data for other genera are included.

The pH optima of the lipases are usually in the alkaline region between 7 and 9[148,149,152-9] but some have considerable or even optimal activity at the pH of milk.[145,148,159] They generally show highest activity at 40–50°C,[153,158,159] although there are reports of higher[148,149,152,155] and lower[145,156,157,160] temperature optima. The apparent optimal temperature may change with the state of purity of the enzyme[149,156] and with the assay conditions used.[159] Many of these lipases show activity at the low temperatures used for storage of dairy products, e.g. 10°C,[161] 1°C[157] and −10°C.[153]

One of the most important properties of these lipases is their heat stability. This varies with the species and strain,[162] and also with the medium in which they are heated.[145,160,163] Their stability decreases with increasing purity.[152,154] Many are sufficiently stable to retain at least some activity after HTST pasteurization,[145,153,155,162,164,165] and even after UHT treatment.[158,161,164,166] Some workers have reported a two-stage inactivation on heating, an initial rapid activity loss followed by a slow or even negligible decline.[155,161,167] A number of the lipases are less stable at temperatures below 70°C than at higher temperatures[160,168] and this suggests that, like the proteases of some psychrotrophic bacteria, they may be susceptible to 'low-temperature inactivation' (i.e. 55°C for 1 h).[169] However, it has been found that only a few are inactivated under these conditions. Moreover, in fat-containing media, considerable lipolysis can occur during prolonged heating at 55°C. Thus such treatment is considered to be of little value for eliminating these lipases from milk products.[162,170] Heating at temperatures >70°C (up to c. 120°C) can cause activation of some lipases.[162,163]

In general, microbial lipases have molecular weights in the region of 25 000–50 000.[148,151] Sugiura and co-workers[148,171] purified a *P. fluorescens* lipase with molecular weight 33 000 and found it to be a single polypeptide chain with no lipid, carbohydrate or disulphide linkage. Some workers have found activity associated with material of molecular weight >100 000,[150,151,156,160] which probably represents aggregates of subunits with molecular weight <50 000.

Unlike the corresponding proteases, the lipases do not appear to contain metal ions,[171,172] but do require metal ions such as Ca^{2+} or Mg^{2+} for activity.[149,159] Excess EDTA causes complete inhibition of most bacterial lipases, which can be reversed by Ca^{2+} or Mg^{2+}. *Acinetobacter* lipases have been found to be irreversibly inactivated by EDTA,[159] while *P. aeruginosa* lipases are exceptional in being almost unaffected by excess EDTA.[159,172] Some heavy metals are inhibitory, in particular zinc, iron, mercury, nickel, copper and cobalt.[148,149,151-3,157,159] These metals are effective at concentrations of less than 10 mM.

Low levels of NaCl (10 mM) may cause activation[173] while high concentrations inhibit the lipases. However, more than half of the activity remains in the presence of 2 M salt, a level similar to that in the aqueous phase of salted butters.[159]

The bacterial lipases act on emulsified triglyceride substrates, although many psychrotrophs also elaborate esterases which prefer soluble substrates[144,168] or short-chain triglycerides, such as tributyrin, to long-chain triglycerides.[146] Unlike milk LPL, they do not require a fatty acid acceptor such as BSA.[70] Blood serum has been found to activate some of these enzymes,[159] including *P. fluorescens* lipases, and these have consequently been designated lipoprotein lipases.[174] The significance of this is unknown.

Since very few of these lipases have been purified, little definitive information is available on their substrate specificity. A purified *P. fluorescens* lipase showed activity against natural vegetable oils and a range of synthetic triglycerides from tributyrin to triolein.[149] Lawrence *et al.*[151] reported some preference for long-chain triglycerides by a *P. fragi* lipase but for short-chain triglycerides by a lipase from *Micrococcus freudenreichii*. Temperature may have an influence on the apparent specificity of lipolysis, with relatively more short-chain and unsaturated fatty acids being released from milk fat at lower temperatures. This appears to be a reflection of the physical state of the substrate.[157,175,176] With regard to positional specificity, most of the lipases have a preference for the primary positions of triglycerides.[172,177,178]

The lipases produced in crude cultures are usually capable of hydrolyzing the triglyceride in intact milk fat globules,[159] a property not exhibited by the natural milk lipase. It is not known whether the lipases *per se* can penetrate the milk fat globule membrane or whether the membrane is disrupted by other enzymes such as phospholipases.[179]

2.6. Lipolytic Enzymes in Milk Product Manufacture

Most lactic starters used in the manufacture of fermented milk products have a weak lipolytic activity[180] due to intracellular lipases and esterases.[181,182] The enzymes, present in the cytoplasm,[181] are released in cheese as the starter cells lyse during maturation.[183] In general, the lipases have pH and temperature optima around 7 and 37°C, respectively.[182,183] They are reported to have a specificity for short-chain fatty acids[184] and to show a preference for partial glycerides over triglycerides.[185] Lipases of adventitious organisms such as yeasts, lactobacilli and micrococci may also contribute to lipolysis during cheese ripening.[186,187]

Moulds involved in the ripening of certain varieties of cheeses (e.g. *Penicillium roqueforti*, *Pe. caseicolum*, *Pe. camemberti*, *Pe. candidum*) usually produce lipases.[188–90] *Pe. roqueforti* lipases have been most extensively studied[191–3] and have been found to show a specificity for the short-chain fatty acids of milk fat.[189,194,195] Lipase preparations from certain other micro-organisms (*Achromobacter lipolyticum*,[173] *Mucor miehei*,[196] *Candida cylindracea*,[197,198] *Candida lipolytica*,[199] *Aspergillus* spp.,[200] *Geotrichum candidum*,[201] *Rhizopus delemar*[202]) have been used to synthesize 'dairy' flavours from milk fat. The diversity of properties of these lipases, such as pH optima and, in particular, substrate specificities,[195,203] enables the selection of an appropriate enzyme for a specific purpose.

Pre-gastric esterases are used in the manufacture of Italian cheeses to produce the characteristic 'picante' flavours. These flavours are due to short-chain fatty acids, especially butyric, which are preferentially released from milk fat by these enzymes.[204] Pre-gastric esterases are produced in the salivary glands and can be obtained from the abomasum of milk-fed calves, lambs and kids. They have been isolated in heterogeneous form and have molecular weights of approximately 172 000 (calf), 168 000 (kid) and 150 000 (lamb).[205] They are optimally active at 32–42 °C at pH 4·8–5·5.[206] An extensive review of pre-gastric esterases was published in 1977.[204]

2.7. Phospholipases

Phospholipases are potentially important in milk and milk products because of their ability to degrade the phospholipids of the milk fat globule membrane, thereby increasing the susceptibility of the milk fat to lipolytic attack.[207,208]

Bovine milk LPL has phospholipase A_1 activity,[84] but its action against milk phospholipids has not been recorded. Freshly secreted goat's milk has been shown to have phospholipase A activity.[209] It is not known whether it can be attributed to the LPL of that milk. No phospholipase C was detected in the goat's milk.[209] An investigation by Chen *et al.*[210] failed to confirm an earlier report[211] that bovine milk contained phospholipase D activity.

Several psychrotrophic bacteria produce extracellular phospholipases, the most prevalent in milk being pseudomonads (particularly *P. fluorescens*), *Alcaligenes*, *Acinetobacter* and *Bacillus* species.[147,207,208] Most of these produce phospholipase C,[207] but some also produce phospholipase

A$_1$ and lysophospholipase C.[212] Phospholipases C from some pseudomonads have been purified and characterized.[213,214] Like the lipases, many of these enzymes have considerable heat stability and are not destroyed by pasteurization.[215]

The phospholipases of *Bacillus* spp., especially *B. cereus*, have received considerable attention because of their association with the 'bitty cream' defect in milk.[216] By partially degrading the milk fat globule membrane, they initiate agglutination of the fat globules into cream flakes or flecks.[217] The degradation is caused by phospholipase C,[218] although *B. cereus* produces a sphingomyelinase[219] which may also be involved. Phospholipase-producing bacteria other than *Bacillus* species do not appear to cause 'bitty cream'.[208,217] *Bacillus cereus* phospholipases have been isolated, purified and extensively characterized.[219-21]

3. CAUSES OF HYDROLYTIC RANCIDITY IN MILK AND MILK PRODUCTS

Raw bovine milk contains a relatively large amount of lipase activity, but seldom lipolyzes sufficiently to acquire an off-flavour. Under optimal conditions, the lipase (milk LPL) can catalyze the hydrolysis of up to $c.$ 5 μmol of triglyceride/ml/min.[77,222] Since milk with $c.$ 2 μmol FFA/ml has a rancid flavour,[223] it is of interest to ascertain why excessive lipolysis occurs in milk only under certain conditions.

Milk when freshly secreted from a healthy udder has $c.$ 0·5 μmol FFA/ml.[138,224-6] These acids result from incomplete synthesis rather than lipolysis. Under proper handling and storage conditions, only small increases in the FFA level should occur. In some cases, however, substantial increases are observed which result from either 'induced' or 'spontaneous' lipolysis. The two types of lipolysis are considered separately here, although some workers maintain that 'spontaneous' lipolysis is just a special case of 'induced' lipolysis.[227]

'Induced' lipolysis results when the milk lipase system is activated by physical or chemical means. 'Spontaneous' lipolysis is defined as that which occurs in milk which has had no treatment other than cooling soon after milking.[11]

Mastitis and microbial contamination can also contribute to hydrolytic rancidity. In general, lipolysis caused by natural milk lipase accounts for most of the rancidity in milk and cream, while in stored milk products lipolysis by microbial lipases is of greatest significance.

3.1. Induced Lipolysis

3.1.1. Agitation and Foaming

Lipolysis in raw milk can be readily initiated by vigorous agitation producing foaming. Such treatment disrupts the milk fat globule membrane and renders the milk triglycerides more accessible to the milk lipase. Incorporation of air (or other gases) and consequent foam formation are essential to the process of milk fat globule damage and to the activation of lipolysis.[228] The damage results from the high interfacial (liquid–air) tension acting on a small region of the fat globule surface.[229]

The amount of lipolysis induced depends on the mode of agitation (e.g. air agitation, pumping, stirring),[230,231] the severity and duration of agitation,[231-3] the amount of lipase present,[234] the content[231] and hardness of the fat,[235] and the vulnerability of the milk fat globule membrane.[236] Milk from cows in late lactation and milk with a tendency to spontaneous lipolysis are more susceptible to agitation-induced lipolysis.[73,237-40]

The temperature of the milk during agitation has a major influence. In general, activation is greatest at 37–40°C and least at cold storage temperatures (<5°C). However, the relationship between temperature and activation is complex, and depends on the conditions of the mechanical treatment and the characteristics of the milk.[231,241,242] The age and previous temperature history of the milk also influence its ease of activation.[228,231,242]

The amount of agitation-induced lipolysis depends on the nature and extent of fat-globule damage. The extent of damage has been estimated by the amount of free fat in the milk,[243] the amount of fat in the skim milk[244,245] and by the amount of milk fat globule associated enzymes (e.g. alkaline phosphatase or xanthine oxidase) released from the fat globules.[246] While good correlations are observed between these parameters and the amount of induced lipolysis for agitation at a given temperature,[244-6] poor correlations are found for treatments at different temperatures. This can be explained in terms of the agitation causing either aggregation or dispersion of the milk fat globules, with the predominant effect being determined by the temperature of the milk and the severity of agitation.[247]

Besides its effect on the integrity of the milk fat globule, agitation causes a redistribution of lipase between the skim-milk and cream phases.[53,231,248,249] Agitation of milk at 5–10°C or 37°C results in a several-fold increase in lipase activity associated with the cream.[231] The amount of lipase transferred to the cream is not reflected in the extent of lipolysis in whole milk, but it is in the lipolysis in cream separated from activated milk. Since the transferred enzyme is bound to the milk fat globule membrane and in this form has enhanced heat stability,[53] the amount of redistribution is of particular relevance in butter manufacture.

The realization that the introduction of mechanized milking and bulk handling and storage of milk was accompanied by increased problems of hydrolytic rancidity prompted investigations of the relationship between the use of milking machines and lipolysis which have extended over the last 30 years. The results of these studies have been previously reviewed.[250,251] Agitation activation normally results from faulty design and installation and inadequate maintenance of milking machines, and is associated with excessive air intake into the system, causing turbulence and frothing.[238,239,252] Considerable scope for activation exists where pipeline milkers are used compared with hand milking or bucket milkers.[253-7] Features such as elbows, joints, in-line fittings, long and narrow pipes and vertical risers in the milking line are mainly responsible for the turbulence.[238,241,258,259] High-line milking machines have come under considerable scrutiny.[73,260]

Bulk milk tanks are rarely implicated in milk lipase activation, although excessive agitation has been reported to increase lipolysis.[252,261-3] Similarly, road tankers have been found to cause little activation.[262] Continuous pumping, particularly with aeration, causes milk fat globule damage and subsequent lipolysis to an extent dependent on the type of pump.[241,260,264,265] Factory separation of cream, where the cream is partially homogenized as it leaves the separator bowl at relatively high pressure, can also promote lipolysis.[229,260] Air agitation of milk in factory silos can cause activation if the air flow rate is excessive or if agitation is used continuously rather than intermittently.[17]

The extent of lipolysis in raw milk or cream following activation is determined by the temperature and duration of storage. Rapid cooling to and storage at low temperatures (without freezing) minimizes lipolysis.[266,267] The rate of lipolysis falls off with time but can be accelerated by further activation treatments.[227,232]

3.1.2. Homogenization

Homogenization of raw milk or cream results in a very strong activation of lipolysis. Milk may become perceptibly rancid within 5 min of treatment.[229] Homogenization produces a large surface area of vulnerable milk fat and permits ready access of the milk lipase. The activity of the enzyme *per se* is not increased by homogenization.[268] Lipolysis in homogenized milk is related to the pressure, time and temperature of homogenization, i.e. to the efficiency of milk fat dispersal.[268,269] The rate of lipolysis in raw milk is greatest immediately after homogenization, then levels off. A second or third homogenization again promotes rapid lipolysis.[269] The slowing and revival of lipolysis is attributable to the accumulation and dissipation of lipolysis products at the interface.[227,269]

Lipolysis proceeds readily when pasteurized homogenized milk is admixed with raw, non-homogenized milk. Here, the amount of lipolysis depends on the ratio of the susceptible substrate and the amount of lipase, and is maximal for an approximately 50/50 mixture.[269,270] This phenomenon is particularly important in the dairy factory since any recirculation of pasteurized homogenized milk back into the raw milk during start-up or close-down of processing plants can cause considerable lipolysis. In commercial practice, the homogenizer is placed immediately before or directly after the pasteurizer so that the milk lipase is heat inactivated before it can cause lipolysis in the homogenized milk.

3.1.3. Temperature Activation

Lipolysis may be induced when fresh milk or cream is subjected to a specific sequence of temperature changes. The optimal amount of 'temperature activation' is promoted by cooling to ≤5°C, warming to 25–35°C, followed by re-cooling to <10°C.[10] Lipolysis proceeds on storage at this low temperature. Re-warming reverses the activation. Milks from individual cows vary widely in their susceptibility to temperature activation, the milks from some cows and even some herds being completely resistant to it. The reasons for the variability are not clear, although it appears that milks susceptible to spontaneous lipolysis are more likely to be affected.[234]

Susceptibility is a property of the lipid phase,[234,271] as has been demonstrated by exchange experiments with cream and skim milk from susceptible and non-susceptible milks. Pasteurized cream can also be activated, and undergoes lipolysis on subsequent mixing with raw (cold) skim milk.[234] The activation treatment facilitates the attachment of the lipase to the surface of the milk fat globule. In contrast to the lipase–milk fat globule membrane interactions which occur in agitation-induced or spontaneous lipolysis, this attachment is reversed by re-warming, but can re-form if the milk is again cooled.[10,234]

In practice, temperature activation can occur if a small amount of cooled milk or cream is topped up with a larger amount of warm milk and then refrigerated.[272,273] Separation of previously cooled milk at temperatures around 30°C can lead to lipolysis if the cream is held in cold storage before pasteurization.

3.1.4. Freezing

Freezing and thawing of milk, which leads to churning of the fat,[229] may induce lipolysis.[274] Slow freezing and repeated freeze–thawing are most effective, but the amount of lipolysis which ensues is less than that promoted by moderate agitation. Freezing of milk in farm bulk vats is the most likely source of this activation.

3.2. Spontaneous Lipolysis

3.2.1. Characteristics

Milk which undergoes spontaneous lipolysis has been referred to as 'naturally active', 'susceptible', 'spontaneously lipolytic' or 'spontaneous', in contrast to 'normal' milk in which no such lipolysis occurs.[11] Spontaneous milk can be produced by most, if not all, cows but because of the individuality of cows in their response to a variety of factors only a percentage of cows in a herd give such milk at any one time. Percentages between 3 and 35 have been reported.[275–7]

Spontaneous lipolysis is defined here as the lipolysis which occurs in some individual milks when cooled to below *c.* 15°C soon after milking. The sooner the milk is cooled, and the lower the temperature to which it is cooled (without freezing), the more lipolysis occurs.[278] If cooling is delayed, the amount of lipolysis is decreased.[5,262,266] Once the milk is cooled, spontaneous lipolysis proceeds at that temperature but its rate increases if the temperature is raised.[278] As with induced lipolysis, the rate of spontaneous lipolysis is initially high but levels off thereafter. FFA levels of up to *c.* 10 meq./litre^{-1} can be attained (in extreme cases) after 24-h storage at 5°C.

A characteristic of spontaneous lipolysis is its inhibition by normal milk. Mixing normal milk and spontaneous milk, before cooling, in the ratio of about 3:1 was reported by Tarassuk and Henderson[279] to prevent lipolysis. The ratio of mixing to cause complete inhibition depends on the properties of both milks. Milks which are highly susceptible require a high ratio of normal to spontaneous milk to prevent lipolysis.[48] Furthermore, admixing of two spontaneous milks can result in more lipolysis than in the two milks incubated separately. This phenomenon can be explained in terms of the lipase activities in the two milks and the relative levels of activators and inhibitors present (see Section 3.2.3).[48] Because of the usual predominance of normal milk over spontaneous milk, lipolysis in herd bulk milks is less than it would be otherwise. Higher proportions of spontaneous milk can result in high lipolysis levels in bulked milks.

3.2.2. Factors Affecting Spontaneous Lipolysis

3.2.2.1. Stage of lactation. The variability in propensity to produce spontaneous milk applies to cows within a herd, cows in different herds, cows in different stages of lactation and even the same cows from day to day and from lactation to lactation.[280,281] Stage of lactation is one of the most important factors responsible for this variability, with cows in late lactation having the greatest tendency to produce spontaneous milk.[240,259,281–9] Milk from cows in any stage of lactation can, however, be susceptible.[280] One of the first flavour defects to be linked with lipolysis was known as 'bitter milk of late lactation' and was described in 1913 (cited in Reference 7). Most workers have found the interval since calving, i.e. length of lactation, to be more important than stage of pregnancy—long lactation being particularly conducive to the production of spontaneous milk.[238,275,281,290,291] However, a few reports have suggested that the gestation stage is most closely related to the incidence of spontaneous lipolysis.[292,293]

3.2.2.2. Feed and nutrition. Both the quality and quantity of feed have an influence on the tendency of a cow to produce spontaneous milk.[281,294] The milk from most cows on a low plane of nutrition has an enhanced susceptibility.[295,296] The effect is particularly marked when the cows are in late lactation.[297] The cow's body condition has not been found to be a reliable indicator of the susceptibility of the milk.[240]

Cows fed dry rations, particularly hay[298] and high-carbohydrate winter feeds,[253,291] are more likely to produce susceptible milk than cows fed green pasture.[281–3,299,300] It has been reported that silage feeding can result in serious spontaneous lipolysis problems.[301,302] Where changeover feeding trials have been conducted, lag times of 4–5 days[297,299] have been observed for corresponding changes in the susceptibility of the milk.

Recently, fat supplements have been fed to cows to increase milk fat yields. Astrup *et al.*[296] observed that supplementing feed with 6% palmitic or myristic acid significantly enhanced spontaneous lipolysis but stearic acid and fatty acids with chain lengths shorter than myristic acid had no effect. Feeding rapeseed oil to underfed cows has been found to reduce the susceptibility of their milk to lipolysis.[296] Protected soybean oil[303] and safflower oil[304] supplements have also been found to cause reduced rancidity in milk. An in-depth review of the effect of feed and nutrition was prepared by Jellema.[294]

3.2.2.3. Season. There have been many reports of seasonal variations in the incidence of spontaneous lipolysis. Most, but not all,[276] have suggested a higher incidence in the coldest months of the year.[5,282,291] However, it appears that the season *per se* is not the determining factor but rather the stage of lactation of the majority of cows[259,266,276,285] and/or the type and availability of feed. The coincidence of the two factors, late lactation and poor feed, is particularly conducive to the production of milk with an enhanced susceptibility to lipolysis, both spontaneous and induced.

3.2.2.4. Milk production. In general, low-yielding cows are more likely to produce spontaneous milk than are high-yielding ones.[73,240,276,305] The influence of production level, like that of season, is usually related to poor feeding and/or advanced lactation, but it is interesting to note that milk obtained at the evening milking is more susceptible than morning milk.[73,255] This is attributed to the shorter intermilking interval prior to the evening collection, which results in lower milk production.[277,289,306]

3.2.2.5. Other factors. Several workers have reported that a cow's hormonal balance can affect the susceptibility of her milk to lipolysis.[224,281,307,308] Bachmann[293] found that the milk from cows with ovarian cysts and from those that had received oestrogen or oestrogen-like hormones tended to produce spontaneously active milk. The oestrus cycle appears to have little effect on spontaneous lipolysis[281] but it may affect the lipase activity in the milk.[309]

Research in The Netherlands has indicated that the type of milking machine system used can affect the physiology of the cow's udder and thereby influence the susceptibility of the milk to lipolysis.[224] Cows milked with a one-line system produced milk which was more susceptible than milk produced by the same cows when milked with a two-line system.

3.2.3. Biochemical Aspects

A priori one might expect the degree of spontaneous lipolysis in a milk to be determined largely by the amount of lipase present. However, both low[51,72,90,236] and high[48,310] correlations between lipase activity and the level of spontaneous lipolysis have been reported. It appears that all raw milks have sufficient lipase activity to cause considerable spontaneous lipolysis,[227] but that this only occurs when other conditions, as outlined below, are conducive to such lipolysis.[48] In addition to the total lipase activity of a milk, the accessibility of the enzyme may also be important. Conditions which favour increased levels of soluble enzyme facilitate access of the substrate to the enzyme and promote lipolysis.[47,62]

The susceptibility of the milk fat globule and the permeability of the milk fat globule membrane have also been considered to be important in spontaneous lipolysis.[48,236,278,293] However, investigations involving intermixing of skim milks and creams from normal and spontaneous milks have indicated that the extent of lipolysis is more dependent on the skim-milk portion than on the cream.[109,234] The state of the milk fat globule membrane may be most significant in late-lactation milk,[48,293] where the amount of phospholipid is believed to be insufficient for the formation of the membrane material required to cover the increased (at least twice that of mid-lactation milks) surface area of the fat globules.[311] A further indication of the importance of the milk fat globule membrane is the increased susceptibility of spontaneous and late-lactation milks to induced lipolysis.[73,237-40]

During spontaneous lipolysis, the milk lipase attaches to the milk fat globule membrane.[11,48,278] The interaction occurs when the milk is cooled and is not disrupted if the milk is re-warmed. The formation of the enzyme–membrane complex appears to be an essential feature of spontaneous lipolysis since little or no lipolysis occurs if the interaction is prevented, e.g. by delayed cooling or addition of NaCl.[48] The role of the milk fat globule membrane may be a rather complex regulatory one since it contains both activating lipoproteins[72] and inhibiting glycoproteins.[75]

Following the discovery that the major lipolytic enzyme in milk was a lipoprotein lipase[20] and that lipolysis could be initiated in any milk by the addition of blood serum or serum lipoproteins,[312] it was suggested that spontaneous milk contained activating cofactors derived from the blood.[90] Some considered that these cofactors might be small (MW c. 10 000) soluble apo-LPs which could leak into the milk from the blood during periods of stress.[42] The blood lipoproteins are much larger and would be less likely to transfer to the milk. However, the work of Clegg[91] and Driessen and Stadhouders[90] indicates that the lipid component is essential for activating lipolysis. Milk has been shown to contain material which is immunologically cross-reactive with bovine serum lipoproteins.[72,313]

The presence of an inhibiting factor (or factors) in milk has been suggested to explain the lack of lipolysis in normal milk and the inhibition of lipolysis when normal milk is mixed with spontaneous milk.[48,236] It has been demonstrated that milk serum from normal milk contains such a factor and that it is heat stable and dialyzable.[48] A proteose-peptone fraction with these properties has been shown to be an effective non-competitive inhibitor of lipolysis in milk[74] but it is not known if this is the only inhibiting factor. The inhibitor prevents lipolysis by blocking the lipase–milk fat globule membrane interaction[48] but does not affect lipolysis of emulsified triglycerides.[91]

Thus four factors have been shown to contribute to the susceptibility of a milk to spontaneous lipolysis: lipase activity, milk fat globule vulnerability, activating factors and inhibiting factors. Since lipolysis can occur with any natural level of lipase activity and with any milk fat globules, it appears that the potential of milk to undergo spontaneous lipolysis is largely determined by the balance between activating and inhibiting substances in the skim milk.

3.3. Mastitis

Mastitis is often considered to be a cause of spontaneous lipolysis[227] because many mastitic milks have elevated levels of FFAs.[138,314–18] The tendency to rancidity increases with increasing somatic cell count. However, many milks which can be classified as mastitic (cell count $>0.5 \times 10^6$ ml^{-1} [319]) do not contain significantly more FFAs than milk from corresponding healthy quarters. In one study, less than 50% of milks with up to $c.\ 3 \times 10^6$ cells ml^{-1} had significantly elevated FFA levels.[138] In contrast, when mastitis is experimentally induced by intramammary infusion of endotoxins or bacteria, increases in the levels of milk FFAs closely correspond to the rises in cell count and other mastitis indices.[138,318,320]

A large proportion of the FFAs in mastitic milk is present when the milk is secreted from the udder, i.e. mastitic milks have high initial FFA levels,[138,315] and, although they show greater increases on storage than normal milk from healthy quarters,[138] the increases are small compared with those observed in spontaneous lipolysis. For these reasons, mastitis is not considered to be a cause of spontaneous lipolysis as defined here.

Surveys of mastitic milks have shown that they usually have lower milk lipase activities than milks from corresponding healthy quarters,[129,138,321,322] probably as a result of the reduced biosynthetic capability of the mammary gland tissue damaged by the infection and inflammation. The lipase in mastitic milk has a different distribution from that in other milk, with a greater proportion being in non-micellar form, either soluble[318] or associated with the cream.[293]

The leucocytes in milk contain a lipase[33] which may be responsible for some of the lipolysis in mastitic milk.[323] When suspensions of these cells are added to milk, lipolysis occurs[44] to produce an increase in FFAs which is almost linear with cell count up to $c.\ 2\times 10^6\ ml^{-1}$. More lipolysis is observed if the cells are disrupted prior to their addition to milk.[323] This lipolysis may be caused directly by the cell lipase activity or indirectly through an effect on the milk lipase system.[44]

3.4. Microbial Lipolysis

Hydrolytic rancidity can be produced in milk and milk products as a result of pre- or post-manufacture contamination by micro-organisms.[17,154] The introduction of bulk cold storage has led to the emergence of psychrotrophic (mainly Gram-negative) bacteria as dominant organisms in raw milk and cream.[324,325] Storage of milk at the farm and factory may extend its age at processing to several days, and allow sufficient growth for spoilage, including lipolysis, to occur. Only a proportion of psychrotrophs (between 0·1 and $c.\ 30\%$,[326,327] depending on the composition of the sample's flora) produces appreciable amounts of lipolytic enzymes, so that there is not necessarily a good correlation between psychrotroph numbers and FFA or rancid flavour.[328]

In cultures of single lipolytic psychrotrophs, a level of approximately 10^6 [329] or $10^7\ ml^{-1}$ [330] is attained before spoilage is apparent. Muir et al.[328] only observed lipolysis in stored farm or factory milks with psychrotroph counts $>5\times 10^6\ ml^{-1}$. Taints may be evident after 4–5 days at $5°C$[331] or after shorter times at a higher temperature.[328]

Heat-stable bacterial lipases have been implicated in the development of rancidity during storage of UHT milk,[166] butter[332] and cheese (Cheddar,[145,333] Dutch,[334] Swiss[335] and Camembert[336]) made from milk containing high levels of psychrotrophs (usually $>10^7\ ml^{-1}$) before UHT or HTST treatment. However, Adams and Brawley[158] reported that UHT milk with counts of 10^4–$10^5\ ml^{-1}$ before processing lipolyzed on storage at 25 or 40°C. The lipases have been found to be concentrated in the cream on separation and in the curd on coagulation.[337,338]

Post-pasteurization contamination is a common cause of microbial lipolysis in milk products.[325] Psychrotrophic yeasts, moulds and bacteria can cause rancidity and surface spoilage in butter,[339] a problem less significant now than previously due to improved factory methods and hygiene. Development of flavour defects in pasteurized milk and cream is associated with the presence of high numbers (e.g. $>10^7\ ml^{-1}$) of psychrotrophs.[147,340,341]

4. DETRIMENTAL EFFECTS OF LIPOLYSIS IN MILK AND MILK PRODUCTS

4.1. Flavour Defects

4.1.1. Milk and Cream

The relationship between the flavour of milk and its FFA content has been examined by numerous workers, and threshold FFA levels for the detection of rancid flavour have been determined. Many of these studies have been reviewed.[225,342,343] The threshold values reported show considerable variation, as the following list testifies: 0·8,[344] 1·25,[345] 1·3,[346] 1·2–1·5,[315] 1·3–1·6,[347] 1·5,[18,238,258,276] 1·5–2·0,[348] 1·8,[349] 1·85–2·05,[350] 2·0,[342] 2·0–2·2,[315] 2·74,[351] meq./100 g fat and 0·7,[349] 1·0,[224] 2·0,[223,226] meq. litre^{-1}. The wide range can be largely attributed to the variation in methods used to obtain the thresholds. Individual people differ considerably in their ability to detect rancid flavour according to their natural flavour perceptions and their degree of training in tasting. This is particularly evident in studies in which taste panellists have been used (e.g. References 225 and 258). The thresholds given by Tallamy and Randolph[315] appear to be reasonable and generally acceptable, i.e. 1·2–1·5 meq./100 g fat for trained experts and 2·0–2·2 meq./100 g fat for the average consumer. Rancidity is detected at approximately the same levels in cream.[352] Similar numerical levels apply for milk where FFA content is measured by an extraction–titration method and expressed in meq.litre^{-1} milk.[226]

The rancid flavour is caused mainly by the fatty acids of chain lengths C-4 to C-12,[353-5] although some of the bitterness may be due to the partial glycerides.[18] The long-chain acids C-14 to C-18 make little contribution to flavour. Al-Shabibi et al.[354] concluded that although the C-4 to C-8 acids constitute part of the rancid flavour, the C-10 and C-12 acids are responsible for most of the unclean, bitter, soapy flavour of lipolyzed milk.

The theoretical concentrations of the C-4–C-12 fatty acids present in milk with a total FFA content of 2 meq.litre^{-1} are shown in Table 1, together with the range for rancid milks found by Kintner and Day[351] and the threshold levels for these individual acids obtained by Scanlan et al.[355] It is clear that the levels of the individual acids in rancid milk can be considerably lower than the threshold values reported by Scanlan et al.,[355] which do not include the levels contributed by the fresh milk, i.e. approximately half the amounts shown in the second column. Thus the flavour of the combination of the acids in rancid milk is apparently sufficient to exceed the threshold.

TABLE 1
C-4 TO C-12 FFA LEVELS IN RANCID MILK

Fatty acid	Theoretical[a] for milk with 2 meq. litre^{-1}	Found[351] for range of rancid[b] milks	Threshold[355] levels of added acids[c]
C-4	26·6	27·5–85·0	46·1
C-6	15·1	16·0–48·7	30·4
C-8	9·1	8·3–27·9	22·5
C-10	17·6	27·6–78·6	28·1
C-12	17·6	26·7–63·3	29·7

[a] Assuming average MW = 228 and FFA profile reported by Day.[356]
[b] 'Threshold rancidity to extremely rancid'.
[c] Amounts added to fresh, pasteurized, homogenized milk to produce a rancid flavour.

In milk, most of the acids are in the salt form and have much less flavour than if they were all in the acid form.[343] In fact, acidification of milk greatly enhances the sensitivity of organoleptic detection of lipolysis in milk.[351,357] The detection of rancidity is reduced by the association of the FFAs with certain proteins in milk[358,359] and by heating of milk.[351] The phase with which the fatty acids associate also influences their flavour threshold, since the short-chain acids have much lower thresholds in fat than in water, while the opposite applies to the long-chain acids.[360]

4.1.2. Butter
Hydrolytic rancidity in butter is characterized by off-flavours variously described as 'bitter', 'unclean', 'wintery', 'butyric', 'rancid' or 'lipase'.[361–3] The defect is sometimes evident at manufacture,[364,365] but may develop during storage.[361,363,366,367]

The FFA concentration in butter is usually expressed as its acid degree value (ADV) in meq./100 g fat (or mg NaOH/100 g fat,[366] 1 meq. = 40 mg NaOH). As in the case of milk, numerous workers have endeavoured to correlate FFA levels with flavour in order to arrive at a threshold for rancidity. In many studies, little or no correlation has been observed between ADV and 'lipase' defects.[225,361,363,364,368,369] However, some investigators have reported thresholds ranging from 0·75 to 2·5 meq./100 g fat.[227,361,370,371] A threshold value of 1·5 meq./100 g fat appears to be a

TABLE 2
C-4 TO C-12 FFA LEVELS IN BUTTER

Fatty acid	FFA level (mg kg^{-1})		Theoretical[b] increases for $\Delta ADV = 0.1$ meq./100 g fat
	Flavour thresholds[a]		
	Added singly	Added in pairs	
C-4	11·4	4– 7	11·3
C-6	51·5	16– 26	6·4
C-8	454·6	148–367	3·9
C-10	161·6	132–173	7·4
C-12	127·9	70– 84	7·4

[a] For acids added to good quality butter.[376]
[b] Based on the FFA profile of rancid milk[356] and assuming average MW of 228 and 82% fat in butter.

realistic guide for the butter manufacturer.[372,373] Besides lipolytic defects, butter with ADV >1·5 is likely to have other defects, e.g. oxidized flavours, and to deteriorate on storage.[373] There is some evidence that free fatty acids oxidize more readily than esterified acids[374] and hence may predispose butter to oxidative rancidity.[17,375]

The difficulty in relating rancid flavours in butter to FFA content arises because the short-chain acids, C-4 and C-6, which are the most flavoursome,[376] are water soluble and hence are mostly lost in the buttermilk and wash water during manufacture. For this reason, even butters made from cream with ADV as high as 2·4 may show little defect.[373,377] On the other hand, butters with quite low ADV can be rancid,[377] particularly if lipolysis occurs after manufacture.[373]

The flavour thresholds for the individual fatty acids are quite different in butter than in milk. Whereas in milk C-10 and C-12 acids are most significant in rancid flavour, in butter C-4 and C-6 are of most interest since they have much lower flavour thresholds in fat than do the longer-chain acids.[360] The reported thresholds[376] of the C-4–C-12 acids added singly and in pairs to butter are shown in Table 2 together with the theoretical amounts for an increase in ADV of 0·1 meq./100 g fat. From these data, it is evident that a small amount of lipolysis in butter produces sufficient butyric acid to exceed its flavour threshold and impart a rancid flavour. Thus measurement of C-4 (and C-6) may give the best indication of hydrolytic rancidity in butter.

The above discussion applies to sweet-cream butter only. Little information is available on cultured butter, but O'Connell et al.[366] found that ripened cream butter was less prone to the development of hydrolytic rancidity than was the corresponding salted or unsalted sweet-cream butter.

4.1.3. Cheese
The typical flavour of Cheddar cheese is due to the combination of a variety of flavour compounds including FFAs.[378] When excessive lipolysis occurs in cheese, or in the cheesemilk prior to manufacture, this balance is upset and rancid flavours result. Because of the high total flavour of cheese, the threshold levels of FFAs are higher than for milk or butter. ADVs of 2·8–3·0 are usually attained before rancidity is evident.[379–81] Various studies have shown rancid Cheddar to have from 2 to 10 times the FFA content of good quality cheese.[145,382,383]

For some non-Cheddar varieties, Kuzdzal-Savoie[342] concluded that measures of caproic acid (C-6) were the best guide to the lipolysis status of a cheese. She determined the following caproic acid levels (in mg/100 g dry matter) beyond which organoleptic defects appeared: 8 for Emmental, 14 for Gruyère de Comté, and 20–25 for Camembert. In some highly flavoured cheeses, very high FFA contents are acceptable, e.g. >66 000 mg kg^{-1} for blue cheese (cf. <4000 for good Cheddar[382]) and for Greek Feta up to 520 mg kg^{-1} of butyric acid[384] (cf. <43[385] or 71–207[382] for Cheddar).

4.2. Technological Consequences
In addition to the flavour aspects of lipolysis in milk products, some technological consequences are of interest.

Difficulties have been experienced in churning cream from spontaneously rancid milk.[7,386] The cream foams excessively and may take up to five-times as long as normal cream to churn. Fat losses in the buttermilk are similar to those for normal milks.[282] However, milk in which lipolysis is induced by agitation at warm temperatures gives high fat losses on separation[244,245,247] due to the partial homogenization of the fat.

The use of lipolyzed milk may slow the manufacture of fermented milk products because of inhibition of starter bacteria by FFAs.[387–9] The amount of inhibition varies with the strain of starter used.[387,390] *Streptococcus lactis* and *S. cremoris* are inhibited by rancid milk.[391–4] Caprylic, capric and lauric are the fatty acids most effective in slowing starter growth and acid production.[390,392,393] The partial glycerides produced by lipolysis may also be involved in the inhibition.[390,394] There are conflicting reports on the ease of manufacture and quality of cheese made from milk containing high levels of psychrotrophs.[325] Proteolysis by the bacterial proteases appears to negate the inhibitory effects of the products of lipase action.[145] The use of rancid milk can also cause weak or delayed rennet clotting[299,395] and decreased yield in cheese manufacture.[396,397]

The determination of the fat content of milk is influenced by the level of lipolysis, to an extent dependent on the assay method used.[398] Only one-third of the FFAs are assayed as fat by the Röse–Gottlieb method while 90% are recovered by the Gerber procedure.[399] The lowering of Milkotester results exceeds the actual loss ascribable to lipolysis because of the surface tension effects produced by the lipolysis products.[400] An FFA increase of 1 meq./100 g fat, corresponding theoretically to c. 0·01% fat in a 4% fat milk, reduces a Röse-Gottlieb fat estimate by 0·01 and Milkotester and IR tests by 0·03%.[400,401]

Lipolysis in milk causes a reduction in surface tension[352,391] and, as a consequence, lipolyzed milk has a low foaming capacity.[402] This is particularly noticeable during the steam frothing of milk used for making Cappuccino-style coffee, where stable foam formation is essential.[262,403] Milks with FFA levels greater than 1·5 meq. litre^{-1} usually have poor foaming properties, while those with levels greater than 2 meq. litre^{-1} exhibit negligible steam frothing.[404] Commercial milk-processing operations, pasteurization and homogenization, markedly enhance the steam frothing capacity of milk. The reduction in surface tension and the consequent detrimental effect on foaming caused by lipolysis is believed to be due to the partial glycerides, particularly the monoglycerides, formed in the lipolysis reaction.[402,403]

5. BENEFICIAL EFFECTS OF LIPOLYSIS IN MILK AND MILK PRODUCTS

5.1. Desirable Flavour Production

Lipolysis plays an important role in providing the characteristic flavour of many milk products. In particular, the ripening of most cheese varieties is accompanied by lipolysis due to micro-organisms or to added enzyme preparations, and, in raw milk cheese, to the milk lipase. Fat breakdown is not extensive, but is more pronounced in some cheeses, e.g. blue-veined and Italian varieties, than in others. It has been estimated that <1% of the fat is degraded during Cheddar-cheese ripening.[405] Excessive lipolysis renders the cheese unacceptable.[381]

In Cheddar cheese, the role of lipolysis, while considered essential, is still not fully understood.[190,378] Of the fatty acids produced in the cheese during ripening, part of the butyric and most of the higher acids are formed by lipolysis of the milk fat,[382,385] mostly due to lipases from lactic acid bacteria.[183] In Swiss varieties, lipases of propionibacteria may also contribute to FFA production.[406] The Italian cheeses such as Provolone, Romano and Parmesan acquire their characteristic flavour from the action of pregastric esterases, traditionally from 'rennet paste' used to curdle the milk but now in the form of commercially prepared oral gland extracts.[204,407] In internal mould-ripened cheeses, e.g. Gorgonzola, Roquefort and Stilton, the free fatty acids produced by the *Penicillium roqueforti* lipase (and milk lipase when raw milk is used) are important both as flavour agents *per se* and as precursors for the methyl ketones which provide the 'peppery' taste of such cheeses.[408] Lipase preparations (microbial and pre-gastric) have been used to enhance the flavour[195,407,409] and accelerate ripening of cheeses,[410] e.g. Cheddar,[383,411] blue[412] and Feta.[413]

Flavour preparations typical of particular varieties of cheese have been produced with the aid of lipases of appropriate specificities.[414-16] Such flavours are used in processed cheeses, dips and spreads.[195] Controlled lipolysis of milk fat is also used to produce creamy and buttery flavours for bakery and cereal products, confectionery (e.g. milk chocolate, fudge), coffee whiteners and other imitation dairy products.[195,204,409] Several reviews on the application of lipolytic enzymes to flavour development have been published.[195,204,407,409,417]

5.2. Digestion of Milk Fat

Fat absorption in the newborn, and particularly in premature infants, is much less efficient than in the adult due to the marked discrepancy between the high fat intake provided by milk and the relatively low output of lipase and bile salts from the pancreas. Intragastric lipolysis by lingual or salivary lipases, whose secretion is stimulated by suckling, appears to augment the pancreatic system in newborn calves, rats and humans.[418]

Although lipoprotein lipase activity is commonly present in the milk of mammalian species, it does not appear to have any function in the digestion of milk fat by the young animal.[34,42] It is unlikely to be active under, or to survive, the acidic conditions of the stomach.

It has been speculated that the bile-salt-stimulated lipase in human milk has evolved as a supplementary milk fat digestive enzyme.[29,100] Many of its characteristics suit it for such a role. BSSL is stable under the physiological conditions of the stomach and small intestine.[108] At bile salt levels similar to those in the newborn's intestine (c. 2 mM), 20–40% of the fat in human milk is hydrolyzed in 2 h by the BSSL present.[29,419] The products of BSSL lipolysis (mainly FFA and glycerol) may be more readily absorbed at low (<critical micelle concentration) bile salt concentrations than are the monoglycerides and FFA produced by pancreatic lipase.[108]

BSSL has been detected in gastric and intestinal contents of infants following feeding of fresh human milk,[28] and average fat absorption has been found to be higher than after feeding cow's milk formulae or heated human milk.[420] This increase is particularly significant in pre-term infants, or those suffering from pancreatic insufficiency.[421,422] Hamosh et al.[420] found high BSSL levels even in the milk of women who delivered prematurely, and suggested that fat digestion and absorption could be improved by feeding fresh breast milk to very small pre-term infants.

BSSL has a wider substrate specificity than does pancreatic lipase. Fredrikzon et al.[28] found that retinol esters were hydrolyzed three-times as rapidly after feeding fresh human milk as after pasteurized milk ingestion. BSSL might therefore be important in the absorption of retinol and hence in vitamin A nutrition in the infant.

6. ANALYTICAL METHODS FOR FREE FATTY ACID DETERMINATION

The quantitative measurement of all the free fatty acids in milk or milk products presents a difficult analytical problem. The acids range in solubility from water soluble (C-4) to almost completely water insoluble (C-18), and are accompanied by other water-soluble acids, particularly lactic, which it is not desired to measure. The methods which have been used can be divided into three types according to the means of isolating the acids prior to their measurement: (1) fat separation; (2) solvent extraction; and (3) solid phase adsorption. A review of the methods was compiled by Kuzdzal-Savoie.[343] Those used for milk are summarized in Table 3.

In general, methods of type 1 measure only the fat-soluble acids, and hence underestimate the total level.[423,424] The most commonly used procedure is the Bureau of Dairy Industry (BDI)[347] procedure or modifications of it.[425] Results are expressed as acid degree values (ADV) in meq./100 g fat.

The solvent extraction methods of type 2 estimate a high proportion of the total FFAs (c. three-times more than the BDI method at ADV = 1^{426}).

TABLE 3
SUMMARY OF METHODS USED FOR DETERMINING THE FFA CONCENTRATION IN MILK

Type of isolation	Isolation reagent/method	Method of quantification	Reference
Type 1, Fat separation	Detergent demulsification (BDI)	Titration—manual —semi-automatic	347 425
	Churning	Titration	445
Type 2, Solvent extraction	Isopropanol–heptane–H_2SO_4 (Dole)	Manual titration Colorimetric, phenol red—barbital, autoanalyser	423 426, 446
	Alcohol–ether–petroleum ether	Titration	447–9
	Ether–hexane–HCl	Titration	450
	Chloroform–heptane–methanol–HCl	Colorimetric, copper diethyldithiocarbamate	451, 452
	Benzene–K oxalate–H_3PO_4	Colorimetric, Rhodamine B Rhodamine 6G	 453 454
Type 3, Solid phase adsorption	Silica gel–5% butanol in chloroform	Titration	429
	Ion-exchange resin	Gas chromatography	351

However, those in which the higher amounts of butyric acid are extracted and measured are the ones in which there is the greatest probability of lactic acid and other acidic components also being measured.

Theoretically, the methods involving an extraction procedure (liquid–liquid or liquid–solid) in which all of the acids are extracted but only the fatty acids are determined, e.g. by gas–liquid chromatography, should be ideal. However, such methods tend to be tedious and time-consuming and are suitable only as reference methods.[343]

For routine analyses, methods of types 1 and 2 are usually employed. According to the replies received by the International Dairy Federation in response to a questionnaire on lipolysis,[2] many different methods are in use throughout the world. For large numbers of analyses, automated procedures based on the Dole extraction[427] are widely used.

For butter, aliquots of fat obtained either by simply melting the fat[428] or by using a BDI procedure[347] are titrated and the results expressed in meq./100 g fat. For cheese, the BDI method can be performed on a slurry prepared in a sodium citrate solution.[385] Cream can be analyzed directly by an extraction–titration procedure[423] or by the BDI method after being diluted with water (1:9, v/v).

Quantitative determination of the individual FFAs involves an initial step in which the free acids are separated from the fat, and a gas chromatographic step for quantifying the acids. The most commonly used separation methods are based on the silicic acid–KOH procedure of Harper *et al.*[429] (cited in References 377, 430 and 431), the anion exchange method of Hornstein *et al.*[432] (cited in References 376, 382 and 433) and solvent extraction, usually involving soap formation.[434,435] A major problem with the solid phase methods is hydrolysis of the fat whilst in contact with the support, especially those involving strong alkalis.[377,436]

Gas chromatography can be performed on the isolated free acids[377,437–9] or on their ester derivatives.[374,431,434,435,440] Because of the risk of loss of short-chain acids during derivatization, determination of the free acids is preferable.

A high-performance liquid chromatographic method for estimating FFAs in butter as their *p*-bromophenacyl esters has recently been devised.[441] This method is attractive because it is much more sensitive for the lower fatty acids owing to the molar response of the ultra violet detector compared with the weight response for the flame ionization detector of gas chromatographs.

Although the above methods have generally been used for butter or milk fat, most are readily adaptable for other products such as cheese[382,383,442] and yoghurt.[443] In addition, steam distillation methods have been used to isolate the volatile FFAs from cheese prior to quantification by gas–liquid chromatography.[385,444]

7. PREVENTION OF HYDROLYTIC RANCIDITY

Although some aspects of the problem are not fully understood, sufficient knowledge now exists to enable dairy industry personnel to take precautions to ensure a low incidence of hydrolytic rancidity in milk and milk products. Detailed lists of recommended precautions have been published elsewhere[250,428] and are not repeated here. A short list is given below which, in the authors' experience, represents situations in which particular care should be taken.

   (i) Large numbers of cows in late lactation, particularly under poor feed conditions.

- (ii) Air intake at teat cups.
- (iii) Centrifugal pumping, especially of warm milk.
- (iv) Agitation of bulk raw milk, particularly by air.
- (v) Mixing of homogenized and raw milk.
- (vi) Time and temperature of storage.
- (vii) Equipment maintenance and hygiene.

The problem of hydrolytic rancidity can be minimized if constant attention is paid to the risks associated with each of these areas.

REFERENCES

1. Downey, W. K., *Ann. Bull., Int. Dairy Fed.*, 1975, Doc. **86**, ii.
2. International Dairy Federation, Results of Questionnaire 282/A, 1982.
3. Rice, F. E. and Markley, A. L., *J. Dairy Sci.*, 1922, **5**, 64.
4. Roahen, D. C. and Sommer, H. H., *J. Dairy Sci.*, 1940, **23**, 831.
5. Dunkley, W. L., *Can. Dairy Ice Cr. J.*, 1946, **25**(6), 27.
6. Herrington, B. L., *J. Dairy Sci.*, 1954, **37**, 775.
7. Palmer, L. S., *J. Dairy Sci.*, 1922, **5**, 201.
8. Krukovsky, V. N. and Sharp, P. F., *J. Dairy Sci.*, 1938, **21**, 671.
9. Dorner, W. and Widmer, A., *Lait*, 1931, **11**, 545.
10. Krukovsky, V. N. and Herrington, B. L., *J. Dairy Sci.*, 1939, **22**, 137.
11. Tarassuk, N. P. and Frankel, E. N., *J. Dairy Sci.*, 1957, **40**, 418.
12. Downey, W. K. and Andrews, P., *Biochem. J.*, 1965, **94**, 642.
13. Downey, W. K. and Andrews, P., *Biochem. J.*, 1969, **112**, 559.
14. Chandan, R. C. and Shahani, K. M., *J. Dairy Sci.*, 1963, **46**, 275.
15. Gaffney, P. J., Harper, W. J. and Gould, I. A., *J. Dairy Sci.*, 1966, **49**, 921.
16. Fox, P. F. and Tarassuk, N. P., *J. Dairy Sci.*, 1968, **51**, 826.
17. Sjöström, G., *Rep. Milk Dairy Res. Alnarp*, 1959, No. **58**.
18. Jensen, R. G., *J. Dairy Sci.*, 1964, **47**, 210.
19. Schwartz, D. P., In *Fundamentals of Dairy Chemistry*, B. H. Webb, A. H. Johnson and J. A. Alford (eds), 2nd edn., 1975, AVI Publishing Coy. Inc., Westport, p. 220.
20. Castberg, H. B., Egelrud, T., Solberg, P. and Olivecrona, T., *J. Dairy Res.*, 1975, **42**, 255.
21. Downey, W. K., *Ann. Bull., Int. Dairy Fed.*, 1975, Doc. **86**, 80.
22. Olivecrona, T., Egelrud, T., Hernell, O., Castberg, H. B. and Solberg, P., *Ann. Bull., Int. Dairy Fed.*, 1975, Doc. **86**, 61.
23. Egelrud, T. and Olivecrona, T., *J. Biol. Chem.*, 1972, **247**, 6212.
24. Iverius, P.-H. and Östlund-Lindqvist, A.-M., *J. Biol. Chem.*, 1976, **251**, 7791.
25. Kinnunen, P. K. J., Huttunen, J. K. and Ehnholm, C., *Biochim. Biophys. Acta*, 1976, **450**, 342.
26. Kinnunen, P. K. J., *Med. Biol.*, 1977, **55**, 187.
27. Hernell, O. and Olivecrona, T., *J. Lip. Res.*, 1974, **15**, 367.

28. FREDRIKZON, B., HERNELL, O., BLÄCKBERG, L. and OLIVECRONA, T., *Pediat. Res.*, 1978, **12**, 1048.
29. HALL, B. and MULLER, D. P. R., *Pediat. Res.*, 1982, **16**, 251.
30. LAW, B. A., *J. Dairy Res.*, 1979, **46**, 573.
31. ROUT, T. P., *Proc. Aust. Biochem. Soc.*, 1970, **3**, 15.
32. MONTGOMERY, M. W. and FORSTER, T. L., *J. Dairy Sci.*, 1961, **44**, 721.
33. GAFFNEY, P. J. and HARPER, W. J., *J. Dairy Sci.*, 1965, **48**, 613.
34. KORN, E. D., *J. Lip. Res.*, 1962, **3**, 246.
35. KORN, E. D., *J. Biol. Chem.*, 1955, **215**, 15.
36. HERNELL, O., EGELRUD, T. and OLIVECRONA, T., *Biochim. Biophys. Acta*, 1975, **381**, 233.
37. CASTBERG, H. B., EGELRUD, T. and OLIVECRONA, T., *19th Int. Dairy Cong.*, 1974, **1E**, 350.
38. FLYNN, A. and FOX, P. F., *Ir. J. Fd Sci. Technol.*, 1980, **4**, 173.
39. JENSEN, R. G. and PITAS, R. E., *J. Dairy Sci.*, 1976, **59**, 1203.
40. SAITO, Z., KITAYA, E. and OKAZAKI, M., *Bull. Fac. Agric. Hirosaki Univ.*, 1978, **29**, 77.
41. ASKEW, E. W., EMERY, R. S. and THOMAS, J. W., *J. Dairy Sci.*, 1970, **53**, 1415.
42. OLIVECRONA, T., *Bull., Int. Dairy Fed.*, 1980, Doc. **118**, 19.
43. BRUMBY, P. E., *Rep. Natn. Inst. Res. Dairy. (Shinfield)*, 1969–70, 121.
44. SALIH, A. M. A. and ANDERSON, M., *20th Int. Dairy Cong.*, 1978, **E**, 32.
45. ZINDER, O., HAMOSH, M., FLECK, T. R. C. and SCOW, R. O., *Am. J. Physiol.*, 1974, **226**, 744.
46. FOX, P. F., YAGUCHI, M. and TARASSUK, N. P., *J. Dairy Sci.*, 1967, **50**, 307.
47. ANDERSON, M., *J. Dairy Res.*, 1982, **49**, 51.
48. DEETH, H. C. and FITZ-GERALD, C. H., *Ann. Bull., Int. Dairy Fed.*, 1975, Doc. **86**, 24.
49. HOYNES, M. C. T. and DOWNEY, W. K., *Biochem. Soc. Trans.*, 1973, **1**, 256.
50. DOWNEY, W. K. and MURPHY, R. F., *Ann. Bull., Int. Dairy Fed.*, 1975, Doc. **86**, 19.
51. CLEGG, R. A., *Ann. Rep. Hannah Res. Inst.*, 1981, 75.
52. SHAHANI, K. M. and KHAN, I. M., *J. Dairy Sci.*, 1968, **51**, 941.
53. FRANKEL, E. N. and TARASSUK, N. P., *J. Dairy Sci.*, 1959, **42**, 409.
54. RAPP, D. and OLIVECRONA, T., *Eur. J. Biochem.*, 1978, **91**, 379.
55. POSNER, I. and BERMÚDEZ, D., *Acta Cient. Venez.*, 1977, **28**, 277.
56. OLIVECRONA, T. and LINDAHL, U., *Acta Chem. Scand.*, 1969, **23**, 3587.
57. IVERIUS, P.-H., LINDAHL, U., EGELRUD, T. and OLIVECRONA, T., *J. Biol. Chem.*, 1972, **247**, 6610.
58. LEBEDEV, A. B. and UMANSKII, M. S., *Appl. Biochem. Micro.*, 1979, **15**, 560.
59. LUHTALA, A. and ANTILA, M., *Fette Seifen AnstrMittel*, 1968, **70**, 280.
60. DOWNEY, W. K., *19th Int. Dairy Cong.*, 1974, **2**, 323.
61. NILSSON, R. and WILLART, S., *Rep. Milk Dairy Res. Alnarp*, 1961, No. **64**.
62. DOWNEY, W. K. and ANDREWS, P., *Biochem. J.*, 1966, **101**, 651.
63. SHIPE, W. F. and SENYK, G. F., *J. Dairy Sci.*, 1981, **64**, 2146.
64. ÖSTLUND-LINDQVIST, A.-M. and IVERIUS, P.-H., *Biochem. Biophys. Res. Commun.*, 1975, **65**, 1447.
65. OWEN, J. S. and MCINTYRE, N., *Trends Biochem. Sci.*, 1982, **7**, 95.
66. LIM, C. T. and SCANU, A. M., *Artery*, 1976, **2**, 483.

67. CLEGG, R. A., *Biochem. Soc. Trans.*, 1978, **6**, 1207.
68. BROWN, W. V. and BAGINSKY, M. L., *Biochem. Biophys. Res. Commun.*, 1972, **46**, 375.
69. EGELRUD, T. and OLIVECRONA, T., *Biochim. Biophys. Acta*, 1973, **306**, 115.
70. BENGTSSON, G. and OLIVECRONA, T., *Eur. J. Biochem.*, 1980, **106**, 557.
71. SHIRAI, K. and JACKSON, R. L., *J. Biol. Chem.*, 1982, **257**, 1253.
72. CASTBERG, H. B. and SOLBERG, P., *Meieriposten*, 1974, **63**, 961.
73. JELLEMA, A. and SCHIPPER, C. J., *Ann. Bull., Int. Dairy Fed.*, 1975, Doc. **86**, 2.
74. ANDERSON, M., *J. Dairy Res.*, 1981, **48**, 247.
75. SHIMIZU, M., MIYAJI, H. and YAMAUCHI, K., *Agric. Biol. Chem.*, 1982, **46**, 795.
76. WAGH, P. V. and OLIVECRONA, T., *Atherosclerosis*, 1978, **29**, 195.
77. BENGTSSON, G. and OLIVECRONA, T., *Biochem. J.*, 1977, **167**, 109.
78. OLIVECRONA, T., BENGTSSON, G. and RAPP, D., In: *Protides of the Biological Fluids*, H. Peeters (ed.), 1978, Pergamon Press, Oxford, p. 201.
79. OLIVECRONA, T. and EGELRUD, T., In: *Lipid Metabolism, Obesity and Diabetes Mellitus: Impact upon Atherosclerosis*, R. Levine and E. F. Pfeffer (eds), 1974, Georg Thieme, Stuttgart, p. 23.
80. OLIVECRONA, T., BENGTSSON, G., MARKLUND, S.-E., LINDAHL, U. and HÖÖK, M., *Fed. Proc.*, 1977, **36**, 60.
81. BENGTSSON, G. and OLIVECRONA, T., *FEBS Letts.*, 1981, **128**, 9.
82. MATSUOKA, N., SHIRAI, K. and JACKSON, R. L., *Biochim. Biophys. Acta*, 1980, **620**, 308.
83. STOCKS, J. and GALTON, D. J., *Lipids*, 1980, **15**, 186.
84. SCOW, R. D. and EGELRUD, T., *Biochim. Biophys. Acta*, 1976, **431**, 538.
85. PORTMAN, O. W. and ALEXANDER, M., *Biochim. Biophys. Acta*, 1976, **450**, 322.
86. WELTZIEN, H. U., *Biochim. Biophys. Acta*, 1979, **559**, 259.
87. BLÄCKBERG, L., HERNELL, O. and OLIVECRONA, T., *J. Clin. Invest.*, 1981, **67**, 1748.
88. LAROSA, J. C., LEVY, R. I., HERBERT, P., LUX, S. E. and FREDRICKSON, D. S., *Biochem. Biophys. Res. Commun.*, 1970, **41**, 57.
89. BLATON, B., VANDAMME, D. and PEETERS, H., *FEBS Letts.*, 1974, **44**, 185.
90. DRIESSEN, F. M. and STADHOUDERS, J., *Neth. Milk Dairy J.*, 1974, **28**, 130.
91. CLEGG, R. A., *J. Dairy Res.*, 1980, **47**, 61.
92. MORLEY, N. and KUKSIS, A., *Biochim. Biophys. Acta*, 1977, **487**, 332.
93. SOMERHARJU, P., KUUSI, T., PALTAUF, F. and KINNUNEN, P. K. J., *FEBS Letts.*, 1978, **96**, 170.
94. NILSSON-EHLE, P., EGELRUD, T., BELFRAGE, P., OLIVECRONA, T. and BORGSTRÖM, B., *J. Biol. Chem.*, 1973, **248**, 6734.
95. HERNELL, O., GEBRE-MEDHIN, M. and OLIVECRONA, T., *Am. J. Clin. Nutr.*, 1977, **30**, 508.
96. TARASSUK, N. P., NICKERSON, T. A. and YAGUCHI, M., *Nature*, 1964, **201**, 298.
97. CASTBERG, H. B. and HERNELL, O., *Milchwissenschaft*, 1975, **30**, 721.
98. WANG, C.-S., ALAUPOVIC, P. and BASS, H. B., *Fed. Proc.*, 1979, **38**, 334.
99. WANG, C.-S., KUKSIS, A. and MANGANARO, F., *Lipids*, 1982, **17**, 278.
100. BLÄCKBERG, L., HERNELL, O., OLIVECRONA, T., DOMELLÖF, L. and MALINOV, M. R., *FEBS Letts.*, 1980, **112**, 51.

101. Freudenberg, E., *Experientia*, 1966, **22**, 317.
102. Bläckberg, L., Lombardo, D., Hernell, O., Guy, O. and Olivecrona, T., *FEBS Letts.*, 1981, **136**, 284.
103. Bläckberg, L. and Hernell, O., *Eur. J. Biochem.*, 1981, **116**, 221.
104. Wang, C.-S., *Anal. Biochem.*, 1980, **105**, 398.
105. Wang, C.-S., *J. Biol. Chem.*, 1981, **256**, 10198.
106. Jubelin, J. and Boyer, J., *Eur. J. Clin. Invest.*, 1972, **2**, 417.
107. Hernell, O. and Olivecrona, T., *Biochim. Biophys. Acta*, 1974, **369**, 234.
108. Hernell, O., *Eur. J. Clin. Invest.*, 1975, **5**, 267.
109. Murphy, J. J., Connolly, J. F. and Headon, D. R., *Ir. J. Fd Sci. Technol.*, 1979, **3**, 131.
110. Chandan, R. C., Parry, R. M. and Shahani, K. M., *J. Dairy Sci.*, 1968, **51**, 606.
111. McBride, O. W. and Korn, E. D., *J. Lip. Res.*, 1963, **4**, 17.
112. Chilliard, Y. and Fehr, P. M., *Ann. Technol. Agric.*, 1976, **25**, 219.
113. Hamosh, M. and Scow, R. O., *Biochim. Biophys. Acta*, 1971, **231**, 283.
114. Bjørke, K. and Castberg, H. B., *Nord-Eur. Mejeri-Tidsskr.*, 1976, **8**, 296.
115. Chilliard, Y. and Fehr, P. M., *Lait*, 1978, **58**, 1.
116. Rifaat, I. D., El-Sadek, G. M., Abdel-Aal, A. T. and Ismail, A. A., *Egypt. J. Dairy Sci.*, 1974, **2**, 137.
117. Ismail, A. A., Rifaat, I. D., Abdel-Aal, A. T. and El-Sadek, G. M., *Egypt. J. Dairy Sci.*, 1975, **3**, 6.
118. Sammanwar, R. D. and Ganguli, N. C., *19th Int. Dairy Cong.*, 1974, **1E**, 349.
119. Hofi, A. A., Mahran, G. A., Abdel-Hamid, L. B. and Osman, S. G., *Egypt. J. Dairy Sci.*, 1976, **4**, 111.
120. Abdel-Hamid, L. B., El Sherif, R., Osman, S. G. and Mahran, G. A., *Egypt. J. Dairy Sci.*, 1977, **5**, 175.
121. Abdel-Hamid, L. B., Mahran, G. A., Shehata, A. E. and Osman, S. G., *Egypt. J. Dairy Sci.*, 1977, **5**, 7.
122. Ribeiro, L. P., *Biochimie*, 1971, **53**, 865.
123. Wallinder, L., Bengtsson, G. and Olivecrona, T., *Biochim. Biophys. Acta*, 1982, **711**, 107.
124. DeFeo, A. A., Dimick, P. S. and Kilara, A., *J. Dairy Sci.*, 1982, **65**, 2308.
125. Desnuelle, P., *Bull. Soc. Chim. Biol.*, 1951, **33**, 909.
126. Nachlas, M. M. and Seligman, A. M., *J. Biol. Chem.*, 1949, **181**, 343.
127. Holmes, R. S. and Masters, C. J., *Biochim. Biophys. Acta*, 1967, **132**, 379.
128. Forster, T. L., Bendixen, H. A. and Montgomery, M. W., *J. Dairy Sci.*, 1959, **42**, 1903.
129. Forster, T. L., Montgomery, M. W. and Montoure, J. E., *J. Dairy Sci.*, 1961, **44**, 1420.
130. Murphy, R. F. and Downey, W. K., *18th Int. Dairy Cong.*, 1970, **1E**, 604.
131. Kitchen, B. J., *J. Dairy Res.*, 1971, **38**, 171.
132. Nakanishi, T. and Tagata, Y., *Jap. J. Dairy Sci.*, 1972, **21**(6), A-207.
133. Deeth, H. C., *20th Int. Dairy Cong.*, 1978, **E**, 364.
134. Purr, A., Mathies, P. and Kotter, L., *Dairy Sci. Abstr.*, 1969, **31**, 404.
135. Marquardt, R. R. and Forster, T. L., *J. Dairy Sci.*, 1965, **48**, 1526.
136. Luedecke, L. O., *J. Dairy Sci.*, 1964, **47**, 696.
137. Kitchen, B. J., MSc Thesis 1971, University of Queensland.

138. FITZ-GERALD, C. H., DEETH, H. C. and KITCHEN, B. J., *J. Dairy Res.*, 1981, **48**, 253.
139. DRIESSEN, F. M., *Neth. Milk Dairy J.*, 1976, **30**, 186.
140. ARBABI, P., *20th Int. Dairy Cong.*, 1978, **E**, 319.
141. MARQUARDT, R. R. and FORSTER, T. L., *J. Dairy Sci.*, 1962, **45**, 653.
142. COGAN, T. M., *Ir. J. Fd Sci. Technol.*, 1977, **1**, 95.
143. VON BOCKELMANN, I., *18th Int. Dairy Cong.*, 1970, **1E**, 106.
144. STEWART, D. B., MURRAY, J. G. and NEILL, S. D., *Ann. Bull., Int. Dairy Fed.*, 1975, Doc. **86**, 38.
145. LAW, B. A., SHARPE, M. E. and CHAPMAN, H. R., *J. Dairy Res.*, 1976, **43**, 459.
146. MUIR, D. D., PHILLIPS, J. D. and DALGLEISH, D. G., *J. Soc. Dairy Technol.*, 1979, **32**, 19.
147. PHILLIPS, J. D., GRIFFITHS, M. W. and MUIR, D. D., *J. Soc. Dairy Technol.*, 1981, **34**, 113.
148. SUGIURA, M., OIKAWA, T., HIRANO, K. and INUKAI, T., *Biochim. Biophys. Acta*, 1977, **488**, 353.
149. SEVERINA, L. O. and BASHKATOVA, N. A., *Biochemistry (USSR)*, 1979, **44**, 96.
150. ANDERSSON, R. E., *Biotechnol. Lett.*, 1980, **2**, 247.
151. LAWRENCE, R. C., FRYER, T. F. and REITER, B., *J. Gen. Microbiol.*, 1967, **48**, 401.
152. LU, J. Y. and LISKA, B. J., *Appl. Microbiol.*, 1969, **18**, 108.
153. NASHIF, S. A. and NELSON, F. E., *J. Dairy Sci.*, 1953, **36**, 459.
154. LAWRENCE, R. C., *Dairy Sci. Abstr.*, 1967, **29**, 1, 59.
155. DRIESSEN, F. M. and STADHOUDERS, J., *Neth. Milk Dairy J.*, 1974, **28**, 10.
156. BREUIL, C. and KUSHNER, D. J., *Can. J. Microbiol.*, 1975, **21**, 434.
157. LANDAAS, A. and SOLBERG, P., *20th Int. Dairy Cong.*, 1978, **E**, 304.
158. ADAMS, D. M. and BRAWLEY, T. G., *J. Dairy Sci.*, 1981, **64**, 1951.
159. FITZ-GERALD, C. H. and DEETH, H. C., *Aust. J. Dairy Technol.*, 1983, **38**, 97.
160. STEPANIAK, L., FOX, P. F. and DALY, C., *Ir. J. Fd Sci. Technol.*, 1981, **5**, 72.
161. TE WHAITI, I. E. and FRYER, T. F., *20th Int. Dairy Cong.*, 1978, **E**, 303.
162. FITZ-GERALD, C. H., DEETH, H. C. and COGHILL, D. M., *Aust. J. Dairy Technol.*, 1982, **37**, 51.
163. ANDERSSON, R. E., HEDLUND, C. B. and JONSSON, U., *J. Dairy Sci.*, 1979, **62**, 361.
164. KISHONTI, E., *Ann. Bull., Int. Dairy Fed.*, 1975, Doc. **86**, 121.
165. KNAUT, T., *20th Int. Dairy Cong.*, 1978, **E**, 305.
166. MOTTAR, J., *Milchwissenschaft*, 1981, **36**, 87.
167. O'DONNELL, E. T., *20th Int. Dairy Cong.*, 1978, **E**, 307.
168. O'DONNELL, E. T., PhD Thesis 1975, University of Strathclyde.
169. BARACH, J. T., ADAMS, D. M. and SPECK, M. L., *J. Dairy Sci.*, 1976, **59**, 391.
170. GRIFFITHS, M. W., PHILLIPS, J. D. and MUIR, D. D., *J. Appl. Bact.*, 1981, **50**, 289.
171. SUGIURA, M. and OIKAWA, T., *Biochim. Biophys. Acta*, 1977, **489**, 262.
172. NADKARNI, S. R., *Enzymologia*, 1971, **40**, 302.
173. KHAN, I. M., DILL, C. W., CHANDAN, R. C. and SHAHANI, K. M., *Biochim. Biophys. Acta*, 1967, **132**, 68.
174. AISAKA, K. and TERADA, O., *Agric. Biol. Chem.*, 1979, **43**, 2125.
175. ALFORD, J. A. and PIERCE, D. A., *J. Fd Sci.*, 1961, **26**, 518.

176. SUGIURA, M. and ISOBE, M., *Chem. Pharm. Bull.*, 1975, **23**, 681.
177. ALFORD, J. A., PIERCE, D. A. and SUGGS, F. G., *J. Lip. Res.*, 1964, **5**, 390.
178. COOKE, B. C., *N.Z. J. Dairy Sci. Technol.*, 1973, **8**, 126.
179. MABBITT, L. A., *Kieler milchw. ForschBer.*, 1981, **33**, 273.
180. FRYER, T. F., REITER, B. and LAWRENCE, R. C., *J. Dairy Sci.*, 1967, **50**, 388.
181. OTERHOLM, A., ORDAL, Z. J. and WITTER, L. D., *Appl. Microbiol.*, 1968, **16**, 524.
182. YU, J. H. and NAKANISHI, T., *Dairy Sci. Abstr.*, 1976, **38**, 855.
183. UMEMOTO, Y. and SATO, Y., *Jap. J. Zootech. Sci.*, 1977, **48**, 731.
184. KADERAVEK, G., CARINI, S. and SACCINTO, I., *Riv. Ital. Sostanze Grasse*, 1973, **50**, 135.
185. STADHOUDERS, J. and VERINGA, H. A., *Neth. Milk Dairy J.*, 1973, **27**, 77.
186. RICHARDSON, G. H., In: *Enzymes in Food Processing*, G. Reed (ed.), 2nd edn., 1975, Academic Press, New York, p. 361.
187. CARINI, S., LODI, R., BRAGHINI, T. and FEDELI, E., *Dairy Sci. Abstr.*, 1976, **38**, 543.
188. LAMBERET, G. and LENOIR, J., *Lait*, 1976, **56**, 119.
189. KORNACKI, K., STEPANIAK, L., ADAMIEC, I., GRABSKA, J. and WRONA, K., *Milchwissenschaft*, 1979, **34**, 340.
190. CHAPMAN, H. R. and SHARPE, M. E., In: *Dairy Microbiology, Vol. 2, The Microbiology of Milk Products*, R. K. Robinson (ed.), 1981, Applied Science Publishers, London, p. 157.
191. MORRIS, H. A. and JEZESKI, J. J., *J. Dairy Sci.*, 1953, **36**, 1285.
192. EITENMILLER, R. R., VAKIL, J. R. and SHAHANI, K. M., *J. Fd Sci.*, 1970, **35**, 130.
193. MENASSA, A. and LAMBERET, G., *Lait*, 1982, **62**, 32.
194. ALIFAX, R., *Lait*, 1975, **55**, 41.
195. ARNOLD, R. G., SHAHANI, K. M. and DWIVEDI, B. K., *J. Dairy Sci.*, 1975, **58**, 1127.
196. MOSKOWITZ, G. J., CASSAIGNE, R., WEST, I. R., SHEN, T. and FELDMAN, L. I., *J. Agric. Fd Chem.*, 1977, **25**, 1146.
197. BENZONANA, G. and ESPOSITO, S., *Biochim. Biophys. Acta*, 1971, **231**, 15.
198. BROCKERHOFF, H. and JENSEN, R. G., *Lipolytic Enzymes*, 1974, Academic Press, New York.
199. DHERBOMEZ, M., LACRAMPE, J.-L., and LARROUQUERE, J., *Revue Fr. Cps Gras*, 1975, **22**, 147.
200. FUKUMOTO, J., IWAI, M. and TSUJISAKA, Y., *J. Gen. Appl. Microbiol.*, 1963, **9**, 353.
201. JENSEN, R. G., *Lipids*, 1974, **9**, 149.
202. FUKUMOTO, J., IWAI, M. and TSUJISAKA, Y., *J. Gen. Appl. Microbiol.*, 1964, **10**, 257.
203. SHAHANI, K. M., In: *Enzymes in Food Processing*, G. Reed (ed.), 2nd edn., 1975, Academic Press, New York, p. 181.
204. NELSON, J. H., JENSEN, R. G. and PITAS, R. E., *J. Dairy Sci.*, 1977, **60**, 327.
205. LEE, H. J., OLSON, N. F. and RYAN, D. S., *J. Dairy Sci.*, 1980, **63**, 1834.
206. RICHARDSON, G. H. and NELSON, J. H., *J. Dairy Sci.*, 1967, **50**, 1061.
207. FOX, C. W., CHRISOPE, G. L. and MARSHALL, R. T., *J. Dairy Sci.*, 1976, **59**, 1857.

208. Owens, J. J., *Process Biochem.*, 1978, **13**(1), 13.
209. Long, C. A. and Patton, S., *J. Dairy Sci.*, 1978, **61**, 124.
210. Chen, C. C. W., Argoudelis, C. J. and Tobias, J., *J. Dairy Sci.*, 1978, **61**, 1691.
211. Shukla, T. P. and Tobias, J., *J. Dairy Sci.*, 1970, **53**, 637.
212. Deeth, H. C., *21st Int. Dairy Cong.*, 1982, **1**, Book 2, 297.
213. Doi, O. and Nojima, S., *Biochim. Biophys. Acta*, 1971, **248**, 234.
214. Sonoki, S. and Ikezawa, H., *Biochim. Biophys. Acta*, 1975, **403**, 412.
215. Owens, J. J., *Process Biochem.*, 1978, **13**(7), 10.
216. Stone, M. J., *J. Dairy Res.*, 1952, **19**, 311.
217. Labots, H. and Galesloot, T. E., *Neth. Milk Dairy J.*, 1959, **13**, 79.
218. Shimizu, M., Yamauchi, K. and Kanno, C., *Milchwissenschaft*, 1980, **35**, 9.
219. Ikezawa, H., Mori, M., Ohyabu, T. and Taguchi, R., *Biochim. Biophys. Acta*, 1978, **528**, 247.
220. Zwaal, R. F. A., Roelofsen, B., Comfurius, P. and Van Deenen, L. L. M., *Biochim. Biophys. Acta*, 1971, **233**, 474.
221. Myrnes, B. J. and Little, C., *Acta Chem. Scand.*, 1980, **34**, 375.
222. Møller-Madsen, A. and Horváth, Z., *Beretn. St. Forsøgsmejeri*, 1980, No. **239**.
223. Lombard, S. H. and Bester, B. H., *S. Afr. J. Dairy Technol.*, 1979, **11**, 163.
224. Schipper, C. J. and Jellema, A., Memo to IDF Lipolysis Group A3, 1972.
225. Connolly, J. F., Murphy, J. J., O'Connor, C. B. and Headon, D. R., *Ir. J. Fd Sci. Technol.*, 1979, **3**, 79.
226. Bråthen, G., *Meieriposten*, 1980, **69**, 345.
227. Downey, W. K., *Bull., Int. Dairy Fed.*, 1980, Doc. **118**, 4.
228. Tarassuk, N. P. and Frankel, E. N., *J. Dairy Sci.*, 1955, **38**, 438.
229. Mulder, H. and Walstra, P., *The Milk Fat Globule*, 1974, Commonwealth Agricultural Bureaux, Farnham Royal.
230. Hlynka, I., Hood, E. G. and Gibson, C. A., *Can. Dairy Ice Cr. J.*, 1944, **23**(3), 26.
231. Deeth, H. C. and Fitz-Gerald, C. H., *J. Dairy Res.*, 1977, **44**, 569.
232. Tarassuk, N. P. and Yaguchi, M., *J. Dairy Sci.*, 1958, **41**, 708.
233. Kitchen, B. J. and Aston, J. W., *Aust. J. Dairy Technol.*, 1970, **25**, 10.
234. Claypool, L. L., PhD Thesis 1965, University of Minnesota.
235. Henningson, R. W. and Adams, J. B., *J. Dairy Sci.*, 1967, **50**, 961.
236. Dunkley, W. L. and Smith, L. M., *J. Dairy Sci.*, 1951, **34**, 940.
237. Herrington, B. L. and Krukovsky, V. N., *J. Dairy Sci.*, 1939, **22**, 149.
238. Olson, J. C., Thomas, E. L. and Nielsen, A. J., *Am. Milk Rev.*, 1956, **18**(10), 98.
239. Whittlestone, W. G. and Lascelles, A. K., *Aust. J. Dairy Technol.*, 1962, **17**, 131.
240. Ortiz, M. J., Kesler, E. M., Watrous, G. H. and Cloninger, W. H., *J. Milk Fd Technol.*, 1970, **33**, 339.
241. Kelley, L. A. and Dunkley, W. L., *J. Milk Fd Technol.*, 1954, **17**, 306.
242. Fitz-Gerald, C. H., *Aust. J. Dairy Technol.*, 1974, **29**, 28.
243. Te Whaiti, I. E. and Fryer, T. F., *N.Z. J. Dairy Sci. Technol.*, 1976, **11**, 273.
244. Hlynka, I., Hood, E. G. and Gibson, C. A., *J. Dairy Sci.*, 1945, **28**, 79.
245. Aule, O. and Worstorff, H., *Ann. Bull., Int. Dairy Fed.*, 1975, Doc. **86**, 116.

246. STANNARD, D. J., *J. Dairy Res.*, 1975, **42**, 241.
247. DEETH, H. C. and FITZ-GERALD, C. H., *J. Dairy Res.*, 1978, **45**, 373.
248. REITER, B., DELLAGLIO, F. and SHARPE, M. E., *Rep. Natn. Inst. Res. Dairy. (Shinfield)*, 1969–70, 151.
249. ROUT, T. P., *Proc. Aust. Biochem. Soc.*, 1971, **4**, 73.
250. WORSTORFF, H., *Ann. Bull., Int. Dairy Fed.*, 1975, Doc. **86**, 156.
251. FLEMING, M. G., *Ir. J. Fd Sci. Technol.*, 1979, **3**, 111.
252. HUNTER, A. C., *Dairy Ind.*, 1966, **31**, 277.
253. CHEN, J. H. S. and BATES, C. R., *J. Milk Fd Technol.*, 1962, **25**, 176.
254. O'HALLORAN, J. C., FLEMING, M. G. and RAFTERY, T. F., *Ann. Bull., Int. Dairy Fed.*, 1975, Doc. **86**, 127.
255. DOODY, K., O'SHEA, J. and RAFTERY, T. F., *Ann. Bull., Int. Dairy Fed.*, 1975, Doc. **86**, 146.
256. TOLLE, A. and HEESCHEN, W., *Ann. Bull., Int. Dairy Fed.*, 1975, Doc. **86**, 134.
257. PILLAY, V. T., MYHR, A. N., GRAY, J. I. and BIGGS, D. A., *J. Dairy Sci.*, 1980, **63**, 1219.
258. MACLEOD, P., ANDERSON, E. O., DOWD, L. R., SMITH, A. C. and GLAZIER, L. R., *J. Milk Fd Technol.*, 1957, **20**, 185.
259. SPEER, J. F., WATROUS, G. H. and KESLER, E. M., *J. Milk Fd Technol.*, 1958, **21**, 33.
260. DOWNES, T. E. H., NIEUWOUDT, J. A. and SLABBERT, E. A., *S. Afr. J. Dairy Technol.*, 1974, **6**, 215.
261. MAGNUSSON, F., *17th Int. Dairy Cong.*, 1966, **B**, 331.
262. KITCHEN, B. J. and CRANSTON, K., *Aust. J. Dairy Technol.*, 1969, **24**, 107.
263. KIRST, E., *Lebensmittel-Ind.*, 1980, **27**, 27.
264. MEIN, G. A., HICKEY, M. W. and BROWN, M. R., *19th Int. Dairy Cong.*, 1974, **1E**, 18.
265. KIRST, E., *Nahrung*, 1980, **24**, 569.
266. KEITH, J. I., *Milk Dealer*, 1939, **28**(4), 82.
267. HERRINGTON, B. L. and KRUKOVSKY, V. N., *J. Dairy Sci.*, 1942, **25**, 241.
268. PARRY, R. M., CHANDAN, R. C. and SHAHANI, K. M., *J. Dairy Sci.*, 1966, **49**, 356.
269. NILSSON, R. and WILLART, S., *Rep. Milk Dairy Res. Alnarp*, 1960, No. **60**.
270. LARSEN, P. B., TROUT, G. M. and GOULD, I. A., *J. Dairy Sci.*, 1941, **24**, 771.
271. WANG, L. and RANDOLPH, H. E., *J. Dairy Sci.*, 1978, **61**, 874.
272. MCDOWELL, A. K. R., *J. Dairy Res.*, 1969, **36**, 225.
273. NIELSEN, J. V., *Nord. Mejeriind.*, 1978, **5**, 9.
274. WILLART, S. and SJÖSTRÖM, G., *17th Int. Dairy Cong.*, 1966, **A**, 287.
275. ROADHOUSE, C. L. and HENDERSON, J. L., *J. Dairy Sci.*, 1936, **19**(Abstr.), 107.
276. HUNTER, A. C., WILSON, J. M. and GREIG, G. W., *J. Soc. Dairy Technol.*, 1968, **21**, 139.
277. CONNOLLY, J. F. and JUDGE, F. J., *Res. Rep. An Foras Talantais*, 1975, 112.
278. TARASSUK, N. P. and RICHARDSON, G. A., *Science*, 1941, **93**(2413), 310.
279. TARASSUK, N. P. and HENDERSON, J. L., *J. Dairy Sci.*, 1942, **25**, 801.
280. KRIENKE, W. A., *J. Dairy Sci.*, 1944, **27**, 683.
281. FREDEEN, H., BOWSTEAD, J. E., DUNKLEY, W. L. and SMITH, L. M., *J. Dairy Sci.*, 1951, **34**, 521.
282. FOUTS, E. L. and WEAVER, E., *J. Dairy Sci.*, 1936, **19**, 482.

283. TARASSUK, N. P. and REGEN, W. M., *J. Dairy Sci.*, 1943, **26**, 987.
284. JENSEN, R. G., *J. Dairy Sci.*, 1959, **42**, 1619.
285. MENGER, J. W., *Ann. Bull., Int. Dairy Fed.*, 1975, Doc. **86**, 108.
286. RENNER, E. and SHAHIN, Y., *20th Int. Dairy Cong.*, 1978, **E**, 274.
287. CARDWELL, J. T. and ASHER, Y. J., *J. Dairy Sci.*, 1980, **63**(Suppl. 1), 46.
288. BROUWER, J., *Zuivelzicht*, 1981, **73**, 714.
289. SAITO, Z., *J. Dairy Sci.*, 1981, **64**(Suppl. 1), 60.
290. HERRINGTON, B. L. and GUTHRIE, E. S., *J. Dairy Sci.*, 1959, **42**, 897.
291. KODGEV, A. and RACHEV, R., *18th Int. Dairy Cong.*, 1970, **1E**, 200.
292. WEAVER, E., *Dairy Sci. Abstr.*, 1939, **1**, 358.
293. BACHMANN, M., *Schweiz. Milchztg*, 1961, **87**, 629.
294. JELLEMA, A., *Bull., Int. Dairy Fed.*, 1980, Doc. **118**, 33.
295. GHOLSON, J. H., SCHEXNAILDER, R. H. and RUSOFF, L. L., *J. Dairy Sci.*, 1966, **49**, 1136.
296. ASTRUP, H. N., BAEVRE, L., VIK-MO, L. and EKERN, A., *J. Dairy Res.*, 1980, **47**, 287.
297. STOBBS, T. H., DEETH, H. C. and FITZ-GERALD, C. H., *Aust. J. Dairy Technol.*, 1973, **28**, 170.
298. TARASSUK, N. P., LABEN, R. C. and YAGUCHI, M., *16th Int. Dairy Cong.*, 1962, **A**, 609.
299. TARASSUK, N. P., *Dairy Sci. Abstr.*, 1940, **2**, 183.
300. JENSEN, R. G., GANDER, G. W. and DUTHIE, A. H., *J. Dairy Sci.*, 1960, **43**, 762.
301. JOHNSON, P. E. and VON GUNTEN, R. L., *J. Dairy Sci.*, 1961, **44**, 969.
302. BRÅTHEN, G., *Meieriposten*, 1975, **64**, 209.
303. KRISTENSEN, V. F. and ANDERSEN, P. E., *Dairy Sci. Abstr.*, 1974, **36**, 668.
304. ASTRUP, H. N., VIK-MO, L., LINDSTAD, P. and EKERN, A., *Milchwissenschaft*, 1979, **34**, 290.
305. PFEFFER, J. C., JACKSON, H. C. and WECKEL, K. G., *J. Dairy Sci.*, 1938, **21**(Abstr.), 143.
306. SUHREN, G., HAMANN, J., HEESCHEN, W. and TOLLE, A., *Milchwissenschaft*, 1981, **36**, 150.
307. KÄSTLI, P., PADRUTT, O. and BAUMGARTNER, H., *Schweiz. Milchztg (Lait. Romand.)*, 1967, **93**, 197.
308. BACHMAN, K. C. and WILCOX, C. J., *J. Dairy Sci.*, 1977, **60**(Suppl. 1), 61.
309. KELLY, P. L., *J. Dairy Sci.*, 1945, **28**, 803.
310. HEMINGWAY, E. B., SMITH, G. H., ROOK, J. A. F. and O'FLANAGAN, N. C., *J. Soc. Dairy Technol.*, 1970, **23**, 44.
311. KINSELLA, J. E. and HOUGHTON, G., *J. Dairy Sci.*, 1975, **58**, 1288.
312. JELLEMA, A., *Neth. Milk Dairy J.*, 1975, **29**, 145.
313. ANDERSON, M., *J. Dairy Sci.*, 1979, **62**, 1380.
314. TARASSUK, N. P. and YAGUCHI, M., *J. Dairy Sci.*, 1958, **41**, 1482.
315. TALLAMY, P. T. and RANDOLPH, H. E., *J. Dairy Sci.*, 1969, **52**, 1569.
316. RANDOLPH, H. E. and ERWIN, R. E., *J. Dairy Sci.*, 1974, **57**, 865.
317. AGARWAL, V. K. and NARAYANAN, K. M., *Ind. J. Dairy Sci.*, 1976, **29**, 83.
318. SALIH, A. M. A. and ANDERSON, M., *J. Dairy Res.*, 1979, **46**, 453.
319. KÄSTLI, P., *Ann. Bull., Int. Dairy Fed.*, 1967, Part 3, 1.
320. BACHMAN, K. C. and GULLER, S. M., *J. Dairy Sci.*, 1980, **63**(Suppl. 1), 161.

321. PETERSON, M. H., JOHNSON, M. J. and PRICE, W. V., *J. Dairy Sci.*, 1943, **26**, 233.
322. LUHTALA, A., *Dairy Sci., Abstr.*, 1970, **32**, 196.
323. JURCZAK, M. E. and SCIUBISZ, A., *Milchwissenschaft*, 1981, **36**, 217.
324. THOMAS, S. B. and THOMAS, B. F., *Dairy Ind. Int.*, 1978, **43**(10), 5.
325. COUSIN, M. A., *J. Fd Prot.*, 1982, **45**, 172.
326. CHAPMAN, H. R., SHARPE, M. E. and LAW, B. A., *Dairy Ind. Int.*, 1976, **41**, 42.
327. MUIR, D. D., KELLY, M. E., PHILLIPS, J. D. and WILSON, A. G., *J. Soc. Dairy Technol.*, 1978, **31**, 137.
328. MUIR, D. D., KELLY, M. E. and PHILLIPS, J. D., *J. Soc. Dairy Technol.*, 1978, **31**, 203.
329. SUHREN, G., HEESCHEN, W. and TOLLE, A., *Milchwissenschaft*, 1977, **32**, 641.
330. OVERCAST, W. W. and SKEAN, J. D., *J. Dairy Sci.*, 1959, **42**, 1479.
331. HAWNEY, S. G. and ROYAL, L., *18th Int. Dairy Cong.*, 1970, **1E**, 502.
332. NASHIF, S. A. and NELSON, F. E., *J. Dairy Sci.*, 1953, **36**, 481.
333. COUSIN, M. A. and MARTH, E. H., *J. Dairy Sci.*, 1977, **60**, 1048.
334. DRIESSEN, F. M. and STADHOUDERS, J., *Neth. Milk Dairy J.*, 1971, **25**, 141.
335. PINHEIRO, A. J. R., LISKA, B. J. and PARMELEE, C. E., *J. Dairy Sci.*, 1965, **48**, 983.
336. DUMANT, J. P., DELESPAUL, G., MIGUOT, B. and ADDA, J., *Lait*, 1977, **57**, 619.
337. KISHONTI, E. and SJÖSTRÖM, G., *18th Int. Dairy Cong.*, 1970, **1E**, 501.
338. DRIESSEN, F. M. and STADHOUDERS, J., *Ann. Bull., Int. Dairy Fed.*, 1975, Doc. **86**, 101.
339. THOMAS, S. B. and DRUCE, R. G., *Dairy Ind.*, 1971, **36**, 75, 145.
340. MIKAWA, K. and ARIMA, S., *Dairy Sci. Abstr.*, 1981, **43**, 647.
341. BANDLER, D. K. *et al.*, *J. Dairy Sci.*, 1981, **64**(Suppl. 1), 56.
342. KUZDZAL-SAVOIE, S., *Ann. Bull., Int. Dairy Fed.*, 1975, Doc. **86**, 165.
343. KUZDZAL-SAVOIE, S., *Bull., Int. Dairy Fed.*, 1980, Doc. **118**, 53.
344. KRUKOVSKY, V. N. and HERRINGTON, B. L., *J. Dairy Sci.*, 1942, **25**, 237.
345. ANDERSON, K. P. and JENSEN, S. G. K., *Beretn. St. Forsøgsmejeri*, 1962, No. **136**.
346. SHIPE, W. F., SENYK, G. F., LEDFORD, R. A., BANDLER, D. K. and WOLFF, E. T., *J. Dairy Sci.*, 1980, **63**(Suppl. 1), 43.
347. THOMAS, E. L., NIELSEN, A. J. and OLSON, J. C., *Am. Milk Rev.*, 1955, **17**(1), 50.
348. GOULD, I. A., *J. Dairy Sci.*, 1944, **27**, 167.
349. MAGNUSSON, F., *19th Int. Dairy Cong.*, 1974, **1E**, 19.
350. PILLAY, V. T., MYHR, A. N. and GRAY, J. I., *J. Dairy Sci.*, 1980, **63**, 1213.
351. KINTNER, J. A. and DAY, E. A., *J. Dairy Sci.*, 1965, **48**, 1575.
352. DUNKLEY, W. L., *J. Dairy Sci.*, 1951, **34**, 515.
353. KOLAR, C. W. and MICKLE, J. B., *J. Dairy Sci.*, 1963, **46**, 569.
354. AL-SHABIBI, M. M. A., LANGNER, E. H., TOBIAS, J., and TUCKEY, S. L., *J. Dairy Sci.*, 1964, **47**, 295.
355. SCANLAN, R. A., SATHER, L. A., and DAY, E. A., *J. Dairy Sci.*, 1965, **48**, 1582.
356. DAY, E. A., In: *Flavor Chemistry*, R. F. Gould (ed.), 1966, American Chemical Society, Washington, p. 94.
357. TUCKEY, S. L. and STADHOUDERS, J., *Neth. Milk Dairy J.*, 1967, **21**, 158.
358. PARKS, O. W. and ALLEN, C., *J. Dairy Sci.*, 1979, **62**, 1045.

359. KEENAN, T. W., HEID, H. W., STADLER, J., JARASCH, E.-D. and FRANKE, W. W., *Eur. J. Cell Biol.*, 1982, **26**, 270.
360. PATTON, S., *J. Fd Sci.*, 1964, **29**, 679.
361. JAMOTTE, P., *18th Int. Dairy Cong.*, 1970, **1E**, 201.
362. NIELSON, V. H., *Am. Dairy Rev.*, 1972, **34**(3), 28.
363. BELL, L. I. and PARSONS, J. G., *J. Dairy Sci.*, 1977, **60**, 117.
364. WHITE, A. H. and GIBSON, C. A., *Can. Dairy Ice Cr. J.*, 1946, **25**(2), 35; (3), 39.
365. STÖRGARDS, T. and MAGNUSSON, F., *17th Int. Dairy Cong.*, 1966, **C**, 173.
366. O'CONNELL, J. M., COGAN, T. M. and DOWNEY, W. K., *Ann. Bull., Int. Dairy Fed.*, 1975, Doc. **86**, 92.
367. ANDERSON, J. A. and PARSONS, J. G., *J. Dairy Sci.*, 1977, **60**(Suppl. 1), 51.
368. FOUTS, E. L., *J. Dairy Sci.*, 1940, **23**, 173.
369. PARMELEE, C. E. and BABEL, F. J., *Dairy Sci. Abstr.*, 1955, **17**, 428.
370. KRUKOVSKY, V. N. and HERRINGTON, B. L., *J. Dairy Sci.*, 1942, **25**, 231.
371. JACK, E. L., TARASSUK, N. P. and SCARAMELLA, E. L., *Dairy Sci. Abstr.*, 1943, **5**, 53.
372. STADHOUDERS, J., *Neth. Milk Dairy J.*, 1972, **26**, 68.
373. DEETH, H. C., FITZ-GERALD, C. H. and WOOD, A. F., *Aust. J. Dairy Technol.*, 1979, **34**, 146.
374. BADINGS, H. T., *Neth. Milk Dairy J.*, 1970, **24**, 145.
375. MUKHERJEE, S., *J. Ind. Chem. Soc.*, 1950, **27**, 557.
376. MCDANIEL, M. R., SATHER, L. A. and LINDSAY, R. C., *J. Fd Sci.*, 1969, **34**, 251.
377. WOO, A. H. and LINDSAY, R. C., *J. Dairy Sci.*, 1980, **63**, 1058.
378. ASTON, J. W. and DULLEY, J. R., *Aust. J. Dairy Technol.*, 1982, **37**, 59.
379. HLYNKA, I. and HOOD, E. G., *J. Dairy Res.*, 1947, **15**, 94.
380. WONG, N. P., ELLIS, R. and LACROIX, D. E., *J. Dairy Sci.*, 1975, **58**, 1437.
381. DEETH, H. C. and FITZ-GERALD, C. H., *Aust. J. Dairy Technol.*, 1975, **30**, 74.
382. BILLS, D. D. and DAY, E. A., *J. Dairy Sci.*, 1964, **47**, 733.
383. OHREN, J. A. and TUCKEY, S. L., *J. Dairy Sci.*, 1969, **52**, 598.
384. HORWOOD, J. F., LLOYD, G. T. and STARK, W., *Aust. J. Dairy Technol.*, 1981, **36**, 34.
385. DULLEY, J. R. and GRIEVE, P. A., *Aust. J. Dairy Technol.*, 1974, **29**, 120.
386. KRUKOVSKY, V. N. and SHARP, P. F., *J. Dairy Sci.*, 1936, **19**, 279.
387. COSTILOW, R. N. and SPECK, M. L., *J. Dairy Sci.*, 1951, **34**, 1119.
388. JOHANSSON, S., *Dairy Sci. Abstr.*, 1965, **27**, 78.
389. SWARTLING, P., *Ann. Bull., Int. Dairy Fed.*, 1967, Part **3**, 6.
390. ANDERS, R. F. and JAGO, G. R., *J. Dairy Res.*, 1964, **31**, 81.
391. TARASSUK, N. P. and SMITH, F. R., *J. Dairy Sci.*, 1940, **23**, 1163.
392. COSTILOW, R. N. and SPECK, M. L., *J. Dairy Sci.*, 1951, **34**, 1104.
393. SCHWAB, H., *Dairy Sci. Abstr.*, 1970, **32**, 430.
394. BITSCH, F. and HOLMSKOV, P., *Maelkeritidende*, 1974, **87**, 876.
395. PALMER, L. S. and HANKINSON, C. L., *J. Dairy Sci.*, 1941, **24**, 429.
396. MCDONALD, S. T., SPURGEON, K. R., PARSONS, J. G. and SEAS, S. W., *J. Dairy Sci.*, 1981, **64**(Suppl. 1), 54.
397. HICKS, C. L., ALLAUDDIN, M., LANGLOIS, B. E. and O'LEARY, J., *J. Fd Prot.*, 1982, **45**, 331.

398. GRAPPIN, R. and JEUNET, R., *J. Dairy Sci.*, 1981, **64**(Suppl. 1), 41.
399. JÖNSSON, H., *Dairy Sci. Abstr.*, 1978, **40**, 122.
400. VAN REUSEL, A., *Ann. Bull., Int. Dairy Fed.*, 1975, Doc. **86**, 185.
401. ROBERTSON, N. H., DIXON, A., NOWERS, J. H. and BRINK, D. P. S., *S. Afr. J. Dairy Technol.*, 1981, **13**, 3.
402. BRUNNER, J. R., *J. Dairy Sci.*, 1950, **33**, 741.
403. BUCHANAN, R. A., *Aust. J. Dairy Technol.*, 1965, **20**, 62.
404. DEETH, H. C. and SMITH, R. A. D., *Aust. J. Dairy Technol.*, 1983, **38**, 14.
405. EVANS, E. W. and MABBITT, L. A., *Bienn. Rev. Natn. Inst. Res. Dairy. (Shinfield)*, 1974, 59.
406. OTERHOLM, A., ORDAL, Z. J. and WITTER, L. D., *J. Dairy Sci.*, 1970, **53**, 592.
407. HUANG, H. T. and DOOLEY, J. G., *Biotechnol. Bioeng.*, 1976, **18**, 909.
408. KINSELLA, J. E. and HWANG, D., *Biotechnol. Bioeng.*, 1976, **18**, 927.
409. FOX, P. F., *J. Soc. Dairy Technol.*, 1980, **33**, 118.
410. LAW, B. A., *Dairy Ind. Int.*, 1980, **45**(5), 15.
411. KOSIKOWSKI, F. V. and IWASAKI, T., *J. Dairy Sci.*, 1975, **58**, 963.
412. JOLLY, R. C. and KOSIKOWSKI, F. V., *J. Dairy Sci.*, 1975, **58**, 846.
413. MAHOUD, M. M. and KOSIKOWSKI, F. V., *J. Dairy Sci.*, 1980, **63**(Suppl. 1), 47.
414. NELSON, J. H., *J. Agric. Fd Chem.*, 1970, **18**, 567.
415. JOLLY, R. and KOSIKOWSKI, F. V., *J. Fd Sci.*, 1975, **40**, 285.
416. KALLE, G. P., DESHPANDE, S. Y. and LASHKARI, B. Z., *J. Fd Sci. Technol.*, 1976, **13**, 124.
417. SEITZ, E. W., *J. Am. Oil Chem. Soc.*, 1974, **51**, 12.
418. HAMOSH, M., *Pediat. Res.*, 1979, **13**, 615.
419. HAYASAWA, H., KIYOSAWA, I. and NAGASAWA, T., *19th Int. Dairy Cong.*, 1974, **1E**, 559.
420. HAMOSH, M. *et al.*, *J. Clin. Invest.*, 1981, **67**, 838.
421. ROSS, C. A. C. and SAMMONS, H. G., *Arch. Dis. Child.*, 1955, **30**, 428.
422. WILLIAMSON, S., FINUCANE, E., ELLIS, H. and GAMSU, H. R., *Arch. Dis. Child.*, 1978, **53**, 555.
423. DEETH, H. C., FITZ-GERALD, C. H. and WOOD, A. F., *Aust. J. Dairy Technol.*, 1975, **30**, 109.
424. HÄNNI, H. and RYCHENER, M., *Mitt. Geb. Lebensmittelunters. Hyg.*, 1980, **71**, 509.
425. DRIESSEN, F. M., JELLEMA, A., VAN LUIN, F. J. P., STADHOUDERS, J. and WOLBERS, G. J. M., *Neth. Milk Dairy J.*, 1977, **31**, 40.
426. LINDQVIST, B., ROOS, T. and FUJITA, H., *Milchwissenschaft*, 1975, **30**, 12.
427. DOLE, V. P., *J. Clin. Invest.*, 1955, **35**, 150.
428. DEETH, H. C. and FITZ-GERALD, C. H., *Aust. J. Dairy Technol.*, 1976, **31**, 53.
429. HARPER, W. J., SCHWARTZ, D. P. and EL-HAGARAWY, I. S., *J. Dairy Sci.*, 1956, **39**, 46.
430. MCCARTHY, R. D. and DUTHIE, A. H., *J. Lip. Res.*, 1962, **3**, 117.
431. IYER, M., RICHARDSON, T., AMUNDSON, C. H. and BOUDREAU, A., *J. Dairy Sci.*, 1967, **50**, 285.
432. HORNSTEIN, I., ALFORD, J. A., ELLIOTT, L. E. and CROWE, P. F., *Anal. Chem.*, 1960, **32**, 540.
433. HUMBERT, E. S. and LINDSAY, R. C., *J. Dairy Sci.*, 1969, **52**, 1862.
434. KUZDZAL-SAVOIE, S., KUZDZAL, W. and LANGLOIS, D., *Lait*, 1971, **5**, 534.

435. KISZA, J., BATURA, K., STANIEWSKI, B. and PANFIL-KUNCEWICZ, H., *20th Int. Dairy Cong.*, 1978, **E**, 871.
436. STARK, W., URBACH, G. and HAMILTON, J. S., *J. Dairy Res.*, 1976, **43**, 469.
437. HRIVNAK, J. and PALO, V., *J. Gas Chromatogr.*, 1967, **5**, 325.
438. SUPELCO, INC., *Chromatography/Lipid Bull.*, 1975, No. **727**.
439. DEETH, H. C., FITZ-GERALD, C. H. and SNOW, A. J., *21st Int. Dairy Cong.*, 1982, **1**, Book 2, 393.
440. LABUSCHAGNE, J. H., *S. Afr. J. Dairy Technol.*, 1975, **7**, 229.
441. REED, A. W., DEETH, H. C. and CLEGG, D. E., (in preparation).
442. NIEUWENHOF, F. F. J. and HUP, G., *Neth. Milk Dairy J.*, 1971, **25**, 175.
443. HULSTKAMP, J., *19th Int. Dairy Cong.*, 1974, **1E**, 499.
444. DIXON, R. P. and DE MAN, J. M., *Can. Inst. Fd Technol. J.*, 1968, **1**, 51.
445. HERRINGTON, B. L. and KRUKOVSKY, V. N., *J. Dairy Sci.*, 1939, **22**, 127.
446. SUHREN, G., HEESCHEN, W. and TOLLE, A., *Ann. Bull., Int. Dairy Fed.*, 1975, Doc. **86**, 51.
447. JOHNSON, B. C. and GOULD, I. A., *J. Dairy Sci.*, 1949, **32**, 435.
448. FRANKEL, E. N. and TARASSUK, N. P., *J. Dairy Sci.*, 1955, **38**, 751.
449. PERRIN, D. R. and PERRIN, D. D., *J. Dairy Res.*, 1958, **25**, 221.
450. SALIH, A. M. A., ANDERSON, M. and TUCKLEY, B., *J. Dairy Res.*, 1977, **44**, 601.
451. KOOPS, J. and KLOMP, H., *Neth. Milk Dairy J.*, 1977, **31**, 56.
452. SHIPE, W. F., SENYK, G. F. and FOUNTAIN, K. B., *J. Dairy Sci.*, 1980, **63**, 193.
453. MACKENZIE, R. D., BLOHM, T. R., AUXIER, E. M. and LUTHER, A. C., *J. Lip. Res.*, 1967, **8**, 589.
454. NAKAI, S., PERRIN, J. J. and WRIGHT, V., *J. Dairy Sci.*, 1970, **53**, 537.

Chapter 7

LIPID OXIDATION

T. RICHARDSON

and

M. KORYCKA-DAHL

University of Wisconsin, Madison, Wisconsin, USA

SUMMARY

Oxidation of lipids involves complex chemical reactions between lipids and oxygen mediated by a variety of catalysts. Active oxygen species such as singlet oxygen, superoxide anion and hydroxyl radical can be generated by various chemical and biochemical reactions to initiate lipid oxidation which can then be sustained as a chain reaction by ground state oxygen. Intermediate peroxides decompose to yield off-flavours and potentially toxic oxidation products. A wide variety of food constituents can participate in the oxidation processes to hinder or accelerate them. The fat globule membrane (FGM) surrounding the triacylglycerol droplets in milk is a focal point for lipid oxidation. Metal ions associated with various ligands in the FGM generally act as pro-oxidants favouring oxidation of unsaturated fatty acid residues in membrane phospholipids. Tocopherols and carotenoids in the FGM oppose pro-oxidant effects of metal ions. Enzymes in FGM and aqueous phases of milk may combine to exert pro-oxidant effects. Thermal denaturation of enzymes might also expose pro-oxidants such as the ferrihaem of lactoperoxidase.

Constituents in the aqueous phase may markedly affect the course of lipid oxidation. Ascorbic acid and thiols can be anti-oxidant or pro-oxidant depending upon the conditions. Photo-excited riboflavin is a very potent, non-discriminating oxidizing agent. The caseins tend to exert an anti-oxidant effect.

A. MECHANISM AND CONTROL OF LIPID OXIDATION

1. INTRODUCTION

The oxidation of lipids in a food results in a number of adverse effects on its quality as shown in Fig. 1. Retardation of lipid oxidation is an important element of food preservation and can be a major factor in retention of palatability, nutritional quality and acceptability of a food. One objective of the food chemist is to retard, as far as possible, oxidation of lipids and the accompanying defects in food quality.

It is often possible to add synthetic compounds to foods to retard the oxidation of lipids. However, legal restrictions in many countries prevent the addition of preservatives to milk and dairy products. Thus preservation of dairy products in general and inhibition of lipid oxidation in particular are often restricted to what can be achieved through processing, packaging and storage parameters. The complexities of the milk system at once present problems and opportunities for protection against oxidation of milk lipids.

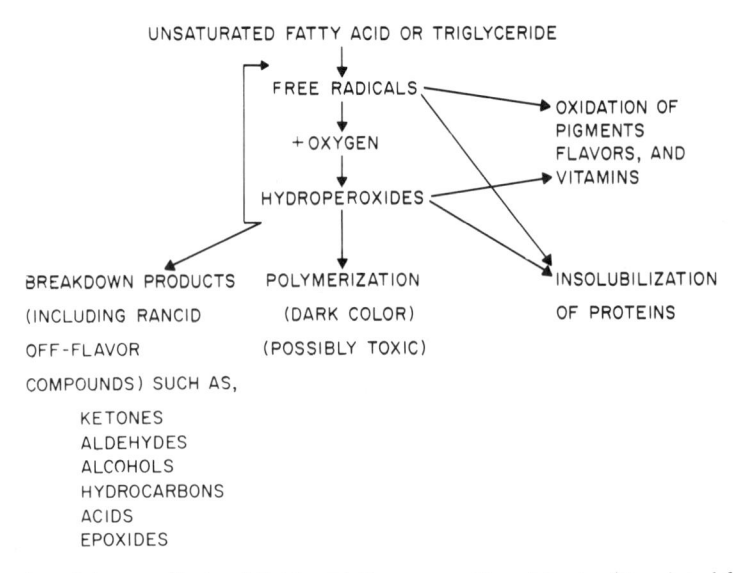

```
           UNSATURATED FATTY ACID OR TRIGLYCERIDE
                          ↓
      ┌──────────→ FREE RADICALS ─────────────→ OXIDATION OF
      │                   ↓                       PIGMENTS
      │               + OXYGEN                    FLAVORS, AND
      │                   ↓                 ────→ VITAMINS
      │            HYDROPEROXIDES ←───
      │                   ↓
      │                   ↓                ↘
BREAKDOWN PRODUCTS   POLYMERIZATION         INSOLUBILIZATION
(INCLUDING RANCID    (DARK COLOR)           OF PROTEINS
OFF-FLAVOR           (POSSIBLY TOXIC)
COMPOUNDS) SUCH AS,
     KETONES
     ALDEHYDES
     ALCOHOLS
     HYDROCARBONS
     ACIDS
     EPOXIDES
```

FIG. 1. Adverse effects of lipid oxidation on quality of foods. (Reprinted from Labuza (1971) with permission. © CRC Press, Inc.)

The susceptibility of milk lipids to oxidation is the net result of complicated interactions and reactions among milk constituents that can favour or oppose oxidation of lipids. In Table 1 are listed the major known factors in milk that may act as anti-oxidants and/or pro-oxidants. Certain characteristics of milk may have profound effects on the course of lipid oxidation. The relative concentrations in milk of the constituents listed in Table 1 may vary depending upon a large number of well-known factors bearing upon the cow, such as plane of nutrition, health, season, stage of lactation (Jenness and Patton, 1959; Johnson, 1974). Subsequent processing of the milk may be anti-oxygenic due to release of latent anti-oxidant activities, removal of oxygen, etc. Pro-oxidant reactions may be catalyzed by indigenous components as well as adventitious factors such as metal ions emanating from equipment (Schwartz and Parks, 1974). Thus, it is small wonder that a complete understanding of lipid oxidation in milk and dairy products has evaded the dairy chemist and, indeed, may never be possible.

It is important to understand the fundamental chemistry of lipid oxidation as fully as possible to provide a basis for developing practical measures of inhibiting the various processes leading to oxidative rancidity in milk and other dairy products. It is well known that lipid oxidation is essentially a free-radical chain reaction involving the familiar initiation, propagation and termination phases (see Section 3.1).

Problems associated with lipid oxidation tend to develop in the free-radical, chain-reaction, propagation phase; intervention with anti-oxidants, for example, is usually also operative at the propagation level. The relentless intrusion of oxygen and the pervasive nature of initiating pro-oxidants, such as preformed peroxides, make it virtually impossible to prevent the oxidative deterioration of foods in general. The obvious corollary is that we must try to minimize the initiation of oxidation as well as retard the propagation reactions in order to extend the useful lifetime of affected dairy products. Although our understanding of the propagation reactions and subsequent events leading to oxidized flavour are generally good, knowledge of the initiation of lipid oxidation is not. In any food, there will always exist preformed peroxides which will serve as sources of initiating radicals upon their homolytic scission. However, we also know that factors such as adventitious copper, light, etc. will accelerate lipid oxidation, thereby emphasizing the importance of a number of factors in the initiating process.

As is well understood, the milk system is a complex mixture of pro-oxidants and anti-oxidants, and the interplay of the various species, as

TABLE 1
MAJOR FACTORS AFFECTING THE OXIDATION OF LIPIDS IN MILK AND DAIRY PRODUCTS[a]

Milk constituents are subject to changes in concentration as a result of seasonal, physiological and nutritional effects on the cow and of processing and storage.

A. *Potential pro-oxidants*
   1. Oxygen and activated oxygen species
      a. Active oxygen system of somatic cells?
   2. Riboflavin and light
   3. Metals (e.g. copper and iron) associated with various ligands
      a. Metallo-proteins
      b. Salts of fatty acids
   4. Metallo-enymes (denatured?)
      a. Xanthine oxidase
      b. Lactoperoxidase, catalase (denatured)
      c. Cytochrome P-420
      d. Cytochrome b_5
      e. Sulphydryl oxidase?
   5. Ascorbate (?) and thiols (?) (via reductive activation of metals?)

B. *Potential anti-oxidants*
   1. Tocopherols
   2. Milk proteins
   3. Carotenoids (β-carotene; bixin in anatto)
   4. Certain ligands for metal pro-oxidants
   5. Ascorbate and thiols
   6. Browning reaction products
   7. Anti-oxidant enzymes (superoxide dismutase, catalase, sulphhydryl oxidase + lactoperoxidase)

C. *Environmental and physical factors*
   1. Inert gas or vacuum packing
   2. Gas permeability and opacity of packaging materials
   3. Light
   4. Temperature
   5. pH
   6. Water activity
   7. Reduction potential
   8. Surface area

D. *Processing and storage*
   1. Homogenization
   2. Thermal treatments
   3. Fermentation
   4. Proteolysis

[a] Many of these factors are necessarily interrelated and may even present paradoxical effects (e.g. ascorbate and thiols) on lipid oxidation.

affected by processing and packaging of dairy products, will determine the eventual oxidative status of the product. This chapter is devoted to brief discussions of the fundamental chemistry of lipid oxidation, some factors affecting it and possibilities for its control including specific illustrations from the dairy literature. The thrust of this chapter is to generalize from the oxidation literature at large to the milk system; the reader is commended to extensive reviews for discussions of the earlier literature (Badings, 1960; Mitchell and Henick, 1962; Patton, 1962; Badings, 1970; Schwartz and Parks, 1974). The exhaustive review on the kinetics of lipid oxidation by Labuza (1971) provides an early attempt to discuss lipid oxidation in foods from a mechanistic perspective. The authors hope that the present discussion will serve as a useful supplement and extension of earlier reviews. Except where reference to an earlier publication is required to further the discussion, the following discourse will be limited largely to developments during the last decade. The intent is to provide the reader with some fundamental information and to ask some questions that may serve as a basis for initiating future studies. A brief review on recent aspects of lipid oxidation in milk systems follows the more general discussion of lipid oxidation and the major factors affecting it.

The common denominator in oxidation of milk lipids is the incorporation of oxygen into these lipids. Some fundamental information on oxygen and so-called 'activated oxygen species' will provide a basis for a better understanding of oxidative processes.

2. PHYSICAL AND CHEMICAL PROPERTIES OF OXYGEN

2.1. Solubilities of Oxygen

Basic to any discussion of oxidation is the concentration of oxygen in the affected system. Oxygen is more soluble in non-polar solvents than in water or aqueous buffers. Consequently, it is not surprising that oxygen is more soluble in liquid milk fat compared to whole raw milk, with the solubility in both phases decreasing as the temperature is increased (Table 2).

Data in Table 2 indicate that a substantial percentage of the total oxygen in whole milk can be in the fat phase with the disparity apparently being greater at higher temperatures. Also, oxygen is apparently excluded from the solid phase of fat as it crystallizes. Thus, low storage temperatures tend to physically exclude oxygen from the fat phase as well as retard many chemical reactions such as the thermal homolytic scission of peroxides.

TABLE 2
PROXIMATE CONTENT OF DISSOLVED OXYGEN IN AIR-SATURATED ANHYDROUS MILK FAT[a,b] AND IN MILK[c] AT VARIOUS TEMPERATURES

Temperature (°C)	Oxygen in total sample mmol kg^{-1} [d]	Solid fat content (%)	Oxygen in liquid phase mmol kg^{-1} [d]	Percentage of total oxygen in fat phase of milk[e]
4	0·67[a]	56·9	1·56[a]	
15	0·95[a], 0·21[c]	33·8	1·44[a]	18[f]
20	0·99[a]	20·0	1·23[a]	
29·7	1·20[a]	4·6	1·26[a]	
35	1·04[a]	0	1·04[a]	
40	1·18[b]	—	—	
45	1·02[a]	0	1·02[a]	
50	0·95[a], 1·00[b]	0	0·95[a]	
60	0·91[b]	—	—	
70	0·78[b]	—	—	
72·8	0·16[c]	—	—	
80	0·96[a], 0·69[b], 0·11[c]	0	0·96[a]	35[f], 25[g]
86	0·08[c]	—	—	
88·5	0·04[c]	—	—	
90	0·66[b]	—	—	
110	0·53[a]	0	0·53[a]	

[a] Calculated from data in Timms *et al.* (1982).
[b] Calculated from data in Jebson *et al.* (1973).
[c] Calculated from data in Noll and Supplee (1941).
[d] Concentrations are millimolal. Density for milk assumed to be approximately 1·04.
[e] Calculated assuming 4% fat in the milk and equilibrium between fat and aqueous phases.
[f] Calculated using data from Timms *et al.* (1982) and Noll and Supplee (1941).
[g] Calculated using data from Jebson *et al.* (1973) and Noll and Supplee (1941).

The overall effect should be to retard initiation of new chain reactions. However, when comparing rates of oxidative reactions as a function of temperature, one might actually observe increased rates of reaction as the temperature is decreased (McWeeney, 1968). These negative temperature coefficients have been observed in a number of food-related reactions involving fats, proteins, carbohydrates and vitamins. The measured reaction rate relating to a single component in a multicomponent system as the temperature is decreased may reflect changes in rates of concurrent reactions, concentrations of reactants, phase conditions, etc. It is important to remember that studies involving alterations in the temperature of food systems may actually be altering the system *per se* rather than the individual reaction being analysed. Thus cold storage of foods may not always be beneficial, and accelerated storage tests can be misleading.

Illustrative of paradoxical situations that may occur are the observations of Betts and Uri (1963). Oxidation of docos-1-ene catalysed by cobalt proceeded as rapidly in the solid phase (at 25 °C) as in the liquid phase (at 50 °C). This was interpreted to result from a sharp reduction in the solid phase of the very rapid, diffusion-controlled rate of termination by free-radical recombination ($RO_2^{\cdot} + RO_2^{\cdot} \rightarrow$ products) in comparison with the rate-determining propagation reaction ($RO_2^{\cdot} + RH \rightarrow ROOH + R^{\cdot}$) which should be favoured because of proximity of reactants. The high rate of hydroperoxide production in irradiated butter at -20 °C was attributed to a similar reduction in the rate of chain-termination reactions. However, if the solubility of oxygen in the solid phase of docos-1-ene approaches zero, as in solidified milk fat (Timms et al., 1982), peroxide formation would be expected at the surface of the powdered hydrocarbon with the rate-limiting propagation reaction also being a function of surface area. Certainly oxygen diffusion is a significant factor superimposed upon the oxidative reactions.

The relatively higher concentrations of oxygen in the liquid fat phase and its exclusion upon crystallization may be relevant to the storage of butter. As the temperature of the butter is lowered, oxygen excluded from the crystallized fat should partition into the liquid fat and aqueous phase of the butter to saturate these compartments in the butter system. Thus oxygen should be available to react with the more unsaturated triacylglycerols in the lipid phase and with pro-oxidants and phospholipids in membrane fragments or lipids at the aqueous–fat interface. The physical distribution of a limiting quantity of liquid milk fat over the surface or within the interstices of mixed crystals may dramatically affect the relative rates of the termination and propagation reactions. The maximum rate of hydroperoxide production in irradiated butter at -20 °C may reflect an increased rate of chain terminations at higher temperatures and decreased propagation reactions at lower temperatures (Hannan and Boag, 1952; Hannan and Shepherd, 1952). Furthermore, as milk fat crystallizes, solutes such as cholesterol and tocopherols may concentrate in the liquid phase together with oxygen and thus affect sterol oxidation reactions.

Since oxygen is excluded from crystalline water (Thompson and Fennema, 1971) as well as crystalline fat, the effects of crystallization and supercooling of both phases on the retention of oxygen cannot be overlooked. For example, the solubility of oxygen in ice is less than one-thousandth its solubility in water (Thompson and Fennema, 1971) and, like other solutes, can become concentrated in the unfrozen phase. Expulsion of gases may occur when they reach their respective saturation levels; however, supersaturation is not uncommon. Hallet (1965) claims that a 30-fold supersaturation with air is possible in aqueous solutions. The complexities of crystallization of milk fat, particularly in the dispersed phase (Mulder, 1953; Brunner, 1974; Sherbon, 1974; Mortensen, Chapter 5, this volume) provide ample possibilities for supercooling and supersaturation of resultant liquid phases with oxygen.

2.2. Chemistry of Oxygen (Ardon, 1965; Taube, 1965)

Kinetically, ground-state or triplet-state oxygen ($^3O_2$) is not very reactive; however, thermodynamically the total reduction of oxygen to water in biological systems is very exothermic, yielding 104 kcal mol^{-1} of oxygen (Hanzlik, 1976). The sluggish kinetic reactivity of oxygen obviously serves as a very useful barrier to hinder the oxidation of food lipids. The reluctance of ground-state oxygen to initiate oxidative reactions directly has been attributed (Hanzlik, 1976) to the following factors:

(i) Incompatibility of triplet-state oxygen in reacting with predominantly singlet-state lipids. This is a spin-forbidden process and is not favoured because of its endothermic nature. Conversely, 'activated', singlet-state oxygen readily reacts directly with double bonds, which are also in the singlet state, in a spin-allowed process to generate peroxides.

(ii) C–H bonds are generally much stronger than the OO–H bond, necessitating the expenditure of substantial energy—about 35–65 kcal mol^{-1}—to initiate the reaction between $^3O_2$ and C–H in the absence of a catalyst (Privett and Blank, 1962; Labuza, 1971). Thus the activation energies for reactions involving $^3O_2$ are relatively high (Ardon, 1965).

(iii) The univalent stepwise reduction of $^3O_2$ is involved in the subsequent 'activation' of oxygen. However, insertion of the first electron into $^3O_2$ to form superoxide (O_2^-) is highly endothermic and thus difficult. Also, direct divalent reduction of $^3O_2$, containing two unpaired electrons with parallel spins, would result in the disallowed situation of two electrons with parallel spins occupying the same orbital (Fridovich, 1977).

Thus, ground-state oxygen ($^3O_2$) requires 'activation' in some way to facilitate oxidative reactions. Three major processes are involved in the 'activation' of oxygen (Fridovich, 1977). One is the photochemical excitation of an electron in $^3O_2$ to a higher energy state thereby generating very reactive singlet oxygen ($^1O_2$). The $^3O_2$ to $^1O_2$ conversion usually requires the intermediate participation of a photosensitizer (Ranby and Rabek, 1978), although it may arise during bimolecular decomposition of peroxy radicals (Nishinaga, 1977) or during microsomal oxidations (Nakano and Noguchi, 1977). Other chemical and biochemical pathways for the formation of $^1O_2$ have also been proposed (Korycka-Dahl and Richardson, 1980). In food systems, the photosensitizer (e.g. riboflavin, chlorophyll, some synthetic food colourants) absorbs light energy in the visible wavelengths and transfers it to $^3O_2$ to generate $^1O_2$. The synthetic food-colourant erythrosine is an efficient photosensitizer; however, nine other synthetic food colourants tested were inactive (Chan, 1975).

In a second mechanism for activation of $^3O_2$, certain metals have been postulated to interact with $^3O_2$ to yield singlet-like O_2 (Hanzlik, 1976; Nishinaga, 1977). In addition, charge–transfer complexes between oxygen and a suitable substrate may also result in a 'singlet-like' state for oxygen to favour oxygenation (Foote, 1978).

The third way for activation of oxygen is in the successive, univalent reduction of oxygen to yield reactive oxygen species. This reduction can be effected photochemically (involving a different mechanism than in the foregoing formation of $^1O_2$), chemically or enzymically (Korycka-Dahl and Richardson, 1980). In general, the activation of $^3O_2$ involves the formation of $^1O_2$ or $^1O_2$-like species and the univalent reduction of $^3O_2$ to yield reactive free radicals or peroxides.

In the latter photochemical process, the photosensitizer (e.g. riboflavin) is activated by absorption of light and in turn abstracts an electron (or H) from an organic molecule to yield an organic free radical which can then propagate subsequent oxidations. The fully reduced photosensitizer can subsequently reduce oxygen univalently to superoxide which subsequently yields more 'activated' oxygen species. Various oxidation reactions provide routes for the return of the resultant riboflavin semiquinone to the oxidized state to initiate a new catalytic cycle. Thus an extremely small amount of photosensitizer can function catalytically to oxidize food constituents with oxygen serving essentially as an electron sink to help drive the reaction but also yielding potentially damaging species of oxygen. A number of enzymes, notably xanthine oxidase from milk, are capable of producing large amounts of superoxide via the univalent reduction of oxygen (Fridovich, 1976).

The implications for riboflavin-catalysed photo-oxidations, possibly involving $^1O_2$ and/or reduced oxygen species in milk and dairy products, are obvious. Thus highly reactive forms of oxygen can arise from various chemical, photochemical and enzymic reactions to initiate oxidation of lipids in foods. A more detailed discussion of these active oxygen species will provide a basis for a better understanding of subsequent sections on oxidation of lipids as affected by milk constituents.

2.3. Active Oxygen Species

2.3.1. Singlet Oxygen

The properties of triplet-state oxygen ($^3O_2$) result from the distribution of its 12 valence electrons upon application of molecular orbital theory. The atomic orbitals of oxygen (O) are combined to form the molecular orbitals of dioxygen containing the appropriate bonding and antibonding orbitals (Fig. 2). The triplet nature (unpaired electrons with the same or parallel spin) of ground-state dioxygen flows necessarily from the distribution of the 12 valence electrons required by application of the Aufbau and Pauli exclusion principles and of Hund's rules (Ardon, 1965). The two unpaired

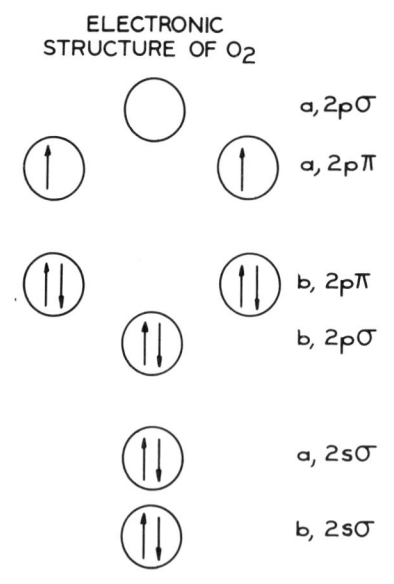

FIG. 2. Electronic configuration of triplet-state oxygen molecular orbitals arising from the linear combination of atomic orbitals of oxygen. (Reprinted from Taube (1965) with permission.)

electrons in degenerate orbitals (equal energy) having parallel spins yield a paramagnetic, triplet-state oxygen (i.e. in a magnetic field *three* possible electronic configurations are observed as triplet absorption bands). Mechanistically, it is useful to think of oxygen as a diradical ($\cdot$O=O$\cdot$) with a bond order of two. The unique behaviour of $^3O_2$ stems from the distribution of electrons whereby the absorption of energy during photo-oxidations results in an inversion of electron spin to yield singlet oxygen ($^1O_2$) in which spins are antiparallel to yield two types of diamagnetic singlet oxygen (only a singlet absorption band is observed in a magnetic field). The first type is a very energetic $^1\Sigma O_2$ (37·5 kcal mol^{-1}) with antiparallel electrons unpaired in different degenerate orbitals, and the second type is of lower energy, $^1\Delta O_2$ (23·4 kcal mol^{-1}) (Fig. 3). The $^1\Sigma$ state is too unstable to be of consequence for initiating oxidative reactions. The

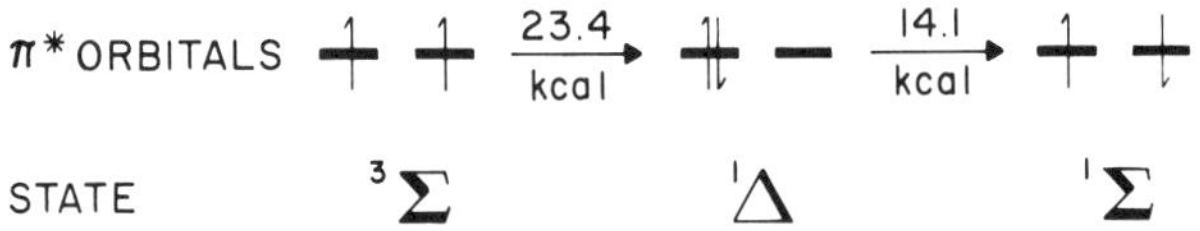

$$\pi^* \text{ORBITALS} \quad \uparrow \ \uparrow \quad \xrightarrow{\frac{23.4}{\text{kcal}}} \quad \uparrow\downarrow \ - \quad \xrightarrow{\frac{14.1}{\text{kcal}}} \quad \uparrow \ \downarrow$$

$$\text{STATE} \qquad \quad ^3\Sigma \qquad\qquad\quad ^1\Delta \qquad\qquad\quad ^1\Sigma$$

FIG. 3. The photochemical excitation of $^3O_2$ ($^3\Sigma$) to yield the two principal species of $^1O_2$ ($^1\Delta$) and ($^1\Sigma$). Electrons in the outer antibonding orbitals, a, $2p\pi$ (Fig. 2) become antiparallel through a spin inversion and are distributed between the two degenerate orbitals in a 'low-energy' ($^1\Delta$) or 'high-energy' ($^1\Sigma$) configuration. (Reprinted from Hanzlik (1976) with permission.)

less energetic $^1\Delta$ state is sufficiently stable to survive relaxation to the ground state long enough to react with singlet organic molecules. In addition to reacting chemically, singlet oxygen can be physically quenched by interaction with solvent or bulk-phase molecules as well as by other solutes such as the tocopherols or carotenoid pigments containing at least nine conjugated double bonds (Foote, 1976; Wilkinson, 1978). β-Carotene is one of the most potent quenching agents known for $^1O_2$, with one molecule capable of quenching, on average, 250 molecules of $^1O_2$ (Foote and Denny, 1968). By comparison, it has been calculated that a molecule of α-tocopherol can quench ~40 molecules of $^1O_2$ before being destroyed (Fragata and Bellemare, 1980). The nature of the bulk phase markedly affects the physical stability of $^1O_2$. The lifetime of $^1O_2$ in water (~2 μs) is markedly less than in non-polar (~45–700 μs) or lipid phases (Young and Brewer, 1978). In aqueous–lipid biphasic systems, the longer half-life of $^1O_2$ in the lipid phase favours oxidation of compounds that partition into the lipids (Suwa *et al.*, 1977). In contrast to triplet oxygen, singlet oxygen is

FIG. 4. Possible reactions between $^1O_2$ and double bonds illustrating the electrophilicity of $^1O_2$ in reacting directly with double bonds to yield various peroxides. (Reprinted from Korycka-Dahl and Richardson (1980) with permission.)

very electrophilic and reacts readily with double bonds in unsaturated lipids as shown in Fig. 4. Consideration of the geometry of possible six-membered transition states in the reaction between $^1O_2$ and linoleate (Fig. 5) leads to the predicted formation of conjugated or non-conjugated *cis–trans* products because of favoured geometry in the *cis–trans* reaction

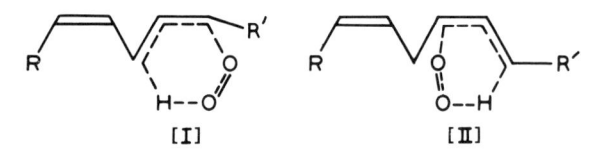

FIG. 5. Two possible, six-membered transition states in the reaction between $^1O_2$ and linoleate. [I] would lead to a peroxide with conjugated double bonds, whereas [II] would yield a peroxide with non-conjugated double bonds. (Reprinted from Clements *et al.* (1973) with permission.)

complex (Fig. 6). This exothermic reaction provides a convenient mechanism whereby the oxidation of lipids may be initiated through the formation of hydroperoxides which can decompose homolytically to initiate new free-radical chain reactions. It is important to remember that an organic free radical is not further constrained by spin conservation and can now react very rapidly with abundant $^3O_2$ to form peroxy radicals. Thus initiation of oxidation by very low levels of $^1O_2$ is sufficient to rapidly propagate large numbers of reaction chains involving ground-state oxygen.

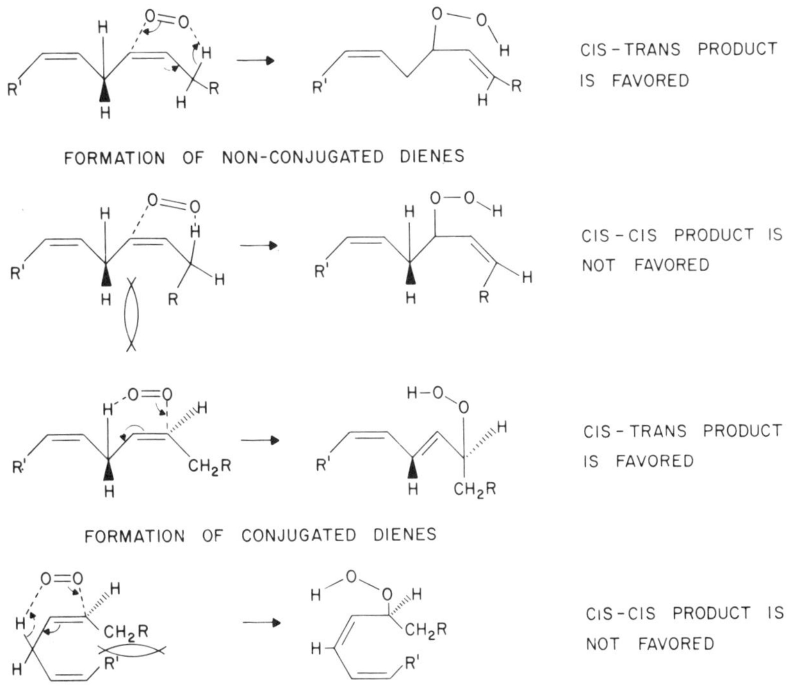

FIG. 6. Sterically favoured formation of conjugated and non-conjugated *cis–trans* double bonds in peroxides resulting from reaction of $^1O_2$ with linoleate. (Reprinted from Thomas and Pryor (1980) with permission.)

In addition, the electrophilicity of $^1O_2$ is exemplified by its rapid reaction with dienes as in substituted furans (Gollnick, 1978). This dienophilic character of $^1O_2$ is exploited in studies where inhibition of oxidative reactions by substituted furans is considered as evidence for participation of $^1O_2$ in promoting the oxidation. Electrophilic $^1O_2$ readily reacts with other nucleophiles, such as amino groups and sulphides, common in food constituents (Gollnick, 1978). In addition, inhibition of oxidative reactions by physical quenchers that 'de-energize' $^1O_2$ (e.g. β-carotene, 1,4-diazabicyclo[2.2.2]octane) is often presented as additional evidence for participation of $^1O_2$ in promoting the oxidations (Ranby and Rabek, 1978). An observed specific reaction between $^1O_2$ and cholesterol to yield the 5-α-hydroperoxide of cholesterol has been proposed as a diagnostic test for $^1O_2$ (Smith, 1981; Gumulka *et al.*, 1982). However, inhibition of oxidative reactions in the presence of chemical traps and physical quenchers for $^1O_2$

may not be taken as absolute proof of its intermediate participation. It would be prudent to provide additional chemical and/or physical evidence. Also, in complex foods such as milk and the products derived from it the various oxidative reactions will probably be a mixture of competing mechanisms.

Although pro-oxidant metals are thought to function primarily in catalysing the homolytic scission of hydroperoxides, they might also interact with $^3O_2$ to generate a singlet-like initiating species (reaction (1)) (Hanzlik, 1976; Nishinaga, 1977):

$$\underset{\overset{\|}{O^\cdot}}{O^\cdot} + M^{n+} \rightarrow \underset{\overset{|}{O^\cdot}}{O^\cdot}\!\!\!\searrow\!\!\!M \tag{1}$$

Based primarily on the effects of chemical trapping agents and physical quenchers, singlet oxygen has been implicated in the oxidation of raw milk lipids (Aurand *et al.*, 1977). However, recent research on the spontaneous dismutation of superoxide (Foote *et al.*, 1980) reveals that $^1O_2$ is not a product. This requires reconsideration of some of the possible mechanisms in the former study.

Five possible ways of generating $^1O_2$ in dairy products as a result of processing and storage come to mind (Korycka-Dahl and Richardson, 1980):

(i) Chemically by reaction of residual hypochlorite with hydrogen peroxide, which may both appear in milk in small quantities:

$$H_2O_2 + {}^-OCl \rightarrow {}^1O_2 + H_2O + Cl^- \tag{2}$$

(ii) Chemically or enzymically by reactions involving metalloproteins:

A. $\circledP - M^{n+} + {}^3O_2 \rightarrow {}^{\prime 1}O_2\text{-like complex'} \tag{3}$

B. $\text{Peroxidase–OCl}^- + H_2O_2 \rightarrow {}^1O_2 \tag{4}$

(iii) Photochemically via riboflavin or another sensitizer:

$$^1\text{sens} \rightarrow {}^1\text{sens*} \rightarrow {}^3\text{sens*} \tag{5}$$

$$^3\text{sens*} + {}^3O_2 \rightarrow {}^1\text{sens} + {}^1O_2$$

(iv) Self-reaction of secondary peroxy radicals (Boveris *et al.*, 1981):

$$2R\text{—HCOO}^\cdot \rightarrow {}^1O_2 + R\text{—HC}\!\!=\!\!O + R\text{—HCOH} \tag{6}$$

(v) Oxidation of superoxide by a restricted number of oxidizing agents can lead to $^1O_2$ (Nanni *et al.*, 1981).

2.3.2. Superoxide

The stepwise univalent four-electron reduction of $^3O_2$ to H_2O is highly exothermic overall. The reduction potentials associated with the various redox reactions involving oxygen are shown in Fig. 7. Although the overall process is exothermic, insertion of the first electron into oxygen to yield superoxide is endothermic and is not favoured. Thus superoxide anion may actually serve to univalently reduce transition metals etc., yielding oxygen and the reduced substrate. From the reduction potentials in Fig. 7, the

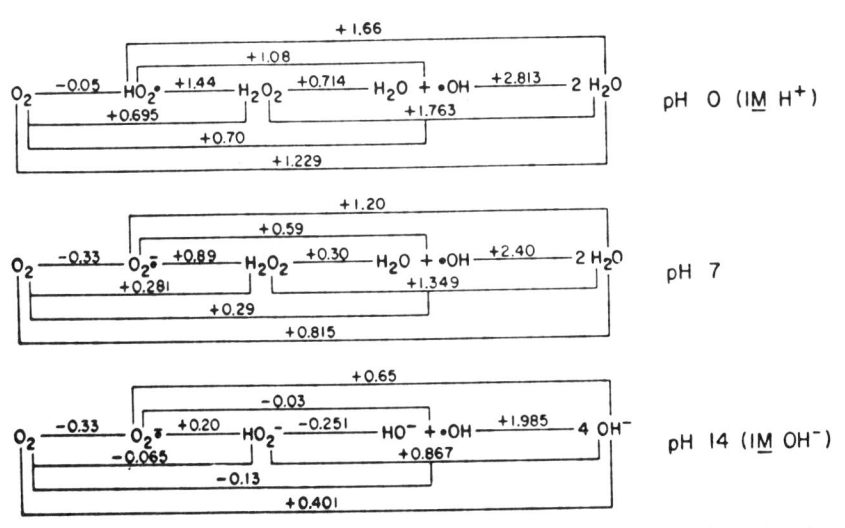

FIG. 7. Standard reduction potentials for dioxygen species in water. (Reprinted from Sawyer and Nanni (1981) with permission.)

addition of each electron generally increases the strength of the oxidizing species until H_2O is formed.

In living organisms, highly reactive oxygen species are usually contained by immobilization on enzymes, by compartmentalization and by various anti-oxidant systems. However, containment of these species that may be generated in dairy products is obviously less exact, thereby providing possibilities for initiation of oxidative reactions.

The superoxide anion is not a strong oxidizing agent and does not oxidize olefinic compounds such as linoleate (Fee and Valentine, 1976); however, the protonated superoxide (perhydroxyl radical) (pKa ~4·8) is a slightly stronger oxidizing agent and reacts slowly with methyl linoleate with a second-order rate constant of 3×10^2 $M^{-1} s^{-1}$ (Gebicki and Bielski, 1981). The electrophilicity of the superoxide radical is possibly enhanced by masking the negative charge on the anion with a proton, which should also favour solubility in non-polar lipids. This begs the question whether divalent cations might form lipophilic salts with the superoxide anion, thereby repressing the dismutation of the anion and yielding a more electrophilic radical soluble in the lipid phase. Although superoxide anion or metal complexes with superoxide are not necessarily strong oxidants, they are apparently capable of oxidizing phenolic compounds such as tocopherols or easily oxidized water-soluble constituents such as thiols and ascorbic acid (Hanzlik, 1976; Nishikimi and Yagi, 1977; Korycka-Dahl and Richardson, 1980). The ascorbyl radical is stabilized by resonance and is thus relatively unreactive (Bielski et al., 1975). Although the superoxide radical may not react readily with unsaturated fatty acids, it may have the practical effect of depleting certain easily oxidized anti-oxidants natural to milk leading to earlier oxidation of lipids.

Despite superoxide not being a strong oxidizing agent (Cotton and Wilkinson, 1972), the products resulting from its subsequent reactions are thought to be important in promoting lipid oxidation (Fridovich, 1976). Superoxide, with a pKa of 4·8, would exist primarily as an anion in most foods and, as such, is highly soluble in water and relatively insoluble in lipids. However, complexing of superoxide with various cations could possibly affect its distribution between the aqueous and lipid phases and, perhaps, its reactivity (more on this later).

Reactions implicating superoxide in damaging oxidative reactions depend primarily on its univalent reduction of various substrates, including itself. Superoxide rapidly dismutates in water, but has a longer lifetime in aprotic solvents (Cotton and Wilkinson, 1972). In the dismutation of superoxide, one molecule oxidizes a second to yield oxygen and potentially damaging hydrogen peroxide:

$$H^+ + HO_2^{\cdot} + O_2^{\cdot -} \xrightarrow[pH\ 4\cdot 8]{} H_2O_2 + {}^3O_2 \qquad (7)$$

Early literature suggests that the resulting oxygen is highly reactive singlet oxygen; however, recent research indicates that ${}^1O_2$ is not generated between pH values of 4 and 8 (Foote et al., 1980). Although ${}^1O_2({}^1\Delta g)$ is not a product in this reaction, the oxidation of superoxide can lead to ${}^1O_2$ when other oxidizing agents are used (Nanni et al., 1981). In general, the formation of ${}^1O_2$ is dependent upon substrates which favour singlet transition states in the electron-transfer process. Reaction by electron transfer to a cation radical from an anion radical such as $O_2^{\cdot -}$ is likely to result in excited-state product molecules which may emit characteristic fluorescence. More specifically, ${}^1O_2$ is formed by the oxidation of superoxide ion with iron and manganese complexes and diacyl peroxides. Interestingly, oxidation of superoxide ion by the Fe^{3+} ion does not result in $^1O_2$, but, since iron in foods is generally complexed, it is not clear as to the significance of this observation for food chemists. In aqueous systems wherein $^1O_2$ should be strongly quenched and the rate of dismutation of O_2^- is generally very rapid, these reactions would probably be of less importance than if they occurred in a non-polar lipid phase.

Although the spontaneous dismutation of superoxide occurs rapidly in water, being fastest at its pKa (Fee and Valentine, 1976), the reaction is catalysed by superoxide dismutase (SOD), known to occur in milk in very low concentrations (Fox and Morrissey, 1981). SOD is thought to provide some anti-oxidant protection in biological systems and in foods; however, the importance of superoxide in oxygen toxicity and SOD in mitigating that toxicity is a matter of some dispute (Fee, 1981; Fridovich, 1981). On the other hand, the generation of hydrogen peroxide as a result of spontaneous or SOD-catalysed dismutation may have additional ramifications in the absence of catalase which would destroy it.

There has been much discussion and argument about the importance of the Haber–Weiss reaction (reaction (8)) whereby superoxide univalently reduces hydrogen peroxide to yield the extremely reactive, electrophilic oxidant, hydroxyl radical (Schaich, 1980):

$$H_2O_2 + O_2^- \rightarrow HO\cdot + {}^-OH + O_2 \tag{8}$$

$$O_2^- + (L)_x M^{(n+1)+} \rightarrow O_2 + (L)_x M^{n+} \tag{9}$$

$$(L)_x M^{n+} + H_2O_2 \rightarrow HO\cdot + {}^-OH + (L)_x M^{(n+1)+} \tag{10}$$

Although the rate of the Haber–Weiss reaction *per se* may be too slow to be of practical significance, the intervention of a metal catalyst could enhance the rate of hydroxyl radical formation *via* a Fenton-type reaction (reactions (9) and (10)). In this case, an appropriate reducing agent (perhaps superoxide, ascorbate, thiols) reduces a metal or metal chelate of sufficiently high reducing potential, which in turn reduces hydrogen peroxide to form the hydroxyl radical (Fee, 1981; Tien *et al.*, 1981*a,b*). Work with iron chelates tends to support this hypothesis. However, chelates of other metals (e.g. protein–Cu chelates) with sufficiently high reduction potentials (Hanzlik, 1976) could possibly be involved in this catalysis. On the other hand, certain synthetic copper complexes have high SOD activity which may catalyse formation of hydrogen peroxide (Fee and Valentine, 1976; Fee, 1981; Leuthauser *et al.*, 1981). Furthermore, reducing agents other than superoxide, such as ascorbate or thiols, could provide the electrons necessary for the production of the reduced metals. The reduction of metals by ascorbate has been proposed as a mechanism for its paradoxical pro-oxidant effects (Uri, 1973). There are indications that the analogous reduction of a lipid hydroperoxide to yield an alkoxy radical proceeds more rapidly than reaction (8) (Schaich, 1980). In a similar vein, reduced metals, depending upon relative reduction potentials and ligands, can reduce oxygen directly via an intermediate perhydroxy radical to hydrogen peroxide (Scott, 1965; Cotton and Wilkinson, 1972; Schaich, 1980) (reactions (11–14)):

$$Cu^+ + O_2 \rightarrow CuO_2^+ \quad (11)$$

$$CuO_2^+ + H^+ \rightarrow Cu^{2+} + HO_2^{\cdot} \quad (12)$$

$$Cu^+ + HO_2^{\cdot} \rightarrow Cu^{2+} + HO_2^- \quad (13)$$

$$H^+ + HO_2^- \rightarrow H_2O_2 \quad (14)$$

It may be relevant to also keep in mind that superoxide adds irreversibly to many copper complexes (Cotton and Wilkinson, 1972) possibly leading to a reactive 'immobilized' peroxy radical.

Superoxide is also a strong nucleophile capable of reacting with esters to yield the potentially damaging peracyl radical (Niehaus, 1978; Gibian *et al.*, 1979):

$$R-CO-OR_1 + O_2^- \rightarrow R-CO-OO\cdot + {}^-OR_1 \quad (15)$$

However, several factors militate against this reaction being of consequence in milk unless, perhaps, in the milk fat globule membrane. The aforementioned dismutation of superoxide in the aqueous phase is very rapid so that its concentration is probably low. The hydration of the anion in the aqueous phase would tend to repress its nucleophilicity (Fee and Valentine, 1976). There would also appear to be a scarcity of water-soluble esters in milk. The anion should not be expected to be very soluble in the lipid phase and would probably partition out of it. On the other hand, lipophilic salts or protons would tend to reduce or prevent the nucleophilic attack of superoxide on triacylglycerol esters. However, any generation of superoxide in the milk fat globule membrane might initiate some oxidative reactions via its nucleophilicity (Niehaus, 1978), provided that excess superoxide does not completely reduce the peracyl radical to the fatty acid (Gibian *et al.*, 1979). Peracyl radicals immobilized on iron porphyrins (e.g. cytochrome P-450) have been proposed as intermediates in oxygen insertion reactions (e.g. epoxide formation) (Sligar *et al.*, 1980).

The significance of superoxide as a pro-oxidant in milk remains to be determined but it can arise from several possible reactions in milk:

(i) Photochemical oxidation of milk constituents by riboflavin followed by univalent reduction of $^3O_2$ by the reduced riboflavin. Thus $^3O_2$ serves as an electron sink and indirectly oxidizes milk constituents univalently to yield free radicals (Korycka-Dahl and Richardson, 1980).

(ii) Xanthine oxidase (XOD) acting upon its substrate is a classical laboratory source of superoxide (Fridovich, 1976) and readily co-oxidizes arachidonic acid (Fridovich and Porter, 1981). However, endogenous substrates for XOD in milk are rather limited in concentration if not missing altogether. Nevertheless, xanthine oxidase has been implicated in the oxidation of milk lipids, perhaps via other mechanisms (Fox and Morrissey, 1981). Certainly XOD contains a number of potential pro-oxidants such as flavins and transition metals.

(iii) Univalent auto-oxidation of thiols (Misra, 1974). Although sulphhydryl groups released as a result of thermal processing of milk have been historically considered as anti-oxidants, their role in the oxidative stability of milk should be assessed very carefully. This will be discussed in more detail in relation to the enzyme, sulphhydryl oxidase which is known to occur in milk (Swaisgood, 1980). Also, if added SOD retards lipid oxidation in biological system it infers involvement of superoxide, but probably indirectly as a source of stronger pro-oxidants.

2.3.3. Hydrogen Peroxide

Hydrogen peroxide might arise in milk as a result of microbial metabolism or by reduction of superoxide by easily oxidizable milk constituents such as ascorbic acid. Recent work (Toyoda et al., 1982a) indicates mean levels of 0·02 mg litre^{-1} hydrogen peroxide in 30 brands of commercial processed fluid milk. The H_2O_2 was thought to have arisen from oxidative reactions in the milk fat. H_2O_2 has been added to milk to facilitate low-temperature pasteurization of milk for cheese making. In the presence of lactoperoxidase, added H_2O_2 oxidizes indigenous thiocyanate to yield highly reactive antimicrobial species (Reiter, 1978; Pruitt et al., 1982). In addition, lactoperoxidase may catalyse the formation of hypochlorite in the presence of H_2O_2 and Cl^- which, in turn, could react with excess H_2O_2 to form $^1O_2$ (Korycka-Dahl and Richardson, 1980). The significance, if any, of these oxidative and antimicrobial agents on lipid oxidation in milk is unclear.

In the presence of various reduced transition metal ions (e.g. Cu^+), the hydroxyl radical is readily generated from H_2O_2 to initiate a multitude of possible free-radical reactions. If added catalase inhibits lipid oxidation, the obvious conclusion is that hydrogen peroxide is somehow involved.

2.3.4. Hydroxyl Radical (Korycka-Dahl and Richardson, 1980)

This very reactive species can result from the radiolysis of water as in the gamma-irradiation of foods. As mentioned previously, the univalent reduction of H_2O_2 also results in the formation of hydroxyl radicals—one of the strongest oxidizing agents released into a food system. It is also very electrophilic and abstracts electrons as well as adding readily to double bonds to create free-radical adducts.

Inhibition of lipid oxidation by addition of mannitol, ethanol or other alcohols has often been presented as evidence for participation of the hydroxyl radical. However, it may be possible that the 'hydroxyl radical' immobilized on metal chelates does not react with mannitol to inhibit lipid oxidation (Tien *et al.*, 1981*b*). In addition, if added SOD and/or catalase inhibit lipid oxidation, the hydroxyl radical is inferred as initiating the oxidations since the enzymes destroy potential precursors.

2.3.5. Ozone (Korycka-Dahl and Richardson, 1980)

Although ozone is an atmospheric pollutant, 6 ppb in the intake air of an urban-sited spray-drier adversely affects the flavour of whole milk powder, apparently resulting from oxidation of lipids (Kurtz *et al.*, 1969*a,b*). Trapping of the ozone with a charcoal filter prevents the flavour defect. Presumably, other atmospheric pollutants such as the oxides of sulphur and nitrogen might also oxidize lipids in spray-dried products. Certainly, existence of substantial amounts of these latter pollutants in the atmosphere has created problems in North America because of effects of resultant 'acid rain' on the ecology of lakes. Obviously, exposure to only 'initiating' amounts of these reactive species is necessary to have a detrimental effect on a highly labile product.

3. FUNDAMENTALS OF LIPID OXIDATION

Over the past 40 years, the familiar hydroperoxide theory of the oxidation of unsaturated fatty acids in lipids has become universally accepted to explain many experimental observations. Largely as the result of early research on auto-oxidation of ethyl linoleate in Great Britain by Farmer, Bolland, Gee, Bateman and their co-workers (Uri, 1961, 1973), it became evident that positions allylic to double bonds were the principal points of attack during the propagation of free-radical chain oxidations of unsaturated fatty acids. Resonating structures involving the resulting allylic free radical with adjacent double bonds provided explanations for the occurrence of various hydroperoxide isomers and a rationale and mechanisms for explaining occurrence of flavourful compounds as predicted decomposition products of these hydroperoxides. It soon became evident that the $-CH_2-$ between the double bonds in polyunsaturated fatty acids such as linoleate, linolenate and arachidonate was particularly vulnerable to auto-oxidation. Hydrogen atoms are easily abstracted from these positions to yield free radicals stabilized, however insufficiently, by resonance, thereby favouring their formation. Consequently, the rate of auto-oxidation of unsaturated fatty acids depends upon the ease with which free radicals result from the foregoing abstraction of hydrogen atoms.

In general, free-radical chain reactions proceed with a very low overall activation energy (Waters, 1971). Consideration of the Arrhenius equation leads one to predict that these free-radical reactions would be little affected by decreasing the temperatures of the reaction. However, in foods where phase changes in water and fat may exclude O_2 and immobilize free radicals, as in frozen foods for example, the situation is obviously more complicated than in pure bulk-phase systems. Since the rate of oxidation can then be diffusion-controlled, the sensitivity of foodstuffs such as butter towards auto-oxidation may be as much a feature of their microscopic structure as of their chemical composition.

3.1. Auto-oxidation of Lipids as a Chain Reaction

The familiar sequence of reactions characteristic of free-radical chain reactions in general, and of auto-oxidation of lipids in particular, is:

Initiation:

$$RH \xrightarrow{k_i} \text{Free radicals (R}^{\cdot}) \quad (16)$$

Propagation:

$$R^{\cdot} + O_2 \xrightarrow{k_O} RO_2^{\cdot} \quad (17)$$

$$RO_2^{\cdot} + RH \xrightarrow{k_P} RO_2H + R^{\cdot} \quad (18)$$

Termination:

$$2RO_2^{\cdot} \xrightarrow{k_t} O_2 + RO_2R \quad (19)$$

$$R^{\cdot} + RO_2^{\cdot} \xrightarrow{k_{t'}} RO_2R \quad (20)$$

$$2R^{\cdot} \xrightarrow{k_{t''}} R\text{—}R \quad (21)$$

Consideration of the rate equation for the overall oxidation of lipids leads to the following generalizations (Schaich, 1980):

(i) The rate of oxidation $(-d[O_2]/dt)$ is directly proportional to the amount of lipid present, to the amount of peroxide produced and, at low oxygen pressures, to the oxygen concentrations.

(ii) A point is reached where the oxygen concentration is not limiting; at atmospheric pressure, the rate of oxidation is independent of the oxygen concentration.

(iii) Except in the early stages of oxidation, the rate of initiation $(R_2^{\cdot})$ depends on (ROOH). The free-radical chain length is proportional to the reciprocal of (ROOH), specifically (constants $\times (O_2)$/ (ROOH)) at low oxygen pressures and (constants $\times$ (RH)/ (ROOH)) at high oxygen pressures.

(iv) Peroxide production mediates the rate of lipid oxidation, and the stability of peroxides as affected by food constituents is a key factor influencing the development of oxidative rancidity.

As mentioned in point (iii), in the very early stages of oxidation, (ROOH) may be very low in concentration. In this case, there probably exist initiating events other than the result of peroxide decomposition. These primordial initiating reactions are poorly understood but are of obvious importance in the stability of food lipids.

Once ROOH decomposition has occurred, the rates of hydrogen abstractions by resultant alkoxy radicals (RO$^{\cdot}$) are in the order of 10^4–10^6 times faster than by peroxy radicals (ROO$^{\cdot}$) generated in the propagation phase.

The autocatalytic oxidation of lipids requires the catalytic decomposition of peroxides as well as other pro-oxidant catalytic events. Schaich (1980) discusses the possible sources of the primordial initiating factors.

At atmospheric oxygen pressures, reaction (17) is very rapid ($>10^6$-fold greater than reaction (18)) which means that the concentrations of alkyl radicals are very low compared to the concentrations of peroxy radicals. Thus reaction (18) is the rate-limiting reaction in the propagation phase of auto-oxidation. Since the levels of peroxy radicals are relatively high compared to alkyl radicals, the principal termination reaction involves reactions between peroxy radicals to yield stable products (reaction (19)). Consequently, the three major reactions that define the steady-state rate of auto-oxidation of RH($-d[O_2]/dt$) are the rate of initiation, R_i (reaction (16)) and the rate-controlling propagation (reaction (18)) and termination (reaction (19)) steps, which are combined to give the rate expression:

$$\frac{-d[O_2]}{dt} = \frac{k_p[RH](R_i)^{0.5}}{(2k_t)^{0.5}}$$

At a constant rate of initiation, $(R_i)^{0.5}$, the oxidizability of a substance, [RH], is proportional to the ratio $k_p/(2k_t)^{0.5}$ and not to k_p alone. Oxidizabilities determined for methyl esters of oleate, linoleate and linolenate are 0·9, 21 and 39, respectively. Although differences between k_p and k_t determine overall oxidizability of a substrate and may differ widely, k_p and k_t for the fatty acid esters tend to increase in parallel (Pryor, 1976). The foregoing provides a rationale based on appropriate rate constants for early observations on the relative rates of oxidation of oleate (1×), linoleate (~28–41×) and linolenate (~77–98×), although some published relative rates do not agree with the calculated oxidizabilities (Holmant, 1954; Swern, 1964; Scott, 1965; Schaich, 1980). Introduction of a second double bond into oleate increases the rate of oxidation 30–40 fold. However, additional double bonds have only a rough additive effect on the rate of oxidation. A major driving force in the enhanced rate of oxidation for linoleate over oleate is the formation of the relatively stable conjugated diene hydroperoxides in the former. It is readily evident that the oxidizability of normal milk fat, where oleate is the principal unsaturated fatty acid, will increase markedly when linoleate is introduced into the milk fat by feeding cows protected vegetable oils.

The much faster rates of oxidation of unsaturated fatty acids as the number of non-conjugated double bonds increases can also be related to markedly increased rates of H˙ abstraction reactions for doubly allylic methylene hydrogens. In linoleate (18:2) there are two doubly allylic hydrogens (C-11) and four singly allylic hydrogens (C-8; 14) compared to equal numbers of each type of hydrogen in linolenate (18:3). Calculated from rate constants for abstraction of H˙ in analogous systems, the ratio of abstraction rates for doubly to singly allylic hydrogens, per hydrogen, is 20:1 at 30°C (Pryor et al., 1976). Abstraction ratios in linolenate, with an equal number of each hydrogen type, should then be 20:1, whereas the ratio for linoleate should be 10:1. Therefore, 5–10% of hydrogens abstracted from singly allylic positions in these fatty acids should yield non-conjugated hydroperoxides whereby $^3O_2$ would attack the resonating radical system at positions 10 or 12 in the case of linoleate, for example. This should be kept in mind when considering the formation of non-conjugated hydroperoxides in the reaction of $^1O_2$ with polyunsaturated fatty acids (Korycka-Dahl and Richardson, 1980).

In the foregoing discussion on substrate oxidizability, $(R_i)^{0.5}$ was constant and oxidizability of substrate was observed to vary. However, it might prove instructive to consider a hypothetical case where $(R_i)^{0.5}$ is varied and oxidizability is a constant.

The dependence of rate of substrate oxidation on the square root of the rate of initiation, $(R_i)^{0.5}$, has been proven for oxidations initiated by light (Uri, 1961) and by homolysis of chemical initiators (e.g. benzoyl peroxide). The square-root dependency for rates of initiation is characteristic of free-radical chain reactions and apparently arises from two free radicals resulting from each initiating event. Consider samples of the same butter product whereby [RH], $k_p/(2k_t)^{0.5}$ are identical. Consider also, exposure of the samples to increasing (or decreasing) intensity of fluorescent lights in a dairy display case. From the well-known inverse-square law whereby light intensity per unit area varies inversely with the square of the distance, D, from a light source, the number of initiating photons impinging on each sample will decrease four-fold when the distance between the product and the light source is doubled. The assumption that the rate of initiation will be proportional to $(1/D^2)^{0.5} = 1/D$ leads one to predict that doubling the distance between the fluorescent light source and the butter will halve the rate of oxidation of the surface lipids. Thus it should be possible to derive relatively simple qualitative relationships that may require experimental verification from the complex kinetics of auto-oxidation. The reviews by Uri (1961) and Labuza (1971) will provide the reader with a basis for further pursuing the kinetics of lipid oxidation in foods.

The development and effectiveness of the anti-oxidant as well as pro-oxidant effects of food constituents can be better understood in terms of the hydroperoxide theory, thus presenting opportunities for asking new questions regarding lipid oxidation. The following examination of the individual steps in the auto-oxidation of lipids should prove useful in the ensuing discussions of factors in milk that may affect the course of the various reactions and in designing additional experiments for retarding and otherwise controlling these largely undesirable reactions.

3.1.1. Initiation

The initiation of the oxidation of unsaturated fatty acids is poorly understood and it has been only in recent years that so-called 'active-oxygen' species have been implicated as possible participants in the initiation process. There are a number of possible primordial primary reactions to generate the initial free radicals; however, once oxidation commences with the production of initiating peroxides, the primordial, priming reactions become minor participants (Schaich, 1980). In complex foods, such as milk and its derivatives, there are undoubtedly complicated interactions that are incompletely understood, often providing contradictory experimental results depending upon concentrations, environmental conditions, etc. The occurrence of the initial free radical to start the oxidation process has been a matter of much speculation and may involve such diverse initiating factors as irradiation directly or in conjunction with photosensitizers, metal complexes, enzymes, active oxygen (perhaps arising via action of the first three factors), and any other sources of sufficiently energetic free radicals. The homolytic scission of initiators (usually lipid hydroperoxides in the auto-oxidation of food lipids) is generally the source of initiating free radicals. These hydroperoxides probably arise from the action of the initiating factors mentioned above.

The decomposition of organic hydroperoxides is extraordinarily complex, varying with structure and reaction conditions. Possible mechanisms for peroxide decomposition include: simple homolytic scission of the O–O bond, oxidation–reduction, induced radical chains, and a number of polar reaction paths (Walling and Heaton, 1965). Decomposition of hydroperoxides can be unimolecular (reaction (22)) or bimolecular (reaction (23)), with the unimolecular reaction predominating in the early stages of lipid oxidations in food (Fig. 8). Although much of the fundamental research on hydroperoxide decomposition employs relatively high peroxide concentrations where the bimolecular process predominates, the food scientist, for obvious reasons (Fig. 8), is interested in the unimolecular phase.

Thermally induced decomposition of hydroperoxides should lead to the unimolecular formation of two free radicals capable of initiating new reaction chains (reaction (22)):

$$R_1CH(OOH)—R_2 \rightarrow R_1—CH(O^{\cdot})—R_2 + {^{\cdot}}OH \qquad (22)$$

$$2ROOH \rightarrow RO_2^{\cdot} + RO\cdot + 2H^{+} \qquad (23)$$

A continued geometric progression of radical formation could, in theory, result in a branching-type chain reaction characteristic of explosions (Stephens, 1965). This can be contrasted with the decomposition of hydroperoxides catalysed by some metals which instead might lead to a linear propagation of the chain where one free radical is required to oxidize or reduce the metal (Kochi, 1973). Possible differences in peroxide scission alluded to in the foregoing discussion may be relevant to accelerated shelf-life studies at higher temperatures whereby different mechanisms of peroxide scission could be operative compared to those at conventional storage temperatures. Details of problems associated with accelerated

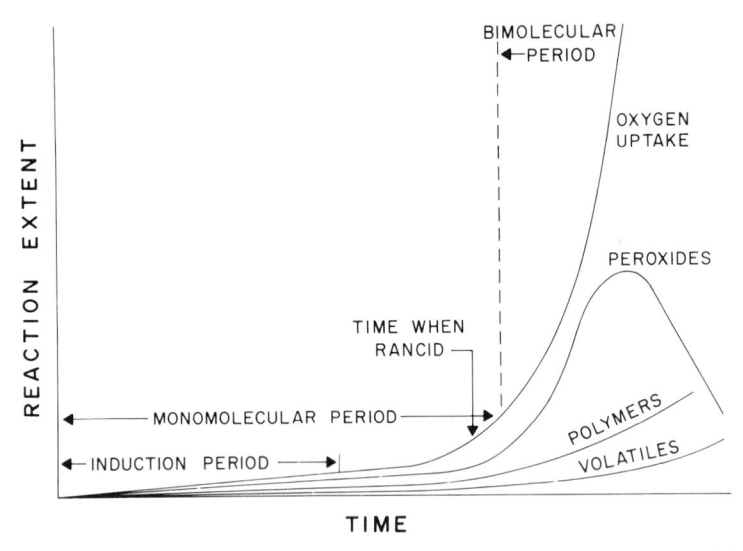

FIG. 8. Various stages in the auto-oxidation of food lipids. (Reprinted from Labuza (1971) with permission. © CRC Press, Inc.)

shelf-life methods can be found in reviews by McWeeny (1968), Labuza (1971) and Ragnarsson and Labuza (1977).

Considerations of thermochemical parameters obtained experimentally for thermal decomposition of various hydroperoxides yield a half-life for RO_2H at 100°C of 27·5 years (Hiatt, 1975) or 10^9 years at 37°C (Pryor, 1976). In practice, however, hydroperoxides usually decompose much faster and thermally generate free radicals in the 60–100°C temperature range between 10^2- and 10^5-times faster than the unimolecular homolysis parameters predict (Hiatt, 1975). The half-life for the decomposition of linoleic acid hydroperoxide at 60°C is 30 min (O'Brien, 1969). It is clear that these decompositions are not simple first-order homolysis reactions (Pryor, 1976). The rapid decomposition of hydroperoxides in a food system may result from a variety of processes:

(i) *Radical-induced decomposition of hydroperoxides:*

$$Q^\cdot + ROOH \rightarrow ROO^\cdot + QH \quad (24)$$

where $Q^\cdot$ is any reactive radical. As written, there is no net gain in numbers of radicals in reaction (24) providing propagation but not initiation. However, allylic hydroperoxides and peroxides are highly susceptible to free-radical-induced decomposition (reactions (25–27)) (Hiatt, 1980):

$$C=C-C-O_2^\cdot + C=C-CO_2H \rightarrow C=C-C-O_2-C-\dot{C}-CO_2H \quad (25)$$

$$C=C-C-O_2-C-\dot{C}-C-O_2H \rightarrow C=C-C-O_2-C-C\overset{O}{\underset{\diagdown}{\diagup}}C + {}^\cdot OH \quad (26)$$

$$C-C-CO_2-C-C\overset{O}{\underset{\diagdown}{\diagup}}C \rightarrow C=C-C-O^\cdot + {}^\cdot O-C-C\overset{O}{\underset{\diagdown}{\diagup}}C \quad (27)$$

The net result in this case is the production of three radicals for one in relatively rapid reactions, but the kinetic expression for the rate of radical production is approximately bimolecular in RO_2H. In general, the free-radical-induced decomposition of hydroperoxides is complicated and can take a number of routes.

(ii) *Molecularly assisted homolysis (MAH) in decomposing lipid hydroperoxides:* hydrogen bonding between hydroperoxides and water has been proposed to inhibit homolysis of peroxides, thereby yielding an anti-oxidant effect (Labuza, 1971; Schaich, 1980). On the other hand, a variety of compounds that can hydrogen bond with hydroperoxides, notably carboxylic acids, can assist in the homolysis of peroxides with decomposition being first-order in substrate and first-order in acid (Pryor, 1976). Since the rate of oxidation of lipids generally increases in the order triacylglycerol < methyl esters < fatty acids (Holmant, 1954; Labuza, 1971), the fatty acids may increase the rate of homolytic scission of hydroperoxides leading to increased rates of oxidation. As Pryor (1976) has suggested, linoleate (and other polyunsaturated fatty acids) hydroperoxides could undergo intramolecular MAH hydrogen atom transfer through a six-membered ring (Fig. 9) to generate initiating radicals. Thus, fatty acids can facilitate lipid oxidation via MAH processes as well as by forming lipid-soluble complexes with pro-oxidant metals (see below). On the other hand, carboxylic acids can also exert an anti-oxidant effect depending upon their structure. In this case, the acids catalyse the heterolytic decomposition of hydroperoxides leading to non-radical products. Butyric

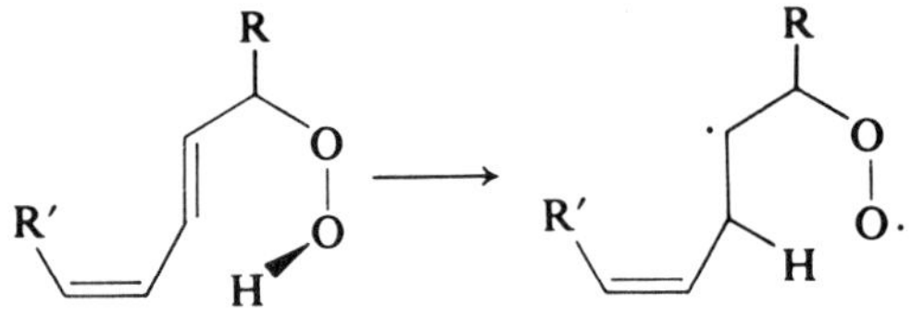

FIG. 9. Molecularly assisted homolysis (MAH) via an intramolecular hydrogen atom transfer through a six-membered ring (Reprinted from Pryor (1976) with permission.)

acid has been shown to exhibit anti-oxidant activity (Scott, 1965). MAH may be operative in the well-known pro-oxidant effect of excessive levels of phenolic anti-oxidants, such as BHT, or in the higher rates of lipid oxidation in the presence of carbohydrates (Labuza, 1971).

In addition, olefins accelerate the decomposition of peroxides by an MAH process which could be either a hydrogen atom transfer (reaction (28)) or an attack on peroxide oxygen (reaction (29)):

$$C=C + ROOH \rightarrow C^\cdot-C-H + ROO^\cdot \qquad (28)$$

$$C=C + ROOH \rightarrow C^\cdot-COH + RO^\cdot \qquad (29)$$

Additional MAH processes are possible (Pryor, 1976), and the superposition of the foregoing reactions on metal-catalysed reactions serves to emphasize the complexities in understanding lipid oxidation in foods.

3.1.2. Propagation
There are five main types of propagation reaction (Pryor, 1976), of which only four (a–d) are of principal concern to the food chemist.

(a) *Atom or group transfer.* This is the most common radical reaction and includes hydrogen atom abstraction (reactions (30) and (31)) and radical displacement on sulphur (reaction (32)):

$$Q^{\cdot} + RH \to QH + R^{\cdot} \quad (30)$$

$$R_1-CH(O^{\cdot})-R_2 + R_3H \to RCH(OH)-R_2 + R_3^{\cdot} \quad (31)$$

$$Q^{\cdot} + R-S-S-R \to QSR + RS^{\cdot} \quad (32)$$

The analogous oxygen compounds do not readily undergo reaction (32) because oxygen cannot accommodate the requisite number of electrons in the transition state compared to sulphur.

Although hydrogen abstractions are dependent upon five factors (Pryor, 1976), the strengths of bonds broken and formed are the two principal determinants of reactivity. Hydrogen abstraction reactions have relatively low activation energies ($\sim$15 kcal mol^{-1}). Thus reactions will occur in which relatively strong bonds are formed and relatively weak bonds are broken to yield exothermic (or slightly endothermic) heats of reaction.

Thiyl radicals can readily abstract hydrogen atoms among organic substrates (reaction (33)). The tendency for thiyl radical is to self-terminate rather than cross-terminate with other radicals. Among other possibilities, thiyl radicals can abstract the α-hydrogen of polypeptides (reaction (33)):

$$RS^{\cdot} + -NH-C(O)-CH-R- \to RSH + -NH-C(O)-C^{\cdot}-R- \quad (33)$$

In addition, low molar ratios (2:1) of cysteine to papain enhance the irradiation-induced inactivation of papain solutions (Pryor, 1976), whereas higher molar ratios (15:1) protect the enzyme from inactivation. From the foregoing, thiol groups are not necessarily a priori anti-oxidants but may participate in radical reactions that could result in oxidative cleavage of polypeptide backbones (reaction (33)) and in paradoxical pro-oxidant events.

(b) *Radical addition.* Radicals add readily to double bonds and this reaction generally competes with hydrogen abstraction:

$$R_1CH(O^{\cdot})-R_2 + -CH=CH- \to \begin{array}{c} R_1-CH-R_2 \\ \diagdown \\ O \\ \diagup \\ -CH-H\dot{C}- \end{array} \quad (34)$$

Thus the ratio of rate constants for addition and abstraction (k_{ad}/k_H) can be used to predict the importance of each pathway in propagation reactions (Pryor, 1976). For example, the methyl radical adds very rapidly to conjugated double bonds (Pryor *et al.*, 1972) leading to the suggestion that radical addition reactions may be a major route to carotenoid destruction although electron abstraction to yield radical cations must also be important (Willson, 1981). Apparently, oxidative attack is favoured at the termini of the conjugated double-bond system (Forss, 1972).

It should be pointed out that alkyl (R˙), alkoxy (RO˙), hydroxy (HO˙) and peroxy (ROO˙) radicals differ greatly in their reactivity (Scott, 1965; Schaich, 1980). The relative rates of atom transfer (H˙ abstraction), addition to double bonds and ease of oxidation will vary with the radical. Oxidizing electrophilic free radicals, such as hydroxy, alkoxy and peroxy, tend to extract electrons from substrates in oxidative processes. The more nucleophilic alkyl radicals tend to be more easily oxidized yielding reactive electrophilic carbonium ions (R^+) which can alkylate various nucleophiles (Norman and West, 1969). Cupric ions oxidize R˙ more readily than Fe^{3+}. These observations probably explain the generally high reactivity of reducing agents (phenols, amines, etc.) toward alkoxy (RO˙) and alkylperoxy (ROO˙) radicals, and of oxidizing agents (e.g. quinones) toward alkyl (R˙) radicals (Scott, 1965).

In the propagation step, reactions of peroxy radicals (ROO˙) are rate-limiting. At atmospheric pressures, peroxy radicals are by far the most common radical species occurring in auto-oxidation. Peroxy radicals may arise from: (1) reaction between R˙ and O_2; (2) one electron oxidation of ROOH; or (3) abstraction of H˙ from ROOH by an appropriate radical, X˙.

The overall concentration of alkoxy radicals is generally much lower than peroxy radicals at high oxygen concentrations. Nevertheless, alkoxy radicals abstract considerably faster and less selectively than peroxy radicals, thus playing a greater role in defining the final oxidation products (Schaich, 1980).

The relative rates of radical additions to double bonds compared to H˙ abstraction reactions will vary with the radical as well as with the structure of the substrate. Alkyl radicals are relatively unreactive in H˙ abstraction reactions and, therefore, tend to add to double bonds (Scott, 1965). Abstraction of an electron from 1,3-conjugated double bonds does not gain from the energetics of conjugation compared to a 1,4-diene system. Thus radical addition reactions are favoured by conjugated double-bond systems (Scott, 1965).

(c) *β-Scission of alkoxy radicals.* This generally yields a carbonyl and a radical which propagates the chain sequence of reaction (35):

$$R_1-CH(O^\cdot)-R_2 \begin{matrix} \nearrow R_1^\cdot + OCH-R_2 \text{ (aldehyde)} \\ \rightarrow \text{Initiation} \\ \searrow R_2^\cdot + OCH-R_1 \text{ (aldehyde)} \end{matrix} \quad (35)$$

Very large facilitation of β-scission of *t*-butoxy radicals by acetic acid suggests that hydrogen bonding is an alternative means of transition-state stabilization (Fig. 10) (Walling and Wagner, 1963). This again may

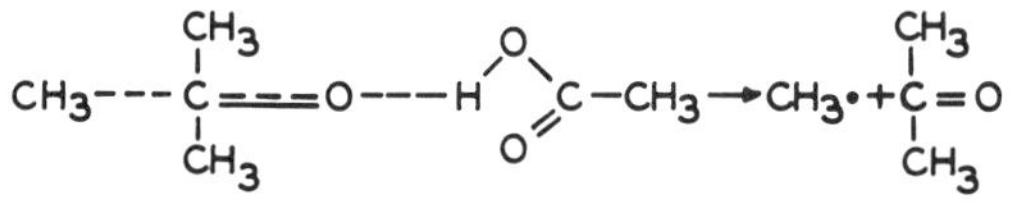

$$CH_3---\underset{\underset{CH_3}{|}}{\overset{\overset{CH_3}{|}}{C}}===O---H\overset{O}{\underset{O}{\diagdown}}C-CH_3 \rightarrow CH_3^\cdot + \underset{\underset{CH_3}{|}}{\overset{\overset{CH_3}{|}}{C}}=O$$

FIG. 10. Stabilization of transition state by acetic acid in β-scission of *t*-butoxy radical to yield methyl radical and acetone. (Reprinted from Walling and Wagner (1963) with permission. © American Chemical Society.)

indicate a detrimental effect of fatty acids, or other hydrogen-bonding agents, leading to accelerated rates of off-flavour development resulting from lipid oxidation. In general, the *t*-butoxy radicals lose much solvation in the transition state for H$^\cdot$ abstraction but have strongly solvated transition states for β-scission (Walling and Wagner, 1964).

(d) *Electron transfer.* Some nucleophiles can decompose certain peroxidic compounds by S_N2 reactions to yield non-radical products. Generally, organic sulphides do not produce free radicals upon nucleophilic decomposition of peroxides and are thus useful in anti-oxidant mixtures. However, certain nucleophiles can react with peroxides to yield *radical intermediates* via electron transfer (ET) reactions as proposed by Pryor (1976):

$$R_3N: + BzOOBz \rightarrow R_3N^{\cdot+} + BzO^\cdot + BzO^- \quad (36)$$

where Bz is a benzoyl group. Reactions such as (36) may be involved in the pro-oxidant effects of certain amino acids, riboflavin, etc. For example, pro-oxidant effects observed in the case of tryptophan (Farag *et al.*, 1978*a,b*; Taylor and Richardson, 1980) could reflect the univalent reduction of a peroxide via electron transfer from the indole nitrogen with the resultant radical being stabilized by resonance with the aromatic ring system (Fig. 11). Traces of Cu^{2+} complexed with various pro-oxidant amino acids may also enhance rates of lipid oxidation (Schaich, 1980).

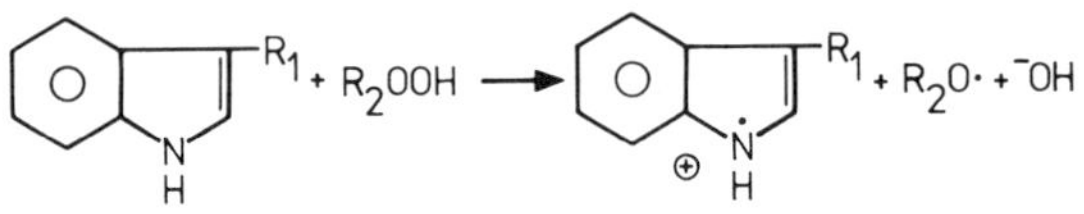

FIG. 11. A possible electron transfer (ET) reaction between tryptophan and a hydroperoxide whereby the formation of the tryptophyl radical would be favoured by resonance stabilization.

3.1.3. Termination

Termination reactions are varied and involve radical–radical reactions between hetero- and/or homo-radicals to yield stable products. In addition, certain anti-oxidants provide reducing equivalents to terminate radical reactions (e.g. reactions (37)–(41)):

A. *Formation of ketones or alcohols*

$$R_1CH(O^{\cdot})R_2 + R_3^{\cdot} \rightarrow R_1CO-R_2 + RH \tag{37}$$

$$R_1CH(O^{\cdot})R_2 + R_3O^{\cdot} \rightarrow R_1CO-R_2 + ROH \tag{38}$$

B. *Polymerization*

$$R_1^{\cdot} + R_2^{\cdot} \rightarrow R_1-R_2 \text{ (polymer)} \tag{39}$$

$$R_1O^{\cdot} + R_2^{\cdot} \rightarrow R_1OR_2 \text{ (ether)} \tag{40}$$

$$R_1O^{\cdot} + R_2O^{\cdot} \rightarrow R_1OOR_2 \text{ (peroxide)} \tag{41}$$

3.2. Peroxidation of Oleate and Linoleate

Examination of the propagation reactions in more detail is essential to understand the development of oxidized flavour. The importance of resulting resonating radicals in defining structures of the eventual carbonyls is illustrated in Figs 12 and 13 for the oxidation of oleate and linoleate, respectively. These classical mechanisms for auto-oxidation of oleate and linoleate have recently been modified slightly to explain small deviations from expected distributions of isomeric hydroperoxides in addition to stereochemical variations (Frankel, 1980). The abstraction (by an initiating radical) of an $H^{\cdot}$ from a carbon α or allylic to a double bond can lead to the depicted resonance structures for the resultant radical. Triplet oxygen reacts very rapidly with this radical to yield the various isomeric peroxy radicals. Usually oxygen attacks the termini of the resonating

LIPID OXIDATION

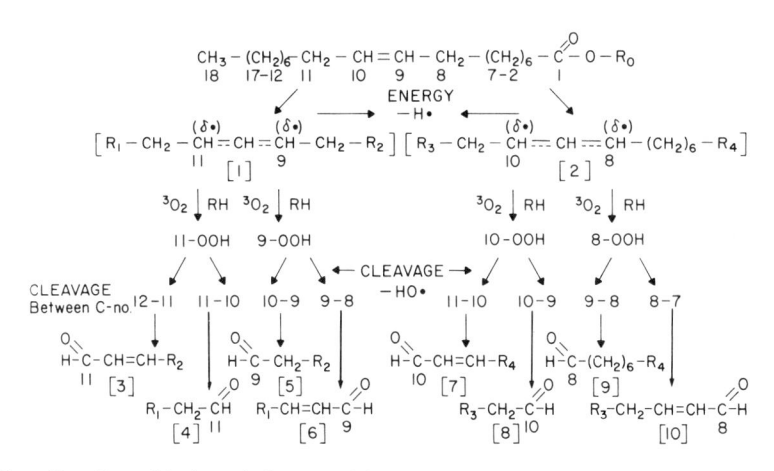

FIG. 12. Peroxidation of oleate and decomposition of the resultant 8, 9, 10 or 11 hydroperoxides to yield specific carbonyls.

FIG. 13. Peroxidation of linoleate and decomposition of resultant 9 or 13 hydroperoxides to yield specific carbonyls.

radical. Presumably, formation of the peroxy radical at C-11 in linoleate (Fig. 13) is not favoured energetically because the product would be non-conjugated requiring intervention of oxygen in the middle of the resonating system. Noted above are the relatively slower rates of H· abstraction of the singly versus doubly allylic hydrogens in polyunsaturated fatty acids.

The double bond that 'moves' into conjugation with the remaining *cis* double bonds tends to be *trans*, again because of favourable energetics and steric factors. However, four isomers of linoleate hydroperoxide can arise via a deoxygenation of the peroxy radical to the pentadienyl radical (Chan *et al.*, 1982) (Fig. 14). This reversible oxygenation of pentadienyl radicals

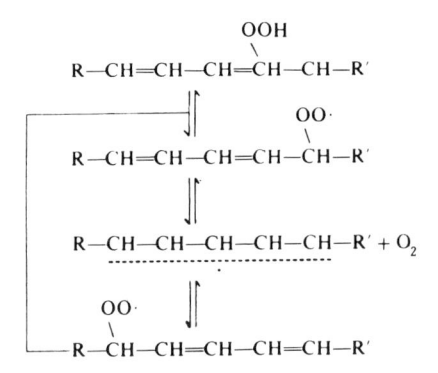

$$\begin{array}{c} \text{OOH} \\ | \\ \text{R—CH=CH—CH=CH—CH—R}' \\ \updownarrow \\ \text{OO·} \\ | \\ \text{R—CH=CH—CH=CH—CH—R}' \\ \updownarrow \\ \text{R—CH—CH—CH—CH—CH—R}' + \text{O}_2 \\ \updownarrow \\ \text{OO·} \\ | \\ \text{R—CH—CH=CH—CH=CH—R}' \end{array}$$

FIG. 14. Isomerization of a linoleate hydroperoxide via reversible deoxygenation of the peroxy radical to yield the pentadienyl radical (Reprinted from Chan *et al.* (1982) with permission.)

has been used to explain the formation of *trans–trans* and *cis–trans* hydroperoxide isomers (Fig. 15).

In the auto-oxidation of linoleate, ~97% of the hydroperoxides will be found as the four possible *trans–cis* (*t,c*) or *trans–trans* (*t,t*) isomers (Porter *et al.*, 1981). The distribution between the *t,c* and *t,t* isomers apparently results from the reversibility of the oxygenation reaction leading to peroxy radicals (Fig. 15) (Porter *et al.*, 1981). The distribution of *t,c* versus *t,t* products depends upon the competition between hydrogen abstraction by peroxy radical conformers [3] and [4] to yield a *t,c* hydroperoxide and formation of radical [6] upon loss of oxygen and subsequent addition of oxygen to the other end of the resonating system to eventually yield a *t,t* hydroperoxide. As expected, the presence of high concentrations of

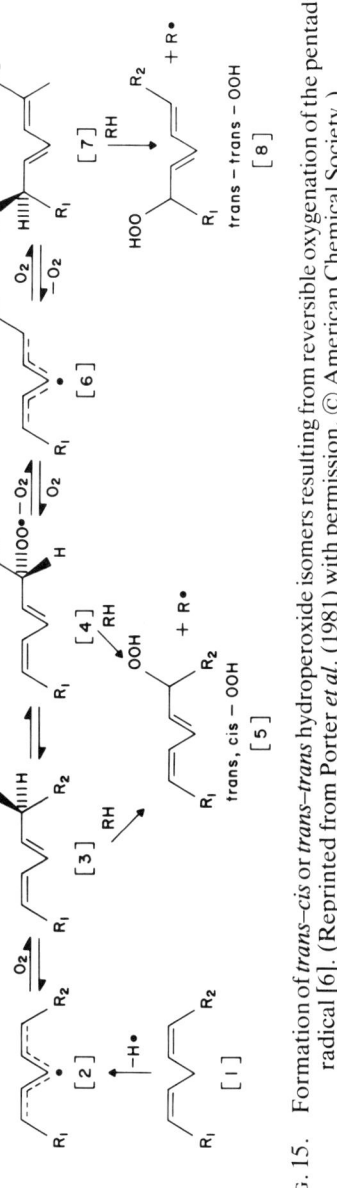

FIG. 15. Formation of *trans–cis* or *trans–trans* hydroperoxide isomers resulting from reversible oxygenation of the pentadienyl radical [6]. (Reprinted from Porter *et al.* (1981) with permission. © American Chemical Society.)

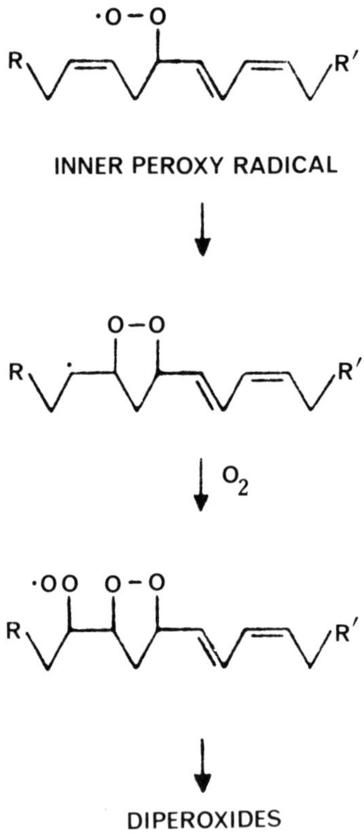

INNER PEROXY RADICAL

DIPEROXIDES

FIG. 16. Cyclization of peroxy radicals to yield cyclic peroxides and diperoxides. (Reprinted from Chan et al. (1982) with permission.)

abstractable H˙ from PUFA, thiols, tocopherols, etc., favours formation of t,c isomers whereas low concentrations of available H˙ yield more of the t,t isomers. As the number of double bonds increase to three and four, possibilities for cyclization of radicals to form cyclic peroxides becomes increasingly important (see below) (Fig. 16).

The non-conjugated double bonds in naturally occurring polyunsaturated fatty acids are quite transparent to the ultraviolet (UV) wavelengths useful in the conventional spectrophotometer; however, the conjugated double bonds resulting from oxidation absorb at wavelengths of 233 nm and above, depending upon the extent of conjugation. This absorption in the far UV has been used to quantify oxidation of polyunsaturated fatty acids.

Consideration of the possible resonating structures resulting from free-radical formation provides a basis for rationalizing structures of isomeric hydroperoxides arising from abstraction of H˙ from an adjacent —C=C—CH$_2$—. It should be pointed out that the carbon bearing the hydroperoxide is asymmetric. Since the oxygen can attack on either side of the plane of the resonating structure, the hydroperoxide is a racemate. On the other hand, hydroperoxides formed enzymically (e.g. by lipoxygenase) are generally optically active because binding to the enzyme is stereospecific requiring stereospecific addition of the oxygen.

Homolytic decomposition of the hydroperoxides yields the alkoxy radical which can undergo the various generic reactions outlined above. However, consideration of the β-scission reaction leads to rationalizations for formation of the carbonyls listed in Table 3. These carbonyls and others

TABLE 3

HYDROPEROXIDES AND ALDEHYDES (WITH SINGLE OXYGEN FUNCTION) THEORETICALLY POSSIBLE IN AUTO-OXIDATION OF SOME UNSATURATED FATTY ACIDS

Fatty acid	Isomeric hydroperoxide	Aldehyde formed
Oleic	8	C_{11}-2-enal
	9	C_{10}-2-enal
	10	C_9-al
	11	C_8-al
Linoleic	9	C_{10}-2,4-dienal
	11	C_8-2-enal
	13	C_6-al
Linolenic	9	C_{10}-2,4,7-trienal
	11	C_8-2,5-dienal
	12	C_7-2,4-dienal
	13	C_6-3-enal
	14	C_5-2-enal
	16	C_3-al
Arachidonic	5	C_{16}-2,4,7,10-tetraenal
	7	C_{14}-2,5,8-trienal
	8	C_{13}-2,4,7-trienal
	9	C_{10}-3,6-dienal
	10	C_{11}-2,5-dienal
	11	C_{10}-2,4-dienal
	12	C_9-3-enal
	13	C_8-2-enal
	15	C_6-al

have been detected in oxidized milk lipids (Badings, 1970; Labuza, 1971; Forss, 1972). The generation of 2,4-decadienal and other carbonyls from the isomeric hydroperoxides of linoleate is illustrated in more detail in Fig. 13. Although the homolytic decomposition of hydroperoxides provides a basis for explaining the formation of many flavourful carbonyls, additional oxidative reactions are also possible. In this regard, the alkoxy radical (R_1—HC(O˙)—R_2) resulting from peroxide homolysis undergoes a variety of reactions as illustrated in the previous discussions. The reader should keep in mind that the R groups in the alkoxy radical may represent the fatty acyl backbone either as the acid or in a triacylglycerol, a sterol ester, etc. and should realize the generic significance of these reactions. As illustrated, these further reactions lead to initiation of new free-radical reaction chains, to dismutations, to fragmentations, to polymerizations and to rearrangements (Howard, 1973). Volatile oxidation products from hydroperoxide decomposition are also potent pro-oxidants (El-Magoli *et al.*, 1979). It becomes readily evident that reactions involving the alkoxy radical are fundamental to the development of oxidative rancidity of foods (Forss, 1972).

3.3. Cyclic Peroxides (Endoperoxides)

Various endoperoxides are intermediates in the enzymic synthesis of hormone-like prostaglandins, etc. However, similar endoperoxides can arise via cyclization of peroxy radicals from the auto-oxidation of linolenate, for example, to yield compounds with possible biological activity (Fig. 17). Endoperoxides, arising from polyunsaturated fatty acids, are apparently involved in the formation of malondialdehyde (MDA) (Pryor and Stanley, 1975) which reacts with thiobarbituric acid (TBA) providing the basis for the well-known TBA test (Fig. 18) (Sinnhuber *et al.*, 1958; Mehlenbacher, 1960; Sinnhuber and Yu, 1977).

In the auto-oxidation of linolenate (18:3) the eight expected monohydroperoxide isomers (*cis–trans* or *trans–trans* 9-, 12-, 13- 16-OOH) result; however, the 'outer' 9- and 16-OOH predominate over the 'inner' 12-, 13-OOH isomers because the latter peroxy radicals cyclize to yield diperoxides (Fig. 16), while the outer peroxy radicals cannot cyclize in this way. Six major diperoxides are obtained by auto-oxidation of methyl linolenate (Chan *et al.*, 1982).

In the auto-oxidation of PUFA, food constituents such as α-tocopherol can bias: (1) the distribution of geometric isomers; (2) the ratio of outer to inner isomers; and (3) the ratio of diperoxides to monohydroperoxides (Chan *et al.*, 1982). For example, good H˙ donors (e.g. α-tocopherol) can

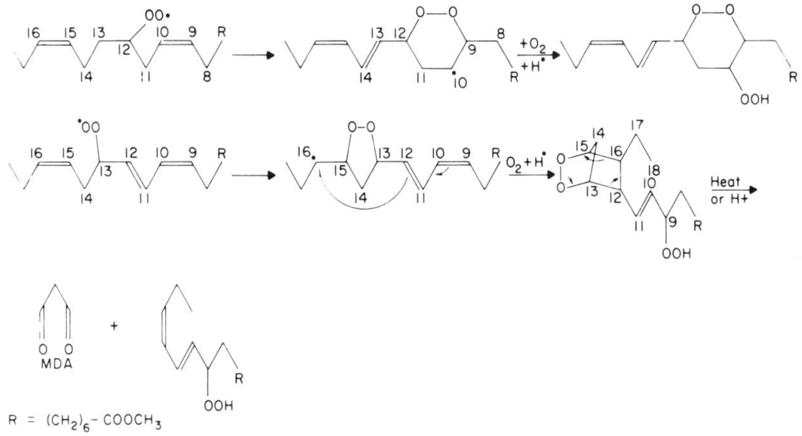

FIG. 17. Production of cyclic endoperoxide upon auto-oxidation of linolenate and the release of malondialdehyde (MDA) upon heating of the endoperoxide under acidic conditions. (Reprinted from Logani and Davies (1980) with permission.)

quench the *cis–trans* peroxy radicals to prevent the formation of diperoxides as well as *trans–trans* monohydroperoxides. Reduced formation of the diperoxides in the presence of α-tocopherol is reflected in the increased proportion of inner isomers (Chan *et al.*, 1982). This selective formation of inner monohydroperoxides may lead to the formation of octa-1,5 diene-3-one responsible for the fishy (metallic) taint of caramels (containing butter) (Swoboda and Peers, 1977).

Since the diperoxides decompose differently from monohydroperoxides, both qualitatively and quantitatively, biochemical and physical factors that alter their relative concentrations may have profound effects on the course of lipid oxidation in foods (Chan *et al.*, 1982). Dihydro-

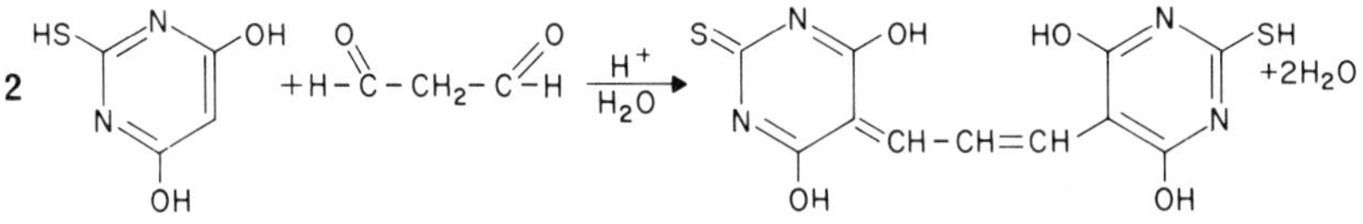

FIG. 18. Reaction between malondialdehyde (MDA) and thiobarbituric acid (TBA) to yield the red TBA pigments. (Reprinted from Sinnhuber *et al.* (1958) with permission. © Institute of Food Technologists.)

peroxides have resulted from the Fe^{2+}/ascorbate-catalysed oxidation of methyl-linolenate in addition to the cyclic monohydroperoxides (Toyoda et al., 1982b).

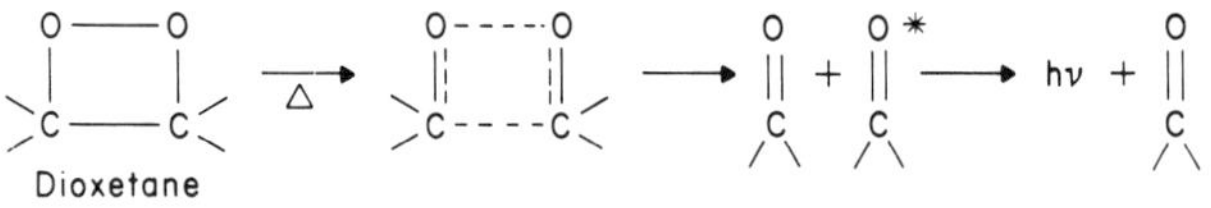

$$\underset{\text{Dioxetane}}{\overset{O\text{—}O}{\underset{>C\text{—}C<}{|\quad\quad|}}} \xrightarrow{\Delta} \overset{O\text{----}O}{\underset{>C\text{----}C<}{||\quad\quad||}} \longrightarrow \overset{O}{\underset{/\backslash}{||}}\overset{O*}{\underset{/\backslash}{||}} \longrightarrow h\nu + \overset{O}{\underset{/\backslash}{||}}$$

FIG. 19. Thermal decomposition of dioxetane yielding an excited-state carbonyl which emits a photon upon return to the ground state. (Reprinted from Huang et al. (1974).)

Dioxetanes (Fig. 19) from the reaction of $^1O_2$ with a double bond might be considered as special cyclic peroxides. Thermal decomposition of a dioxetane leads to the formation of two carbonyl compounds, one of which must be in the excited state thus emitting a photon as it returns to the ground state. Other reactions leading to luminescence might include the generation of $^1O_2$ from the decomposition of peroxy radicals (Ingold, 1973; Logani and Davies, 1980; Boveris et al., 1981) or oxidation of superoxide (Sawyer and Nanni, 1981). Reactions such as the foregoing are probably involved in the observed weak chemiluminescence phenomena used primarily by Russian investigators as a method for predicting storage stability of butter and other dairy products.

3.4. Epoxidation

Epoxidation of a double bond in lipids to yield epoxides (oxiranes) can occur *via* addition of a peroxy radical to the double bond (Fig. 20) or by a polar reaction between the hydroperoxide and a double bond (Fig. 21). The latter reaction can be catalysed by transition metals, notably molybdenum complexes (Sobczak and Ziolkowski, 1981). Complexation of the peroxide by molybdenum apparently increases the electrophilicity of the peroxide thereby facilitating its reaction with nucleophilic double bonds (Fig. 21). A peroxidized metal complex $(MoO(O_2)_2 L)$ intermediate results in epoxidation (Hiatt, 1980). Also, double bonds α to a carbonyl are very susceptible to base-catalysed epoxidation (Sobczak and Ziolkowski,

$$ROO\bullet + \underset{\underset{/\backslash}{C}}{\overset{\backslash/}{\underset{||}{C}}} \longrightarrow \overset{O}{\underset{>C-C<}{/\backslash}} + RO\bullet$$

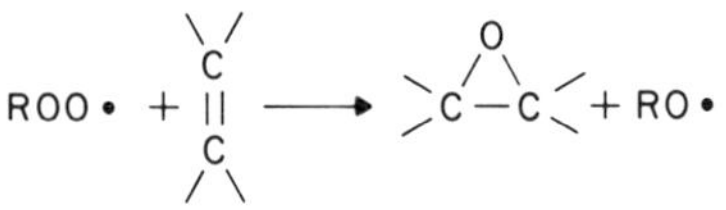

FIG. 20. Addition of a peroxy radical to a double bond yielding an epoxide (oxirane). (Reprinted from Sobczack and Ziolkowski (1981) with permission.)

FIG. 21. Epoxidation of a double bond by a hydroperoxide as catalysed by molybdenum (Mo). (Reprinted from Sobczack and Ziolkowski (1981) with permission.)

1981). Note that epoxidation via the free-radical reaction results in a chain-carrying alkoxy radical, and that the ionic reaction is stoichiometric.

The epoxidation of double bonds in lipids may have important health implications. Epoxides can be electrophilic alkylating agents by reacting with various nucleophiles. This electrophilicity is exploited to quantify epoxides using an HBr-based, back-titration procedure (Mehlenbacher, 1960) or a colorimetric picration method (Frioriti *et al.*, 1966a,b).

Upon hydrolysis, epoxides tend to yield the corresponding *trans*-diol (Fig. 22). However, reactions with other nucleophiles *in vitro* or *in vivo* result in essentially an alkylation reaction which might have biomedical ramifications if epoxides in tissues alkylate sensitive biological macromolecules thereby covalently modifying them and perhaps altering their function. Double bonds, whether in fatty acids, cholesterol or carotenoids, may be subject to epoxidation. However, it is likely that epoxides in foods are hydrolysed or otherwise inactivated during processing or digestion.

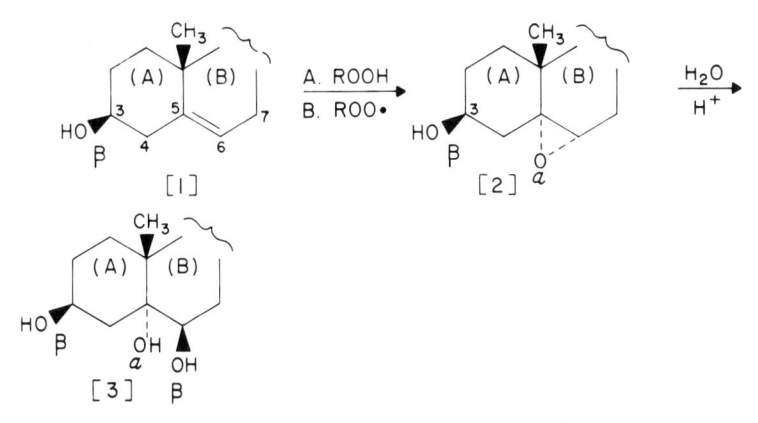

FIG. 22. Formation and hydrolysis of the 5-6-α-epoxide [2] of cholesterol to yield the 5α-6β-*trans*-diol.

3.5. Oxidation of Cholesterol

Cholesterol is generally thought of as relatively stable to oxidation; however, approximately 60 oxidation products of cholesterol have been observed (Smith, 1980, 1981). Oxidized cholesterol species have been implicated in a variety of diseases and conditions. Although oxidation products of cholesterol have been found in foods including dairy products (Table 4), their biological significance remains to be determined (Smith, 1980, 1981). Experience in the author's laboratory (Daly et al., 1983), indicates that the principal oxidation products of cholesterol, as well as those from structurally related phytosterols, are the epimeric 5,6 α- or β-epoxysterols (and the corresponding triol), the epimeric 7α- or β-hydroxysterols and three major ketones. The epoxides arise from aforementioned epoxidation reactions with the α- or β-epimers resulting from the epoxidation occurring on either side of the plane of the cholesterol molecule observed edge-on (Fig. 22). In the case of cholesterol, free-radical epoxidation may proceed rapidly since addition of a peroxy radical at position 6 should yield a relatively stable tertiary radical at position 5. Hydrolysis of the 5,6-epoxides yields the corresponding

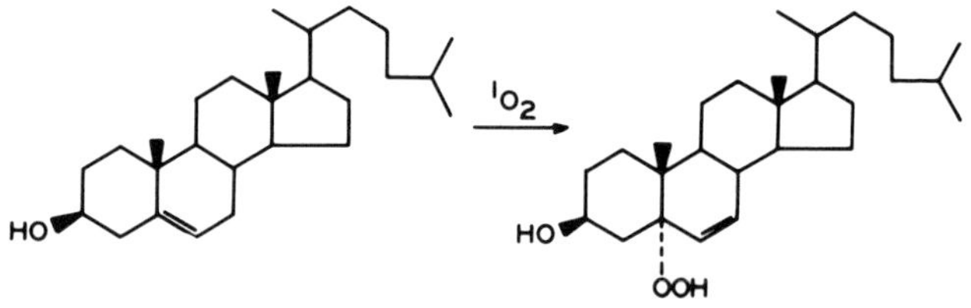

FIG. 23. Reaction of cholesterol with $^1O_2$ to yield the 5α-hydroperoxide considered to be characteristic for $^1O_2$. (Reprinted from Schenck et al. (1957) with permission.)

3,5,6-triol which has been implicated in atherogenesis. Although the biological effects of the α-epoxide have been studied in some detail, little is known about the biological properties of the β-epimer. By the same token, the 7α- or β-hydroxy-epimers result from reduction of epimeric hydroperoxides at the 7 position allylic to the double bond or from dismutation reactions also yielding sterol ketones which have been detected in foods.

Cholesterol has been proposed as a reagent for the specific detection of $^1O_2$ (Fig. 23) (Smith, 1980; Gumulka et al., 1982). A unique 5α-hydroperoxide is reported as the product of the cisoid attack of $^1O_2$ on the 5,6 double bond of cholesterol to distinguish this type of initiator from the more random oxidation expected from free-radical initiation. In the latter

TABLE 4
OXIDIZED STEROLS IN FOODSTUFFS[a]

Food	Sterols found
Egg products	
Egg yolk	5-Cholestane-3β,5,6β-triol
	Cholesta-3,5-dien-7-one
Egg dough[b]	Cholest-5-ene-3β-7-diols
	Cholesterol hydroperoxides
Spray-dried egg[b]	Cholest-5-ene-3β-7-diols
	3β-Hydroxycholest-5-en-7-one
	5-,6β-Epoxy-5β-cholestan-3β-ol
	5-Cholestane-3β-5,6β-triol
Heat-dried egg	5,6-Epoxy-5-cholestan-3β,ol
	5,6β-Epoxy-5β-cholestan-3β-ol
Dried-egg mix	5,6-Epoxycholestan-3β-ols
	3β-Hydroxycholest-5-en-7-one
	5-Cholestane-3β,5,6β-triol
Milk products	
Anhydrous milk fat	Cholest-4-en-3-one
	Cholesta-3,5-dien-7-one
Non-fat dry milk	Cholest-4-en-3-one
	Cholesta-3,5-dien-7-one
	Campest-2-ene[c]
	Stigmast-2-ene[c]
Butterfat	5-Cholestan-3-one
Butter oil	Cholesta-3,5-diene
Other products	
Pork fat	3β-Hydroxycholest-5-en-7-one
Brewer's yeast	Ergosterol 5,8-peroxide
	Cerevisterol
	5,8-Ergosta-6,E-22-diene-3β,5,8-triol
Baker's yeast	Cholesta-3,5-dien-8-one
Beef[b]	Cholestatriene[c]
	Cholesta-3,5-dien-7-one
Edible oils	3β-Hydroxystigmast-5-en-7-one
	Stigmasta-3,5-dien-7-one
	Steroid hydrocarbons[c]

[a] Reproduced from Smith (1981) with permission. Original references in Smith (1980, 1981).
[b] Foodstuffs variously irradiated.
[c] Elimination (not oxidation) products.

instance, one might expect the 5α- and β-hydroperoxide resulting from a resonating radical as a consequence of radical formation at the allylic 7 position. It is puzzling that the 5α-hydroperoxide is unique for the reaction of $^1O_2$ with cholesterol. Although free radical β-peroxidation may be sterically hindered by the presence of the axial β-methyl group (C-19), one could argue that the 5α-hydroperoxide might also arise via a free-radical process in addition to a $^1O_2$ reaction.

3.6. Oxidation of β-Carotene and Vitamin A

Compounds containing conjugated double bonds generally oxidize at a much faster rate than those with the same number of non-conjugated double bonds (Holmant, 1954). The activation energy for the autoxidation of β-carotene in model systems of low A_w is 10·3 kcal mol^{-1} compared to 15·2–17·2 kcal mol^{-1} for linoleic acid or its esters, suggesting oxidation of β-carotene is more favoured than linoleate (Balock et al., 1977). Studies on the oxidation of β-carotene, beyond the observations that it is bleached as a result, are rather limited. As expected, more polar oxidation products result including many of the carbonyl compounds observed in the oxidation of unsaturated fatty acids (Forss, 1972; Teixeira Neto et al., 1981). Epoxides are readily formed at the double bond in the β-ionone ring and such epoxidized carotenoids are commonly found in naturally occurring pigments and as a metabolite of retinoids (McCormick et al., 1978). Oxidation of β-carotene is very complex and may take several paths. In general, conjugated double-bond systems favour radical addition reactions rather than abstraction reactions (Scott, 1965). Carotenoids undergo radical addition reactions leading to bleaching (Bors et al., 1981). Compared to abstraction reactions, addition of the methyl radical to conjugated double bounds is very rapid (Pryor et al., 1972). However, very electrophilic oxidizing free radicals can abstract an electron from β-carotene to yield a radical cation (Willson, 1981):

$$Cl_3\text{—}CO_2^{\cdot} + \beta\text{-Car} \rightarrow Cl_3\text{—}CO_2^{-} + \beta\text{-car}^{\cdot+} \tag{42}$$

This is a very rapid reaction approaching diffusion control, and has called into question the specificity of β-carotene as a scavenger for $^1O_2$ and its use as a diagnostic reagent for initiation of lipid oxidation by $^1O_2$ (Willson, 1981). Oxidation of the related vitamin A should follow the same patterns as β-carotene. Although carotenoids and retinoids, such as the various species of vitamin A, are sensitive to oxidative destruction, little is known about the mechanisms of their oxidation. This is a field ripe for studies relating to oxidation of conjugated double bonds in general and those in carotenoids, retinoids and the mould-inhibitor sorbic acid. Carotenoids and retinoids are known to retard tumourigenesis, and their oxidation *in vitro* and *in vivo* may have profound implications in the aetiology of this disease (Sporn *et al.*, 1976; Lotan, 1980).

The oxidative destruction of the fat-soluble vitamins A, D, E and K has important and obvious nutritional implications. The oxidation of vitamin E is understood in some detail (discussed in Section 5) with much less being known about the oxidation of vitamin A and its precursors. Although the conjugated diene in vitamin D reacts readily with photogenerated $^1O_2$ to form endoperoxides (Moriuchi *et al.*, 1979), virtually nothing is known about the oxidation of vitamins D and K in milk fat or dairy products.

3.7. Oxidation of Amino Acids by Oxidizing Lipids

Many free amino acids and those in peptides and proteins are readily oxidized in association with oxidizing lipids (Fig. 24) (Schaich, 1980; Karel and Yong, 1981). This may provide, in part, a basis for the proposed use of certain amino acids, peptides, protein hydrolysates and proteins as antioxidants (Dugan, 1980; Eriksson, 1982).

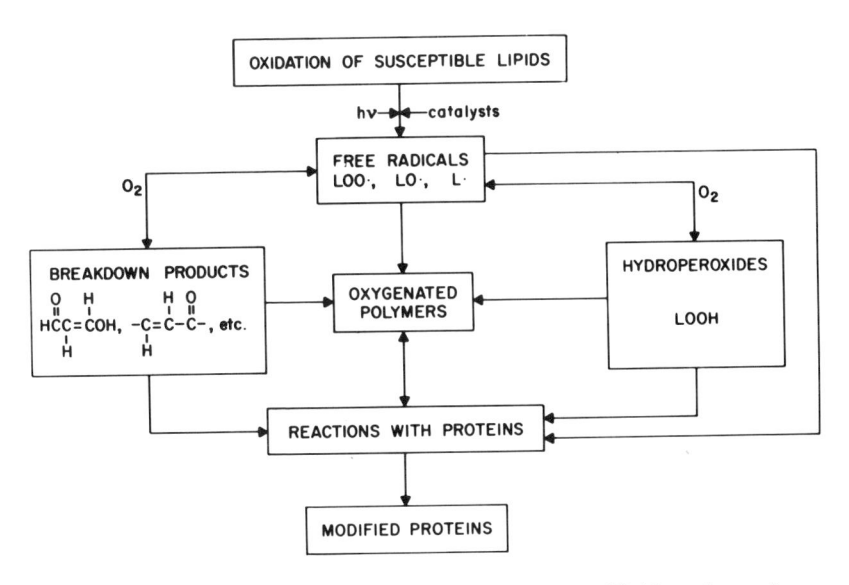

FIG. 24. Possible reactions between oxidation products of lipids and proteins to yield modified proteins. (Reprinted from Karel (1980) with permission.)

4. PRO-OXIDANTS

In general, pro-oxidants (e.g. metals, irradiation, enzymes) lower the activation energy for initiating lipid oxidation (Labuza, 1971). The initial rate of lipid oxidation has been calculated to be about 10^5-10^6 times slower than for the monomolecular decomposition of peroxides at 25 °C, implying the presence of a low concentration of an initiator (e.g. $^1O_2$) before monomolecular peroxide decomposition becomes a major factor. Monomolecular decomposition of single hydroperoxides occurs up to 0·5–1% oxidation on a molar basis (Fig. 8). The food becomes organoleptically unacceptable during the monomolecular phase of hydroperoxide decomposition (Labuza, 1971).

4.1. Metal Ions

Milk (Table 5) and other dairy products contain a wide variety of metal ions, including pro-oxidant transition metal ions such as Cu, Fe, Mo (Johnson, 1974; Jarrett, 1979; Mitchell, 1981). Metal ions capable of undergoing reversible one-electron reductions are very important in catalysed lipid oxidation. Metals are distributed among the various phases of milk with relatively high concentrations of pro-oxidant copper being

TABLE 5
SELECTED TRACE ELEMENTS IN COW'S MILK[a,b]

Element	Cow receiving normal ration (μg litre^{-1}) Mean(s) (Range)	Cow receiving supplement of element in ration[a] (μg litre^{-1})
Copper	130 (20–200)[a] 58–100 (20–400)[b]	NI[c]
Iron	450 (100–900)[a] 370 (210–730)[b]	NI[c]
Manganese	22 (20–30)[a] 25 (10–48)[b]	64
Molybdenum	73 (18–120)[a,b]	371
Nickel	27 (18–36)[a] 30 (0–50)[b]	NI[c]

[a] Johnson (1974).
[b] Jarrett (1979).
[c] NI, no increase.

associated with the fat globule membrane in juxtaposition with the membrane lipids (Haase and Dunkley, 1970).

Metals are thought to function primarily by decomposing hydroperoxides to generate new reaction chains (Labuza, 1971). Either the oxidized or the reduced metal ion can decompose lipid hydroperoxides to allow the following catalytic cycle to increase the rate of lipid oxidation:

$$M^{n+} + ROOH \rightarrow M^{(n+1)+} + {}^-OH + RO^\bullet \qquad (43)$$

$$M^{(n+1)+} + ROOH \rightarrow M^{n+} + {}^+H + RO_2^\bullet \qquad (44)$$

Thus only small quantities of an appropriate metal ion can generate large numbers of reaction chains by cycling between the oxidized and reduced forms.

The reduction of peroxides by metals to generate free radicals is well established (reaction (43)); however, reaction (44) requires that the metal be a sufficiently strong oxidizing agent if it is to be a significant factor in catalysis. Certain iron complexes can cycle effectively between reactions (43) and (44) (Hiatt, 1975). Although iron (with most ligands) and copper in their reduced states are effective reducers of ROOH (reaction (43)), they are inefficient oxidizers in their higher oxidation states (reaction (44)). The oxidation step might be rate-limiting in the catalytic cycle unless alternative mechanisms are available for regenerating the reduced metal (Kochi, 1973). Apparently, alkyl hydroperoxides are relatively more inert to oxidation by Fe^{3+} than to oxidation by Cu^{2+} (Kochi, 1973). Indeed, Fe^{2+}-catalysed decomposition of linoleate hydroperoxides is about tenfold faster than that with Fe^{3+}. Constituents of foods that reduce the Fe^{3+} (or Cu^{2+}?) may accelerate the breakdown of peroxides, perhaps yielding the more reactive Fe^{2+} (O'Brien, 1969; Gardner and Jursinic, 1981). In this way, a redox reaction between a reductant and the peroxide can be catalysed by Fe^{2+}/Fe^{3+} or other transition metal ion couples. Ascorbic acid, thiols and transient superoxide may provide reducing equivalents to accelerate these reactions. Interactions between metal catalysts and reducing groups (e.g. sulphydryl groups) in milk are undoubtedly complex and may be responsible for inconsistencies and paradoxes in the literature (Yee and Shipe, 1982).

In a comprehensive study on the decomposition of linoleic acid hydroperoxide (LAHPO), O'Brien (1969) observed that the effectiveness of metals at decomposing LAHPO was in the order $Fe^{2+} > Fe^{3+} > Cu^{2+}$. EDTA ($10^{-5}$ M) and cyanide (10^{-2} M) completely inhibited the decomposition of LAHPO by these metals. The decomposition of LAHPO by

Cu^{2+} and Fe^{3+} was markedly stimulated upon addition of ascorbate or thiols which probably reduced the ions to facilitate reduction of the peroxide. Ascorbate was about 32-fold more effective than cysteine in enhancing the rate of the reaction of Cu^{2+} with LAHPO at pH 5·5. Also, Fe^{3+} in the presence of ascorbate was only about 8% as effective at pH 5·5 as the Cu^{2+}–ascorbate system in decomposing LAHPO. A copper–ascorbate complex is probably responsible for the degradation of LAHPO with maximal activity at about pH 6·0. The greater effectiveness of the Cu–ascorbate system for decomposing LAHPO compared to the others tested may be relevant to the oxidation of lipids in milk.

Copper rapidly oxidizes carbon-centred radicals (reaction (45)) (Kochi, 1973; Hiatt, 1975) to provide an alternative route to reduced copper which can also sustain the catalytic cycle of reactions (43) and (44):

$$R_2\dot{C}O + Cu^{2+} \rightarrow Cu^+ + R_2CO \qquad (45)$$

It is thus possible that mixtures of metals can be synergistic pro-oxidants by analogy with the enhanced anti-oxygenesis observed for mixed antioxidants (Kochi, 1973; Uri, 1973; Scott, 1965). Trace quantities of copper could perhaps catalyse decomposition of alkyl hydroperoxides. In this case, Cu^+ regenerated through oxidation of carbon-centred radicals by Cu^{2+} could reduce Fe^{3+} to Fe^{2+} to facilitate this decomposition of alkyl hydroperoxides. However, it should be borne in mind that Cu^{2+} is apparently more effective in the oxidation of ROOH (Kochi, 1973) (reaction (44)) to accelerate decomposition of hydroperoxides. The situation is further complicated because metal ions in their lower valence state can rapidly reduce alkoxy (reaction (46)) and peroxy (reaction (47)) radicals and in their higher oxidation states (e.g. Cu^{2+}) can oxidize alkyl radicals (reaction (48)) (Hiatt, 1975):

$$RO^\cdot + M^{n+} \rightarrow RO^- + M^{(n+1)+} \qquad (46)$$

$$RO_2^\cdot + M^{n+} \rightarrow RO_2^- + M^{(n+1)+} \qquad (47)$$

$$R^\cdot + Cu^{2+}(OH^-)(O\bar{A}c) \rightarrow ROAc + Cu^+(OH^-) \qquad (48)$$

It may be important that various reagents (e.g. Cu^{2+}, H_2O_2, $^3O_2$) can oxidize alkyl radicals ($R^\cdot$) to generate a reactive electrophilic carbonium ion (R^+). These latter species alkylate various susceptible nucleophiles (Scott, 1965; Norman and West, 1969).

The sum of reactions (43) and (44) involving metal-catalysed decomposition of hydroperoxides results in a stoichiometry of one initiating radical from the decomposition of one hydroperoxide, i.e., a linear propagation. Possible differences in the mechanism of hydroperoxide decomposition involving metal-catalysed versus MAH thermally induced homolysis may be exaggerated at higher temperatures thereby yielding differences in the flavour stability of foods subjected to accelerated storage tests at higher temperatures compared to lower temperatures where metal catalysis may be relatively more important. It is known that accelerated storage tests at high temperatures may not correlate with the observed shelf-life of a product at lower temperatures (Labuza, 1971; Ragnarsson and Labuza, 1977).

Metals may act as pro-oxidants via other mechanisms as well. For example, metals may actually react with substrate to yield free radicals to initiate oxidation (Kochi, 1973; Uri, 1973).

$$M^{(n+1)+} + RH \rightarrow M^{n+} + R^{\cdot} + H^{+} \qquad (49)$$

The rate of this direct initiation reaction may be 50-fold less than the previously mentioned decomposition of peroxides (Uri, 1973). In addition, this proposed reaction has been criticized on kinetic grounds (Kamiya et al., 1963; Labuza, 1971). Metals under certain conditions (probably in some complexed form) can 'activate' oxygen to provide a pro-oxidant effect (Cotton and Wilkinson, 1972; Hanzlik, 1976; Keevil and Mason, 1978):

$$(L)_x M^{(n+1)+} + {}^3O_2 \rightarrow (L)_x M^{n+} ----O^{-} \qquad (50)$$

As mentioned previously, the reduced form of the metal is probably a better 'activating' agent than the oxidized form as in reaction (50) (Scott, 1965; Cotton and Wilkinson, 1972; Schaich, 1980). The foregoing discussion is predicated on the existence of transition metal ions as essentially free (solvated) in solution, which is rarely the case in foods. Metal ions co-ordinated by water molecules in aqueous solution, for example, are not stabilized in higher oxidation states, as is possible with some other ligands, to catalyse oxidative reactions. Ligands associated with transition metals can have profound effects on the catalytic properties of the bound metal (Cotton and Wilkinson, 1972; Hanzlik, 1976). Ligands associated with pro-oxidant metals can facilitate electron transfer and stabilize lower or higher oxidation states of the metals to allow rationalization of a branching-type free-radical chain reaction involving complex formation between the metal chelate and hydroperoxide:

$$(L)_x M^{n+}-OH + {}^{-}OOR \rightarrow (L)_x M^{n+}---OOR \rightarrow (L)_x M-O^{\cdot} + {}^{\cdot}OR$$
$$\xrightarrow{2R'H} 2R'^{\cdot} + (L)_x M^{n+}-OH + HOR \qquad (51)$$

This reaction sequence is generalized from the formation of compounds I and II in the proposed mechanisms for peroxidases (Walsh, 1979; Sligar *et al.*, 1980).

Ligand groups surrounding a metal may participate in its oxidation or reduction by facilitating the extraction or addition of electrons (Waters, 1971). The reduction of a transition metal in the chelate by an outer sphere mechanism resulting in the direct oxidation of an organic radical or olefin can readily proceed if a ligand (e.g. water) possesses an unshared electron pair to aid electron flow and provides a linear path for electron transfer. Replacement of H_2O with other ligands (e.g. NH_3) which have no free electron pair to serve as a conductor, or anions (e.g. F^-) that are tightly bound, blocks electron flow.

Inner sphere reactions are two-stage processes in which the attacking group displaces a ligand group (e.g. ROO^- displacing HO^- in reaction (51)). Subsequent one-electron transfers within the complex lead to the release of $RO^{\cdot}$ with the metal in a transient higher formal valence state such as the ferryl ion, Fe^{4+}. Since an intermediate complex is formed which decomposes to products analogous to an enzyme reaction, these inner sphere reactions may exhibit saturation 'Michaelis–Menten' kinetics.

Although other possibilities may exist for metal catalysis in lipid oxidation (Kochi, 1973; Uri, 1973), transition metals might act as pro-oxidants in the reduced state (co-called 'reduction activation') or in higher transient oxidation states stabilized by ligands:

$$(L)_x M^{(n-1)+} \underset{}{\overset{+e^-}{\rightleftharpoons}} (L)_x M^{n+} \underset{}{\overset{-e^-}{\rightleftharpoons}} (L)_x M^{(n+1)+} \tag{52}$$

Iron associated with various ligands provides a good example of the foregoing possibilities in progressing from reduced Fe^{2+} to Fe^{4+} oxidation states via one-electron oxidations. Transient higher oxidation states of iron chelated in porphyrins are important in the enzymic activity of peroxidases, mono-oxygenases (Walsh, 1979; Sligar *et al.*, 1980) and possibly in the homolytic decomposition of hydroperoxides by haem pigments in lipid oxidation (Waters, 1971; Schaich, 1980).

Thus ligands can alter catalytic reactivity by controlling the electron density at the metallic centre of a complex ion and so its reduction potential. A higher positive charge and readily available electrons from metals in the high-spin state may favour binding of ROO^- and ready donation of an electron to the hydroperoxide (Waters, 1971).

Since it is likely that complex formation between a metal (or a preformed complex) and a peroxide anion involving co-ordination sites in the metal is a precondition for homolysis of peroxides, other co-ordinating species may actually inhibit lipid oxidation via steric hindrance and/or by forming inert complexes more stable than the peroxide complex. More polar solvents tend to retard lipid oxidation catalysed by metals (Uri, 1970; Labuza, 1971), perhaps by interaction with the metal to form inert complexes. Water may retard oxidation by such interactions with metals or by hydrogen bonding to the hydroperoxide at low water activities (A_w). However, higher A_w's may serve to mobilize catalysts, thereby enhancing the rate of lipid oxidation (Fig. 25).

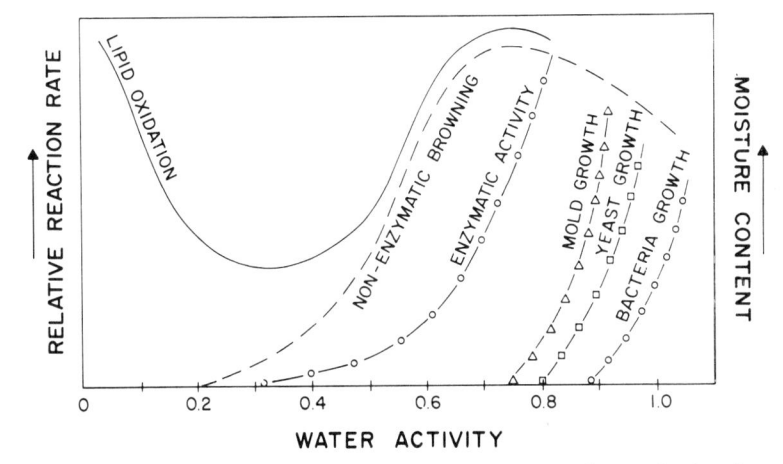

FIG. 25. Effect of water activity in foods on lipid oxidation and other factors affecting the stability of foods (Reprinted from Labuza (1971) with permission. © CRC Press, Inc.)

The principal metals of historical concern in milk products have been ions of copper and iron. Consideration of only the standard reduction potential for Fe^{3+} ($E_0 = 0.77$ V) compared to that of Cu^{2+} ($E_0 = 0.19$ V) indicates that Fe^{3+} is a much stronger oxidizing agent than Cu^{2+}. However, experience indicates that copper in milk is more pro-oxidant than iron, which, under normal circumstances, is of little consequence in catalysing lipid oxidation in milk (Haase and Dunkley, 1970; Jarrett, 1979). This difference probably reflects differences between the interactions of the two metals with other milk constituents (e.g. ascorbic acid, thiols, serine phosphate residues). Since it is unlikely that these ions exist freely to any significant extent in milk systems, the ligands associated with them must play an important role in defining their reactivities as well as their distribution in the three phases of milk, which in turn may influence their reactivity in a secondary way. Possibly reflecting this situation, citric acid is totally ineffective in retarding Cu^{2+}-catalysed oxidations (Scott, 1965).

Differences in reactivities between iron and copper ions might be associated with practical differences in their usual co-ordination numbers, i.e. $Fe^{3+} = (6)$ and $Cu^{2+} = (4)$ (Freifelder, 1982). Differences in the geometry of their respective chelates, and whether these might change upon reduction, may play a crucial role in limiting their pro-oxidant activities (Scott, 1965). In addition to changes in spatial arrangements of members of the chelates, the distribution of electrons in the d-orbitals of the metal in relation to the electronegativity of the donor ligand may profoundly affect the reduction potential of the metal in a chelate. Furthermore, the net charge on a chelate can attract or repel electrons, thereby affecting the apparent reduction potential of the metal. For example, a negatively charged metal chelate should not attract electrons as avidly as a neutral or positively charged complex. Also, bulky ligands around a potential pro-oxidant can sterically hinder the transfer of electrons.

Secondary effects may arise simply from alterations in solubilities of metal ions dictated by the relative formation constants of their chelates. Partitioning of pro-oxidant metals into the lipid phase, or the aqueous phase, would have obvious implications for the scope and extent of oxidative reactions. Copper and iron soaps, for example, are 'lipid soluble', perhaps orienting at interfaces to effectively catalyse the oxidation of fats and oils. On a gram-atom equivalent basis, copper soaps are about 14-fold more pro-oxidant than iron soaps (Ohlson, 1973). Thus low levels of lipolysis of milk lipids may provide ligands for metals to facilitate their solubilization in milk lipids leading to subsequent oxidation. Cupric stearate is a potent pro-oxidant (Scott, 1965). Recall also the possible effects of fatty acids in promoting heterolytic or homolytic peroxide decomposition and in facilitating the β-scission of alkoxy radicals. Alternatively, metals bound by proteins can concentrate at lipid interfaces by virtue of the surface activity of the protein (Waters, 1971). Ligands are known to alter the interaction of metals with oxygen yielding 'immobilized' oxygen species that may be relatively innocuous, as in the case of oxyhaemoglobin, or quite reactive, as with cytochrome P-450, peroxidases (Walsh, 1979; Sligar et al., 1980) and ferric–ADP complexes (Tien et al., 1981a).

Thus metals may have no effect, may promote or in some cases may actually retard oxidative reactions acting through a variety of physico-chemical mechanisms (Uri, 1970, 1973; Howard, 1980). The importance of the milk fat globule membrane in promoting the oxidation of milk lipids no doubt reflects the importance of juxtaposition of lipids with various pro-oxidants.

The reducibility and reduction potential of a metal is determined by two major factors: (1) its position in the periodic table; and (2) the effects of ligands bound to the metal ion. The periodic effect is fixed, but ligands may lead to profound differences in the reduction potential of metal ions and their reactivity (Cotton and Wilkinson, 1972; Hanzlik, 1976).

4.2. Ligand Effects

Indigenous and added metals may exist in milk as insoluble species (e.g. as phosphates) in equilibrium with the soluble ions or bound to ligands as chelates soluble in lipid or water, or associated with the various lipoproteins and proteins. Of course, some metals are naturally associated with milk constituents, such as molybdenum with xanthine oxidase (Johnson, 1974). Thus the distribution of metals in milk and dairy products will be some complex function of relative solubility products and formation constants of the various metal–ligand species. Superimposed upon this will be the effects of changes in this distribution as a result of processing, storage and season of the year. In some cases, anti-oxidant effects will occur, in others, pro-oxidant effects will occur depending upon the associated ligands. Although the two principal metals of concern in the oxidation of milk lipids are copper and iron, other transition metals in dairy products (Johnson, 1974; Jarrett, 1979) (Table 5) cannot be ignored since they may also catalyse decomposition of hydroperoxides (reactions (43) and (44)).

Donor ligands associated with metal are generally anions of oxygen, nitrogen and sulphur (Hanzlik, 1976). Certain ligands can stabilize metal ions in higher oxidation states than normal (Cotton and Wilkinson, 1972). Nitrogen and sulphur anion donors are usually more valuable in this regard than oxygen donors because they retain metal ions more strongly and can also bias the spin-state equilibria (e.g. in haem proteins, see below). In protein–metal complexes with an open co-ordination site (no ligand from protein), various dioxygen species may combine with the metal to yield potential 'immobilized active oxygen species'.

Protein ligands that bind metal ions can alter the reduction potential of the metal dramatically, thereby changing its affinity for electrons. In Table 6 are listed ligands in proteins which favour the principal oxidation states of copper, iron and other metals. Much of our knowledge about metalloproteins is available from basic studies on metalloenzymes. A good

TABLE 6
PREFERRED BINDING SITES FOR BIOLOGICALLY IMPORTANT METALS[a]

Metal	Donors at site
Na^+	Neutral oxygen donors, crown ethers, peptide carbonyls, possibly one negative charge at site, small site
K^+	Large site of neutral donors, possibly one negative charge at site
Mg^{2+}	Prefers basic (hard) nitrogen donors, hard oxyanions (carboxylate, phosphate, phenols, and catechols)
Ca^{2+}	Hard oxyanions >nitrogen donors
Mn^{2+}	Similar to Mg^{2+}, often interchangeable but Mn^{2+} decidedly more octahedral
Fe^{2+}	—SH >—NH_2 >—SCH_3 >carboxylates
Fe^{3+}	Serine OH, phosphate, hydroxamate, hydroxide, often polymeric with —O— bridges
Cu^+	Soft donors, —SH >—NH_2 >>others
Cu^{2+}	Imidazole and amines, peptide —CON=
Zn^{2+}	Imidazole, —SH >carboxylates

[a] Hanzlik (1976).

understanding of ligand effects in the activities of metalloenzymes should provide the food scientist with a better understanding of any pro-oxidant effects of these proteins when denatured, as well as non-enzymic metal–protein complexes. Although the pro-oxidant effects of copper in the presence of proteins (Tappel, 1955; Yee and Shipe, 1981) and haemoproteins (Tappel, 1962) are documented, the authors are unaware of any systematic comprehensive studies on the binding of metals to individual demetalated milk proteins and the effects of these chelates on lipid oxidation. A limited number of studies using equilibrium dialysis have demonstrated that skim-milk proteins and soluble proteins of the fat globule membrane bind Cu^{2+} to various degrees (Aulakh and Stine, 1971).

To summarize, ligands may have profound effects on the oxidizing activities of bound metal ions in foods by:

(i) Altering the physical distribution of the metal ion thereby hindering or enhancing access of redox reagents (and lipids) to the metal ion.

(ii) Facilitating electron transfer and increasing or decreasing the reduction potential of the metal ion.

(iii) Stabilizing metals in lower (e.g. Cu^+) or higher oxidation states (e.g. Fe^{4+}) and altering equilibria toward higher or lower oxidation states.

(iv) Favouring the formation of complexes with various oxygen species to 'activate the oxygen'.
(v) Biasing spin-state equilibria of certain metals (e.g. high-spin or low-spin states of iron complexes).

Although these factors are often interrelated, primarily in altering the reduction potential of the metal, a brief comment on each factor may provide some insight into metal pro-oxidant effects.

4.2.1. Redistribution

Redistribution of metal pro-oxidants resulting from homogenization of milk serves as a prime example in which a process may retard lipid oxidation (Schwartz and Parks, 1974). Also, the thermally induced dissociation of haemoproteins may allow the partitioning of the very strong pro-oxidant iron porphyrin into the lipid phase. Furthermore, a change in the axial ligand associated with the fifth co-ordination site of haem as a result of any altered distribution may markedly affect its pro-oxidant effects. The haem in certain haemoproteins (e.g. cytochrome b_5) is sequestered in a hydrophobic cleft in the protein and may become available for reaction with hydroperoxides only after thermally induced unfolding. Solubilized cytochrome b_5 from rabbit liver microsomes undergoes thermal unfolding between 57 and 60°C with the eventual loss of iron protoporphyrin IX (Kitagawa et al., 1982). Non-enzymic lipid oxidation activity of thermally denatured lactoperoxidase increases about 14-fold when heated at 70°C for 56 min (Eriksson, 1974). This is accompanied by a decrease in actual enzymic activity. Apparently, the haem moiety becomes available for reaction upon removal of steric constraints. Certainly, free haemin exerts a much more pro-oxidant effect when compared to various haem proteins on a haem-equivalent basis (Kendrick and Watts, 1969). Haemoproteins in order of effectiveness in decomposing linoleic acid hydroperoxide (LAHPO) are methaemglobin >cytochrome c >oxy-haemoglobin >myoglobin (O'Brien, 1969). Haematin is more active than any of the haemoproteins and is about 430-fold more active in decomposing LAHPO than ionic Fe^{3+}. Native catalase and horse-radish peroxidase are essentially inactive. This is to be contrasted with the pro-oxidant activities of denatured catalase and horse-radish peroxidase (Eriksson, 1974, 1982). Surfactants and tryptic digestion markedly enhance the rate at which cytochrome c degrades LAHPO, presumably by exposing reactive haeme sites (O'Brien, 1969).

Ligands, such as cyanide (10^{-2} M), completely inhibit decomposition of

LAHPO by haematin and the haemoproteins. At pH 7, the Fe^{3+} methaemoglobin was about six-fold more effective than Fe^{2+}-oxyhaemoglobin in catalysing the decomposition of LAHPO. Although a change in the valence state of haem iron may not be essential, the more effective Fe^{3+} oxidation state may have a greater charge affinity for ROO^- thereby enhancing the reaction rate by effectively reducing the 'K_m' for forming the reaction complex. Indeed, some metal complexes are thought to behave in this enzyme-like manner as pro-oxidants (Waters, 1971). Alternatively, the O_2 of the oxyhaemoglobin may have sterically interfered with its interaction with LAHPO. However, it is likely that haemoproteins in the Fe^{2+}-state are eventually oxidized to generate a pro-oxidant Fe^{3+}-haemoprotein (Brown et al., 1963; Morey et al., 1973, and references cited therein).

Thus the availability of the metal to the substrate as affected by competing ligands, as a result of processing etc., can have a marked effect on lipid oxidation in a food.

4.2.2. Effects on Electron Transfer and Reduction Potential

The effects of various ligands on the reduction potential (oxidizing power) of iron are given in Fig. 26. Note that chelation of iron by o-phenanthroline enhances the oxidizing power of Fe^{3+} whereas complexing it with EDTA or cyanide results in a substantial decrease in the affinity of Fe^{3+} for electrons.

Anionic chelates of Fe^{3+} tend to repel electrons, thereby contributing to a decrease in the reduction potential of the complex. On the other hand, the nitrogen ligands in o-phenanthroline favour the attraction of electrons

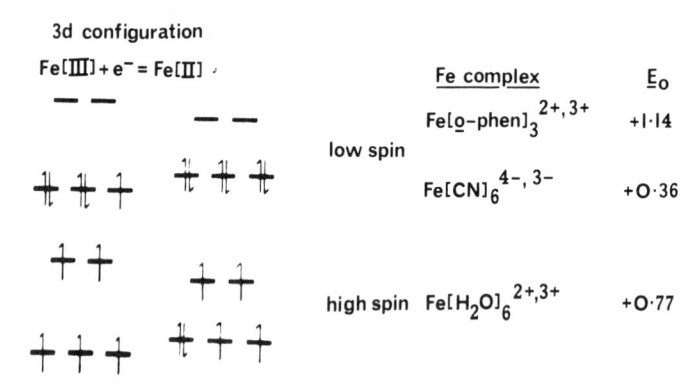

FIG. 26. Effect of ligands on the reduction potential and spin states of iron in chelates. (Reprinted from Hanzlik (1976) with permission.)

to yield higher oxidation states of iron. Ferric chelates of o-phenanthroline are potent pro-oxidants for lipids (Labuza, 1971). Although chelates between iron or copper and EDTA can slowly oxidize ascorbic acid (Khan and Martell, 1967a,b), it is known that the lipid pro-oxidant effects of these metals in foods can be greatly retarded by the addition of EDTA. Thus chelating agents such as EDTA and cyanide form water-soluble iron complexes with low reduction potentials, factors favouring an anti-oxidant effect. Consequently, the electron repelling or attracting effects of ligands may profoundly affect the oxidizing properties of metals.

4.2.3. Stabilization of Oxidation States

Anionic ligands such as nitrogen (in iron porphyrins) or sulphur can stabilize metals in transient higher oxidation states (Cotton and Wilkinson, 1972; Hanzlik, 1976; Williams, 1981). It is probable that this property favours the strong pro-oxidant activity of iron in haem proteins compared to Fe^{3+} $(H_2O)_6$ and which has been attributed to peroxide decomposition (Tappel, 1962; Kendrick and Watts, 1969; Labuza, 1971; Schaich, 1980; Griffin and Ramsey, 1981). Transient higher oxidation states for the iron in haem (e.g. Fe^{4+}, Fe^{5+}) are considered essential to explain oxidative reactions catalysed by peroxidases and cytochrome P-450 (Fig. 27). In these reactions, a common $[FeO]^{3+}$ intermediate, with the iron formally as Fe^{5+}, is proposed (Sligar *et al.*, 1980). The two pathways each require two electrons to 'activate' the oxygen and these are provided by the peroxide in the case of peroxidase and by exogenous reductants in the case of cytochrome P-450.

The radical-like 'oxenoid' oxygen in the $[FeO]^{3+}$ complex is inserted into the substrate by P-450 to effect the oxidation. This insertion reaction is probably favoured by the electron-donating effects of a thiol group as the axial fifth ligand in P-450, thereby easing release of the 'O' to oxygenate the substrate (Ullrich and Kuthan, 1981). This again illustrates the marked effects that ligands may have on the oxidative characteristics of metals.

In comparison, peroxidases utilize iron in the higher oxidation states ($[FeO]^{3+}$–Compound I; $[FeO]^{2+}$–Compound II) to univalently oxidize appropriate substrates by, essentially, abstracting two electrons to eventually yield iron in the Fe^{3+} state. The mechanism proposed by Tappel (1962) (Fig. 28) to explain the decomposition of hydroperoxides by haem is peroxidatic in the sense that a transient higher oxidation state for iron might be invoked and that two radicals might result to initiate new reaction chains.

As depicted in Fig. 28, ROO^- serves as an axial ligand to haem iron

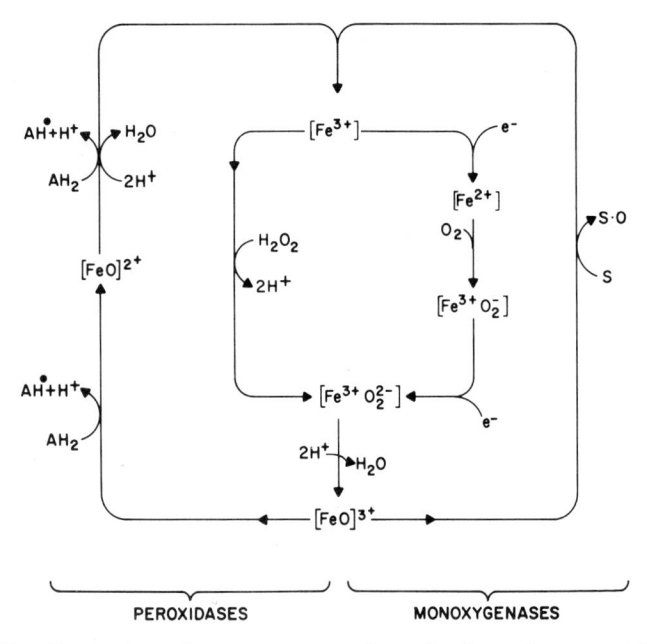

FIG. 27. Comparison between proposed mechanisms for peroxidases and monoxygenases (e.g. cytochrome P-450). (Reprinted from Sligar et al. (1981) with permission.)

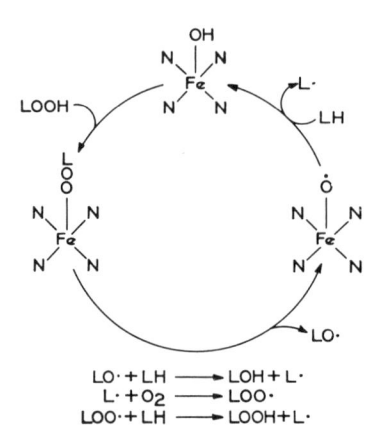

FIG. 28. Proposed mechanisms for haem-catalysed lipid peroxidation. (Reprinted from Tappel (1962) with permission.)

(Fe^{3+}). The 'donation' of an electron to an antibonding orbital in the ROO–Fe^{3+}-(porphyrin) complex could result in a transient Fe^{4+} state and a bond order of 1/2 between the oxygens of the bound peroxide. Energy of solvation of a leaving RO$^\cdot$ may be sufficient to favour the homolytic scission of the peroxide (Taube, 1965). The resulting $^-$O–Fe^{4+}-(porphyrin), in which the transient Fe^{4+} oxidation state is stabilized by the porphyrin ring, is analogous to Compound II in the mechanism for peroxidase activity (Walsh, 1979). Abstraction of an electron from RH by $^-$O–Fe^{4+}-(porphyrin) should initiate new reaction chains via R$^\cdot$ and renew the HO–Fe^{3+}-(porphyrin) catalyst for another cycle (Fig. 28). The existence of a higher oxidation state for porphyrin iron may be inferred from the transient red colour observed during haem-catalysed oxidation of linoleate (Kendrick and Watts, 1969), analogous to that attributed to Compound II in peroxidase catalysis (Walsh, 1979) and observed by George and Irvine (1952) during the decomposition of hydrogen peroxide catalysed by myoglobin. This contrasts with the observations of Griffin and Ramsey (1981) who found that metmyoglobin readily decomposed cumene hydroperoxide by a radical mechanism common to that for the ferrous ion or for several ferric haem compounds. These workers postulated that the electron-rich haem group supplied an electron for the reduction of cumene hydroperoxide to the cumyloxyl radical and $^-$OH analogous to that formed from reaction of metmyoglobin with H_2O_2 (George and Irvine, 1952). However, numerous attempts to detect the transient red oxidized haem species were unsuccessful. Although formal valence changes may not be necessary in the haemoprotein-catalysed decomposition of hydroperoxides the bound metal might still serve as a conduit for electron transfer (Walsh, 1979).

The transient Fe^{4+} oxidation state is termed the 'ferryl' state and is apparently involved in peroxidase reactions. Thus the proposed foregoing mechanism is essentially a 'non-specific' peroxidatic reaction. On the other hand, other mechanisms of peroxide decomposition by iron porphyrins have been envisioned but are less well defined (Labuza, 1971; Schaich, 1980). It is worth noting that each turn of the catalytic cycle (Fig. 28) produces a branching-type chain reaction as with the induced decomposition of peroxides, compared to the linear propagation of the chain reaction in the generally conceived pro-oxidant effects of metals (reactions (43) and (44)). This may provide a rationale for the extremely rapid rate of lipid oxidation catalysed by haem compounds compared to ionic metals (Tappel, 1962; O'Brien, 1969; Schaich, 1980). While direct initiation reactions by Fe^{3+}-haematin may be possible, the extremely rapid rate of lipid oxidation in the presence of these catalysts probably depends upon decomposition of hydroperoxide (Paulson et al., 1974; Schaich, 1980).

On the other hand, high concentrations of haem compounds (low ratio of lipid to haem) may actually be anti-oxidant via some unknown mechanism (Lewis and Wills, 1963; Kendrick and Watts, 1969; Hirana and Olcott, 1971). Strongly bound axial ligands in haem (e.g. CO and NO) may inhibit the pro-oxidant effects of haem or may actually provide anti-oxidant protection to lipids (Brown et al., 1963; Kanner et al., 1980). In the case of nitric oxide myoglobin (MbNO), an anti-oxidant effect was attributed, in part, to the quenching of substrate free radicals by a nitroxide radical, MbNO˙ (Waters, 1971; Kanner et al., 1980).

The foregoing discussions on the pro-oxidant effects of haem compounds represent extreme examples of how ligands can be important in oxidative reactions of lipids. Certainly, many of these observations can be pertinent to other metalloprotein systems. Stabilization of the Fe^{3+} oxidation state in the O-phenanthroline chelate may provide a partial explanation for its effectiveness as a lipid pro-oxidant. The formation constant for $Fe(O\text{-phen})^{3+}$ is much smaller than that for $Fe(O\text{-phen})^{2+}$ (Clark, 1960) thereby favouring reduction of the former to the latter (by ROOH?), contributing to a higher reduction potential for the Fe^{3+} complex compared to $Fe(H_2O)_6^{3+}$. On the other hand, the formation constant for Fe^{3+} $(EDTA)^{-1}$ is much greater than that for Fe^{2+} $(EDTA)^{2-}$ to favour formation of the ferric chelate leading to a lower reduction potential for $Fe(EDTA)^-$ ($E_0 = +0\cdot12$) compared to $Fe(H_2O)_6^{3+}$ (Clark, 1960). The energetics of the redox system as a function of the most stable complex thus contribute to the reduction potential of a metal. Consequently, the reduction potential of a given redox system will reflect the oxidation state of the metal in the most stable complex.

In some cases, metals may be 'frozen' into a particular oxidation state by constraints imposed by the 'stiffness' of the ligands which could prevent reversible univalent redox reactions. For example, cuprous ion has a tetrahedral configuration whereas cupric ion tends to be square planar (Fig. 29) (Hanzlik, 1976). If the ligands cannot accommodate the required rearrangement (Fig. 29) upon oxidation or reduction of the metal, the chelate may not be a good pro-oxidant. Thus this type of binding might yield an anti-oxidant effect. Certain ligands in proteins binding copper may favour a cuprous geometry thereby helping to 'poise' a system and favour reversible redox reactions. In addition, polar ligands (ascorbate?) should retain insoluble cuprous ions in solution. One can envisage that a bidentate ligand co-ordinated with only two sites on copper could 'allow' the conversions depicted in Fig. 29. On the other hand, rigid chelates wherein three or four co-ordination sites are occupied would retard or prevent redox reactions from occurring (Scott, 1965).

Note in Table 6 that amine donors preferentially co-ordinate with both Cu^+ and Cu^{2+} and might provide sufficiently flexible ligands to allow catalytic cycling of the copper ions without requiring an intermediate dissociation. In an essential dissociation of the reduced ion from a chelate followed by its univalent oxidation by oxygen (to yield superoxide) and rebinding of the Cu^{2+} to possibly yield a catalyst with a higher reduction potential, the relative formation constants of the Cu^{2+}, Cu^+ chelates would play a crucial role in defining the catalytic system.

Knowledge of ligands in copper redox enzymes is meagre (Walsh, 1979). Details of the ligands in proteins binding Cu should provide information applicable to an understanding of copper catalysis in the oxidation of milk

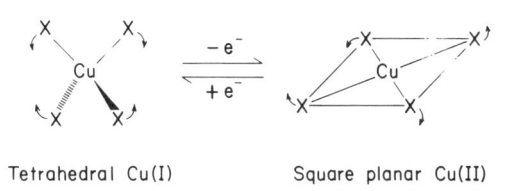

FIG. 29. Change in electronic configuration of copper as a result of reversible oxidation–reduction. (Reprinted from Hanzlik (1976) with permission.)

lipids. For example, metal complexes that favour the decomposition of H_2O_2 must contain 'free' co-ordination sites to complex with H_2O_2 or HO_2^- (Sigel, 1969). As the co-ordination sites on copper are filled (with amino ligands), the rate of decomposition of H_2O_2 decreases, being essentially nil in the case of a four-co-ordinate complex. Copper complexes with two free co-ordination sites are active in the pH range 5·5–7·5, whereas complexes with only one available co-ordination site possess catalytic activity between pH values of 8 and 10.

Apparently, H_2O_2 is decomposed upon formation of a ternary Cu–ligand–peroxo complex without free radicals necessarily being liberated (Sigel, 1969). However, these studies clearly indicate the importance of ligands (and electron donors) in directing the course of the reactions and that saturation of the principal four-co-ordination sites on copper prevents formation of the catalytic complex. This latter effect begs the question whether saturation (with low molecular weight thiols?) of the co-ordination sites of copper in a putative, pro-oxidant, bidentate copper–protein complex might protect foods from oxidation.

As early as 1955, Tappel demonstrated the enhanced rate of lipid oxidation with copper in the presence of proteins, including caseins, compared to copper alone. Copper–serum albumin is more effective in catalysing oxidation of ammonium linoleate at pH 7·0 than either copper–conalbumin or copper–caseinate. An approximation from initial rates of oxidation over the first 10 h of oxidation in a Warburg respirometer indicates that 10^{-3} M copper in the presence of 10^{-4} M conalbumin is over twice as pro-oxidant as 10^{-3} M copper alone.

Tappel (1955) represented the overall reaction as follows:

$$Cu^{2+}\text{–Protein} + ROOH \rightleftharpoons ROO^- - Cu^{2+}\text{–Protein} \rightarrow$$

$$RO^{\cdot} + {}^{\cdot}OH + Cu^{2+}\text{–Protein} \qquad (53)$$

A valence change in copper during the reaction was considered as doubtful. It was thought that the formation of four-co-ordinate, square-planar copper complexes analogous to the planar iron porphyrin ring might favour oxidation in an analogous reaction. However, copper in nature is usually four-co-ordinate (Cotton and Wilkinson, 1972; Hanzlik, 1976; Freifelder, 1982) and does not readily form six-co-ordinate, octahedral, bipyramidal complexes involving axial ligands, as in the iron–porphyrin system. Since free-co-ordination sites for complexation of the peroxide are probably necessary in these types of reaction (Sigel, 1969), it is likely that the four-co-ordination sites in the copper–protein complex are not saturated. The inhibition of copper-catalysed fat oxidation observed at high protein levels may reflect the complete co-ordination of copper in this case compared to catalysis when the molecular ratio of copper to protein approaches unity (Tappel, 1955). However, Smith and Dunkley (1962) point out that Tappel's observations may not be relevant to oxidized flavour in milk since ascorbic acid was absent and the copper concentrations were far in excess of the 10^{-6}–10^{-7} M normally present in uncontaminated milk. Furthermore, the caseins tend to be anti-oxidative (Ericksson, 1982).

Alternatively, stabilization of bound copper in a transient Cu^{3+} state by various protein ligands (analogous to iron in Compounds I and II in peroxidases) may facilitate homolysis of lipid peroxides. A higher oxidation state of Cu^{3+} has been suggested in the oxidation of galactose by galactose oxidase (Walsh, 1979).

Although changes in the oxidation state of metals bound to enzymes (or other proteins?) may not be observable, they may still serve as conduits for electrons during redox reactions (Walsh, 1979), particularly if ligands such as the phenolate ion (and possibly imidazole, sulphur ligands, etc.) provide a 'resonance' stabilization of intermediates. Little is known about ligands and their effects on oxidative reactions catalysed by cupro-proteins (Walsh, 1979). However, the copper in plastocyanin is apparently associated with two histidine imidazole nitrogens and two sulphurs from cysteine and methionine (Walsh, 1979). Likewise, little is known about the association of copper with proteins to form complexes that catalyse lipid peroxidation.

In contrast to transient stabilization of higher oxidation states of copper in peroxide decomposition, copper co-ordinated in only two sites may easily cycle between Cu^+ and Cu^{2+} since the conformational constraints imposed by four-co-ordination complexes (Fig. 29) would not be operative (Scott, 1965).

More recently, research on the pro-oxidant effects of copper–protein complexes has been extended to include observations on the progressive decline in oxidation rates of methyl linoleate in the presence of Cu^{2+} upon addition of peptic digests of soy protein isolate, egg albumin or sodium caseinate (Yee and Shipe, 1981, and references cited therein). Addition of peptides resulting from 1–16 h of enzymic treatment reduces the pro-oxidant effects of Cu^{2+} to near that of the control containing Cu^{2+} without the proteins or their hydrolysates. The decrease in pro-oxidant effects of Cu^{2+}–protein complexes upon peptic hydrolysis is attributed to the formation of less-reactive complexes between Cu^{2+} and amino acids/peptides resulting from proteolysis (Yee and Shipe, 1981). Any conformational constraints imposed on the complex by the three-dimensional structure of the protein would be released upon hydrolysis. Formation of square-planar complexes between amino acids and Cu^{2+} would tend to satisfy the four-co-ordination sites of copper thereby preventing formation of complexes with hydroperoxides necessary for propagation of the oxidation via homolysis of the hydroperoxides. Certainly, the observation of Tappel (1962) that four-co-ordinate, cupro–porphyrin complexes cannot catalyse lipid oxidation is consistent with this interpretation. On the other hand, certain amino acids in the presence of Cu^{2+} readily catalyse the oxidation of linoleic acid in model systems (Farag et al., 1978a,b).

4.2.4. Oxygen Complexes
Oxygenase enzymes contain four prosthetic groups (Cu, Fe, flavins, haem) involved in the activation of oxygen (Keevil and Mason, 1978). Non-

enzymic proteins or denatured enzymes containing these functionalities may also be involved in pro-oxidant reactions in foods via an enzyme-like activation of oxygen. However, since little is actually known about the oxidative mechanisms involved in the highly studied oxidase enzymes, one can only speculate on the non-enzymic oxidations based on what is known about oxidases.

Restrictions imposed on reactivity of triplet-state oxygen, as mentioned previously, can be circumvented by complexation with bound transition metal ions in which spin-allowed oxidations become possible because the number of unpaired electrons on the overall metal ion complex does not change. Alternatively, intermediate formation of stable radicals (e.g. flavin semiquinone) may provide mechanisms for univalent oxidations at physiological temperatures. The bound dioxygen may be co-ordinated in potentially reactive superoxo (not superoxide) or peroxo (not peroxide) forms (Fig. 30). Although not identical to superoxide or peroxide, the bound dioxygen forms resemble, in many respects, either of the two ionic species.

FIG. 30. Oxygen bound to a transition element exhibits or approaches the superoxo or peroxo states, which do not necessarily exhibit properties identical to superoxide or peroxide ions, respectively. (Reprinted from Keevil and Mason (1978) with permission.)

Additional points on the reactivity of bound dioxygen include the following possibilities (Keevil and Mason, 1978):

'there are three possible explanations for the increase in reactivity of molecular oxygen upon coordination. (1) Coordinated dioxygen is in general diamagnetic; therefore reactions with diamagnetic substrates to form diamagnetic products are not limited by the requirement for spin conservation; (2) the metal may hold dioxygen and the substrate in cis positions, lowering the activation energy for oxidation of the substrate; and (3) coordinated dioxygen is, in most cases, partially reduced; increased electron density on the O_2 may activate it. An additional possibility cannot be ruled out: that the superoxo ligand in such complexes is chemically equivalent to an excited state of the oxygen molecule, since bending the molecule may lift the degeneracy of the antibonding orbitals and permit a spin-paired couple to occupy one of them in analogy to the $^1\Delta_g$ singlet state of oxygen.'

Various forms of 'immobilized' reactive oxygen species have been proposed for oxidizing enzymes such as cytochrome P-450 (Sligar *et al.*, 1980) and in model systems whereby the ferryl oxygen complex is considered to be an 'immobilized' but highly reactive hydroxyl radical (Tien *et al.*, 1981*a*,*b*).

Obviously, ligands will markedly affect any 'activation' of oxygen that might result in low levels of non-enzymic oxidation thus providing potentially pro-oxidant products.

Comparison of the rates of decomposition of linoleic acid hydroperoxide by the transition metal ions with their abilities to catalyse oxidation of unsaturated fatty acids indicates that metals may vary in their pro-oxidant mechanism (O'Brien, 1969). As mentioned, perhaps substrate and/or oxygen are activated by complexing with the metals to initiate lipid oxidation.

As seen in the foregoing discussions concerning peroxide decomposition by metals, co-ordination of a metal with a reduced species of oxygen is markedly affected by ligands. As mentioned, the relatively unreactive superoxide radical anion becomes slightly more pro-oxidant upon prior neutralization with a proton (Gebicki and Bielski, 1981) or interaction with complex metal ions (Cotton and Wilkinson, 1972; Hanzlik, 1976).

It is not known what effect, if any, salts of superoxide with divalent metal ions (e.g. Ca^{2+}, Zn^{2+}) in milk may have on lipid oxidation. Electrolytic univalent reduction of oxygen in the presence of Ca^{2+}, Zn^{2+} or other divalent cations results in the production of superoxide salts which subsequently decompose into oxygen and peroxide salts (reactions (54) and (55)) analogous to a dismutation reaction:

$$2O_2^- + M^{2+} \rightarrow M^{2+}(O_2^-)_2 \quad (54)$$

$$M^{2+}(O_2^-)_2 \rightarrow O_2 + MO_2^+ \quad (55)$$

Although the significance of these reactions, if they occur in milk, is unknown, it has been postulated that superoxide bound to calcium in the minerals of bone may present an oxidative hazard (Bray *et al.*, 1977). In this regard, the amorphous apatitic structure of calcium phosphate in milk is thought to be similar to the precursor of bone mineral (McGann, 1979). Furthermore, divalent salts of peroxide are considered to be strong oxidizing agents (Cotton and Wilkinson, 1972), and calcium peroxide is used in the United States as an oxidant for improving bread-dough rheology (Pyler, 1982).

4.2.5. Spin State

Ligands can affect the electronic configuration in the *d*-orbitals of transition metals such as iron in porphyrins (Hanzlik, 1976; Walsh, 1979; Livingston and Brown, 1981). In addition to differences in the oxidation state (e.g. Fe^{2+}, Fe^{3+}, Fe^{4+}, Fe^{5+} etc.), metals such as iron can exist in more than one spin state according to ligand field theory (Hanzlik, 1976; Colton and Wilkinson, 1972).

Ligand field theory (LFT), an outgrowth of crystal field theory (CFT), provides a relatively simple system for rationalizing electronic transitions of metals as affected by various ligands (Cotton and Wilkinson, 1972). However, one should keep in mind that LFT cannot be intermixed with the more sophisticated molecular orbital theory (MOT) to visualize how the electrons of ligand donors may actually form bonds involving the various orbitals in the outer valence shells of metals. For example, if one apportions the electrons available from the nitrogen ligands in the porphyrin ring to form two covalent bonds and two co-ordinate covalent bonds with the 3*d* electrons of iron to form haem compounds, understanding of the subsequent electronic transitions for high-spin versus low-spin haem species is obscured.

LFT treats the electrons in the anionic ligands associated with the metal as point charges and makes no attempt to involve them physically in bonding, as with MOT. This allows a relatively simple discussion of low-spin–high-spin electronic conversions of metals in chelates, uncomplicated by molecular orbitals involving ligands. At the same time, the reader should be aware of the power and importance of MOT and how this can be used to understand the bonding of ligands with metals to provide mechanistic insights (Livingston and Brown, 1981, and references cited therein).

Iron and copper each have electrons in their 3*d* shells which can accommodate ten electrons in five orbitals to assume a stable noble electronic configuration. However, Fe^{2+} has only six e^- in the 3*d* shell ($3d^6$), and Fe^{3+} only five e^- ($3d^5$), whereas Cu^+ has ten e^- ($3d^{10}$) and Cu^{2+} has nine e^- ($3d^9$) (Hanzlik, 1976). These differences in electronic configuration between iron and copper may result in profound differences in their pro-oxidant capabilities as affected by associated ligands.

Unlike the Cu^+/Cu^{2+} system, Fe^{2+} and Fe^{3+} can exist chelated within the porphyrin conjugated ring system with relatively slight dislocations in the octahedral complex resulting from univalent redox reactions (Scott, 1965; Hanzlik, 1976; Walsh, 1979). Furthermore, the conjugated porphyrin ring would appear to support the higher iron oxidation states (Fe^{4+}, Fe^{5+})

necessary for some biochemical/chemical reactivities. Iron with a co-ordination number of six can form octahedral complexes, in the case of haem, associated with four nitrogenous ligands in the plane of the ring with two additional sites for *trans* axial ligands. In addition, the electrons in the Fe^{2+} and Fe^{3+} chelates can distribute themselves to occupy the $3d$ orbitals in two principal configurations, i.e. high spin or low spin (see below). By comparison, metals with $3d^{10}$ or $3d^9$ electronic configurations, i.e. Cu^+ or Cu^{2+}, are constrained to interact differently with ligands, partly because the filled or nearly filled $3d$ shell prevents expansion of the shell to accommodate necessary ligands. Furthermore, octahedral configurations for metals with these electronic configurations are not possible. Instead, Cu^+ exists in a tetrahedral configuration whereas Cu^{2+} favours a square-planar complex. Any interconversion entails redistribution of orbital energy levels (Cotton and Wilkinson, 1972). Consequently, Cu^{2+} chelated into a porphyrin ring would resist reversible redox reactions requiring a transient tetrahedral configuration. Furthermore, the four available ligands in the porphyrin ring would satisfy the sites on copper thereby preventing the necessary co-ordination with peroxides to catalyse homolytic scission to generate free radicals. Pro-oxidant copper complexes, therefore, would probably require ligands that would not satisfy all of their co-ordination sites and which would facilitate any essential changes in electronic and spatial configurations (Fig. 29).

A description of the electronic configuration of haem iron according to LFT will serve to illustrate possible electronic transitions available to the haem iron. In the gas phase, the five $3d$ orbitals of iron are of equal energy (degenerate) so that each orbital in Fe^{3+} contains one electron. Addition of an electron to yield Fe^{2+} requires spin-pairing energy (P) to fill one of these degenerate orbitals with an electron of opposite spin. However, the binding of iron by ligands, as in haem, results in ligand field splitting, and, in the case of haem, the orbitals are no longer degenerate but are split into two discrete groups, three low-lying and two higher-lying orbitals separated by a field-splitting energy, Δ (Fig. 31). When Δ is small relative to P, the higher spin-pairing energy favours the population of all five $3d$ orbitals to give high-spin iron. Fe^{3+} has all five $3d$ electrons unpaired whereas Fe^{2+} has four unpaired electrons. However, if Δ is larger than P, the two higher orbitals are not populated and the electrons will distribute themselves in the three lower-lying orbitals to yield low-spin iron. In the latter case, Fe^{3+} has only one unpaired electron and Fe^{2+} has none and is, therefore, electron paramagnetic resonance (EPR) silent. Low-spin iron (Fe^{2+}, Fe^{3+}) porphyrin is more compact than high-spin iron. Thus low-spin iron rests in the plane of the porphyrin ring system, but the larger high-spin iron cannot be accommodated and must lie out of the plane of the macrocyclic ring. Axial ligands may determine the equilibrium between the high-spin and low-spin states in haem groups.

The reduction potential of a metal can be altered dramatically by its association with a given ligand (Figs 26 and 31). Since the energies of the d orbitals in iron, for example, are split differently by various ligands, the type of orbital an electron leaves or enters during redox reactions will affect the reduction potential of the metal. In Fig. 26 Fe(o-phen)$_3^{3+}$ has a much higher reduction potential than Fe(H$_2$O)$_6^{3+}$ and should be a much

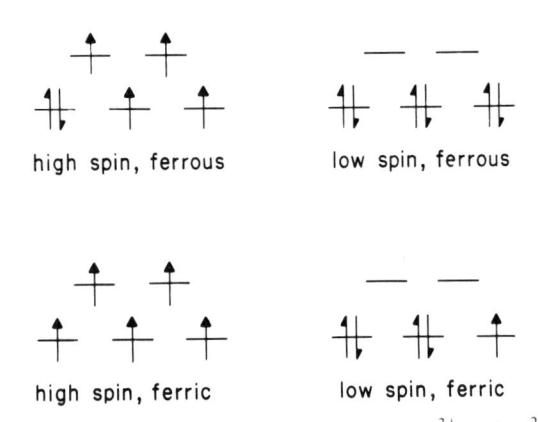

FIG. 31. Electronic configurations in the 3d orbitals of Fe^{3+} and Fe^{2+} in high and low spin states. Various ligands influence the spin state as well as the oxidation state. (Reprinted from Livingston and Brown (1981) with permission.)

stronger oxidizing agent. In the low-spin state, Fe(o-phen)$_3^{3+}$ can accept an electron much easier for maximum electron pairing than Fe(H$_2$O)$_6^{3+}$ because Δ is much larger than P in the former yielding the low-spin state rather than the high-spin state in the latter. Although the net charge on each iron species is the same, the electron entering an orbital of lower energy in the Fe(o-phen)$_3^{3+}$ will be more stable, thereby favouring the reduction of Fe(o-phen)$_3^{3+}$. On the other hand, the low-spin Fe(CN)$_6^{4-,3-}$ complex has a very low reduction potential since the net negative charge on the complex tends to repel electrons making reduction more difficult. Considerations such as the foregoing can be useful in developing a qualitative appreciation for ligand effects on the redox potential of metals and on possible pro-oxidant properties of metals in dairy products. As mentioned previously, the abilities of transition metals to catalyse otherwise forbidden reactions lies, in part, in their facilitation of overcoming restrictions due to electronic configurations as in the activation of oxygen by oxygenases (Keevil and Mason, 1978).

It is worth noting that lactoperoxidase (Walsh, 1979), metmyoglobin (Livingston and Brown, 1981), methaemoglobin (Weissbluth, 1974) and denatured cytochrome b_5 (Kitagawa *et al.*, 1982) all contain ferric haem with the iron in the high-spin state. Because metmyoglobin and methaemoglobin are potent lipid pro-oxidants, one wonders whether one could generalize to include the ferric, high-spin states in exposed haem as favouring lipid oxidation. Electron exchanges involving the high-spin state with unpaired electrons in outer $3d$ orbitals might favour transient electron donation to a peroxide ligand with the higher oxidation state of iron stabilized by electron donation from the conjugated porphyrin ring system. In this sense, denatured haem proteins, such as cytochrome b_5 in juxtaposition with lipids in the fat globule membrane, may serve as a focal point for catalysing lipid oxidation. In contrast, the relatively inert carbonmonoxy myoglobin and the anti-oxidative nitric oxide myoglobin (Kanner *et al.*, 1980) are in the low-spin, ferrous state (Livingston and Brown, 1981). The low-spin ferric state of the phenanthroline chelate (Fig. 26), on the other hand, might have a sufficiently high reduction potential to oxidize a hydroperoxide or substrate directly.

5. ANTI-OXIDANTS

In general, anti-oxidants have been classified into three major groups:
  (i) Inhibitors of free-radical chain reactions (e.g. hindered phenolic anti-oxidants (such as the tocopherols, BHT, BHA), thiols or ascorbic acid).
 (ii) Peroxide decomposers such as the thioethers, methionine and thiodipropionic acid and its esters. These sulphides decompose peroxides by nucleophilic, heterolytic rather than homolytic mechanisms thereby avoiding the formation of free radicals for initiation of new chain reactions (Hiatt, 1975). Although the non-radical nucleophilic decomposition of linoleic acid hydroperoxide at pH 7, 23°C by cysteine and ascorbate is about 15- to 17-fold that for methionine (O'Brien, 1979), the amino acid residues in proteins constrained at a lipid interface should be relatively more effective than their counterparts in solution. Furthermore, certain nucleo-

philes may be pro-oxidant as a result of one-electron-transfer reactions (Pryor, 1976).

(iii) Ligands that de-activate metallic pro-oxidants, presumably resulting from a decrease in their reduction potential and by favouring the partitioning of the chelate away from lipids. Some chelates are actually excellent anti-oxidants (Uri, 1970).

Thus metals can, depending upon the circumstances, stabilize or markedly inhibit oxidation (Schaich, 1980). For example, at very low oxygen concentrations, certain metals (e.g. Cu^{2+} and Fe^{3+}) can actually exert an anti-oxidant effect not evident when these metals, in their lower oxidation states, were present. Such anti-oxidant functions of Cu^{2+} and Fe^{3+} have been attributed to the oxidation of $R^{\cdot}$ to R^{+} leading to chain termination and also to formation of unreactive radical–metal complexes. In addition, other food constituents such as enzymes (e.g. sulphydryl oxidase, lactoperoxidase, superoxide dismutase, catalase, and glutathione peroxidase), singlet oxygen quenchers (e.g., β-carotene, α-tocopherol) and browning reaction products represent additional specific groups of anti-oxidants that may or may not function according to the three aforementioned mechanisms. The interplay among these various factors and their abundance relative to the pro-oxidants as affected by physiology and health of the cow, and by the processing and storage of milk and its products will obviously help to define the oxidative stability of dairy products. Labuza (1971) also considers favourable environmental conditions for a food as a class of anti-oxidant.

5.1. Tocopherols

Apparently, the tocopherols (α-tocopherol) serve an anti-oxidant function by: (1) physically quenching $^{1}O_{2}$; (2) chemically reacting with $^{1}O_{2}$; and (3) univalently reducing free radicals to break chain reactions followed by a reaction between the resultant tocopheryl radical and a peroxy radical to yield a peroxide. Most information in the literature relates to the function of the four isomers of tocopherol (α, β, γ and δ) as hindered phenolic anti-oxidants. The hindered phenolic anti-oxidants (AH) suffer from several distinct disadvantages (Howard, 1980). Firstly, they have no effect on the rate of chain initiation occurring by homolysis of peroxidic impurities and, secondly, a hydroperoxide is formed during the slow rate-controlling termination reaction between AH and $ROO^{\cdot}$ and a peroxide is formed during the rapid chain-terminating reaction between $A^{\cdot}$ and $ROO^{\cdot}$. Although each phenolic anti-oxidant can react with two peroxy radicals (stoichiometric factor, $n = 2$), the resultant peroxides may eventually undergo homolytic scission to initiate new reaction chains (Scott, 1965; Pryor, 1976). Furthermore, the hindered phenolic group may undergo direct reaction with molecular oxygen at high temperatures (~100°C) to render these types of anti-oxidant ineffective. Anti-oxidant activity among the four isomers is thought to decrease in the order $\alpha > \beta > \gamma > \delta$ (Fukuzawa et al., 1982). The chemistry of the oxidation of the tocopherols is very complex with a large variety of oxidation products possible (Korycka-Dahl and Richardson, 1980). Initially, at least, α-tocopherol can reduce peroxy radicals univalently to yield the corresponding hydroperoxide and a phenolic free radical sufficiently stabilized by resonance to prevent the initiation of new free-radical reaction chains. The reaction of R˙ with $^3O_2$ is extremely rapid so that phenolic anti-oxidants are compelled to react with ROO˙ to break the chain reaction rather than reduce R˙ to an innocuous RH. Unfortunately, a hydroperoxide is produced as a result of this univalent reduction. However, the tocopherols can also reduce the hydroxy and alkoxy radicals from homolytic scission of this peroxide to retard initiation of further oxidative chains. Of course, the tocopherols are eventually destroyed. Usually, the phenolic radical formed as a result of the foregoing univalent reduction of a peroxy radical can readily react via a radical–radical reaction with a peroxyl radical to yield a peroxide adduct (Fig. 32). Thus, the stoichiometry of reactions between phenolic anti-oxidants, such as BHT, and peroxy radicals is considered to be two radicals per anti-oxidant.

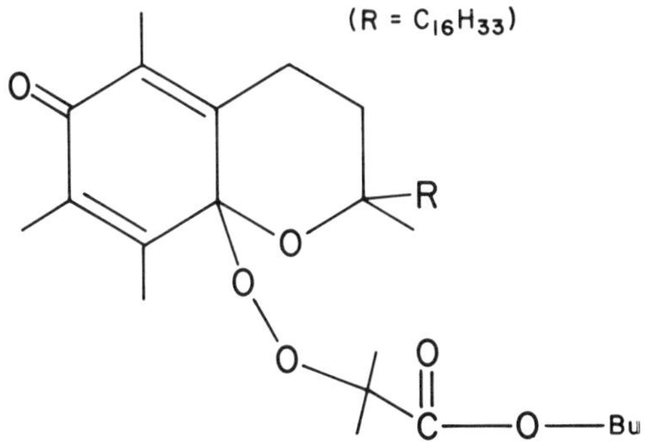

FIG. 32. A peroxide adduct resulting from a radical–radical reaction between the tocopheryl radical and a peroxy radical. (Reprinted from Winterle and Mill (1981) with permission.)

Although α-tocopherol readily forms analogous peroxy adducts (Fig. 32), the overall stoichiometry may be less than two; 1·6 + 0·1 in hexane compared to 1·3 + 0·3 in a liposomal membrane (Winterle and Mill, 1981).

By analogy with synthetic phenolic anti-oxidants, α-tocopherol is pro-oxidant at higher concentrations (Cillard et al., 1980a,b). Reduced effectiveness and/or pro-oxidant effects of α-tocopherol may be related to the formation of unstable peroxides (Fig. 33), or carbon-centred radicals

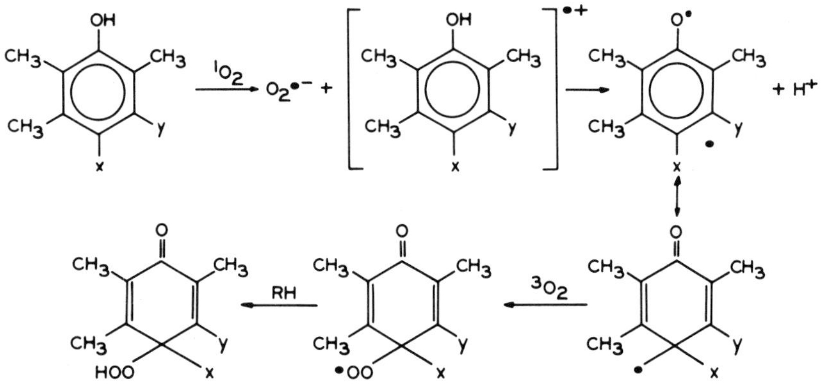

FIG. 33. Some proposed reactions for the formation of unstable peroxides in the oxidation of the tocopherols. (Reprinted from Carlsson et al. (1976) with permission.)

from scission of the chroman ring (Fig. 34) from α-tocopherol, or some type of MAH or electron transfer (ET) reaction with other peroxides (reaction (56)), all of which may initiate new reaction chains:

$$ROOH + AH \rightarrow A^{\cdot} + RO^{\cdot} + H_2O \qquad (56)$$

Regeneration of α-tocopherol from the phenolic radical by donation of electrons from thiols and ascorbate may, on the other hand, enhance the anti-oxidant effectiveness of α-tocopherol. Also, nucleophilic hydroperoxide decomposers that destroy the peroxides formed in the radical termination steps involving hindered phenolic anti-oxidants can give synergistic effects (Uri, 1970; Wright, 1975).

In some cases, α-tocopherol is considered to be inferior to synthetic, phenolic anti-oxidants in retarding lipid oxidation. However, a consideration of stereo-electronic effects (Burton et al., 1980) indicates that α-

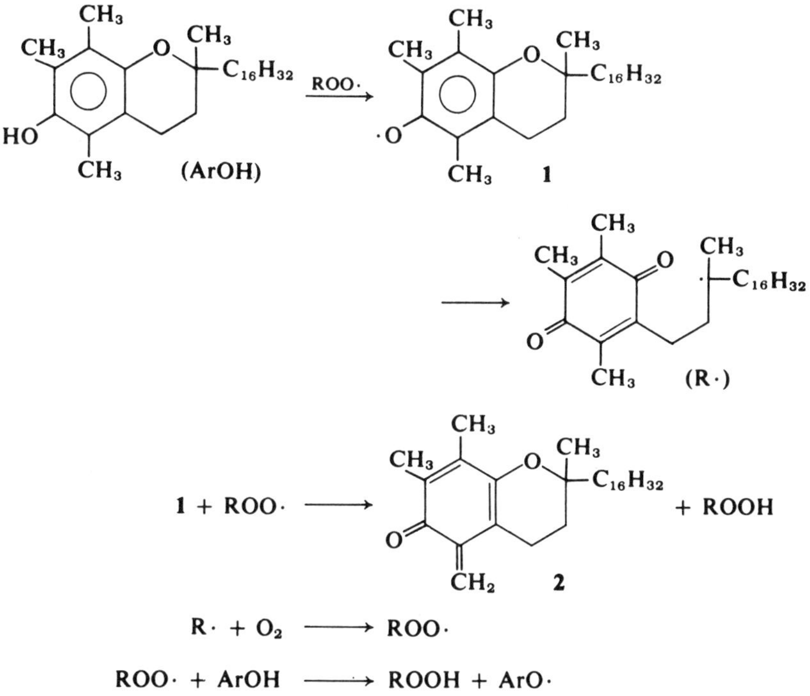

FIG. 34. Some potential pro-oxidant reactions between α-tocopherol and peroxy radicals. The (R·) resulting from scission of the chroman ring can form unstable hydroperoxides. (Reprinted from Pryor (1976) with permission.)

tocopherol possesses an excellent structure for a chain-breaking anti-oxidant (Fig. 35).

Maximum anti-oxidant activity is generally observed with those phenols hindered by methyl substituents at the 2, 3, 5 and 6 positions and electron-donating groups, such as methoxyl groups, in position 4. Apparently the predicted effectiveness of α-tocopherol does not reside in the isoprenoid side-chain but in the fused chroman ring which represents an 'immobilized' methoxy group. Thus the p-type lone-pair electrons of the oxygen in the chroman ring are constrained to be more-or-less perpendicular to the aromatic ring thereby stabilizing the phenoxyl radical via delocalization of the unpaired electron (Fig. 35). The combination of this near optimal orientation of the ethereal oxygen p-type lone pair with respect to the aromatic ring and the alkyl substitutions at the other four ring positions

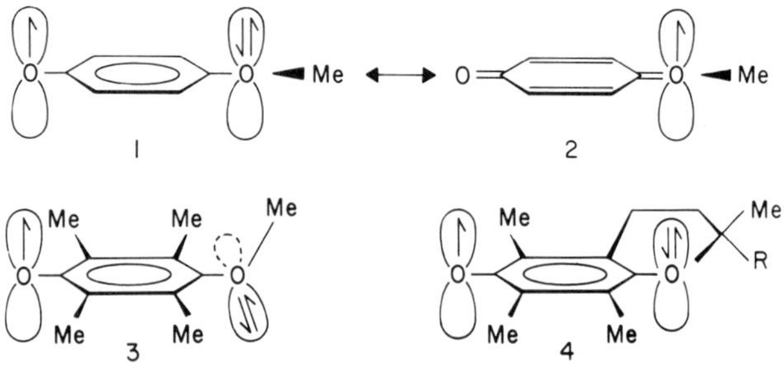

FIG. 35. Stereo-electronic factors favouring increased anti-oxidant effectiveness of α-tocopherol (4). In (1) and (2) the methoxy substituent is free to orient so that the *p*-type orbital on oxygen can participate in resonance stabilization of the phenolic radical. This favourable orientation of the methoxy group is prevented in (3) as a result of steric interference from *o*-methyl groups. In contrast, the chroman ring in α-tocopherol (4) requires that the *p*-type orbitals of the ring oxygen be nearly normal to the aromatic ring. (Reprinted from Burton *et al*. (1980) with permission. © American Chemical Society.)

should yield superior chain-breaking anti-oxidant properties for α-tocopherol. In addition, α-tocopherol can react with singlet oxygen to yield endoperoxides or can also physically quench $^1O_2$.

It might be appropriate at this point to discuss some observed differences between lipid oxidation in the bulk phase compared to that within synthetic membranes and at interfaces and to consider the effectiveness of α-tocopherol as an anti-oxidant in this context.

The catalysis of lipid oxidation by pro-oxidants acting at the lipid–water interface (Morita *et al*., 1976) is of obvious importance in food systems. Apparently, polar lipid hydroperoxides orient at the interface and compete with surfactants in model systems for interfacial area and exposure to peroxide decomposers and other anti-oxidants (Marita *et al*., 1976). As Uri (1961) and other investigators (Walling and Wagner, 1963, 1964; Taube, 1965) have pointed out, energy of solvation of products during homolysis of peroxides can be an important driving force favouring homolysis. For example, univalent reduction of H_2O_2 leads to an intermediate with essentially 1/2 of a covalent bond with the energy of solvation for the products facilitating scission of the bond (Taube, 1965). By the same token, solvation of homolytic products of organic peroxides is important in their decomposition (Walling and Wagner, 1964). One might postulate that at a lipid–water interface homolysis of lipid hydroperoxides might be greatly accelerated by co-operative effects involving solvation by water of a leaving ⁻OH or ˙OH group coupled with solvation of RO˙ by the lipid phase. The combined solvation energies may accelerate peroxide decomposition at the interface catalysed by various components that exist in the aqueous phase functioning through various mechanisms.

The net result of the foregoing considerations is a general perception that the tocopherols in milk are effective anti-oxidants. Consequently, there has been substantial research on increasing the tocopherol content of milk by feeding various supplements of these agents to dairy cows. Apparently, however, this has not become a general part of the feeding regimen for enhancement of the oxidative stability of milk, perhaps because of unfavourable economics compared to adding anti-oxidants directly to milk.

Peroxides can be decomposed by various nucleophiles with the effectiveness increasing in the order (Hiatt, 1975):

$$\text{\textbackslash}C=C\text{\textbackslash} < RNH_2 < R_2SO < R_2S$$

In principle, if heterolytic peroxide decomposers, such as thioethers (Fig. 36), were available to decompose peroxides formed in reactions between hindered phenols and peroxy radicals an additive or synergistic anti-oxidant effect should result (Uri, 1970). Indeed, this synergism has been

$$R_2S \quad O\underset{H}{\overset{O-R}{\diagup}} \longrightarrow R_2SO + ROH$$

FIG. 36. Nucleophilic decomposition of a hydroperoxide by a sulphide to obviate free-radical intermediates. (Reprinted from Hiatt (1975) with permission. © CRC Press, Inc.)

observed in certain patented anti-oxidant formulations (Wright, 1975). Consequently, commercial anti-oxidants may contain a mixture of the three aforementioned anti-oxidant types which are generally more effective than any single anti-oxidant.

The oxidizability of egg lecithin at 30°C, under 760 torr of O_2, in liposomal membranes is only 2·7% of that for the homogeneous material in chlorobenzene (Barclay and Ingold, 1981). A reduced efficiency in initiation of lipid oxidation in the membrane bilayer is probably the result of a reduction in the fraction of t-butoxyl radicals, used in initiating the oxidation, that escape from the solvent cage which in turn is due to the high microviscosity within the bilayer membrane. The reduced oxidizability of the phosphatidyl choline also reflects the partitioning of polar peroxy radicals away from auto-oxidizable acyl residues into the more polar non-auto-oxidizable surface regions of the bilayers. This would retard propagation and accelerate termination reactions. These observations suggest that orientation of membrane components plays an important role in lipid oxidations which may be altered upon destruction of membrane integrity with associated changes in these orientational effects. Perhaps alterations in the structure of the fat globule membrane in milk induced by processing would be reflected in changes in the oxidation of membrane components. In addition, reactions at lipid–membrane–water interfaces cannot be ignored. Well known is the resurfacing of milk fat droplets with caseins upon homogenization of milk resulting in retardation of lipid oxidation (Brunner, 1974). Along the same lines, water-soluble free-radical initiators, such as peroxy-chromate, fail to penetrate multi-bilayers of phosphatidyl choline liposomes to catalyse oxidation of polyenoic fatty acids within the interior of the membranes (Edwards and Quinn, 1982). Disruption or enhanced dispersion of the phospholipids exposed the otherwise hindered fatty acid residues to oxidation. The foregoing studies clearly indicate fundamental differences between auto-oxidation of lipids in the bulk phase compared to that at interfaces including membranous dispersions, especially as affected by the relative polarity of the initiating radical species.

In a reverse sense, reduction of tocopheryl radicals or tocopherones by other reducing agents in milk to regenerate the tocopherol as a lipid-soluble anti-oxidant could be an important interfacial reaction. For example, ascorbic acid or thiols in the aqueous phase (where they might be less effective as anti-oxidants for lipids) could reduce tocopheryl radicals or tocopherones oriented at the lipid–water interface thereby contributing reducing equivalents to the lipid phase (Parker et al., 1979; Fragata and Bellemare, 1980; Simic, 1981). Furthermore, competition of amphipathic molecules such as tocopherols and hydroperoxides at the interface for reducing equivalents may play a role in defining the balance between anti- and pro-oxidant effects.

Although vitamin E is considered to be less effective than BHT in scavenging peroxy radicals in lipid bilayers (Winterle and Mill, 1981), tocopherols in the milk fat globule membrane compared to those in the bulk phase of the milk fat may play an especially important role in retarding oxidation of milk lipids and in preventing oxidized flavours.

In micellar systems, for example, the chromanoxy part of an α-tocopherol radical is close to or at the surface of the micelle with the hydrocarbon tail of vitamin E buried within the micelle (Simic, 1981). By analogy, if α-tocopherol is to seek ROO˙ within a membrane, large entropy changes are required for it to move in the membrane thereby reducing the rate of reaction. However, interception of free radicals at the interface coupled with regeneration of α-tocopherol via interfacial reactions with polar reductants can combine to increase the efficiency of α-tocopherol as an anti-oxidant in milk.

Furthermore, if one assumes an intermolecular hydrogen atom transfer between polyunsaturated fatty acids (PUFA), radicals can be transferred within a membrane as follows (Pryor, 1981):

$$(\text{PUFA radical})_1 + (\text{PUFA})_2 \rightarrow (\text{PUFA})_1 + (\text{PUFA radical})_2 \rightarrow \text{etc.} \quad (57)$$

Thus a small amount of α-tocopherol in a membrane may have many opportunities to intercept the radical. This 'transport' of radicals in the lipid phase probably plays a crucial role in the oxidation of food lipids, particularly in the dry or frozen state (Schaich, 1980).

In addition, the rate constant for the reaction of $RO_2^{\cdot}$ with a polyunsaturated fatty acid molecule is about 50 $M^{-1} s^{-1}$ while the rate constant for the reaction between α-tocopherol and an $RO_2^{\cdot}$ in the lipid medium is about 10^4–10^5 $M^{-1} s^{-1}$. Therefore, one α-tocopherol molecule can protect about 10^4 PUFAs (Patterson, 1981).

In the scavenging of $^1O_2$ by α-tocopherol the rate of physical quenching (k_Q) is several orders of magnitude higher than the rate for chemical reactions of $^1O_2$ with α-tocopherol (k_R) (reaction (58)) (Fragata and Bellemare, 1980):

$$^1O_2 + \text{tocopherol} \underset{k_R}{\overset{k_Q}{\rightleftarrows}} \begin{array}{l} ^3O_2 + \text{tocopherol} \\ \text{Reaction products} \end{array} \quad (58)$$

Further, $^1O_2$ reacts chemically with tocopherol about 90-fold faster than with methyl linoleate (Yamauchi and Matsushita, 1977). The ratio k_Q/k_R increases with decreasing polarity of the medium, and each molecule of α-tocopherol oriented in a relatively non-polar portion of a synthetic, phospholipid membrane may quench 40 or more molecules of $^1O_2$ before it is destroyed (Fragata and Bellemare, 1980). If the chromanol moiety resides in a more hydrophobic environment within the membrane, it may quench even more $^1O_2$ molecules. Thus polarity and reactions at the lipid interface with the aqueous phase will affect the anti-oxidant efficiency of the tocopherols.

5.2. Ascorbic Acid

Ascorbic acid can be a very effective anti-oxidant and is known to be one of the most efficient scavengers of alkoxy radicals, outperforming thiols and phenols (Bors *et al.*, 1981a,b; Swartz and Dodd, 1981). Milk contains only small quantities (about 20 mg litre^{-1}) of ascorbic acid (Johnson, 1974) which can be readily oxidized to dehydroascorbic acid. This oxidation can be catalysed by agents such as the copper ion via a free-radical or an ionic mechanism (Fig. 37). In the one-electron oxidation, the relatively unreactive ascorbyl radical (Bielski *et al.*, 1975) is apparently stabilized by resonance (Fig. 38). The ease with which ascorbate can donate a single electron to other free radicals provides one basis for anti-oxidant function.

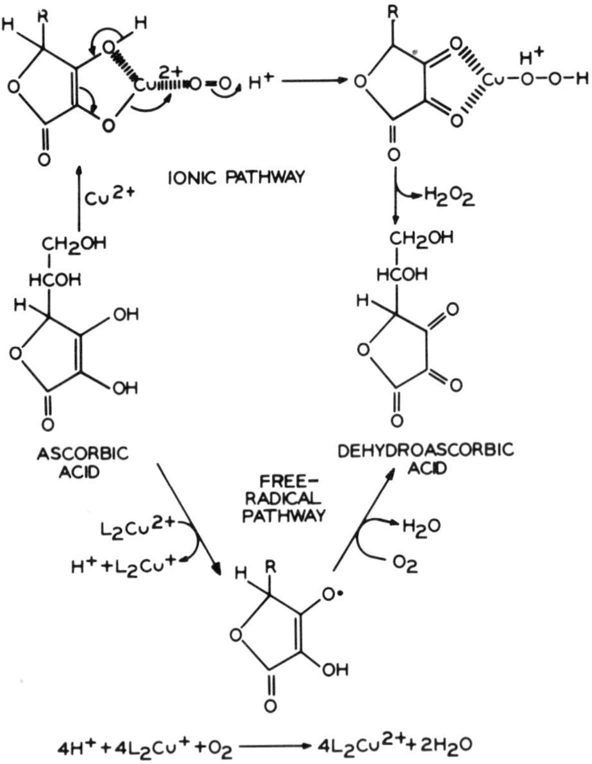

FIG. 37. Copper-catalysed oxidation of ascorbic acid to dehydroascorbic acid via a free-radical or an ionic pathway. Notice that H_2O_2 can be produced and that L_2Cu^{2+} can be univalently reduced to L_2Cu^+. The latter could reduce the H_2O_2 to yield strongly oxidizing ·OH and $^-$OH. (Reprinted from Hanzlik (1976) with permission.)

Two ascorbyl radicals can efficiently disproportionate to yield ascorbate and dehydroascorbate to effectively provide the second electron from ascorbate *per se* rather than necessarily from the ascorbyl radical (Bielski *et al.*, 1975). On the other hand, the one-electron reduction of a transition metal (e.g. Fe^{3+}) by ascorbate may be instrumental in yielding a pro-oxidant effect from resultant reduction of peroxides as mentioned previously (Haase and Dunkley, 1969c; Uri, 1973; Kanner *et al.*, 1977) (Fig. 37).

This univalent reduction of metal ions, such as copper, by ascorbate followed by the reduction of hydroperoxides to yield hydroxyl radical

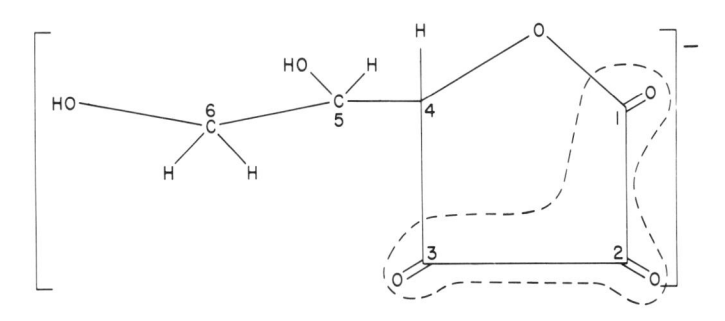

FIG. 38. Ascorbyl free radical stabilized by resonance among positions 1, 2 and 3. (Reprinted from Bielski *et al.* (1975) with permission.)

provides a straightforward rationalization for the pro-oxidant effects of ascorbate. In the case of the ascorbate–Fe^{3+} couple, reductive activation of the iron may indeed be operative (Haase and Dunkley, 1969c; Kanner *et al.*, 1977).

However, combinations of ascorbate and copper can be pro-oxidant or anti-oxidant depending upon their relative concentrations (Haase and Dunkley, 1969a,b,c; Allan and Wood, 1970; Kanner *et al.*, 1977, and references cited therein). Pro-oxidant effects of the ascorbate–Cu^{2+} couple catalysing the oxidation of linoleate are observed at low levels of ascorbate (up to $\sim 1.8 \times 10^{-4}$ M) in the presence of Cu^{2+} (up to $\sim 1.3 \times 10^{-5}$ M) (Table 7). However, at higher concentrations of ascorbate ($>10^{-4}$ M) an anti-oxidant effect is observed when the Cu^{2+} concentration becomes $\sim 1.3 \times 10^{-5}$ M. At this point the molar concentration of ascorbate is more than 10-fold that of the Cu^{2+}. A number of possible mechanisms have been suggested for this apparent paradox, including: (1) reductive activation of Cu^{2+}; (2) increased levels of the supposed pro-oxidant ascorbyl radical;

TABLE 7
RATE CONSTANTS FOR THE FORMATION OF CONJUGATED DIENES DURING THE OXIDATION OF LINOLEATE CATALYSED BY COMBINATIONS OF ASCORBIC ACID AND COPPER

Ascorbic acid concentration (M)	Copper concentration, $1 \cdot 3 \times$ (M)	$k_i \times 10^{6b}$ (mol litre^{-1} min^{-1})
A. $1 \cdot 8 \times 10^{-6}$	10^{-7}	1·02
	10^{-6}	1·73
	10^{-5}	3·09
	10^{-4}	4·60
	10^{-3}	6·59
B. $1 \cdot 8 \times 10^{-5}$	10^{-7}	5·27
	10^{-6}	7·37
	10^{-5}	8·80
	10^{-4}	9·10
	10^{-3}	9·78
C. $1 \cdot 8 \times 10^{-4}$	10^{-7}	11·4
	10^{-6}	12·8
	10^{-5}	0
	10^{-4}	0
	10^{-3}	0
D. $1 \cdot 8 \times 10^{-3}$	10^{-7}	13·8
	10^{-6}	0
	10^{-5}	0
	10^{-4}	0
	10^{-3}	0

[a] Reprinted from Haase and Dunkley (1969c) with permission.
[b] Calculated from the first 100 min of the oxidation.

and (3) formation of unspecified ascorbate–Cu^{2+} complexes that may differentially affect propagation and termination reactions in lipid oxidation. Mechanisms 1 and 2 are less attractive than mechanism 3 or a variant of it.

In metal-catalysed oxidation of ascorbate, copper was more effective than iron at higher pH values, presumably because a Cu^{2+}–ascorbate catalytic complex formed more readily than an iron complex (Khan and Martell, 1967a). The oxidation of ascorbate was postulated to proceed through a ternary complex comprised of Cu^{2+}–ascorbate–O_2 (Fig. 39). An electron from ascorbate is transferred through the Cu^{2+} to O_2. In this case, there is no need for a formal reduction of the Cu^{2+} and, as mentioned previously, it is essentially serving as a conduit for electron flow. However, subsequent to the electron transfer, dissociation of the complex leads to superoxide (perhydroxyl), Cu^{2+} and ascorbyl radicals. The ascorbyl radical can disproportionate or be oxidized univalently by Cu^{2+} to reductively activate the copper. Superoxide, of course, can undergo the various reactions discussed in the foregoing sections to yield pro-oxidants. Additional possibilities relate to the activation of oxygen alluded to in the foregoing discussion of Fig. 37 wherein reduced metals and H_2O_2 resulting from ascorbate oxidation might combine to lead to a marked pro-oxidant effect.

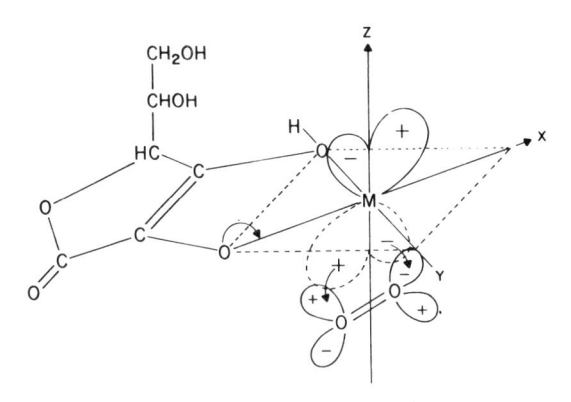

Fig. 39. Proposed metal–ascorbate–oxygen complex. (Reprinted from Khan and Martel (1967a) with permission. © American Chemical Society.)

At this point, it might prove useful to comment briefly on copper catalysis of lipid oxidation in relation to ascorbate and to the change in electronic configuration of Cu^{2+} when it is reduced to Cu^+ (Fig. 29). As discussed previously, particular bidentate chelates of Cu^{2+} might allow any required electronic transitions. However, in the presence of a large excess of ascorbate, Cu^{2+} might co-ordinate with two ascorbates similar to the hybrid ascorbate–Cu^{2+}–EDTA complex proposed by Khan and Martell (1967a). Alternatively, hybrid chelates involving ascorbate and a second food constituent as donors (by analogy with the ascorbate–EDTA complex) could provide additional means for binding Cu^{2+} into essentially unproductive complexes that might actually be anti-oxidant.

Consider the formation of such hypothetical chelates between ascorbate and Cu^{2+}:

$$AA + Cu^{2+} \rightleftharpoons AACu^{2+} + AA \rightleftharpoons AACu^{2+}AA \quad (59)$$
$$[I] \qquad\qquad [II]$$

If one assumes that chelate [I] is pro-oxidant and that chelate [II] is anti-oxidant, low concentrations of both AA and Cu^{2+} would favour formation of [I] leading to enhanced lipid oxidation (Table 7). Even with an excess of Cu^{2+} relative to ascorbate, a sufficiently high formation constant for [II] might yield enough of (II) to provide an anti-oxidant effect at high reactant concentration (Table 7, C and D).

Although the foregoing is highly speculative, there are suggestions in the literature that some chelates of Cu^{2+} may be effective lipid anti-oxidants (Uri, 1970; Kanner et al., 1977). A square-planar complex of nickel is an excellent peroxide decomposer and free-radical scavenger (Uri, 1970). This combination of anti-oxidant effects in a single complex was termed 'homosynergism' by Uri (1970) and autosynergism by Scott (1965). Square-planar Cu^{2+} chelates are moderately effective in protecting polyolefins from oxidative degradation (Uri, 1970) suggesting that such compounds could behave as anti-oxidants (Kanner et al., 1977). The planar ene-diol double-bond system of ascorbate would favour the formation of a square-planar chelate between two ascorbates and one Cu^{2+}. In this regard, one might predict that the previously mentioned cupro–porphyrin chelate could behave as an anti-oxidant. Thus excess ascorbate in a food system may not only provide electrons to satisfy free radicals but may also saturate the co-ordination sites of Cu^{2+}, thereby inhibiting complex formation with O_2 necessary for its activation and actually yielding anti-oxidant activity. Dehydroascorbate exhibits effects similar to those of ascorbic acid in the presence of copper, but to a lesser degree (Allan and Wood, 1970; Kanner et al., 1977).

Indeed, the behaviour of ascorbic acid in milk reflects the paradox evident in the foregoing discussion on possible anti-oxidant or pro-oxidant effects of ascorbic acid. Numerous studies in this area have not succeeded in defining the complex role of ascorbate in the oxidative stability of milk.

5.3. Thiols

It is well known that thiols are generated during thermal processing of milk, and they are generally thought to serve an anti-oxidant function, presumably as univalent reducing agents, peroxide decomposers or possibly as ligands for heavy metals. However, the univalent auto-oxidation of thiols yielding thiyl radicals, superoxide and hydrogen peroxide may provide the basis for additional paradoxical situations as with ascorbate (Yee and Shipe, 1982). These products of thiol oxidation may actually serve as pro-oxidants in certain cases. At pH values above 7 the auto-oxidation of thiols yields very high levels of superoxide (Misra, 1974). Thiyl radicals also add readily to double bonds generating a radical which might form peroxides upon reaction with $^3O_2$ (Abel, 1973). On the other hand, more likely reactions of thiyl radicals with oxygen can be very rapid, approaching diffusion control in the oxidation of cysteine, to yield a variety of stable oxidation products (Fig. 40) (Jocelyn, 1972; Asmus *et al.*, 1981). However, one should not overlook the possible formation of unstable thioperoxy (sulpho-peroxy) radicals, particularly in dry systems

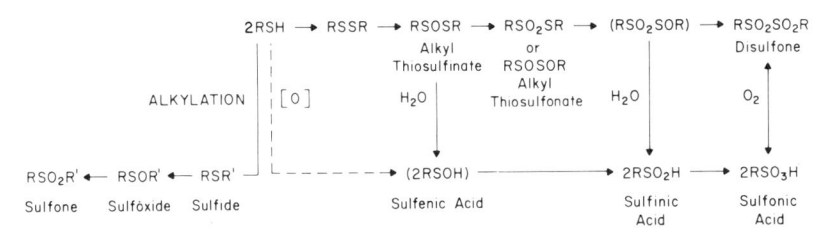

```
              2RSH → RSSR → RSOSR → RSO₂SR → (RSO₂SOR) → RSO₂SO₂R
                            Alkyl         or                      Disulfone
                         Thiosulfinate  RSOSOR
                                         Alkyl
 ALKYLATION      [O]       H₂O       Thiosulfonate    H₂O         O₂

RSO₂R' ← RSOR' ← RSR'  ┄┄┄→ (2RSOH) ──────────→ 2RSO₂H ────→ 2RSO₃H
Sulfone Sulfoxide Sulfide    Sulfenic Acid      Sulfinic       Sulfonic
                                                  Acid           Acid
```

FIG. 40. Summary of some reactions between oxygen and thiol groups. Compounds in parentheses are unstable. (Reprinted from Jocelyn (1972) with permission.)

(Schaich, 1980). Free-radical and nucleophilic displacement reactions involving such oxygenated sulphur species are discussed in detail by O'Brien (1980). Furthermore, thiols in the aqueous phase might also reduce tocopheryl radicals or tocopherone at the lipid–water interface as proposed for ascorbic acid to provide an anti-oxidant function.

Sulphydryl oxidase, indigenous to milk, has been proposed for the oxidation of thiols in UHT-processed milk to reduce cooked flavour and also thereby to serve as an anti-oxidant, in conjunction with lactoperoxidase, by obviating pro-oxidants resulting from auto-oxidation of thiols (Swaisgood and Abraham, 1980). The auto-oxidation of thiols may also be catalysed by a variety of agents, many of which are known to occur in milk. Yee and Shipe (1982) propose that copper-catalysed oxidation of sulphydryl groups may generate pro-oxidants in milk leading to paradoxical observations on their effectiveness as anti-oxidants.

In addition to possibilities for participating in one-electron reductions, ascorbate (dehydroascorbate?) and thiols may also bind certain pro-oxidant metals to prevent them from further catalysing pro-oxidant reactions, to yield anti-oxidants or conversely to form active catalysts.

5.4. Carotenoids

β-carotene, bixin and other conjugated systems with more than nine double bonds are effective physical quenchers of singlet oxygen (Foote, 1976). The highly conjugated double-bond systems favour low-energy

$\pi \to \pi^*$ transitions which may be important in thermally dissipating the energy in various electronically excited species upon collision with the carotenoid. β-Carotene is known to be one of the most potent quenchers of singlet oxygen with one molecule estimated to quench 250 molecules of $^1O_2$ (Foote and Denny, 1968). Bixin in anatto is probably also effective as a singlet-oxygen quencher (Hicks and Draper, 1981). In addition to quenching $^1O_2$, β-carotene readily forms adducts through radical addition reactions, is easily epoxidized and otherwise oxidized to yield radical cations and a variety of compounds characteristic of lipid oxidation in general (Chiba, 1967; Forss, 1972; Ramakrishnan and Francis, 1980; Bors et al., 1981a; Teixeiro Neto et al., 1981; Willson, 1981). These additional reactions of β-carotene have called into question the extensive use of the carotenoid as a probe to implicate $^1O_2$ as an initiator of lipid oxidation (Willson, 1981).

Novel oxidation products containing the β-ionone moiety have been implicated in violet-like off-flavour development in dairy products (Forss, 1972). Bleaching of added or natural carotenoid pigments in dairy products is, of course, a natural consequence of their oxidation with associated destruction of the chromogenic double-bond system. Surface bleaching of cheese and butter may result from, as yet, poorly understood oxidative reactions which may, in the case of cheese, be catalysed by photochemical activation of residual riboflavin by fluorescent lights in display cabinets. However, one would think that the carotenoid would act as a 'molecular filter' and absorb light at the wavelengths necessary to activate riboflavin.

Whether β-carotene serves a useful function as an anti-oxidant in milk and dairy products is controversial with more than one study indicating that its addition to foods does not provide anti-oxidant protection (Dugan, 1980; Eriksson, 1982).

5.5. Phospholipids

It is common knowledge that phospholipids can have an anti-oxidant effect in foods with some species being more effective than others (Dugan, 1980; Eriksson, 1982). Although the mechanism of this anti-oxidant action is not well understood, phospholipids may bind metals, may regenerate other anti-oxidants (Dugan, 1980), and may provide a synergism with phenolic anti-oxidants via the nitrogen function (Olcott and Van der Veen, 1963). On the other hand, unsaturated fatty acids usually associated with phospholipids in juxtaposition with pro-oxidants in membranes are quite susceptible to auto-oxidation (Schaich, 1980).

5.6. Proteins

Proteins, partial hydrolysates of proteins, synthetic peptides and amino acids (Dugan, 1980; Taylor and Richardson, 1980; Eriksson, 1982; Yee and Shipe, 1982) can be anti-oxidant via a number of possible mechanisms, including: (1) binding of pro-oxidant metals to prevent their participation in oxidative reactions; (2) orientation at a lipid interface providing a physical barrier to possible macromolecular pro-oxidants; and (3) providing functionalities to break free-radical chain reactions, to heterolytically decompose peroxides and to scavenge various reactive oxidizing species. Little is known about the anti-oxidant mechanisms of proteins, peptides and amino acids. However, the auto-oxidation of various amino acids as well as the reactions they undergo upon exposure to ionizing radiation should provide some clues as to how individual amino-acid residues might function as anti-oxidants. Some possible reactions include:

(i) Heterolytic decomposition of peroxides by thiol groups of cysteine and thioether groups of methionine.
(ii) Breaking free-radical chain reactions via donation of electrons (or H) by thiol groups of cysteine, by the phenolic function of tyrosine, by the indole N of tryptophan or by the α-C of peptide bonds.
(iii) Scavenging reactive oxygen species (e.g. $^1O_2$).
(iv) Chelating pro-oxidant metals and binding other pro-oxidants.

Of course, the selective oxidations of certain amino acids in proteins may also create special problems of their own including off-flavours. In some cases, however, amino acids or proteins, especially in the presence of copper ions, can be pro-oxidants (Tappel, 1955; Farag et al., 1978a,b; Yee and Shipe, 1982). The association of peroxidizing lipids with proteins can be especially damaging to the amino acids in proteins affecting their physical properties (Schaich, 1980). Perhaps the formation of unstable peroxides at the β-carbon of amino acids may be favoured in those residues allowing delocalization of the radical via appropriate functionalities.

The caseins seem to be particularly effective as anti-oxidants (Taylor and Richardson, 1980; Eriksson, 1982). Their known high hydrophobicity should favour orientation at the lipid interface maximizing the anti-oxidant functions of the various side-chains.

5.7. Browning Reaction Products

Carbonyl-amine reactions, such as those between lactose and milk proteins, are known to produce potent anti-oxidants (Dugan, 1980; Eichner, 1980; Eriksson, 1982). Products of the browning reaction can stabilize milk fat from oxidative deterioration for long periods (Wyatt and Day, 1965). However, it should be pointed out that some compounds resulting from the browning reaction may have adverse biological effects (Adrian, 1974; Tanaka et al., 1977). In recent years, substantial research efforts have been expended to characterize the effective agents. Recently, the adduct formed between ascorbate and tryptophan has been characterized as the first such product of a browning-type reaction (Namiki et al., 1982) (Fig. 41).

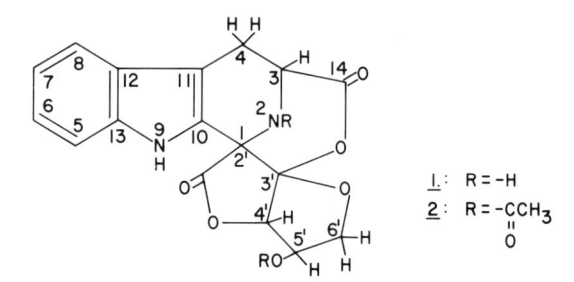

FIG. 41. Adduct from carbonyl–amine reaction between tryptophan and ascorbic acid. (Reprinted from Namiki et al. (1982) with permission.)

5.8. Anti-oxidant Enzymes

There are a few enzymes which can apparently function as anti-oxidants in biological systems. These enzymes will be discussed briefly; however, they are apparently only present in milk at low concentrations. Furthermore, thermal processing may inactivate them and actually convert them into non-specific metallo-protein pro-oxidants, since most of them are metalloenzymes.

(i) *Catalase*. This well-known enzyme destroys H_2O_2 via the disproportionation reaction:

$$2H_2O_2 \rightarrow {}^3O_2 + 2H_2O \tag{60}$$

The resultant oxygen is not an activated form but is in the ground state.

(ii) *Superoxide dismutase (SOD)*. As its name implies, SOD catalyses the dismutation of superoxide:

$$O_2^- + HO_2^{\cdot} \xrightarrow{H^+} {}^3O_2 + H_2O_2 \tag{61}$$

Although the product oxygen is in the ground state, SOD does generate potentially damaging H_2O_2. SOD has been patented as an anti-oxidant for foods (Michelson and Monod, 1975; Michelson, 1977), but should be more effective in conjunction with catalase to destroy the H_2O_2. Indeed, the co-operative effect between these enzymes has been observed when both are added to a linoleate–Fe^{2+} system as anti-oxidants (Valenzuela *et al.*, 1981). SOD occurs at very low levels in milk, presumably insufficient to be of practical significance, and is quite susceptible to thermal inactivation (Fox and Morrissey, 1981). Certain low molecular weight copper complexes possess substantial SOD activity (Leuthauser *et al.*, 1981).

(iii) *Glutathione peroxidase*. This seleno-enzyme catalyses the reduction of peroxides by reduced glutathione:

$$2GSH + ROOH \rightarrow ROH + H_2O + GSSG \qquad (62)$$

Although of considerable importance for the destruction of peroxides *in vivo*, this enzyme has not been reported to be in milk (Fox and Morrissey, 1981).

(iv) *Sulphydryl oxidase*. This enzyme is bound to membranes in milk and occurs as hydrophobic aggregates (Swaisgood, 1980). It catalyses the oxidation of thiols to yield disulphides and H_2O_2:

$$2R\text{—}SH + O_2 \rightarrow R\text{—}S\text{—}S\text{—}R + H_2O_2 \qquad (63)$$

The resultant H_2O_2 may be pro-oxidant; however, it can apparently be utilized by lactoperoxidase in a non-pro-oxidant fashion (Swaisgood and Abraham, 1980) with this enzymic combination actually exerting a possible anti-oxidant effect in milk.

These anti-oxidant enzymes are apparently not present in milk in sufficient quantities to provide protection against oxidative changes. If anything, the pro-oxidant effects of enzymes (denatured) such as lactoperoxidase, cytochrome P-420, and xanthine oxidase override possible beneficial effects of the former enzymes.

6. SOME ENVIRONMENTAL EFFECTS

6.1. Light, Photo-oxidations and Riboflavin

The photochemical oxidation of foods has been extensively reviewed recently by Sattar and deMan (1975) and by Spikes (1981). Although highly energetic photons in the ultraviolet spectrum can directly initiate free-radical reactions in lipids, the photochemical oxidation of food constituents under fluorescent illumination in display cases is usually mediated by photosensitizers such as riboflavin. In addition to the inherent light sensitivity of a food, other variables that affect photochemical oxidation of foods include: light intensity (spectrum, distance from foods, packaging), duration of exposure, oxygen concentration, temperature, surface-to-volume ratio, composition, and mixing.

Nutrients considered to be 'light-sensitive' include: vitamin A, carotenes, cyanocobalamin (vitamin B_{12}), vitamin D, folic acid, vitamin K, vitamin B_6, riboflavin, tocopherols, tryptophan and unsaturated fatty acid residues in oils, solid fats and phospholipids (Spikes, 1981). The opacity of most foods restricts photochemical oxidations to surface regions. However, these surface oxidations can be sufficient in foods such as meat, butter, milk and even cheese to create major colour and flavour problems.

Although riboflavin is the principal photosensitizer in dairy products, other natural and synthetic food colourants (e.g. chlorophyll, erythrosin) can also serve as photosensitizers. Riboflavin mediates photochemical oxidations by producing singlet oxygen and/or by generating superoxide anion in addition to free radicals in susceptible substrate molecules. The free-radical reactions probably predominate in the milk system.

Although much has been learned about some of the photo-oxidations of biological model compounds, applications of this knowledge to complex mixtures of biochemicals as in milk is difficult. In these cases, there exist large numbers of competing reactions, and the relative importance of a given pathway will depend upon the complexities imposed by concentrations of reactants, environmental effects, etc. in the food.

The photochemistry of riboflavin is complicated and incompletely understood (Muller, 1981). As shown in Fig. 42, a photosensitizer such as riboflavin (F_0) is 'activated' upon absorption of the appropriate wavelength of light to eventually yield an excited triplet state which can react by Type I (C, D) and/or Type II (B) reactions depending largely upon relative concentrations of potential reactants. At relatively high concentrations of oxygen, a Type II reaction favours the formation of $^1O_2$ which can react with double bonds and other electron-rich centres in various substrates to yield oxygenated products which can then decompose to yield free radicals, carbonyls, etc. The sensitizer then reverts to the ground state to absorb an additional quantum of light to initiate a new catalytic cycle. Under appropriate conditions, photo-excited riboflavin can generate $^1O_2$ upon interaction with $^3O_2$ (Song and Moore, 1968; Moore and Song, 1969).

An alternative reaction pathway involves the abstraction of an electron or a hydrogen atom from a substrate (RH) by the sensitizer to yield a free radical (R·) which can then react with ground-state oxygen ($^3O_2$) to form an initial peroxy radical to initiate new free-radical chain reactions (Fig. 42). If the sensitizer is riboflavin, it apparently is univalently reduced in the Type I reaction to yield a stable flavin semiquinone (FH·) central to a variety of ensuing reactions (Fig. 42). Although the flavin semiquinone radical is relatively stable, it can abstract an electron from an easily

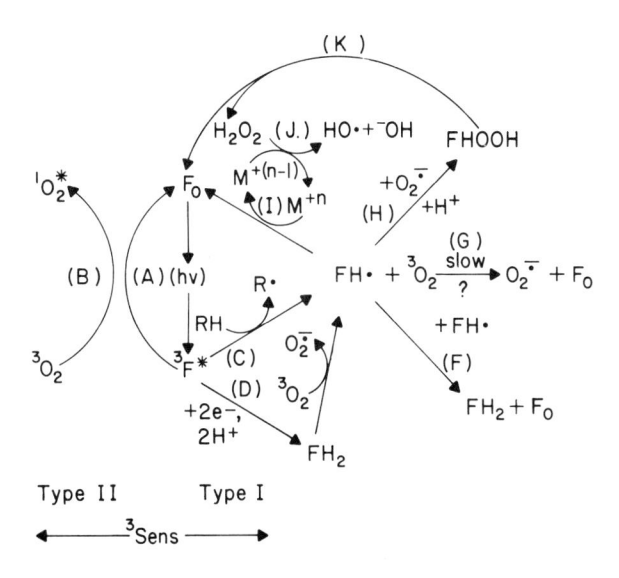

FIG. 42. Possible photocatalysed reactions involving riboflavin as a photosensitizer. Note the central role of the semiquinone radical (FH·) resulting from a Type I reaction of $^3F^*$ with RH (C).

oxidizable substance such as a phenol (tyrosine?) (Vaish and Tollin, 1970, 1971) to become fully reduced (FH_2). The flavin semiquinone can rapidly disproportionate to yield reduced and oxidized riboflavins (Fig. 42). Because the flavin semiquinone radical is stabilized by resonance, it forms readily from the univalent auto-oxidation of the fully reduced flavin to also yield superoxide anion and H_2O_2 (Fig. 43). The superoxide radical can react with the riboflavin semiquinone radical to give a peroxide which can decompose to yield the oxidized flavin and hydrogen peroxide (Fig. 44). These flavin peroxides are thought to be intermediates in enzymic oxidations in which they serve as prosthetic groups (Walsh, 1979). Decomposition of such flavin peroxides via scission reactions may be involved in the eventual oxidative destruction of riboflavin observed in milk.

Apparently, the neutral semiquinone radical (FH˙) reacts only slowly to reduce $^3O_2$ to yield superoxide (Fig. 42(G)) (Vaish and Tollin, 1971;

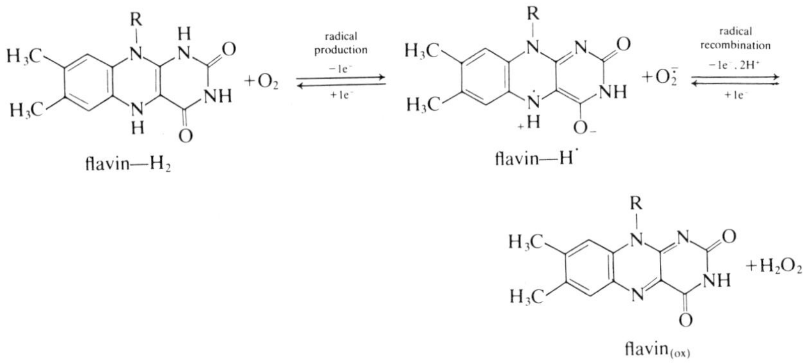

FIG. 43. The stepwise, univalent reduction of oxygen to superoxide ($O_2^{\bar{\cdot}}$) thence to H_2O_2 starting with reduced flavin and proceeding via the intermediate semiquinone (Walsh, 1979).

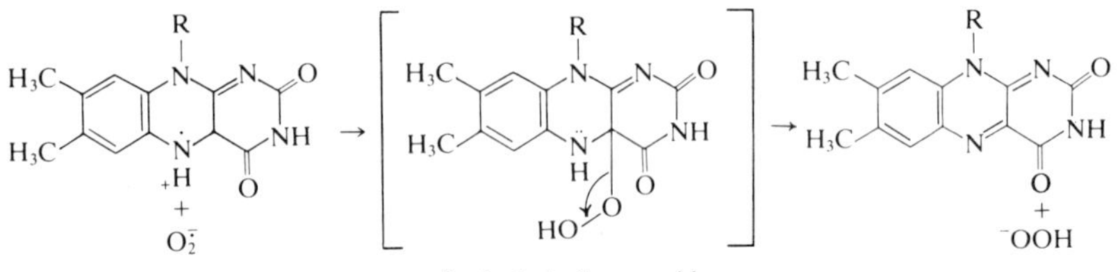

FIG. 44. A proposed flavin-4a-hydroperoxide as an intermediate in the enzymic reduction of oxygen to hydrogen peroxide (Walsh, 1979).

Anderson, 1981). However, the flavin radical anion ($F^{\bar{\cdot}}$) formed at a pH $>\sim 8\cdot 5$ reacts rapidly with $^3O_2$ to yield oxidized F_0 and superoxide (Vaish and Tollin, 1971; Anderson, 1981). Since the pKa of FH˙ is about 8·5, the concentration of F˙ at the pH of milk is very low. Thus it would appear that a major source of superoxide upon irradiation of milk is the univalent auto-oxidation of fully reduced riboflavin (FH_2) (Korycka-Dahl and Richardson, 1980). On the other hand, metal chelates (e.g. ferricyanide) can effectively interact with the flavin semiquinone (FH˙) to become reduced (reductive activation?) by abstracting an electron from the resonating radical system (Fig. 42 (I), (J)) (Vaish and Tollin, 1971).

The substrates for riboflavin-catalysed photo-oxidations in milk no doubt are many and varied including ascorbic acid, thiols and various amino acids in proteins. Riboflavin is a very potent catalyst in the photo-oxidation of ascorbic acid (Heelis *et al.*, 1981).

It is generally thought that the immobilization of a photosensitizer on an insoluble substrate favours a Type I photo-oxidation (Hanzlik, 1976; Foote, 1978). The preferential binding of one form of the photosensitizer over another (e.g. F_0, FH˙, F⁻, FH_2) while in equilibrium with an insoluble material suspended in a fluid could profoundly affect the course of the foregoing reactions. On the other hand, in dried high-fat products such as whole milk, where the riboflavin may be immobilized on protein, lactose and lipid, reaction with substrate is favoured because of proximity to form free radicals on the substrate (e.g. proteins, lipids) yielding peroxides and other active oxygen species. It is interesting to note, however, that the yield of excited triplet state upon irradiation of lumiflavin in a non-polar environment (as in dried high-fat products?) is low because of reduced intersystem crossing from the excited singlet state but increases with added water (Vaish and Tollin, 1970). This would suggest that a situation analogous to that observed for metals in catalysing lipid oxidations as a function of water activity (see below) might hold for riboflavin.

6.2. Water Activity

The quantity of water in a food affects rates of lipid oxidation and other deteriorative processes in a complex manner (Fig. 25). To explain the oxidation minima at intermediate levels of moisture, Labuza (1971), Karel (1980) and Schaich (1980) have proposed that an extremely dry environment promotes lipid oxidation because small amounts of water (perhaps a monolayer) are not available to mask pro-oxidants or to hydrogen-bond to hydroperoxides and retard their decomposition. As the amount of water is increased to form a monolayer, pro-oxidants may be masked via hydration, as in the co-ordination of water by metals, and otherwise serve as a barrier to oxidation. However, higher water activities promote oxidation perhaps by mobilizing pro-oxidants allowing them to diffuse through the food (Labuza, 1971). Sufficiently high A_w, however, could then retard oxidation by diluting the reactants.

6.3. pH
The acidity of a food may exert large effects on lipid oxidations in foods. Protons compete with metal ions for ligands thereby releasing metal ions to become more or less pro-oxidant. It should be pointed out that in the iodometric determination of peroxide values the reduction of peroxide by iodide ion is favoured if protons are present to help promote the 'decomposition':

$$3I^- + ROOH \rightarrow RO^- + {^-}OH \xrightarrow{2H^+} ROH + H_2O + I_3^- \qquad (64)$$

The protonation of superoxide (pKa of 4·8) yields perhydroxy radicals which can initiate the oxidation of methyllinoleate (Gebicki and Bielski, 1981). In general, fermentation of dairy products stabilizes them against oxidative deterioration which may be the result of the decline in pH, lower oxygen tensions and/or other factors such as anti-oxidants produced by micro-organisms (Eriksson, 1982).

7. MEASURING LIPID OXIDATION

Quantifying the rate and extent of lipid oxidation in foods is important in relation to development of oxidized flavour and in trying to predict the oxidative stability of foods. Numerous accelerated tests have been devised for various foods in attempting to correlate oxidation of lipids with storage stability (Ragnarsson and Labuza, 1977). Simple, reliable, sensitive methods for quantifying or estimating lipid oxidation are essential to these efforts. A brief discussion follows on some methods for quantifying lipid oxidation that have served as a basis for accelerated storage tests, for quality control and for research purposes. The various methods for measuring lipid oxidation have been reviewed by Holmant (1954), Mehlenbacher (1960) and Gray (1978).

7.1. Peroxide Values
Measurement of peroxides is the classical method for quantifying lipid oxidation (Mehlenbacher, 1960; Barthel and Grosch, 1974; Timmen, 1975). Determination of peroxide values relies on the oxidation of iodide ion to iodine in an acidic medium by lipid peroxides (reaction (64)). The released iodine can be quantified iodometrically or spectrophotometrically. Auto-oxidation of iodide by oxygen can lead to serious errors in micromethods and must be avoided (Heaton and Uri, 1958). Since two electrons are provided by the iodide to reduce 1 mol of peroxide, peroxide

values are reported as meq./kg fat or mmol (meq./2)/kg fat. Colorimetric micromethods (Heaton and Uri, 1958; Asakawa and Matsushita, 1980) are useful for measuring oxidation of phospholipids, often avoiding problems with solubilizing these compounds.

7.2. Thiobarbituric Acid (TBA) Test
This is a very sensitive and useful method for quantifying lipid oxidation. The lipid oxidation product, malondialdehyde (MDA), reacts with thiobarbituric acid to yield a red pigment which is quantified at 532 nm (Sinnhuber et al., 1958) (Fig. 18). The commercially available acetal tetraethoxypropane (TEP) hydrolyses under the acidic conditions of the test to generate in situ a source of the unstable malondialdehyde. Thus TEP has been used as a standard to prepare calibration curves (Sinnhuber et al., 1958). Although the TBA test is very popular, compounds other than malondialdehyde may react to form artifacts which may interfere (Marcuse, 1973). Recently, high-performance liquid chromatography has been used to determine TBA numbers (Kakuda et al., 1981).

7.3. Ultraviolet Absorption
As mentioned previously, the conjugated double bonds resulting from oxidation of polyunsaturated fatty acids can be observed in the ultraviolet (UV) region at 233 nm and at higher wavelengths depending on the extent of conjugation. Increased absorption of lipid solutions in the UV region has been used to follow oxidation of milk lipids.

7.4. Ferric Thiocyanate
The oxidation of ferrous to ferric iron by peroxides in the presence of thiocyanate to produce ferric thiocyanate has been used extensively to study the oxidation of milk lipids (Loftus-Hills and Thiel, 1946). Measurement of the red colour at 500–510 nm of the ferric thiocyanate provides a simple rapid method for estimating lipid oxidation.

7.5. Kreis Test
This test (Mehlenbacher, 1960) is one of the best known for the detection of oxidative rancidity in foods. It depends upon the reaction between phloroglucinol and some constituent (an epihydrin aldehyde?) in the oxidized lipid to yield a red colour. Non-rancid fats, however, may also give intense Kreis reactions. The Kreis test has been judged to be inadequate for assessing the oxidative state of food lipids (Gray, 1978).

7.6. Oxygen Uptake

Reaction of oxygen with lipids allows measurements of lipid oxidation by sensing a decrease in the partial pressure of oxygen in the system. This can be done manometrically or polarographically using an oxygen electrode (Tappel, 1955; Hamilton and Tappel, 1963).

7.7. Chemiluminescence

Oxidizing lipids emit photons upon the decomposition of peroxy radicals and of dioxetane (Ingold, 1973; Boveris *et al.*, 1981). Apparently, these decompositions yield molecular species in excited states that release photons upon reversion to the ground state. Counting the photons using special photomultiplying equipment (Usuki *et al.*, 1979) has been used to study lipid oxidation and to predict the oxidative stability of dairy products. This technique has been developed and used largely by Russian researchers (Shlyapintokh *et al.*, 1968).

7.8. Induction Period

Food lipids have an inherent stability to oxidation depending upon concentrations of natural anti-oxidants, pro-oxidants, etc. Consequently, the use of the foregoing methods to monitor the course of food-lipid oxidation usually results in the types of curve shown in Fig. 8. After a period of relative stability (induction period), oxidation of the lipids becomes autocatalytic at about the time for rancidity development. The longer the induction period, the more stable the food lipids to oxidation. This can serve as a basis for prediction of oxidative stability of a food as well as for evaluating anti-oxidants which increase the induction period.

REFERENCES

ABEL, P. I. (1973). In: *Free Radicals*, Vol. 2, J. K. Kochi (ed.), John Wiley & Sons, New York, p. 63.
ABEL, E. W., PRATT, J. M., WHELAN, R. and WILKINSON, P. J. (1974). *J. Am. Chem. Soc.*, **96**, 7119.
ADRIAN, J. (1974). *World Rev. Nutr. Diet.*, **19**, 71.
ALLAN, W. A. and WOOD, H. L. (1970). *J. Sci. Food Agric.*, **21**, 282.
ANDERSON, R. F. (1981). In: *Oxygen and Oxy-Radicals in Chemistry and Biology*, M. A. J. Rodgers and E. L. Powers (eds), Academic Press, New York, p. 597.
ARDON, M. (1965). *Oxygen: Elementary forms and hydrogen peroxide*, W. A. Benjamin, New York.
ASAKAWA, T. and MATSUSHITA, S. (1980). *Lipids*, **15**, 965.
ASMUS, K.-D., BAHNEMANN, D., BONIFACIC, M. and SCHAFER, K. (1981). In:

Oxygen and Oxy-Radicals in Chemistry and Biology, M. A. J. Rodgers and E. L. Powers (eds), Academic Press, New York, p. 69.
AUKLAKH, J. S. and STINE, C. M. (1971). *J. Dairy Sci.*, **54**, 1605.
AURAND, L. W., BOONE, N. H. and GIDDINGS, G. G. (1977). *J. Dairy Sci.*, **60**, 363.
BADINGS, H. T. (1960). *Neth. Milk Dairy J.*, **14**, 214.
BADINGS, H. T. (1970). PhD Thesis, Agricultural University, Wageningen, Netherlands.
BALOCK, A. K., BUCKLE, K. A. and EDWARDS, R. A. (1977). *J. Food Technol.*, **12**, 309.
BARCLAY, L. R. C. and INGOLD, K. U. (1981). *J. Am. Chem. Soc.*, **103**, 6487.
BARTHEL, G. and GROSCH, W. (1974). *J. Am. Oil Chem. Soc.*, **51**, 540.
BETTS, A. T. and URI, N. (1963). *Nature*, **199**, 568.
BIELSKI, B. H. J., RICHTER, H. W. and CHAN, P. G. (1975). *Ann. N.Y. Acad. Sci.*, **258**, 231.
BORS, W., MICHEL, C. and SARAN, M. (1981*a*). In: *Oxygen and Oxy-Radicals in Chemistry and Biology*, M. A. J. Rodgers and E. L. Powers (eds), Academic Press, New York, p. 75.
BORS, W., MICHEL, C. and SARAN, M. (1981*b*). *Bull. Eur. Physiopathol. Respir.*, **17**, Suppl., 13.
BOVERIS, A., CADENAS, E. and CHANCE, B. (1981). *Fed. Proc.*, **40**, 195.
BRAY, R. C., MAUTNER, G. N., FELDEN, E. M. and CARLE, C. I. (1977). In: *Superoxide and Superoxide Dismutases*, A. M. Michelson, J. M. McCord and I. Fridovich (eds), p. 61.
BROWN, W. D., HARRIS, L. S. and OLCOTT, H. S. (1963). *Arch. Biochem. Biophys.*, **101**, 14.
BRUNNER, J. R. (1974). In: *Fundamentals of Dairy Chemistry*, B. H. Webb, A. H. Johnson and J. A. Alford (eds), AVI Publ. Co., Westport, p. 497.
BURTON, G. W., LEPAGE, Y., GABE, E. J. and INGOLD, K. U. (1980). *J. Am. Chem. Soc.*, **102**, 7792.
CARLSSON, D. J., SUPRANCHUK, T. and WILES, D. M. (1976). *J. Am. Oil Chem. Soc.*, **53**, 656.
CHAN, H. W. S. (1975). *Chem. Ind.*, 612.
CHAN, H. W. S., COXON, D. T., PEERS, K. E. and PRICE, K. R. (1982). *Food Chem.*, **9**, 21.
CHIBA, N. (1967). MSc Thesis, Oregon State University, Corvallis.
CILLARD, J., CILLARD, P. and CORMIER, M. (1980*a*). *J. Am. Oil Chem. Soc.*, **57**, 255.
CILLARD, J., CILLARD, P., CORMIER, M. and GIRRE, L. (1980*b*). *J. Am. Oil Chem. Soc.*, **57**, 252.
CLARK, W. M. (1960). *Oxidation–Reduction Potentials of Organic Systems*, Williams and Wilkins Co., Baltimore, p. 466.
CLEMENTS, R. H., VAN DER ENGH, R. H., FROST, D., HOOGENHOUT, K. and NOOI, J. R. (1973). *J. Am. Oil Chem. Soc.*, **50**, 325.
COTTON, F. A. and WILKINSON, G. (1972) *Advanced Inorganic Chemistry*, 4th edn. John Wiley & Sons, New York.
DALY, G. G., FINOCCHIARO, E. T. and RICHARDSON, T. (1983). *J. Agric. Food Chem.*, **31**, 46.
DUGAN, L. R. (1980). In: *Autoxidation in Food and Biological Systems*, M. G. Simic and M. Karrel (eds), Plenum Press, New York, p. 261.

EDWARDS, J. C. and QUINN, P. J. (1982). *Biochim. Biophys. Acta*, **710**, 502.
EICHNER, K. (1980). In: *Autoxidation in Food and Biological Systems*, M. G. Simic and M. Karel (eds), Plenum Press, New York, p. 367.
EL-MAGOLI, S. B., KAREL, M. and YONG, S. (1979). *J. Food Biochem.*, **3**, 111.
ERIKSSON, C. (1974). In: *Industrial Aspects of Biochemistry*, B. Spencer (ed.), Fed. Eur. Biochem Soc., p. 865.
ERIKSSON, C. E. (1982). *Food Chem.*, **9**, 3.
FARAG, R. S., OSMAN, S. A., HALLABO, S. A. S. and NASR, A. A. (1978a). *J. Am. Oil Chem. Soc.*, **55**, 703.
FARAG, R. S., OSMAN, S. A., HALLABO, S. A. S., GIRGIS, A. N. and NASR, A. A. (1978b). *J. Am. Oil Chem. Soc.*, **55**, 708.
FEE, J. A. (1981). In: *Oxygen and Oxy-Radicals in Chemistry and Biology*, M. A. J. Rodgers and E. L. Powers (eds), Academic Press, New York, p. 205.
FEE, J. A. and VALENTINE, J. S. (1976). In: *Superoxide and Superoxide Dismutases*, A. M. Michelson, J. M. McCord and I. Fridovich (eds), Academic Press, New York, p. 19.
FIORITI, J. A., BENTZ, A. P. and SIMS, R. J. (1966a). *J. Am. Oil Chem. Soc.*, **43**, 37.
FIORITI, J. A., BENTZ, A. P. and SIMS, R. J. (1966b). *J. Am. Oil Chem. Soc.*, **43**, 487.
FOOTE, C. S. (1976). In: *Free Radicals in Biology*, Vol. II, W. A. Pryor (ed.), Academic Press, New York, p. 85.
FOOTE, C. S. (1978). In: *Singlet Oxygen: Reactions with Organic Compounds and Polymers*, B. Ranby and J. F. Rabek (eds), John Wiley & Sons, New York, p. 137.
FOOTE, C. S. and DENNY, R. W. (1968). *J. Am. Chem. Soc.*, **90**, 6233.
FOOTE, C. S., SHOOK, F. C. and ABAKERLI, R. A. (1980). *J. Am. Chem. Soc.*, **102**, 2503.
FORSS, D. A. (1972). *Progress in the Chemistry of Fats and Other Lipids*, **13**, p. 181.
FOX, P. F. and MORRISSEY, P. A. (1981). In: *Enzymes and Food Processing*, G. G. Birch, N. Blakebrough and K. J. Parker (eds), Applied Science Publishers Ltd, London, p. 213.
FRAGATA, M. and BELLEMARE, F. (1980). *Chem. Phys. Lipids*, **27**, 93.
FRANKEL, E. N. (1980). In: *Autoxidation in Food and Biological Systems*, M. G. Simic and M. Karel (eds), Plenum Press, New York, p. 141.
FREIFELDER, D. (1982). *Physical Chemistry for Students of Biology and Chemistry*, Science Books International, Boston, p. 308.
FRIDOVICH, I. (1976). In: *Free Radicals in Biology*, Vol. I, W. A. Pryor (ed.), Academic Press, New York, p. 239.
FRIDOVICH, I. (1977). In: *Biochemical and Medical Aspects of Active Oxygen*, O. Hayaishi and K. Asada (eds), University Park Press, Baltimore, p. 3.
FRIDOVICH, I. (1981). In: *Oxygen and Oxy-Radicals in Chemistry and Biology*, M. A. J. Rodgers and E. L. Powers (eds), Academic Press, New York, p. 197.
FRIDOVICH, S. E. and PORTER, N. A. (1981). *J. Biol. Chem.*, **256**, 260.
FUKUZAWA, K., TOKUMARA, A., OUCHI, S. and TSUKATANI, H. (1982). *Lipids*, **17**, 511.
GARDNER, H. W. and JURSINIC, P. A. (1981). *Biochim. Biophys. Acta*, **665**, 100.
GEBICKI, J. M. and BIELSKI, B. H. J. (1981). *J. Am. Chem. Soc.*, **103**, 7020.
GEORGE, P. and IRVINE, D. H. (1952). *Biochem. J.*, **52**, 511.
GIBIAN, M. J., SAWYER, D. T., UNGERMAN, T., TANGPOONPHOLVIVAT, R. and MORRISON, M. M. (1979). *J. Am. Chem. Soc.*, **101**, 640.

GOLLNICK, K. (1978) In: *Singlet Oxygen: Reaction with Organic Compounds and Polymers*, R. Ranby and J. F. Rabek (eds), John Wiley & Sons, New York, p. 111.
GRAY, J. I. (1978). *J. Am. Oil Chem. Soc.*, **55**, 539.
GRIFFIN, B. W. and RAMSEY, D. (1981). *Bio-organic Chem.*, **10**, 177.
GUMULKA, J., PYREK, J. St. and SMITH, L. L. (1982). *Lipids*, **17**, 197.
HAASE, G. and DUNKLEY, W. L. (1969a). *J. Lipid Res.*, **10**, 555.
HAASE, G. and DUNKLEY, W. L. (1969b). *J. Lipid Res.*, **10**, 561.
HAASE, G. and DUNKLEY, W. L. (1969c). *J. Lipid Res.*, **10**, 568.
HAASE, G. and DUNKLEY, W. L. (1970). *Milchwissenschaft*, **25**, 656.
HALLET, J. (1965). *Fed. Proc.*, **24**(2) Suppl. 15, S-42.
HAMILTON, J. W. and TAPPEL, A. L. (1963). *J. Am. Oil Chem. Soc.*, **40**, 52.
HANNAN, R. S. and BOAG, J. W. (1952).*Nature*, **169**, 152.
HANNAN, R. S. and SHEPHERD, H. J. (1952). *Nature*, **170**, 1021.
HANZLIK, R. P. (1976). *Inorganic Aspects of Biological and Organic Chemistry*, Academic Press, New York.
HEATON, F. W. and URI, N. (1958). *J. Sci. Food Agric.*, **9**, 781.
HEELIS, P. F., PARSONS, B. J., PHILLIPS, G. O. and MCKELLAR, A. (1981). *Photochem. Photobiol.*, **33**, 7.
HIATT, R. R. (1975). *Crit. Rev. Food Sci. and Nutr.*, **7**, 1.
HIATT, R. R. (1980). In: *Frontiers of Free Radical Chemistry*, W. A. Pryor (ed.), Academic Press, New York, p. 225.
HICKS, C. L. and DRAPER, J. (1981). In: *Oxygen and Oxy-Radicals in Chemistry and Biology*, M. A. J. Rodgers and E. L. Powers (eds), Academic Press, New York, p. 663.
HIRANA, Y. and OLCOTT, H. S. (1971). *J. Am. Oil Chem. Soc.*, **48**, 528.
HOLMANT, R. T. (1954). In: *Progress in the Chemistry of Fats and other Lipids*, Vol. 2, R. T. Holman, W. O. Lundberg, T. Malkin (eds), Academic Press, New York, p. 51.
HOWARD, J. A. (1973). In: *Free Radicals*, Vol. 2, J. J. Kochi (ed.), John Wiley & Sons, New York, p. 3.
HOWARD, J. A. (1980). In: *Frontiers of Free Radical Chemistry*, W. A. Pryor (ed.), Academic Press, New York, p. 237.
HUANG, R. L., GOH, S. H. and ONG, S. H. (1974). *The Chemistry of Free Radicals*, Edward Arnold, London, p. 89.
INGOLD, K. U. (1973). In: *Free Radicals*, Vol. 1, J. J. Kochi (ed.), John Wiley & Sons, New York, p. 37.
JARRETT, W. D. (1979). *Aust. J. Dairy Tech.*, **34**, 28.
JEBSON, R. S., EVANS, A. A. and COOKE, D. (1973). *N.Z. J. Dairy Sci. Tech.*, **8**, 60.
JENNESS, R., PATTON, S. (1959). *Principles of Dairy Chemistry*, John Wiley & Sons, New York, p. 446.
JOCELYN, P. C. (1972). *Biochemistry of the SH Group*, Academic Press, New York, p. 108.
JOHNSON, A. H. (1974). In: *Fundamentals of Dairy Chemistry*, B. H. Webb, A. H. Johnson and J. A. Alford (eds), AVI Publ. Corp., Westport, p. 1.
KAKUDA, Y., STANLEY, D. W. and VAN DE VOART, F. R. (1981). *J. Am. Oil Chem. Soc.*, **58**, 773.
KAMIYA, Y., BEATON, S., LAFORTUNE, A. and INGOLD, K. U. (1963). *Can. J. Chem.*, **41**, 2034.

KANNER, J., MENDEL, H. and BUDOWSKI, P. (1977). *J. Food Sci.*, **42**, 60.
KANNER, J., BEN-GERA, I. and BERMAN, S. (1980). *Lipids*, **15**, 944.
KAREL, M. (1980). In: *Autoxidation in Foods and Biological Systems*, M. G. Simic and M. Karel (eds), Plenum Press, New York, p. 191.
KEEVIL, T. and MASON, H. S. (1978). *Meth. Enzymol.*, **52**, 3.
KENDRICK, J. and WATTS, B. M. (1969). *Lipids*, **4**, 454.
KHAN, M. M. T. and MARTELL, A. E. (1967a). *J. Am. Chem. Soc.*, **89**, 4176.
KHAN, M. M. T. and MARTELL, A. E. (1967b). *J. Am. Chem. Soc.*, **89**, 7104.
KITAGAWA, T., SUGIYAMA, T. and YAMANO, T. (1982). *Biochemistry*, **21**, 1680.
KOCHI, J. (1973). In: *Free Radicals*, J. K. Kochi (ed.), John Wiley & Sons, New York, p. 591.
KORYCKA-DAHL, M. and RICHARDSON, T. (1980). *J. Dairy Sci.*, **63**, 1181.
KURTZ, F. E., TAMSMA, A. and PALLANSCH, M. J. (1969a). *J. Dairy Sci.*, **52**, 425.
KURTZ, F. E., TAMSMA, A., SELMAN, R. L. and PALLANSCH, M. J. (1969b). *J. Dairy Sci.*, **52**, 158.
LABUZA, T. P. (1971). In: *CRC Critical Reviews in Food Technology*, T. E. Furia (ed.), CRC Press, Boca Raton, p. 355.
LEUTHAUSER, S. W. C., OBERLEY, L. W., OBERLEY, T. D., SORENSON, J. R. J. and BUETTNER, G. R. (1981). In: *Oxygen and Oxy-Radicals in Chemistry and Biology*, M. A. J. Rodgers and E. L. Powers (eds), Academic Press, New York, p. 679.
LEWIS, S. E. and WILLS, E. D. (1963). *Biochim. Biophys. Acta*, **70**, 336.
LIVINGSTON, D. J. and BROWN, W. D. (1981). *Food Tech.*, 244.
LOFTUS-HILLS, G. and THIEL, C. C. (1946). *J. Dairy Res.*, **14**, 340.
LOGANI, M. K. and DAVIES, R. E. (1980). *Lipids*, **15**, 485.
LOTAN, R. (1980). *Biochim. Biophys. Acta*, **605**, 33.
MARCUSE, R. (1973). *J. Am. Oil Chem. Soc.*, **50**, 387.
MCCORMICK, A. M., NAPOLI, J. L., SCHNOES, H. K. and DELUCA, F. H. (1978). *Biochemistry*, **17**, 4085.
MCGANN, T. C. A. (1979). *Biochem. Soc. Trans.*, **7**, 51.
MCWEENY, D. J. (1968). *J. Food. Technol.*, **3**, 15.
MEHLENBACHER, V. C. (1960). *The Analysis of Fats and Oils*, The Garrand Press, Champaigne, p. 191.
MICHELSON, A. M. (1977). US Patent 4,029,819, June 14.
MICHELSON, A. M. and MONOD, J. (1975). US Patent 3,920,521, November 18.
MISRA, H. P. (1974). *J. Biol. Chem.*, 249.
MITCHELL, G. E. (1981). *Aust. J. Dairy Tech.*, **36**, 70.
MITCHELL, J. H., JR. and HENICK, A. S. (1962). In: *Autoxidation and Antioxidants*, Vol. II, W. O. Lundberg (ed.), Interscience–John Wiley & Sons, New York, p. 558.
MOORE, T. A. and SONG, P.-S. (1969). *Photochem. Photobiol.*, **10**, 13.
MOREY, K. S., HANSEN, S. P. and BROWN, W. D. (1973). *J. Food Sci.*, **38**, 1104.
MORITA, M., MUKUNOKI, M., OKUBO, F. and TADOKORO, S. (1976). *J. Am. Oil Chem. Soc.*, **53**, 489.
MORIUCHI, S. *et al.* (1979). *J. Nutr. Sci. Vitaminol.*, **25**, 455.
MULDER, H. (1953). *Neth. Milk Dairy J.*, **7**, 149.
MULLER, F. (1981). *Photochem. and Photobiol.*, **34**, 753.
NAKANO, M. and NOGUCHI, T. (1977). In: *Biochemical and Medical Aspects of*

Active Oxygen, O. Hayaishi and K. Asada (eds), University Park Press, Baltimore, p. 29.
NAMIKI, M., HAYASHI, T. and SHIGETA, A. (1982). *Agric. Biol. Chem.*, **46**, 1207.
NANNI, E. J., JR, BIRGE, R. R., HUBBARD, L. M., MORRISON, M. M. and SAWYER, D. T. (1981). *Inorg. Chem.*, **20**, 737.
NIEHAUS, W. G., JR (1978). *Bioorg. Chem.*, **7**, 77.
NISHIKIMI, M. and YAGI, K. (1977). In: *Biochemical and Medical Aspects of Active Oxygen*, O. Hayaishi and K. Asada (eds), University Park Press, Baltimore, p. 79.
NISHINAGA, A. (1977). In: *Biochemical and Medical Aspects of Active Oxygen*, O. Hayaishi and K. Asada (eds), University Park Press, Baltimore, p. 13.
NOLL, C. I. and SUPPLEE, G. C. (1941). *J. Dairy Sci.*, **24**, 993.
NORMAN, R. O. C. and WEST, P. R. (1969). *J. Chem. Soc.*, **B**, 389.
O'BRIEN, P. J. (1969). *Can. J. Biochem.*, **47**, 485.
O'BRIEN, P. J. (1980). In: *Autoxidation in Foods and Biological Systems*, M. G. Simic and M. Karel (eds), Plenum Press, New York, p. 563.
OHLSON, R. (1973). Cited in Berger, K. G. (1975). *Chem. Ind.*, **5**, 194.
OLCOTT, H. S. and VAN DER VEEN, J. (1963). *J. Food Sci.*, **28**, 313.
PARKER, J. E., SLATER, T. F. and WILSON, R. L. (1979). *Nature*, **278**, 737.
PATTERSON, L. K. (1981). In: *Oxygen and Oxy-Radicals in Chemistry and Biology*, M. A. J. Rodgers and E. L. Powers (eds), Academic Press, New York, p. 89.
PATTON, S. (1962). In: *Lipids and Their Oxidation*, H. Schultz, E. A. Day and R. Sinnhuber (eds), AVI Publ. Corp., Westport, p. 190.
PAULSON, D. R., ULLMAN, R. and SLOANE, R. B. (1974). *J. Chem. Soc. Chem. Com.*, 186.
PORTER, N. A., LEHMAN, L. A., WEBER, B. A. and SMITH, K. J. (1981). *J. Am. Chem. Soc.*, **103**, 6447.
PRIVETT, D. S. and BLANK, M. L. (1962). *J. Am. Oil Chem. Soc.*, **39**, 465.
PRUITT, K. M., TENOVIVO, J., ANDREWS, R. W. and MCKANE, T. (1982). *Biochem.*, **21**, 562.
PRYOR, W. A. (1976). In: *Free Radicals in Biology*, Vol. I, W. A. Pryor (ed.), Academic Press, New York, p. 1.
PRYOR, W. A. (1981). In: *Oxygen and Oxy-Radicals in Chemistry and Biology*, M. A. J. Rodgers and E. L. Powers (eds), Academic Press, New York, p. 133.
PRYOR, W. A. and STANLEY, J. P. (1975). *J. Org. Chem.*, **40**, 3615.
PRYOR, W. A., FULLER, D. L. and STANLEY, J. P. (1972). *J. Am. Chem. Soc.*, **94**, 1632.
PRYOR, W. A., STANLEY, J. P., BLAIR, E., CULLEN, G. B. (1976). *Arch. Environ. Health*, **28**, 201.
PYLER, E. J. (1982). *Baker's Digest*, **56**, 23.
RAGNARSSON, J. O. and LABUZA, T. P. (1977). *Food Chem.*, **2**, 291.
RAMAKRISHNAN, T. V. and FRANCIS, F. J. (1980). *J. Food Qual.*, **3**, 25.
RANBY, B. and RABEK, J. F. (1978). *Singlet Oxygen: Reactions With Organic Compounds and Polymers*, John Wiley & Sons, New York.
REITER, B. (1978). *J. Dairy Res.*, **45**, 131.
ROONEY, M. L. (1981). *J. Food Sci.*, **47**, 291.
SATTAR, A. and DEMAN, J. M. (1975). *CRC Crit. Rev. Food Sci. Nutr.*, **7**, 13.
SAWYER, D. T. and NANNI, E. G., JR (1981). In: *Oxygen and Oxy-Radicals in*

Chemistry and Biology, M. A. J. Rodgers and E. J. Powers (eds), Academic Press, New York, p. 15.
SCHAICH, K. M. (1980). *CRC Crit. Rev. Food Sci. Nutr.*, 189.
SCHENCK, G. O., GOLLNICK, K. and NEUMULLER, O. A. (1957). *Justus Liebigs Ann. Chem.*, **603**, 46.
SCHWARTZ, D. P. and PARKS, O. W. (1974). In: *Fundamentals of Dairy Chemistry*, B. H. Webb, A. H. Johnson and J. A. Alford (eds), AVI Publishing Co., Westport, p. 240.
SCOTT, G. (1965). *Atmospheric Oxidation and Antioxidants*, Elsevier Publishing Co., New York.
SHERBON, J. W. (1974). *J. Am. Oil Chem. Soc.*, **51**, 22.
SHLYAPINTOKH, V. YA. et al. (1968). *Chemiluminescence Techniques in Chemical Reactions*, Consultants Bureau, New York. (Cited by Ingold, 1973.)
SIGEL, H. (1969). *Angew. Chem. Internat. Edit.*, **8**, 167.
SIMIC, M. G. (1981). In: *Oxygen and Oxy-Radicals in Chemistry and Biology*, M. A. J. Rodgers and E. L. Powers (eds), Academic Press, New York, p. 109.
SINNHUBER, R. O. and YU, T.-C. (1977). *J. Jap. Oil Chem. Soc.*, **25**, 259.
SINNHUBER, R. O., YU, T. C. and YU, T.-C. (1958). *Food Res.*, **23**, 626.
SLIGAR, S. G., KENNEDY, K. A., PEARSON, D. C. (1980). *Proc. Natl. Acad. Sci. USA*, **77**, 1240.
SMITH, L. L. (1980). In: *Autoxidation in Food and Biological Systems*, M. G. Simic and M. Karel (eds), Plenum Press, New York, p. 119.
SMITH, L. L. (1981). *Cholesterol Autoxidation*, Plenum Press, New York.
SMITH, G. J. and DUNKLEY, W. L. (1962). *J. Food Sci.*, **27**, 127.
SOBCZAK, J. and ZIOLKOWSKI, J. J. (1981). *J. Mol. Catal.*, **13**, 11.
SONG, P.-S. and MOORE, T. A. (1968). *J. Am. Chem. Soc.*, **90**, 6507.
SPIKES, J. D. (1981). In: *Photochemical and Photobiological Reviews*, Vol. 6, K. C. Smith (ed.), p. 39.
SPORN, M. B., DUNLAP, N. H., NEWTON, D. L. and SMITH, J. M. (1976). *Fed. Proc.*, **35**, 1332.
STEVENS, B. (1965). *Chemical Kinetics*, Chapman and Hall, London, p. 97.
SUWA, K., KIMURA, T. and SCHAO, A. P. (1977). *Biochem. Biophys. Res. Commun.*, **75**, 785.
SWAISGOOD, H. (1980). *Enzyme Microb. Tech.*, **2**, 265.
SWAISGOOD, H. E. and ABRAHAM, P. (1980). *J. Dairy Sci.*, **63**, 1205.
SWARTZ, H. M. and DODD, N. J. F. (1981). In: *Oxygen and Oxy-Radicals in Chemistry and Biology*, M. A. J. Rodgers and E. L. Powers (eds), Academic Press, New York, p. 161.
SWERN, D. (ed.), (1964). *Bailey's Industrial Oil and Fat Products*, Interscience, New York, p. 74.
SWOBODA, P. A. T. and PEERS, K. E. (1977). *J. Sci. Food Agric.*, **28**, 1019.
TANAKA, M., KIMIAGAR, M., LEE, T.-C. and CHICHESTER, C. O. (1977). In: *Protein Cross-Linking—Nutritional and Medical Consequences*, (Adv. Exp. Med. Biol. 86B), M. Friedman (ed.), Plenum Press, New York, p. 321.
TAPPEL, A. L. (1955). *J. Am. Oil Chem. Soc.*, **32**, 252.
TAPPEL, A. L. (1962). In: *Lipids and Their Oxidation*, H. W. Schultz, E. A. Day and R. O. Sinnhuber (eds), AVI Publ. Corp., Westport, p. 122.
TAUBE, H. (1965). In: *Proc. New York Heart Assoc. Symposium on Oxygen*, A. P. Fishman (ed.), Little, Brown and Co., Boston, p. 29.

TAYLOR, M. J. and RICHARDSON, T. (1980). *J. Dairy Sci.*, **63**, 1783.
TEIXEIRA NETO, R. O., KAREL, M., SAGUY, I. and MIZRAHI, S. (1981). *J. Food Sci.*, **46**, 665, 676.
THOMAS, M. J. and PRYOR, W. A. (1980). *Lipids*, **15**, 544.
THOMPSON, L. U. and FENNEMA, O. (1971). *J. Agric. Food Chem.*, **19**, 121.
TIEN, M., SVINGEN, B. A. and AUST, S. D. (1981a). *Fed. Proc.*, **40**, 179.
TIEN, M., SVINGEN, B. A. and AUST, S. D. (1981b). In: *Oxygen and Oxy-Radicals in Chemistry and Biology*, M. A. J. Rodgers and E. L. Powers (eds), Academic Press, New York, p. 147.
TIMMEN, H. (1975). *Milchwissenschaft*, **30**, 329.
TIMMS, R. E., ROUPAS, P. and ROGERS, W. P. (1982). *Aust. J. Dairy Tech.*, **37**, 39.
TOYODA, M., ITO, Y. U., IWAIDA, M., ITSUGI, Y., OHASHI, M. and FUJU, T. (1982a). *N.Z. J. Dairy Sci. and Tech.*, **17**, 41.
TOYADA, I., TERAO, J. and MATSUSHITA, S. (1982b). *Lipids*, **17**, 84–90.
ULLRICH, V. and KUTHAN, H. (1981). In: *Oxygen and Oxy-Radicals in Chemistry and Biology*, M. A. J. Rodgers and E. L. Powers (eds), Academic Press, New York, p. 497.
URI, N. (1961). In: *Autoxidation and Antioxidants*, W. O. Lundberg (ed.), Interscience–John Wiley & Sons, New York, p. 55.
URI, N. (1970). *Is. J. Chem.*, **8**, 125.
URI, N. (1973). *IFST Proceedings*, **6**, 179.
USUKI, R., KANEDA, T., YAMAGISHI, A., TAKYU, C. and INABA, H. (1979). *J. Food Sci.*, **44**, 1573.
VAISH, S. P. and TOLLIN, G. (1970). *J. Bioenergetics*, **1**, 181.
VAISH, S. P. and TOLLIN, G. (1971). *Bioenergetics*, **2**, 61.
VALENZUELA, A., ADARMES, H. and GUERRA, R. (1981). *Alimentos*, **6**, 5; *Dairy Sci. Abstr.* (1982), **44**, 4328.
WALLING, C. and HEATON, L. (1965). *J. Am. Chem. Soc.*, **87**, 38.
WALLING, C. and WAGNER, P. (1963). *J. Am. Chem. Soc.*, **85**, 2333.
WALLING, C. and WAGNER, P. J. (1964). *J. Am. Chem. Soc.*, **86**, 3368.
WALSH, C. (1979). *Enzymatic Reaction Mechanisms*, W. H. Freeman and Co., San Francisco, p. 371, 464.
WATERS, W. A. (1971). *J. Am. Oil Chem. Soc.*, **48**, 428.
WEISSBLUTH, M. (1974). *Hemoglobin: Cooperativity and Electronic Properties*. Springer-Verlag, New York.
WILKINSON, F. (1978). In: *Singlet Oxygen: Reaction with Organic Compounds and Polymers*, R. Ranby and J. F. Rabek (eds), John Wiley & Sons, New York, p. 27.
WILLIAMS, R. J. P. (1981). In: *Oxygen and Life: Second BOC Priestly Conference*, Royal Society of Chemistry, London, p. 18.
WILLSON, R. (1981). In: *Oxygen and Oxy-Radicals in Chemistry and Biology*, M. A. J. Rodgers and E. L. Powers (eds), Academic Press, New York, p. 117.
WINTERLE, J. S. and MILL, T. (1981). In: *Oxygen and Oxy-Radicals in Chemistry and Biology*, M. A. J. Rodgers and E. L. Powers (eds), Academic Press, New York, p. 779.
WRIGHT, R. B. (1975). US Patent 3,873,466.
WYATT, J. C. and DAY, E. A. (1965). *J. Dairy Sci.*, **48**, 682.
YAMAUCHI, R. and MATSUSHITA, S. (1977). *Agric. Biol. Chem.*, **41**, 1425.
YEE, J. J. and SHIPE, W. F. (1981). *J. Food Sci.*, **46**, 966.
YEE, J. J. and SHIPE, W. F. (1982). *J. Dairy Sci.*, **65**, 1414.

Young, R. H. and Brewer, D. R. (1978). In: *Singlet Oxygen: Reaction with Organic Compounds and Polymers*, R. Ranby and J. F. Rabek (eds), John Wiley & Sons, New York, p. 36.

B. OXIDATION OF MILK LIPIDS

1. INTRODUCTION

Milk lipids easily undergo auto-oxidation, which leads to development of oxidized flavour defects that limit the shelf-life of dairy products. In the presence of oxygen and under the usual processing and storage conditions for dairy foods, the saturated fatty acids in milk fat are considered to remain stable, whereas unsaturated fatty acids undergo auto-oxidation and form odourless and tasteless hydroperoxides. However, these are unstable and readily decompose to yield flavourful carbonyl compounds. Carbonyls often have an unpleasant flavour which can be detected even at concentrations below 1 μg/g of fat. Since unstable, unsaturated carbonyls are also formed during auto-oxidation of lipids, they will be further oxidized to yield a new group of flavour compounds. Furthermore, small quantities of polyperoxides arising from the oxidation of polyunsaturated fatty acid residues of lipids (PUFA) can yield novel flavour compounds upon decomposition (Badings, 1970; Chan *et al.*, 1982).

Therefore, the off-flavour formed during unsaturated lipid auto-oxidation is actually composed of many flavours often referred to as 'oxidized flavour'—this is a cumulative term describing flavours composed of many characteristics such as 'metallic', 'fishy', 'cardboard', 'tallowy', 'fatty' (Table 8) which appear in dairy products in different intensity and order during the progress of lipid oxidation (Badings, 1970). Since peroxide values of less than 1 meq. kg^{-1} in many products containing milk fat result in distinct off-flavours (Badings, 1970), the importance of oxidative reactions in limiting the shelf-life of dairy products cannot be over-estimated. It should be noted, however, that not all oxidized flavour characteristics appear to be undesirable. Mushroom and creamy flavours (Table 8) can be appreciated in certain products, for example, cheeses.

TABLE 8
COMPOUNDS CONTRIBUTING TO TYPICAL OXIDIZED FLAVOUR [a]

Compounds	Flavours
Alkanals C_6-C_{11}	Green tallowy
2-Alkenals C_6-C_{10}	Green fatty
2,4-Alkadienals C_7-C_{10}	Oily deep-fried
3-cis-Hexenal	Green
4-cis-Heptenal	Creamy/putty
2,6- and 3,6-Nonadienal	Cucumber
2,4,7-Decatrienal	Fishy, sliced beans
1-Octen-3-one	Metallic
1,5 cis-Octadien-3-one	Metallic
1-Octen-3-ol	Mushroom

[a] Reproduced from Badings and Neeter (1980) with permission.

Milk is a very complex system and many of its components have been shown to participate in the induction, propagation and inhibition of lipid auto-oxidation and the subsequent generation of oxidized flavours. The mechanisms of these reactions, however, are not fully understood. For obvious reasons, it is impossible to divorce the various interactions and reactions in milk that combine to yield organoleptically unacceptable flavours in dairy products as a result of lipid oxidation. Various approaches have been taken by investigators in studying oxidation of milk lipids. Often it has proven useful and, in many cases, necessary to study model systems employing purified milk constituents; for example, in examining larger structural units of the milk system in isolation (e.g. the milk fat globule membrane) or in correlating levels of suspected pro-oxidants or anti-oxidants with the natural susceptibility of lipids in certain milk samples to rapid oxidation (e.g. in spontaneous oxidation; in milks high in PUFA). Of course, difficulties always arise in trying to interpret how the individual entities in question combine to give a net effect as well as in assessing their relative importance in the oxidation of milk lipids. Nonetheless, dairy chemists have experienced much success since the turn of the century in establishing many of the molecular mechanisms involved in the oxidation of milk lipids. From the early appreciation of the importance of metals, especially copper, to the current interest in 'active oxygen' as pro-oxidants for milk lipids, there has been a steady increase in the level of understanding of this important area of dairy research.

In order to facilitate discussion, it will be necessary to arbitrarily group the major factors that may combine to promote or inhibit oxidative rancidity in milk. Attempts will be made to interrelate some of the disparate observations that have been made in recent years.

From the work of Aurand *et al.* (1977), Hill (1979), Korycka-Dahl and Richardson (1980), Allen and Wrieden (1982*b*) and Yee and Shipe (1982) an appreciation is developing for the initiation of lipid oxidation in dairy products by so-called 'active oxygen' species. Since these apparently arise from a variety of sources in milk, any discussion of them will be integrated into the appropriate sections that follow.

2. MILK COMPONENTS AND ENVIRONMENTAL FACTORS AFFECTING LIPID OXIDATION

2.1. Spontaneous Oxidation

It is well established that lipids in a substantial percentage (12–20%) of raw milk samples undergo rapid oxidation (Bruhn *et al.*, 1976). Although this has been a practical problem for the dairy industry, it has also provided a vehicle whereby identification of the factors contributing to this type of oxidative rancidity has led to a better generic understanding of lipid oxidation in dairy products. This has produced a number of suggestions on ways of inhibiting lipid oxidation in milk (Shipe, 1964).

Raw milk has been classified into three categories depending upon the ease with which it undergoes oxidation, namely:

(i) *Spontaneous milk*. Milk capable of developing oxidized flavour without the presence of iron or copper as a contaminant.
(ii) *Susceptible milk*. Milk which does not develop oxidized flavour spontaneously but is susceptible if copper or iron is added.
(iii) *Non-susceptible milk*. Milk which does not become oxidized even in the presence of iron or copper.

Although spontaneous oxidation of milk lipids seems to be affected by heredity and stage of lactation of the cow as well as feeding practices (Shipe, 1964), two dominant themes have developed over the last 25 years to explain this phenomenon at the molecular level. Aurand and Woods (1959) proposed that a positive direct correlation between xanthine oxidase (XOD) activity and spontaneous oxidation in milk indicated that XOD is a primary pro-oxidant in milk. There are many attractive features about XOD that would favour it as a major pro-oxidant in milk, i.e. its essential iron, molybdenum and flavin cofactors and its association with the labile lipids in the fat globule membrane. Furthermore, XOD possesses a rather broad substrate specificity and in the presence of an appropriate substrate will generate H_2O_2 and O_2^- (Walsh, 1979). However, there are few, if any, substrates for XOD that occur normally in milk which tends to becloud the hypothesis that XOD is a principal determinant in spontaneous oxidation. Nonetheless, recent research suggests that XOD is involved to some extent in the peroxidation of milk lipids (Allen and Humphries, 1977; Hill, 1979; Allen and Wrieden, 1982b). Its high concentration in milk ($\sim$120 mg litre^{-1}) and the proposed coupling of its activity with that of lactoperoxidase (LP) in milk have led to indications that XOD plays a role not only in lipid oxidation (Hill, 1979) but also as part of a bactericidal system in milk that relies upon the oxidation of endogenous thiocyanate (Björck and Claesson, 1979).

In opposition to the XOD hypothesis for spontaneous oxidation, Smith and Dunkley (1960, 1961) could find no relationship between the XOD content of milk and spontaneous oxidation of milk lipids. A second major hypothesis on molecular mechanisms for spontaneous oxidation focusses on the copper–ascorbic acid system in milk and has been developed largely by Dunkley and co-workers. Although the chemistry is complicated, the concept has emerged that a copper–ascorbate complex is intimately involved in the peroxidation of milk lipids. Initially it was suggested that ascorbic acid served to reduce Cu^{2+} to Cu^+ which was the ultimate pro-oxidant, possibly as a complex with ascorbate (Smith and Dunkley, 1961). Certainly, copper in the presence of ascorbic acid is a strong oxidizing agent and may catalyse the formation of active oxygen species which can attack the milk lipids to yield flavour defects (Fig. 37).

For example, Cu^+ (as an ascorbate complex) can reduce $^3O_2$ to O_2^- or H_2O_2 (Fig. 37) to eventually yield the very reactive ·OH as an oxidant. Thus copper serves as a conduit to facilitate the transfer of electrons from ascorbic acid to generate damaging oxygen species. There can be little doubt that the copper–ascorbate system in milk is a primary pro-oxidant in milk. The relative levels of anti-oxidant tocopherols and pro-oxidant copper in milk may play a dominant role in defining the oxidative stability of individual milk samples (Bruhn *et al.*, 1976).

Obviously, the dynamics of pro-oxidants in milk are complex and cannot be considered as self-contained separate entities. It is apparent that interactions exist that may have additional profound consequences for the oxidative stability of milk lipids. For example, interesting proposals (discussed later) have been advanced for a coupling between XOD in the fat globule membrane and lactoperoxidase in the milk serum (Hill, 1979) or between contaminating copper and denatured XOD (Allen and Wrieden, 1982b) to promote the oxidation of milk lipids. Although these and undoubtedly other unknown interactions exist in milk, the following discussions will be simplified by grouping the major components in milk that affect lipid oxidation. Further studies on spontaneous oxidation of milk lipids will no doubt continue to provide us with basic as well as practical information.

2.2. Metals

Metals in milk are central to any understanding of lipid oxidation in dairy products. Various metals in milk are loosely combined according to those that occur naturally and those that arise from exogenous sources as contaminants. Although a wide variety of metals has been quantified in milk (Jarrett, 1979), attention has focussed on iron and copper with respect to problems involving lipid oxidation. Interest in these two metals stems from studies in which the addition of various metal salts to milk, including those of copper and iron, catalysed lipid oxidation (Shipe, 1964; Schwartz and Parks, 1974).

Since milk is a complex dynamic system, metals involved in the oxidation of milk lipids cannot be studied in isolation and must be considered within the overall context of the various factors in milk that may physically or chemically alter their behaviour. Metals, whether naturally occurring or contaminants, are generally associated to varying degrees with all of the milk constituents. Metals that occur naturally in milk tend to be associated with identifiable components of milk in a fixed stoichiometry, such as the 8 mol of iron, the 2 mol of molybdenum and the 2 mol of flavin adenine dinucleotide (FAD) found in XOD (Walsh, 1979) associated with the milk fat globule membrane or the haem iron of lactoperoxidase in the milk serum and cytochromes of the milk fat globule membrane. On the other hand, metal contaminants will distribute amongst the various phases of milk according to poorly understood factors that determine their affinities for available ligands as binding sites. Although binding studies employing excess Cu^{2+} in equilibrium with various milk macromolecules indicate that these compounds have a large capacity to bind Cu^{2+} (Aulakh and Stine, 1971), a limited amount of contaminating Cu^{2+} should seek sites of strongest affinity (Freifelder, 1982). In the latter more realistic case, small amounts of radioactive $^{64}Cu^{2+}$ or $^{59}Fe^{3+}$ added to raw whole milk or administered to lactating cows have been used to determine the distribution of added or natural copper or iron in milk (King et al., 1959). The natural copper and iron favoured the fat globules, but only 2–3% of the total added copper and none of the added iron associated with the fat globules. The naturally occurring metals were not dialysable whereas added metals in the skim milk were slightly dialysable. Added copper associated with the fat globules was non-dialysable. The small amount of added copper rather tightly bound to the fat globule may play a crucial role in the development of oxidized flavour in milk.

Certainly, increased chelation capacity resulting either from the treatment of milk with trypsin or from ageing has been amply demonstrated by Gregory and Shipe (1975) and has been proposed as a means for inhibiting development of oxidized flavour in milk (Shipe, 1977). Presumably, enzymic hydrolysis of milk proteins in the serum, as well as in the milk fat globule membrane, increases available sites for chelating metals, particularly Cu^{2+}, into non-pro-oxidant complexes.

Administration of copper in the diet or by injection as various complexes while attempting to study the effects of the administered copper on the oxidation of milk lipids is also complicated by the variety of ligands available to bind heavy metals *in vivo* in relation to the stability constants of the administered chelates. In general, administration of copper complexes to cows has not provided a clear-cut answer to the problem of spontaneous oxidation of lipids in milk (Dunkley *et al.*, 1968).

It is well known that the distribution of the naturally occurring metallo-protein and other metallo-organic entities in milk, as well as those formed with metal contaminants, is affected by a wide array of environmental factors and processing treatments (McPherson and Kitchen, 1983). Nonetheless, a substantial portion of the milk fat globule membrane, including pro-oxidant XOD, remains with the fat globules after homogenization of milk (Keenan *et al.*, 1983).

The migration of metals (however associated) within the milk system and in dairy products as affected by handling, processing, storage and marketing can have profound effects on the oxidative stability of their lipids (Downey, 1969). Depending upon the circumstances, an inhibition or an enhancement of lipid oxidation may occur. From a combination of empirical studies and observations with those of a more fundamental nature, copper has emerged as the dominant metal for catalysing the oxidation of milk lipids and the balance between levels of tocopherol and copper in milk plays a crucial role in defining the spontaneous oxidation of raw whole milk (Bruhn *et al.*, 1976). Only in recent years, however, has a better understanding of the interplay among copper and other milk constituents, such as ascorbic acid, been developed at the molecular level.

Nonetheless, much more must be learned before a clear knowledge of the complexities involved in metal-catalyzed lipid oxidation will emerge. Where possible, metals will be integrated into the following discussions of various factors known to affect the oxidation of lipids in milk.

In the pro-oxidant metal system there are four basic crucial components: (1) a metal that undergoes univalent redox reactions (e.g., copper, iron); (2) appropriate ligands to facilitate these reactions; (3) electron donors (which may also be the substrate) to provide electrons (usually to oxygen via the complexed metal); and (4) oxygen which, upon reduction, is 'activated' and which also serves as a virtually limitless electron sink to drive the oxidative reactions. In the previous discussions in Part A, we have seen how these components combine to catalyse numerous oxidative reactions; subsequently, we hope to provide some insight as to how these basic components relate to other factors in the milk system to enhance or retard lipid oxidation.

2.3. Electron Donors/Acceptors

The flow of electrons from a donor to an appropriate acceptor in the milk system has profound effects, particularly if one-electron redox reactions are involved. Although reactions involving the transfer of two electrons concurrently may be favoured in metabolic reactions (Walsh, 1979), it is the univalent transfer of electrons that characterizes the oxidation of lipids. The consideration of particular milk constituents as electron donors or acceptors is necessarily arbitrary. However, ultimately the unsaturated fatty acids of milk lipids are the electron donors and oxygen is the principal acceptor. Along the reaction path, however, other milk constituents intervene to play crucial roles in enhancing or retarding the rate of the overall oxidative reactions.

In this section we shall briefly consider only tocopherols, thiols and ascorbic acid as major electron donors in milk. Where necessary, oxygen and active oxygen species as electron acceptors will be discussed subsequently.

2.3.1. Tocopherols

It is common knowledge that the tocopherols generally serve an antioxidant function in lipids, although under certain circumstances (probably at high concentrations) the tocopherols can be pro-oxidant (see Part A). However, under the normal conditions existing in milk, a pro-oxidant effect of tocopherol is highly unlikely.

Tocopherol concentrations are at least three-fold higher in the lipids of the milk fat globule membrane than inside the fat globule (Erickson et al., 1964). Tocopherol concentrations in fractions obtained from washed cream range from 17–21 µg/g butteroil to 170 µg/g lipid in the milk fat globule membrane. During storage (35 h) of gravity cream containing added Cu^{2+} and ascorbic acid, total destruction of tocopherol in the milk fat globule membrane was observed compared to ~30% destruction in the butteroil. Thus the proximity of tocopherols in the milk fat globule membrane to pro-oxidants and highly oxidizable phospholipids favours lipid oxidation.

Attempts to increase levels of tocopherols in milk by feeding tocopherol derivatives have been disappointing because of the low incorporation of administered tocopherols into the milk lipids. Consequently, the economics associated with feeding tocopherol supplements to lactating dairy cows are unfavourable (Bruhn et al., 1976).

Other phenolic compounds that exist in milk may also contribute electrons to perform an anti-oxidant function. The well-known antioxygenesis of the caseins (discussed in Section 2.6.1) could partially result from available phenolic side-chains of tyrosine associated with the lipids of milk and favoured by the hydrophobicities of the casein fractions.

2.3.2. Thiols

Historically, thiols resulting from thermal treatments of milk have been considered as contributing an anti-oxidant function in milk which can be rationalized from possible univalent reduction of free radicals. It is worthy of note that the principal sources of thiols in milk are the milk fat globule membrane (McPherson and Kitchen, 1983) and the serum proteins, particularly β-lactoglobulin (Schwartz and Parks, 1974).

However, recent research suggests that thiols occurring in milk may actually play a somewhat paradoxical dual role, reminiscent of ascorbic acid. Certainly, as potential one-electron donors, thiols may provide reducing equivalents to 'activate' oxygen via the reduction of metals. Auto-oxidation of thiols is known to produce substantial amounts of superoxide (Misra, 1974). Recent research (Rowley and Halliwell, 1982) indicates that in the presence of iron salts and H_2O_2, thiol compounds may facilitate the formation of $\cdot OH$ via the following reactions:

$$RSH + {}^3O_2 \xrightarrow[\text{Salts}]{\text{Iron}} O_2^- \qquad (65)$$

$$O_2^- + Fe^{3+} \longrightarrow Fe^{2+} + O_2 \qquad (66)$$

$$Fe^{2+} + H_2O_2 \longrightarrow {}^{\cdot}OH + {}^-OH + Fe^{3+} \qquad (67)$$

The production of ˙OH is dependent on thiol concentration—being favoured at low levels but with scavenging the ˙OH at high levels. Interestingly, the H_2O_2 is generated from the oxidation of hypoxanthine by XOD. These data suggest concentration-dependent anomalous effects for thiol compounds in the oxidation of milk lipids.

A potential source of thiol groups from the milk fat globule membrane is XOD, an iron–sulphur enzyme (Walsh, 1979). The inactivation of XOD upon storage has been attributed to the decomposition of a persulphide, proposed as part of the active site:

$$E—S—S^- \rightarrow E—S^- + {}^-SH \quad (68)$$

Nucleophiles can displace the sulphur to yield a variety of sulphur compounds that should be flavourful. It is interesting that CN^- reacts with the proposed persulphide to yield thiocyanate (Walsh, 1979). This has numerous possible ramifications for the proposed bactericidal system in milk that relies on the oxidation of endogenous thiocyanate to yield antimicrobial species (Reiter, 1978; Björck and Claesson, 1979). Additionally, thermal treatments of milk known to generate thiol groups should enhance the rate of decomposition of the persulphide.

Taylor and Richardson (1980), studying the oxidation of lipids in model systems, questioned whether thiol groups generated in milk are obligatory anti-oxidants. Consideration of reactions (65)–(68) could actually provide a basis for rationalizing at least a portion of the pro-oxidant effect on milk lipids ascribed to XOD in the milk fat globule membrane (Allen and Humphries, 1977; Hill, 1979; Allen and Wrieden, 1982a). Yee and Shipe (1982) compared the effects of cysteine (10^{-5} M) and glutathione on copper (10^{-5} M) or haem-catalysed oxidation of emulsified methyl linoleate. In general, free thiol groups were pro-oxidative in the presence of copper but not so with haem. Blocking of the thiol groups, using iodoacetate or by previous oxidation, nullified the pro-oxidant effects of the thiols in the copper-catalysed oxidations. On the other hand, free thiols actually behaved as anti-oxidants in the haem-catalysed oxidation of methyl linoleate. Thiols in the presence of Cu^{2+} promoted the formation of superoxide, whereas haem did not. Interestingly, the addition of cysteine (100–1000 μM) to milk did not inhibit either copper- or haem-induced lipid oxidation. Yee and Shipe (1982) concluded that thiol groups in milk may be pro-oxidant or anti-oxidant in milk depending upon the conditions. Perhaps more lipophilic thiols partition into the lipids, especially in the milk fat globule membrane, to exert effects different than those observed with water-soluble cysteine.

Swaisgood and Abraham (1980) suggested that a sulphydryl oxidase indigenous to milk, which couples with lactoperoxidase in the serum to destroy the resultant H_2O_2, may serve an anti-oxidant function. One would expect that lactoperoxidase in the presence of the H_2O_2 would also destroy residual ascorbic acid in milk to provide additional protection against oxidation of milk lipids (see below).

2.3.3. Ascorbic Acid

Early research, reviewed by Shipe (1964) and by Schwartz and Parks (1974), indicated that ascorbic acid in association with copper is an important pro-oxidant in milk.

The behaviour of the ascorbic acid in relation to oxidized flavour in milk is anomalous. Similar to the anomaly proposed above for thiol groups, concentrations of ascorbic acid above those in normal milk provide anti-oxidant protection; however, at lower concentrations ascorbic acid is a pro-oxidant. Since oxidized flavour can be prevented by rapid destruction of ascorbic acid in milk, it is crucial to the development of oxidized flavour.

In early work, Brown and Olson (1942) postulated that ascorbic acid reduces Cu^{2+} to Cu^+, which in turn reduces molecular oxygen to hydrogen peroxide which oxidizes lipids in the fat globule membrane. Dunkley's group firmly established the copper–ascorbic acid system as a principal cause of lipid oxidation in milk. The relative concentration-dependence between copper and ascorbate required for enhanced lipid oxidation was discussed in Part A. In a series of papers (King and Dunkley, 1959; Smith and Dunkley, 1960, 1961, 1962; Haase and Dunkley, 1969*a*, *b*, *c*) that involved addition of copper ions to milk and model systems, judicious application of specific chelators for cuprous ions, addition or specific destruction of ascorbic acid in milk and model systems, and use of reducing agents other than ascorbic acid, the primary function of copper and ascorbic acid in catalysing lipid oxidation was carefully documented. Among reducing agents tested, ascorbic acid was the most effective promoter of lipid oxidation in the presence of copper, prompting Smith and Dunkley (1961) to propose a specific association between copper and ascorbate as the ultimate pro-oxidant. As discussed in Part A, a copper–ascorbate chelate has subsequently been proposed as a potent oxidizing agent. It would appear that such a chelate can activate oxygen to yield pro-oxidant species (Fig. 37).

On the other hand, addition of adequate levels of surface-active ascorbyl palmitate to dairy products may lead to an orientation of this anti-oxidant at the lipid–aqueous interface to intercept damaging radicals and otherwise retard lipid oxidation (Badings, 1970; Badings and Neeter, 1980).

2.4. Light and Riboflavin

Riboflavin serves as a potent photosensitizer in milk to catalyse a variety of oxidative reactions, not only of lipids but also of proteins, ascorbic acid, etc. (Sattar and déMan, 1975; Spikes, 1981). As indicated in Fig. 42, the photo-oxidation reactions catalysed by riboflavin are extremely varied and complex. This complexity is evident in the dual roles that riboflavin apparently plays in the flavour defects that develop in milk and dairy products upon exposure to light, particularly fluorescent light and sunlight (White and Bulthaus, 1982). The so-called 'light-induced (or activated) flavour' arises primarily from the formation of 3-methyl thiopropanal (methional) as a result of the photo-oxidation of methionine mediated by riboflavin (Badings and Neeter, 1980). The photogeneration of methional is thought to proceed via a Type I mechanism (Fig. 42) (Foote, 1976).

The low concentration of free amino acids in milk available as substrates for photo-oxidations (Aston, 1975) suggests that methionine residues in the primary sequence of the milk proteins may somehow serve as electron donors in the univalent reduction of riboflavin. This would, of course, necessitate chain scission with the release of methional.

Superimposed upon the generation of the light-induced flavour related to methional formation is the oxidation of milk lipids. This could result also from a Type I mechanism (Fig. 42) wherein an excited riboflavin triplet abstracts an H$^\cdot$ from an appropriate fatty acid residue to initiate lipid oxidation. However, the generation of active oxygen species, particularly $^1O_2$ and $O_2^{\bar{\cdot}}$, cannot be ignored as initiators of lipid oxidation, perhaps concurrently with the univalent oxidation of unsaturated fatty acid residues. The work of Aurand *et al.* (1977), Hill (1979), Korycka-Dahl and Richardson (1980), Yee and Shipe (1982) and Allen and Wrieden (1982*b*) suggests that active oxygen species are involved in the oxidation of milk constituents in general and of lipids in particular. Aurand *et al.* (1977) used quenching and scavenging agents to implicate $^1O_2$ in the oxidation of milk lipids as catalysed by copper ions, enzymes and light. The presence of a singlet-oxygen quencher (1,4-diazabicyclo[2-2-2] octane) or scavenger (1,3-diphenylisobenzofuran) inhibited lipid oxygen in those systems. They concluded that singlet oxygen was the immediate source of the hydroperoxides that initiate lipid oxidation as catalyzed by the three agents above. Although these studies represent some of the first attempts to implicate active oxygen species in the oxidation of milk lipids, the participation of $^1O_2$ in initiating the oxidative processes should be judged with caution. In the absence of confirmatory evidence, the trapping and quenching experiments offer only a preliminary indication of $^1O_2$ involvement. Further, as mentioned previously, it is now known that the spontaneous dismutation of O_2^- proposed as a likely source of the $^1O_2$ in the oxidations mediated by copper and enzymes does not yield this species (Foote *et al.*, 1980). Nevertheless, under appropriate conditions, the oxidation of O_2^- to O_2 can yield $^1O_2$ (Nanni *et al.*, 1981). We do not have sufficient knowledge of the oxidizing agents in milk that may be capable of oxidizing O_2^- to $^1O_2$ (see Part A). This could be a fruitful field of study to characterize possible oxidizing species in milk that might yield $^1O_2$ upon oxidation of O_2^-.

In milk systems containing riboflavin and irradiated with a 15 W fluorescent light generating 5200 lx, substantial amounts of O_2^- are produced (Korycka-Dahl and Richardson, 1980). Bovine milk serum proteins (β-lactoglobulin, α-lactalbumin and bovine serum albumin) and free amino acids (cysteine, methionine, histidine, tyrosine and tryptophan) can serve as photo-oxidizable substrates in the reduction of excited, triplet-state riboflavin ($^3F^*$). Fully reduced riboflavin (FH_2) can subsequently univalently reduce oxygen to yield O_2^- and the flavin semiquinone (FH˙) (Fig. 42). These reductions imply a univalent oxidation of milk serum proteins which could lead to their polymerization and aggregation or chain scission as demonstrated by Gilmore and Dimick (1979). Apparently, the univalent reduction of O_2 to O_2^- by FH˙ (Fig. 42) proceeds slowly, being much more rapid near the pKa for FH˙ ($\sim$8·5) (Vaish and Tollin, 1970, 1971). The major photodegradation product of riboflavin found in milk is lumichrome (Parks and Allen, 1977).

There are ample opportunities for the generation of active oxygen species and reactive radicals in milk and dairy products. This is particularly true in photo-oxidative reactions catalysed by riboflavin. In summary, highly reactive damaging species involved in initiating many of the adverse reactions in dairy products include excited triplet-state riboflavin, $^1O_2$, O_2^-, H_2O_2 and resultant free radicals. The genesis of these agents in dairy products can be countered, in part, by opaque packaging, inert-gas packing and generation of anti-oxidants *in situ* by appropriate processing techniques (Shipe, 1964).

2.5. Milk Fat Globule Membrane (MFGM)

This complex lipoprotein membrane that surrounds the fat globule is derived largely from the plasma membrane of the secretory cell and entrapped cytoplasmic membranes (e.g. endoplasmic reticulum) (Mulder and Walstra, 1974; Keenan *et al.*, Chapter 3, this book). The phospholipids in the MFGM contain ~6% linoleic acid, and their juxtaposition with various pro-oxidants in the lipoprotein matrix makes the MFGM a focal point for oxidation of milk lipids (Badings, 1970; Mulder and Walstra, 1974). King (1962) used the thiobarbituric acid reaction to demonstrate the rapid oxidation of isolated MFGM. Purified phospholipids, especially phosphatidyl ethanolamine, readily undergo oxidation in the presence of Cu^{2+} (O'Mahony and Shipe, 1970; Morita and Fujmaki, 1972a, b).

OXIDIZED FLAVORS

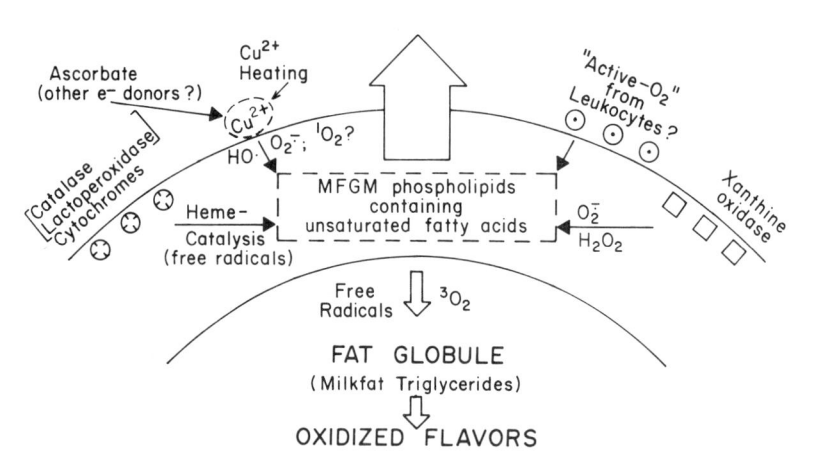

FIG. 45. Factors associated with the fat globule membrane of milk that may favour the oxidation for milk lipids (modified from McPherson and Kitchen, 1983).

However, once oxidation starts, diffusion of the propagating chain reaction radicals into the more saturated fat globule core from the fat–plasma interface results in generalized oxidation of milk fat triglycerides (Fig. 45). Indeed, this was observed (Badings, 1970) in 'artificial butter' made from butterfat and milk cream devoid of MFGM and, therefore, also of phospholipids. In this butter, oxidized flavours developed, albeit slowly, provided that copper was also present in the associated butter serum.

About one-third of the milk phospholipids in freshly drawn milk is located in the milk serum as MFGM. These small lipoprotein particles are sometimes referred to as 'milk microsomes', and their proportion in milk serum can be increased in processed milk as a result of disruption of the MFGM and release of membrane phospholipids into the aqueous phase (Mulder and Walstra, 1974; McPherson and Kitchen, 1983). Thus many of the following reactions may occur in MFGM lipoproteins associated with the fat globule as well as those dispersed in the serum phase.

The major factors that affect lipid oxidation in the MFGM are illustrated in Fig. 45. Exogenous copper bound to membrane constituents probably catalyses oxidation of membrane lipids, especially the phosphatidyl ethanolamine (O'Mahony and Shipe, 1970; Morita and Fujimaki, 1972a, b); endogenous copper does not appear to be as damaging (Badings, 1970). Modification of the membrane by processing treatments that may alter the distribution of pro- and anti-oxidants can markedly affect the oxidative stability of milk (McPherson and Kitchen, 1983).

2.5.1. Xanthine Oxidase (XOD)

XOD plays an important role in the peroxidation of membrane lipids and those lipids in the triglyceride core of the fat globule (Allen and Humphries, 1977; Bruder et al., 1980). Although controversial, XOD has been implicated in the spontaneous oxidation of milk lipids (Aurand and Woods, 1959). The mechanisms whereby XOD exerts its pro-oxidant effects are not completely understood. In the simplest case, H_2O_2 resulting from the oxidation of an apparent substrate by XOD could oxidize the milk lipids. However, there is little, if any, known substrate available for XOD in normal milk, which complicates the situation. Perhaps varying levels of superoxide dismutase (SOD) in milk may have confounded some of the earlier observations implicating XOD in the spontaneous oxidation of milk lipids (Holbrook and Hicks, 1978) (see below).

Recent research by Hill (1979, and references cited therein) and by Allen and Wrieden (1982b) indicates possibilities for interactions between native and denatured XOD in MFGM and lactoperoxidase (LP) or copper from the serum phase to yield pro-oxidant effects.

Hill (1979) studied lipid oxidation in milks high in linoleic acid that were obtained by feeding vegetable oils protected from biohydrogenation in the rumen. Milk thus obtained contains up to 35% linoleic acid (w/w) in the milkfat (Hill, 1979) and is easily oxidized (Sidhu et al., 1976). XOD possesses a broad substrate specificity and generates O_2^- and H_2O_2 upon oxidizing suitable substrates (Walsh, 1979). Since the concentration of XOD in milk is high (~ 120 mg litre^{-1}), it is fortunate that there is a dearth of substrates for this enzyme in milk. Nonetheless, aldehydes that may arise from incipient lipid oxidation could serve as substrates for XOD to provide H_2O_2 which, in turn, might be utilized by lactoperoxidase (~30 mg litre^{-1}) from the milk serum to catalyse further peroxidation reactions. Thus Hill (1979) proposes a coupled enzyme system in which lactoperoxidase catalyses lipid oxidation generating aldehyde substrates for XOD that in turn furnishes H_2O_2 for use by lactoperoxidase as an oxidizing agent. A similar enzyme couple has been proposed by Björck and Claesson (1979) as a means for oxidizing endogenous thiocyanate in milk to generate a bactericidal agent.

Milks high in linoleic acid pasteurized at 80 °C/15 s are much more stable to oxidation than those heated at 72 °C/15 s. The increased stability is attributed to thermal inactivation of the oxidative enzymes at the higher temperature (Hill, 1979). After treatment of the milk at 80 °C/15 s, addition of XOD or LP to more readily oxidizable milks causes more rapid development of oxidized flavours. However, more stable milks tend to remain stable after addition of XOD or LP. When substrate for XOD is added, oxidative processes are accelerated. Apparently the XOD generates H_2O_2 from substrate ordinarily absent from milk, implying that a 'substrate' is present in unstable but not in stable milks. Addition of as little as 1 mg litre^{-1} of SOD (and a small amount of catalase to remove resultant H_2O_2) improves the stability of these unstable milks considerably, suggesting that O_2^- is involved in the enzymatic peroxidation.

On the other hand, oxidation of milk lipids after addition of low levels (~0·1 ppm) of Cu^{2+} to the milk is not inhibited by SOD and catalase. However, the ˙OH scavenger formate does inhibit lipid oxidation, suggesting that ˙OH is the active pro-oxidant. Copper added before pasteurization at 80 °C/15 s results in slight acceleration of lipid oxidation, but when added after pasteurization results in rapid development of oxidized flavour. In comparison, milk pasteurized at 72 °C/15 s undergoes lipid oxidation irrespective of when the copper is added. The higher temperatures may possibly result in the unfolding of casein with the consequent binding of copper available during the heat treatment. Hill (1979) proposes that two major systems in milk catalyse lipid oxidation via: (1) the copper–ascorbate generation of ˙OH; and (2) the XOD–LP enzymic generation of O_2^- and $^1O_2$. In milk, however, these two systems are coupled so that aldehydes from the copper-based system are oxidized by XOD to furnish LP with the H_2O_2 to initiate lipid oxidation. The more effective higher pasteurization temperature was thought to protect milk lipids from oxidation by inactivating XOD and LP and by inhibiting Cu^{2+}-induced lipid oxidation.

In model studies using trilinolein emulsions, Allen and Wrieden (1982b) confirmed the strong pro-oxidant effects of LP and their reduction upon heating the enzyme at 80 °C/20 s. They proposed that LP was pro-oxidant by virtue of generalized haem catalysis in addition to its enzymic effects. They also suggested that the two lipid-oxidizing systems in milk (Hill, 1979) be modified, since the generation of O_2^- is not attributable only to oxidases, and that O_2^- itself has no activity in lipid oxidation as proposed by Hill (1979).

In the presence of 10 μM added Cu^{2+}, XOD rapidly oxidized trilinolein in the absence of a substrate for XOD (Allen and Wrieden, 1982b). Since the added Cu^{2+} inactivated the XOD, this rate enhancement was considered analogous to that observed when XOD from MFGM was thermally denatured (Allen and Humphries, 1977). Any O_2^- that might be produced could possibly be converted to very strong oxidizing species by the Cu^{2+} bound to the inactive enzyme. Thus Cu^{2+} bound to denatured XOD may yield a very strong pro-oxidant which can utilize O_2^- or other active oxygen species to catalyse lipid oxidation. One should not ignore the potential for photocatalysed reactions involving the FAD associated with XOD as a source of O_2^-. Korycka-Dahl and Richardson (1978) demonstrated that FAD and native XOD were about 26 and 4%, respectively, as effective as riboflavin in the photogeneration of O_2^- on an equivalent riboflavin basis. Furthermore, the potentially longer lifetime for O_2^- in the membrane lipids in association with bound Cu^{2+} could make these reactions of some significance. In addition, any resultant semiquinone radical (FH·) might reduce metals (e.g. Cu^{2+}, Fe^{3+}) to react with available H_2O_2 (Fig. 42). Consider also reactions (65)–(68) as a potential pro-oxidant system involving XOD.

2.5.2 Cytochromes

Cytochromes have been observed in the MFGM by several workers (Bailie and Morton, 1958a, b; Plantz et al., 1973; Gregory et al., 1975; Bernstein, 1977; Jarasch et al., 1977). Because of their low concentrations in the MFGM, these pigments would appear to be of minor importance in catalysing lipid oxidations. However, the powerful pro-oxidant properties of ferri–porphyrin proteins as catalysts for lipid oxidation (Tappel, 1962; Kendrick and Watts, 1969) and their juxtaposition with MFGM lipids, coupled with the longer lifetimes of active oxygen species in non-polar environments (see Part A), suggest a role for these pigments as focal points for oxidation of MFGM lipids.

A b_5-type cytochrome and a CO-binding cytochrome have been observed in MFGM material by Bernstein (1977). He showed that the CO-binding cytochrome was as active as haemoglobin in promoting lipid oxidation, whereas the native b_5-type cytochrome was inactive. Thermal processing may affect lipid oxidation by exposing or masking ferri–porphyrin groups in the membrane material. Gregory *et al.* (1975) concluded that haem proteins in the MFGM are involved in lipid oxidation, but are not the only factors.

MFGM is known to be an important source of thiol groups released upon heating (McPherson and Kitchen, 1983). Univalent reduction of Cu^{2+} and Fe^{3+} by lipid-soluble thiols in the MFGM coupled with their auto-oxidation may actually result, in some cases, in an enhanced rate of lipid oxidation, as discussed in Section 2.3.2 on thiols.

2.6. Serum Enzymes and Other Proteins

From the foregoing discussions, it is evident that complex interactions between a number of constituents, including proteins, in the serum phase and lipids associated with the fat globule (and lipoproteins in the serum phase) can markedly affect the oxidation of milk lipids. In the following sections, effects of the major known serum proteins and enzymes on oxidation of milk lipids will be considered briefly.

2.6.1. Serum Proteins

The caseins possess a remarkable anti-oxidant activity (El-Negoumy, 1965; Taylor and Richardson, 1980; Allen and Wrieden, 1982*a*; Ericksson, 1982), which may be related in part to their hydrophobic nature. This could constrain proper orientation of potential anti-oxidant side-chains of constituent amino acids at the lipid interface. Caseins also strongly bind Cu^{2+} (Aulakh and Stine, 1971), which could provide anti-oxidant protection, particularly if the caseins are in excess. On the other hand, the major whey proteins are substantially less effective as anti-oxidants (Taylor and Richardson, 1980; Allen and Wrieden, 1982*a*). Among the minor serum proteins, lactoferrin with available binding sites for Fe^{3+} can inhibit peroxidation induced by Fe^{2+} (Gutteridge *et al.*, 1981; Allen and Wrieden, 1982*b*).

2.6.2. Serum Enzymes

We have seen that lactoperoxidase (LP) is strongly pro-oxidant in the presence or absence of added Cu^{2+} or Fe^{3+} when studied in a trilinolein emulsion (Allen and Wrieden, 1982*a*) or in milk high in linoleate (Hill, 1979). Heat treatments of LP or milk at 72°C/15–20 s had little effect on lipid oxidation, but after 80 °C/15–20 s attendant lipid oxidation was greatly reduced. These findings tend to conflict with the observations of Ericksson's group discussed in Part A (Ericksson, 1982) which found that increased heat treatments of lactoperoxidase in model systems increased the rate of lipid oxidation. It has recently been demonstrated (Kanner and Kinsella, 1983b) that the lactoperoxidase/H_2O_2/halide system has the capacity to initiate lipid peroxidation in model systems.

Superoxide dismutase (and catalase) exerts a strong anti-oxidant effect when added to milk at a level of 1 mg litre^{-1} (Hill, 1979) or to model emulsion systems (Allen and Wrieden, 1982b). However, Cu^{2+} (10 μM) added to the emulsion with SOD was pro-oxidant and might compete for O_2^- to convert it to pro-oxidant species (Allen and Wrieden, 1982b). SOD has been detected and isolated as a constituent of milk (Hill, 1975; Asada, 1976; Korycka-Dahl et al., 1979), but it is apparently present at insufficient levels to provide substantial anti-oxidant protection (Holbrook and Hicks, 1978). However, the observed inhibition by SOD of lipid oxidation catalysed by XOD (Holbrook and Hicks, 1978) may have confounded earlier observations on the importance of XOD in spontaneous oxidation of milk lipids (Aurand and Woods, 1959; Smith and Dunkley, 1961). On the other hand, attempts to correlate the SOD content of milk with resistance to spontaneous oxidation were unsuccessful (Holbrook and Hicks, 1978). Partially purified SOD studied in milk serum is incompletely inactivated by heat treatments greater than 75 °C and not at all by a minimal pasteurization at 71·7 °C/15 s (Hicks et al., 1979).

2.7. Oxygen

Removal of dissolved oxygen from fluid milk or its replacement by inert gases is expected to reduce the incidence and intensity of oxidized flavour (Schwartz and Parks, 1974; Schroder, 1982). However, not all oxidized flavours are equally reduced by lowering the oxygen concentration. Schroder (1982) reported that light-induced, but not copper-induced, oxidized flavour development in milk could be controlled by oxygen removal. To prevent copper-induced oxidized flavour, milk would have to be de-aerated to a very low oxygen level. One wonders whether the greater oxygen-dependence of the light-induced oxidized flavour signals an interaction between excited triplet-state riboflavin and the higher oxygen concentrations to yield $^1O_2$. Aurand et al. (1977) have proposed that $^1O_2$ is generated directly in photo-oxidative reactions catalysed by riboflavin in milk.

In fermented dairy products, such as certain types of cheese and yoghurt, oxidized flavour was reported to be of minor importance (Wong et al., 1973; Czulak et al., 1974; Korycka-Dahl et al., 1983). Apparently, the oxygen content within these products was lowered due to its utilization for growth by the lactic acid starter bacteria. However, other factors, such as the acidic pH conditions prevailing in these products, the presence of peptides from protein proteolysis, higher protein:copper molar ratios or the formation of anti-oxidative compounds by micro-organisms, could also add to their oxidative stabilities.

As expected, high-fat products such as dried whole milk and butter are quite susceptible to lipid oxidation. In addition to the foregoing factors affecting lipid oxidation, available oxygen may be differentially distributed as a result of physical constraints placed upon the products.

It is evident from Table 2 that oxygen is excluded from crystallized milk fat, presumably because closely packed crystals cannot physically accommodate the oxygen. The question remains whether amorphous lipids such as those that are prevented from orienting into crystalline structures (possibly in the MFGM or resulting from rapid cooling of milk fat) can retain oxygen physically dissolved or entrapped in the 'liquid' phase. The interstices within lipids associated with the MFGM or within supercooled or amorphous milk fat triglycerides surrounded by case-hardened fats could occlude substantial quantities of oxygen that would be retained upon storage at sufficiently low temperatures.

Thus rapid surface cooling of high-fat products (e.g. dried whole milk) after drying may actually entrap small quantities of oxygen that may have redissolved in the fat phase thereby favouring long-term oxidation of lipids. It might be that slow-cooling (tempering) to maximize fat crystallization (in an inert atmosphere) would be more beneficial to the shelf-stability of high-fat dairy products.

A final comment on the possible pro-oxidant effects of the oxygen-activating system of phagocytes is worth considering (Fig. 45). Milk leucocytes can adhere to the MFGM (Peters and Trout, 1945a, b). The enzymic generation of 'active oxygen' species resulting from phagocytic activity (Salin and McCord, 1977; Kanner and Kinsella, 1983a) should not be overlooked as a possible source of pro-oxidants.

3. CONCLUSIONS

One can only conclude that the diverse nature of dairy products and the multitude of factors that affect the oxidation of lipids in these products will continue to make this a fertile field of research for some time. A knowledge of fundamental reaction mechanisms gleaned from chemistry and the life sciences will continue to provide the dairy chemist with new insights into methods of controlling lipid oxidation. The current interest in 'active oxygen species' in initiating lipid oxidation and the observations on the potential use of superoxide dismutase as an anti-oxidant in milk are sufficient to justify optimism for continued progress in mitigating the detrimental effects of lipid oxidation in dairy products.

ACKNOWLEDGEMENTS

This review was made possible by support from the College of Agricultural and Life Sciences, University of Wisconsin–Madison, Madison, USA. The authors are grateful to Professors W. Hoekstra, W. A. Pryor, W. F. Shipe and L. M. Smith for reviewing portions or all of the manuscript.

REFERENCES

References not found below may be found in the reference section to Part A of this chapter.

ALLEN, J. C. and HUMPHRIES, C. (1977). *J. Dairy Res.*, **44**, 495.
ALLEN, J. C. and WRIEDEN, W. L. (1982*a*). *J. Dairy Res.*, **42**, 239.
ALLEN, J. C. and WRIEDEN, W. L. (1982*b*). *J. Dairy Res.*, **49**, 249.
ASADA, K. (1976). *Agr. Biol. Chem.*, **40**, 1659.
ASTON, J. W. (1975). *Aust. J. Dairy Technol.*, **30**, 55.
AURAND, L. W. and WOODS, A. E. (1959). *J. Dairy Sci.*, **42**, 1111.
AURAND, L. W., BOONE, N. H. and GIDDINGS, G. G. (1977). *J. Dairy Sci.*, **60**, 363.
BADINGS, H. T. and NEETER, R. (1980). *Netherlands Milk and Dairy J.*, **34**, 9.
BAILIE, M. J. and MORTON, R. K. (1958*a*). *Biochem. J.*, **69**, 35.
BAILIE, M. J. and MORTON, R. K. (1958*b*). *Biochem. J.*, **69**, 44.
BERNSTEIN, A. (1977). PhD Thesis, University of Wisconsin, Madison.
BJÖRCK, L. and CLAESSON, O. (1979). *J. Dairy Sci.*, **62**, 1211.
BROWN, W. C. and OLSON, F. C. (1942). *J. Dairy Sci.*, **25**, 1041.
BRUDER, G., HEID, H. W., JARASCH, E.-D. and KEENAN, T. W. (1980). *Eur. J. Cell Biol.*, **22**, 271.
BRUHN, J. C., FRANKE, A. A. and GOBEL, G. S. (1976). *J. Dairy Sci.*, **59**, 828.
CZULAK, J., HAMMOND, L. A. and HORWOOD, J. F. (1974). *Aust. J. Dairy Technol.*, **29**, 124.
DOWNEY, W. K. (1969). *J. Soc. Dairy Technol.*, **22**, 154.
DUNKLEY, W. L., FRANKE, A. A., ROBB, J. and RONNING, M. (1968). *J. Dairy Sci.*, **51**, 863.

EL-NEGOUMY, A. M. (1965). *J. Dairy Sci.*, **48**, 1406.
ERICKSON, D. R., DUNKLEY, W. L. and SMITH, L. M. (1964). *J. Food Sci.*, **29**, 269.
FREIFELDER, D. (1982). *Physical Chemistry for Students of Biology and Chemistry*, Science Books International, Boston, p. 564.
GILMORE, T. M. and DIMICK, P. S. (1979). *J. Dairy Sci.*, **62**, 189.
GREGORY, J. F. and SHIPE, W. F. (1975). *J. Dairy Sci.*, **58**, 1263.
GREGORY, J. F., BABISH, J. G. and SHIPE, W. F. (1975). *J. Dairy Sci.*, **59**, 364.
GUTTERIDGE, J. M. C., PATTERSON, S. K., SEGAL, A. W. and HALLIVELL, B. (1981). *Biochem. J.*, **199**, 259.
HICKS, C. L., BUCY, J. and STOFER, W. (1979). *J. Dairy Sci.*, **62**, 529.
HILL, R. D. (1975). *Aust. J. Dairy Technol.*, **30**, 26.
HILL, R. D. (1979). *CSIRO Fd. Res. Q.*, **39**, 33.
HOLBROOK, J. and HICKS, C. L. (1978). *J. Dairy Sci.*, **61**, 1072.
JARASCH, E-D., BRUDER, G., KEENAN, T. W. and FRANKE, W. (1977). *J. Cell Biol.*, **73**, 223.
KANNER, J. and KINSELLA, J. E. (1983a). *J. Agric. Food Chem.*, **31**, 370.
KANNER, J. and KINSELLA, J. E. (1983b). *Lipids*, **18**, 204.
KEENAN, T. W., MOON, T.-W., and DYLEWSKI, D. P. (1983). *J. Dairy Sci.*, **66**, 196.
KING, R. L. (1962). *J. Dairy Sci.*, **45**, 1165.
KING, R. L. and DUNKLEY, W. L. (1959). *J. Dairy Sci.*, **42**, 420.
KING, R. L., LUICK, J. R., LIBMAN, I. I., JENNINGS, W. G. and DUNKLEY, W. L. (1959). *J. Dairy Sci.*, **42**, 780.
KORYCKA-DAHL, M.and RICHARDSON, T. (1978). *J. Dairy Sci.*, **61**, 400.
KORYCKA-DAHL, M., RICHARDSON, T. and HICKS, C. L. (1979). *J. Food Prot.*, **42**, 867.
KORYCKA-DAHL, M., VASSAL, L., RIBADEAU DUMAS, B. and MOCQUOT, G. (1983). *Sci. Aliments*, in press.
MCPHERSON, A. V. and KITCHEN, B. J. (1983). *J. Dairy Res.*, **50**, 127.
MORITA, M. and FUJIMAKI, M. (1972a). *Agr. Biol. Chem.*, **36**, 1751.
MORITA, M. and FUJIMAKI, M. (1972b). *Agr. Biol. Chem.*, **36**, 1163.
MULDER, H. and WALSTRA, P. (1974). *The Milk Fat Globule: Emulsion Science as Applied to Food Products and Comparable Foods*, Farnham Royal: Commonwealth Agricultural Bureaux.
O'MAHONEY, J. P. and SHIPE, W. F. (1970). *J. Dairy Sci.*, **53**, 636.
PARKS, O. W. and ALLEN, C. (1977). *J. Dairy Sci.*, **60**, 1038.
PETERS, I. I. and TROUT, G. M. (1945a). *J. Dairy Sci.*, **28**, 277.
PETERS, I. I. and TROUT, G. M. (1945b). *J. Dairy Sci.*, **28**, 283.
PLANTZ, P. E., PATTON, S. and KEENAN, T. W. (1973). *J. Dairy Sci.*, **56**, 978.
ROWLEY, D. A. and HALLIWELL, B. (1982). *FEBS Letts.*, **138**, 33.
SALIN, M. and MCCORD, J. M. (1977). In: *Superoxide and Superoxide Dismutases*, A. M. Michelson, J. M. McCord and I. Fridovich (eds), Academic Press, New York, p. 257.
SCHRODER, M. J. A. (1982). *J. Dairy Res.*, **49**, 407.
SHIPE, W. F. (1964). *J. Dairy Sci.*, **47**, 221.
SHIPE, W. F. (1977). In: *Enzymes in Food and Beverage Processing*, R. L. Ory and A. J. St. Angelo (eds), Am. Chem. Soc., Washington, DC, p. 57.
SIDHU, G. S., BROWN, M. A. and JOHNSON, A. R. (1976). *J. Dairy Res.*, **43**, 239.
SMITH, G. J. and DUNKLEY, W. L. (1960). *J. Dairy Sci.*, **43**, 278.

SMITH, G. J. and DUNKLEY, W. L. (1961). *J. Dairy Sci.*, **44**, 115.
WHITE, C. H. and BULTHAUS, M. (1982). *J. Dairy Sci.*, **65**, 489.
WONG, N. P., WALTER, H. E., VESTAL, J. H., LACROIX, D. E. and ALFORD, J. A. (1973). *J. Dairy Sci.*, **56**, 1271.

Chapter 8

THE NUTRITIONAL SIGNIFICANCE OF LIPIDS

M. I. GURR

National Institute for Research in Dairying, University of Reading, UK

SUMMARY

Humans obtain their dietary fats from several sources. The milk and adipose tissue of farm animals and the seed oils of plants supply triacylglycerols (storage lipids) while the flesh of animals and plant leaves are a source of structural lipids. Different types of fatty acid are contributed by these different classes of dietary lipid. Storage lipids are major suppliers of dietary energy. One-third of all fat in the average UK diet comes from dairy products. Fats also provide a number of dietary essentials: the fat-soluble vitamins, A, D, E and K and the essential fatty acids. Food is nutritious only in so far as it is good to eat and the presence of fat contributes substantially to the palatability of food. Dairy fats in particular contribute to palatability through their unique flavour and textural properties.

1. INTRODUCTION

Lipids contribute about 42% of dietary energy in the UK and similar figures apply to many other affluent societies. Of these, lipids originating from milk, in the form of dairy products, contribute about 33% of dietary lipids in the UK or 14% of total energy intake. This chapter describes the various roles of dietary lipids in general and sets the nutritional significance of dairy lipids in perspective.

2. TYPES OF LIPID IN THE BODY AND THEIR FUNCTION

2.1. Definitions

Lipids form a chemically heterogeneous group of substances, having in common the property of poor solubility in water but appreciable solubility in non-polar solvents such as chloroform, hydrocarbons and alcohols. Many lipids are 'amphiphilic', having within the same molecule moieties that confer some ability to associate with water as well as a hydrophobic moiety. This amphiphilic nature has immense importance in biological structures and foods. In general, the hydrophobic nature of lipids derives from either the *fatty acids* esterified with glycerol or in amide linkage with sphingosine, or from the *steroid* ring system. The wide variety of lipid structures present in nature is due mainly to the large number of types of fatty acid (Christie, Chapter 1).[1,2]

2.2. Structural, Metabolic and Storage Lipids

Although individual lipids may have many separate roles in the body, these functions can be conveniently divided into three types: structural, metabolic and storage.

2.2.1. Structural Lipids

Lipids play an important part in biological structures that form barriers against the environment. One such barrier is the skin which is covered by a protective layer of 'surface lipids' comprised of tri-, di- and mono-acylglycerols, wax esters, sterols, sterol esters, hydrocarbons and non-esterified fatty acids. The lipids in this layer differ from almost all others in humans in containing significant quantities of fatty acids with odd-numbered and branched chains and with double bonds in unusual positions.[3]

Lipids form an integral part of biological membranes, the structures that provide a barrier between the living cell and its environment or, within the cell, provide a matrix on which much of the cell's complex chemistry takes place. In mammals, the lipids involved in membrane structures are mainly phospholipids and unesterified cholesterol. Glycolipids are also components of membranes but their concentration depends very much on the type of membrane and the tissue in which it is located; brain and nervous tissues are particularly rich in glycolipids.

Current theories of biological membrane structure envisage that most of the lipid is present as a bimolecular sheet with the fatty acid chains of the phospholipids in the interior of the bilayer. Membrane proteins are located at intervals on the internal or external face of the membrane, or projecting through from one side to the other (Fig. 1). There may be polar interactions between phospholipid head-groups and ionic-groups on the proteins as well as hydrophobic interactions between fatty acid chains and hydrophobic amino-acid sequences. Lipid molecules are quite mobile along the plane of the membrane and there may be limited movement across the membrane. The fatty acid chains are in constant motion and the degree of molecular motion within the membrane (often referred to as 'fluidity') is influenced by the nature of the fatty acid chains, interactions between fatty acid chains and cholesterol, and lipid–protein interactions.

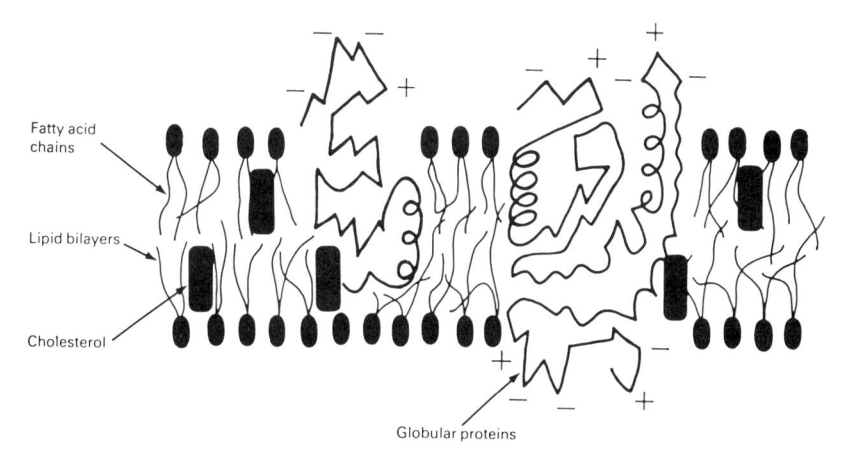

FIG. 1. Schematic illustration of the modern concept of biological membrane structure.

Cholesterol plays a vital role in stabilizing hydrophobic interactions within the membrane. The lipid serves to provide an insulating environment for the many metabolic activities of the membrane that involve proteins. These proteins may be enzymes, transporters of small molecules across the membrane or 'receptors' for substances such as hormones, antigens and nutrients. The property of 'fluidity' seems to be important in as far as it is regulated in the face of different dietary intakes by subtle changes in the proportions of phospholipids, cholesterol and fatty acids.

Because of the similarity of the major metabolic processes in different mammals, structural lipids differ little from one to another, although there may be minor differences in fatty acids and phospholipid polar head-groups between ruminants and non-ruminants.

2.2.2. Metabolic Lipids

In providing a stabilizing environment for metabolic reactions in membranes, the important features of lipids are their physical properties that allow interactions between large numbers of lipid molecules, and between lipid and protein molecules. However, equally important may be the capacity of individual lipid molecules to undergo biochemical transformations to produce specific substances of metabolic importance.

Polyunsaturated fatty acids are released from membrane lipids and transformed by specific enzymes into hormone-like substances called prostaglandins (Section 5.3).[1,4] Fatty acid molecules are also released from storage lipids to be oxidized for the provision of cellular energy (Section 4.4.2).

Cholesterol is metabolized in the adrenal gland to form a series of steroid hormones, and in the liver to form a variety of bile acids that are secreted into bile and are subsequently involved in the processes of digestion and absorption (Section 4.1). These processes appear to be remarkably similar throughout the mammalian kingdom, although the patterns of prostaglandins, steroid hormones and bile salts may differ between species.

The fat-soluble vitamins, in so far as they participate in metabolic processes (in ways not yet completely defined), can also be included in the metabolic lipids (Section 5.4). However, they may also be stored in the liver or adipose tissue (Section 2.2.3) for some time before being used in these processes.

2.2.3. Storage Lipids

Fatty acids in the form of simple glycerides constitute the major source of energy in mammals. There tends to be a distinction between the types of fatty acid fulfilling a storage role, i.e. esterified in simple glycerides, and those fulfilling a structural role, i.e. esterified in 'polar lipids'. Storage lipids tend to contain more saturated or monounsaturated fatty acids, although the composition of storage lipids is influenced by the fatty acid composition of the diet in non-ruminant animals or by the fermentative activities of the rumen micro-organisms in ruminants, as discussed below.

Adipose tissue is the principal reservoir of fatty acids to supply the long-term needs of the animal; mobilization of fatty acids from this tissue to meet the energy demands of the animal at times when dietary energy is limiting is under nutritional and hormonal control. Other tissues, such as the liver, can accommodate fat in the form of small globules, but only in the short-term. The excessive accumulation of fat in the liver is a pathological condition. Milk fat can also be regarded as an energy store for the benefit of the newborn animal and, like adipose-tissue fat, is composed mainly of triacylglycerols.

Storage fat may be derived directly from the diet or synthesized from carbohydrate precursors in the adipose tissue or in milk fat in the mammary gland. The capacity of these tissues to synthesize fatty acids is geared to the needs of the animal and is under dietary and hormonal control. The range of fatty acids synthesized in mammalian adipose tissue

TABLE 1
THE FATTY ACID COMPOSITION OF SOME ANIMAL STORAGE LIPIDS OF IMPORTANCE IN FOODS

Fatty acid	Fatty acid composition (g/100 g total fatty acids)				
	Beef suet	Butter	Lard[a]	Lard[b]	Cod liver oil
4:0[c]	0	3	0	0	0
6:0	0	2	0	0	0
8:0	0	1	0	0	0
10:0	0	3	0	0	0
12:0	tr[d]	4	0	0	0
14:0	3	12	2	1	6
16:0	28	26	27	19	13
18:0	26	11	16	9	3
16:1	2	3	3	2	13
18:1	34	28	41	37	20
20:1	0	0	1	0	12
22:1	0	0	0	0	6
18:2	1	1	9	32	2
18:3	0	2	1	0	0
20:4	0	0	0	0	0
20:5	0	0	0	0	9
22:5	0	0	0	0	2
22:6	0	0	0	0	9
Others	6[e]	4[f]	0	0	5[g]

[a] Composition in pigs on a standard low-fat diet.
[b] Composition in pigs on a diet containing 10% of a vegetable oil such as soybean.
[c] Some figures may represent a mixture of isomers; thus 18:1 is mainly oleic acid but may also contain some elaidic acid (*trans* isomer). All figures have been rounded to the nearest whole number and figures represent average values for a large number of analyses; there is considerable variation in composition as discussed in the text.
[d] tr = trace.
[e,f] Includes some odd-chain and branched-chain acids.
[g] The sum of a large number of minor, mostly polyunsaturated, components.

is limited (normally palmitic (16:0), stearic (18:0) and oleic (18:1) acids), but ruminants may also synthesize odd- and branched-chain fatty acids from precursors arising from rumen fermentation.[5] Mammary gland is unique in synthesizing short- and medium-chain length fatty acids but the pattern of milk fatty acids is characteristic of each species (Christie, Chapter 1).[6] On a fat-free diet, the pattern of storage lipids (in milk or adipose tissue) is entirely dependent on the biosynthetic capacity of the tissue. However, the introduction of fat into the diet suppresses, to varying extents, the synthetic activity of the tissue and mechanisms operate to incorporate dietary fatty acids into storage lipids. Hence in non-ruminant adipose tissue, fatty acids other than 16:0, 18:0 and 18:1 are likely to be of dietary origin and can often give an accurate picture of the animal's recent dietary intake (Table 1). This applies particularly to polyunsaturated fatty acids of plant origin although the characteristic mammalian polyunsaturated fatty acid, arachidonic acid (20:4), is rarely found in more than minute concentrations in storage lipids. Although ruminants habitually eat a diet

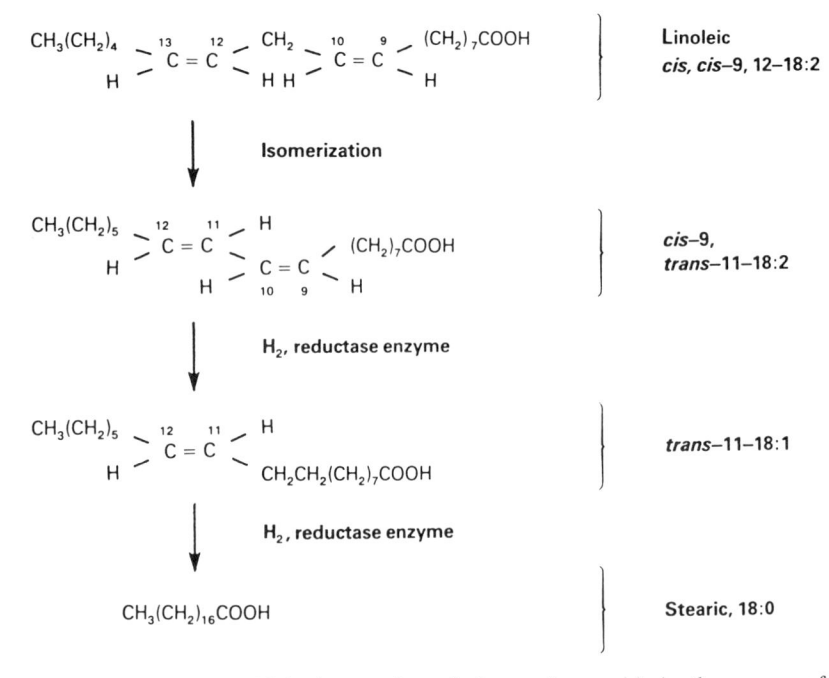

FIG. 2. Steps in the biohydrogenation of dietary fatty acids in the rumen of ruminant animals.

in which plant polyunsaturated fatty acids form a high proportion of the total fatty acids, the adipose tissue in no way reflects this because these acids are converted, before absorption, into saturated, and *cis* and *trans* mono-unsaturated fatty acids by rumen micro-organisms (Table 1, Fig. 2).

2.3. Lipids in Transit

Body lipids, even when they are part of the 'structural' or 'storage' pool, are in a continuously dynamic state. There is continuous exchange of fatty acids in membranes or adipose tissue with fatty acids in the blood supply. Lipids are also transferred from tissues where they are synthesized to storage sites or from storage pools to sites of metabolism. Lipids absorbed from the gut after digestion of the diet are transported to sites of storage or metabolism, depending on the current energy needs of the animal. The route of transport is the bloodstream and the predominantly hydrophobic lipids are carried in a form, i.e. associated with proteins as lipoproteins, that stabilizes them in an aqueous environment (Section 4.2).

3. TYPES OF LIPID IN FOOD

3.1. Source of Dietary Lipids

Omnivorous humans derive their dietary lipids from plant and animal tissues. Just as these tissues contain 'structural' and 'storage' lipids, so can the dietary lipids of humans be considered in these terms. Food in general contains the 'visible fats' that derive from the adipose tissue and the milk of animals or the seed oils, which are, in effect, the storage lipids of plants, and the 'hidden fats' that derive from the membranes of animal tissues or plant leaves. Completely fabricated fats, such as margarines, are themselves derived from plant or animal storage fats but these may be modified by processing.

3.2. Composition of Food Lipids

3.2.1. Adipose Tissue

For detailed information, the reader is referred to Gurr and James[1] and to Paul and Southgate.[7]

The adipose tissue of ruminants is less variable than that of monogastric animals because about 90% of the unsaturated dietary lipids are hydrogenated by rumen micro-organisms.[8] It contains a larger proportion of saturated and mono-unsaturated fatty acids and a lower proportion of polyunsaturated acids than the storage fats of monogastric animals. It also contains *trans* and branched-chain fatty acids in amounts and types dependent upon the species and the way in which diet affects the activities of the rumen microflora. The fatty acid composition of the adipose tissue of monogastric animals, such as pigs and poultry, depends on their diet. The adipose tissue of all animals also contains cholesterol and variable amounts of fat-soluble vitamins. Lard and suet (tallow) are common food fats derived from pig and beef adipose tissue, respectively; their fatty acid compositions are shown in Table 1. The storage fats of fish are found not in adipose tissue but mainly in the liver. Their compositions are markedly different from those of other animals and are characteristically rich in polyunsaturated fatty acids containing five or six double bonds. A large proportion of these have the *n*-3 structure (related to all-*cis* 9,12,15-linolenic acid) rather than *n*-6 (related to *cis cis* 9,12-linoleic acid) because of the consumption of phytoplankton by fish. The contribution of cholesterol to the diet by crustaceans can be considerable.

3.2.2. Milk

The composition of milk fat has been reviewed by Christie (Chapter 1) and by Morrison,[6] Moore,[8] and Jensen *et al.*[9] The only milk of quantitative importance in the adult human diet in the UK and most western countries is cow's milk. However, in some countries, the milks of goats and sheep are of considerable importance, and there is increasing interest in goat's milk in the UK. Comparative fatty acid compositions are given in Table 2. Goat's milk is reported[10] to possess smaller fat globules than cow's milk and it has been inferred that this enables better digestion of the fat by humans. The author is not aware that this has been subjected to rigorous scientific investigation.

Butter is a common food fat containing ~16% water as an emulsion in oil. The fat is derived entirely from cow's milk and therefore its composition does not vary greatly, in contrast to that of margarines (Section 3.2.6). However, the fatty acid composition of milk fat and butter is susceptible to dietary modification if the fat in the ruminant's diet is protected from hydrogenation in the rumen. This subject has been reviewed by McDonald and Scott,[11] and Hawke and Taylor (Chapter 2).

Human milk fat has a composition very different from that of cow's milk (Table 2), a fact that is of great importance in infant nutrition. The reader is referred to Jensen *et al.*[9] for further details of the composition of human milk fat, and to Gurr[12] for a detailed discussion of the roles of human and cow's milks in infant feeding. Apart from differences in the fatty acid composition of the two milks, differences in structure of the triacylglycerols are important to the digestion of milk fat (Section 4.1). Human milk triacylglycerols resemble pig storage fat in that a high proportion (70%) of palmitic acid is esterified at the glycerol sn-2 position, while the primary positions contain a high proportion of unsaturated fatty acids. Cow's milk triacylglycerols, however, have a much smaller proportion of 16:0 (40%) at position 2 and relatively large proportions of saturated fatty

TABLE 2
FATTY ACID COMPOSITION OF MILKS USED IN HUMAN FOODS

Fatty acid	Fatty acid composition (g/100 g total fatty acids)			
	Cow	Goat	Sheep	Human
4:0	3	2	4	0
6:0	2	2	3	0
8:0	1	3	3	0
10:0	3	9	9	1
12:0	4	5	5	5
14:0	12	11	12	7
16:0	26	27	25	27
18:0	11	10	9	10
16:1	3	2	3	4
18:1	28	26	20	35
18:2	1	2	2	7
18:3	2	0	1	1
20:4	0	0	0	tr
Others	4[a]	1[a]	4[a]	3

[a] Includes odd-chain and branched-chain acids.

acids at position 1 (16:0, 18:0) and position 3 (4:0, 6:0) (see Christie, Chapter 1; Hawke and Taylor, Chapter 2).

In comparing the milk fat intakes of breast-fed babies with that of babies fed on formulae based on cow's milk, it is important to note that the total concentration of fat in breast milk increases markedly during a single feed,[12,13] whereas the intake from a formula is constant during feeding.

3.2.3. Plant Oils

The lipids of seed oils are composed mainly of triacylglycerols, the fatty acid compositions of which are highly specific for the plant species (Table

TABLE 3
THE FATTY ACID COMPOSITION OF SOME VEGETABLE OILS USED IN HUMAN FOODS

Fatty acid	Fatty acid composition (g/100 g total fatty acids)							
	Coconut	Corn	Olive	Palm	Peanut	Rape (Low erucic)	Rape (High erucic)	Soybean	Sunflower
8:0	8	0	0	0	0	0	0	0	0
10:0	7	0	0	0	0	0	0	0	0
12:0	48	0	0	tr	tr	0	0	tr	tr
14:0	16	1	tr	1	1	tr	tr	tr	tr
16:0	9	14	12	42	11	4	4	10	6
18:0	2	2	2	4	3	1	1	4	6
20:0	1	tr	tr	tr	1	1	1	tr	tr
22:0	0	tr	0	0	3	tr	tr	tr	tr
16:1	tr	tr	1	tr	tr	2	tr	tr	tr
18:1	7	30	72	43	49	54	24	25	33
18:2	2	50	11	8	29	23	16	52	52
18:3	0	2	1	tr	1	10	11	7	tr
Others	0	1	1	2	2	5	10 (20:1) 33 (22:1)	2	3

3). As can be seen, not all vegetable oils are rich in polyunsaturated fatty acids as the lipids of coconut and palm oils are predominantly saturated.

Seed oils also contain variable amounts of tocopherols and plant sterols, such as β-sitosterol, although the latter is not absorbed from the human gut. In addition, some may contain unusual fatty acids which, if ingested in large concentrations, may have toxic or otherwise undesirable metabolic effects (see Sections 6.1 and 6.4).

3.2.4. Muscle Lipids and Other Animal Structural Lipids

Lipids ingested from lean meat comprise mainly phospholipids and cholesterol, the fatty acid compositions of which are less variable than that of the storage lipids. Muscle lipids are the main sources of dietary arachidonic acid. Liver lipids contain variable amounts of triacylglycerols, the fatty acid compositions of which are intermediate between those of storage and muscle lipids. Brain, although not an important form of meat in the UK diet, contains a high concentration of structural fat (8%) (Table 4).

TABLE 4
FATTY ACID COMPOSITION OF SOME STRUCTURAL LIPIDS OCCURRING IN FOODS

Fatty acid	Fatty acid composition (g/100 g total fatty acids)				
	Beef muscle	Chicken liver	Cod	Lamb brain	Green leaves
14:0	3	1	1	1	0
16:0	13	25	22	22	13
18:0	16	17	4	18	tr
16:1	2	3	2	1	3
18:1	21	26	11	28	7
18:2	20	15	1	1	16
18:3	2	1	tr	0	56
20:4	19	6	4	4	0
20:5	0	0	17	1	0
22:5	0	1	2	3	0
22:6	0	5	33	10	0
Others	4	0	3	11	5

3.2.5. Plant Leaves

Lipids ingested from green leaves are comprised mainly of the galactolipids, the fatty acid composition of which varies little from one plant to another. Although the fat content of leaves is low (0·5%), they make a contribution to the intake of polyunsaturated fatty acids by humans, particularly α-linolenic acid (all-*cis*-9,12,15-octadecatrienoic acid, 18:3) (Table 4).

3.2.6. Manufactured Fats

Margarines, like butter, are water-in-oil emulsions. The aqueous component is skimmed milk, slightly soured by micro-organisms to enhance flavour, while the fat phase is a blend of several fats and oils from vegetable and animal sources used in proportions determined by the supply and cost of oils at a given time. The most commonly used vegetable fats are palm, soybean, groundnut, coconut, sunflower, cottonseed and rapeseed oils, while the animal fats are mainly beef and mutton tallow, and herring and pilchard oils. Therefore, even within a single brand, the fat blend may change from batch to batch and the fatty acid composition may vary between certain limits designed to maintain the physical properties of that brand. In addition, manufacturing processes designed to modify the physical properties of the oils, so as to achieve the desired properties of the product, alter the fatty acid composition of the oils in the blend. These include catalytic hydrogenation (which results in a decrease in the number of double bonds, an increase in the proportion of *trans* isomers and a

TABLE 5
THE APPROXIMATE FATTY ACID COMPOSITIONS EXPECTED IN MARGARINES OF DIFFERENT TYPES

Fatty acid	\multicolumn{5}{c}{Fatty acid composition (g/100 g total fatty acids)}
	Hard A[a]	Hard B[b]	Soft A[a]	Soft B[b]	Polyunsaturated[c]
12:0	tr	tr	1	2	3
14:0	6	1	5	1	1
16:0	20	28	16	24	11
18:0	8	7	5	5	9
20:0	2	1	2	1	1
22:0	2	1	3	1	1
16:1	6	1	6	1	tr
18:1	22	42	25	37	18
20:1	9	2	7	1	1
22:1	9	4	9	4	1
18:2	5	10	9	21	53
18:3	tr	1	tr	2	1
20:4 + 20:5	7	1	7	tr	tr
22:5 + 22:6	4	1	5	tr	tr
Others	0	0	0	0	0

[a] Manufactured from a blend of animal and vegetable oils.
[b] Manufactured from a blend of vegetable oils only.

randomization of double-bond positions along the chain); interesterification (which randomizes the positions of the fatty acids on the triacylglycerol molecules); and fractionation (which separates fats of different melting points). The fatty acid compositions of different margarines, listed in Table 5, should therefore be regarded as examples of the type of fatty acid spectrum found in margarines of different types and should not be regarded as absolute.

3.3. Modification of Food Lipids in Processing

3.3.1. Heating

In some processes, such as frying, used either in the industrial preparation of foods or in household cooking, fats may be subjected to elevated temperatures. During heating, changes may occur in the structure of fats, the nature of which depends on several factors such as temperature, the duration of heating and the amount of air to which the fat is exposed. At the temperature of deep-fat frying (~180°C), evolution of water vapour from the food results in steam-stripping of volatiles from the fat which act as a barrier to limit the entrance of air.[14]

If air is present, the first products formed at low temperatures are the hydroperoxides of unsaturated fatty acids (see Section 3.4) which break down at higher temperatures to a range of polar fatty acids. At higher temperatures, in the absence of oxygen, cyclic monomers of triacylglycerols are formed, which later polymerize. Fats heated under household conditions may contain 10–20% of polymerized material, but the functional properties of such oils are not noticeably worse and they are not regarded as harmful.[14,15]

3.3.2. Hydrogenation

Highly unsaturated oils are unsuitable for many food uses because they have very low melting points[16] and they are more susceptible to oxidative deterioration (Section 3.4). The objective of hydrogenation, which is probably the margarine manufacturer's most important tool, is to reduce the degree of unsaturation thereby increasing the melting point of the oil and extending its range of food applications.[16] By careful choice of catalyst and temperature, the oil can be selectively hydrogenated so as to achieve a product with precisely the desired characteristics. Hydrogenation is carried out in an enclosed tank in the presence of 0·05–0·20% of finely powdered catalyst (usually nickel) at temperatures up to 180°C, after which all traces of catalyst are removed by filtration. The process is seldom taken to completion since the melting points of totally saturated fats are much too high. In the course of the hydrogenation, a proportion of the *cis* double bonds in natural oils are isomerized to *trans* configurations and there is a migration of double bonds along the chain. Hence, whereas a natural fat contains a high proportion of double bonds at position 9, the products of partial hydrogenation contain randomly distributed double bonds. In principle, there is little difference between the chemically catalysed reduction of double bonds employed by the food industry and the enzymically catalysed reduction of double bonds by rumen microorganisms.

3.3.3. Interesterification

In addition to the nature of the fatty acids esterified in triacylglycerols, the distribution of the three fatty acids among the different positions on glycerol also influences the physical properties of a fat. Interesterification, a method of altering the melting point of a fat by randomizing the positions of its fatty acids, is achieved by heating the fat in the presence of a catalyst (usually sodium, sodium ethoxide or sodium methoxide) to 110–160°C. Positions may be exchanged between fatty acids of the same triacylglycerol molecule (intramolecular exchange) or between fatty acids of different molecules (intermolecular exchange).

3.4. Spoilage of Food Lipids

Deterioration in the quality of food lipids generally arises from hydrolytic release of fatty acids from glyceride molecules, oxidation of the fatty acids, or both.

Hydrolysis of ester bonds gives rise to increases in the concentration of non-esterified fatty acids in the oil, resulting in off-flavours. This process may be catalysed by alkaline conditions and can occur during storage or processing of an oil. In addition, most biological tissues contain enzymes that catalyse the release of fatty acids from lipids.

The more highly unsaturated the fatty acid, the greater its susceptibility to oxidation. Vitamin A is also particularly susceptible to peroxidation damage. Compounds that react quickly with free radicals (anti-oxidants), such as the naturally occurring tocopherols (vitamin E) or synthetic antioxidants such as butylated hydroxytoluene (BHT), are useful in retarding peroxidation damage. As well as resulting in physical deterioration of the food and generating potentially toxic compounds (see Section 6.5), peroxidation reduces the organoleptic properties of foods and hence their value in nutrition. The subject of hydrolytic rancidity in plant lipids is reviewed by Galliard[17] and in milk and dairy products by Deeth and Fitz-Gerald (Chapter 6); oxidative rancidity has been reviewed by Frankel,[18] and by Richardson and Korycka-Dahl (Chapter 7).

4. METABOLISM OF LIPIDS

This section is intended as a brief background only and, for more detail, the reader is referred to Gurr and James.[1]

4.1. Digestion and Absorption of Food Lipids

After ingestion of food (Fig. 3), the first process, occurring in the stomach, is the formation of an oil-in-water emulsion brought about by mechanical movements in the stomach. Lipoproteins are broken down by proteolysis, liberating the lipids. Little or no lipolysis or absorption of fat is thought to occur in the stomach of adult animals. However, newborn animals, including human babies, secrete a lipase (pregastric esterase) from glands around the tongue which is carried to the stomach (~pH 5) where it hydrolyses glycerides without the aid of bile salts. The secretion of the enzyme is stimulated by sucking and by milk, a diet rich in fat.

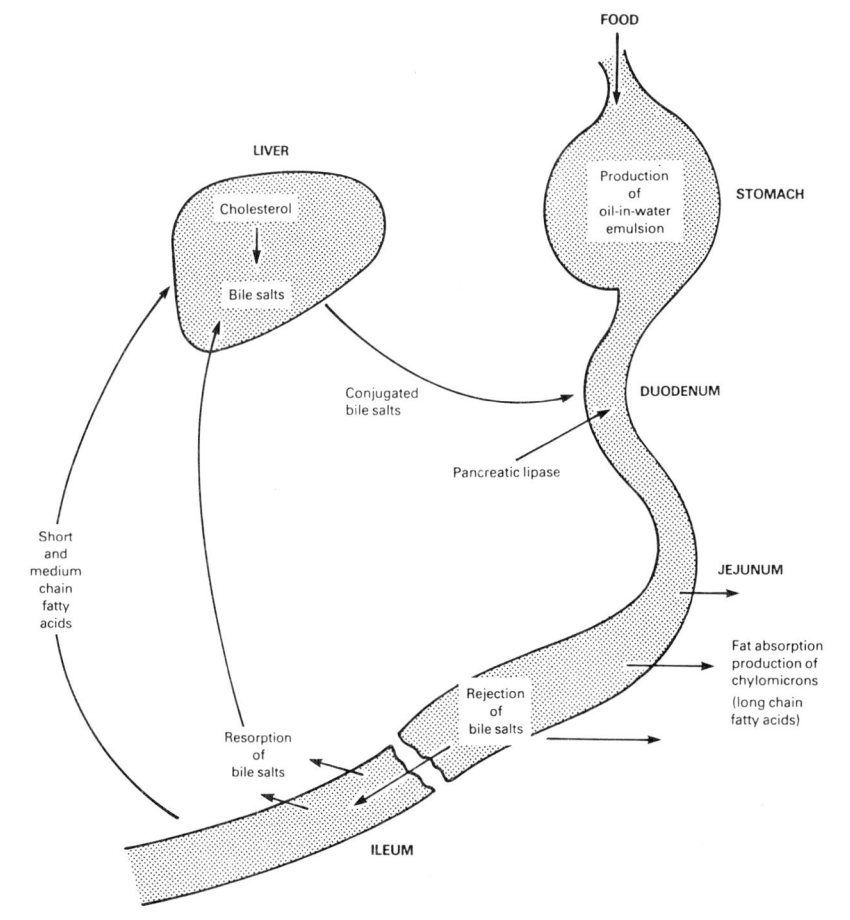

FIG. 3. Schematic diagram of the processes involved in the first stage of fat digestion and absorption.

In grown animals, the secretion of pancreatic juice and bile into the duodenum initiates lipid digestion through the action of pancreatic lipase and phospholipase A and the marked ability of the conjugated bile acids to solubilize partial glycerides and fatty acids. Attack by pancreatic lipase on triacylglycerols at the oil–water interface of the fat particles is mainly at the 1 and 3 positions and generates two types of surface-active agent: fatty acids and 2-monoacylglycerols. Any phospholipids present are degraded by phospholipase A to lysophospholipids which are also powerful detergents. Little or no work has been done on the intestinal digestion of plant galactolipids in monogastric animals.

The molecular species primarily absorbed by the brush border are 2-monoacylglycerols and non-esterified fatty acids. The bile salts themselves are not absorbed in the proximal small intestine but pass on into the ileum where they are absorbed and recirculated via the portal blood to the liver and then to the bile for re-entry at the duodenum. This is generally called the 'entero-hepatic circulation'. In the enterocytes, fatty acids are converted into their acyl-coenzyme A thiol esters and in this form are re-esterified at the 1 and 3 positions of 2-monoacylglycerols. Thus, water-insoluble triacylglycerols are regenerated in the intestinal epithelial cells and are stabilized and converted into a form that enables them to be readily transported in body fluids by surrounding the fat droplets with a layer of protein and phospholipids. The resulting particles, known as *chylomicrons*, pass from the cells to the intercellular spaces, thence to the lacteals, the lymphatic channels and the bloodstream.

The chylomicrons, irrespective of the fat eaten, consist mainly of long-chain triacylglycerols, with fatty acids longer than 14 carbon atoms. Short-chain fatty acids (from dairy products) and medium-chain fatty acids (from milk and coconut oil) are selectively transported into mesenteric portal blood as non-esterified fatty acids and are *not* absorbed into the epithelial cells as components of mixed micelles. For this reason, medium-chain triacylglycerols (largely refined from coconut oil) are used therapeutically in diets for patients who are unable to absorb long-chain fatty acids.

The steroids and fat-soluble vitamins are also absorbed by incorporation into mixed micelles in the intestinal lumen and into the core of the chylomicrons in the intracellular phase of fat absorption.

Fatty acids may undergo some modification by the micro-organisms in the gut of monogastric animals before absorption takes place. The greatest concentration of intestinal micro-organisms in monogastric animals occurs in the lower gut, i.e. after the normal sites of absorption. These carry out many transformations of lipids that result in a different composition of faecal lipids compared with those in the diet.There are, however, significant numbers of organisms in the jejunum which may make minor modifications, such as biohydrogenation of unsaturated fatty acids. The tissues of germ-free animals, for example, have higher proportions of linoleic acid than animals with a conventional flora. It is unlikely, however, that this process has much nutritional significance.

The processes of fat digestion and absorption in ruminants differ somewhat from those in monogastric animals. In the rumen, a complex population of micro-organisms hydrolyses all types of dietary lipid with almost complete release of free fatty acids. The glycerol moiety is fermented and the unsaturated acids undergo isomerization and biohydrogenation to produce saturated and *trans*-unsaturated isomers. The fermentative activities generate large amounts of short-chain carboxylic acids, acetic and propionic, which are absorbed extensively in the rumen and carried to the liver where they are substrates for gluconeogenesis (propionic) and lipid biosynthesis (acetic). Longer-chain fatty acids, including saturated, *trans*-unsaturated, branched- and odd-chain acids, pass through the abomasum into the small intestine where an emulsion of non-esterified fatty acids, conjugated bile acids and lysophospholipids is formed and absorption of fatty acids occurs. In contrast to monogastrics, there is little or no absorption of glycerol or monoacylglycerol which are fermented or hydrolysed, respectively, in the rumen. The triacylglycerols formed in the intestinal epithelial cells are derived largely from the absorbed saturated acids and *trans*-unsaturated acids which accounts for the differences in composition of the depot fats between ruminants and non-ruminants.

4.2. Transport of Absorbed Lipids

The chylomicrons formed in intestinal epithelial cells during the intracellular phase of fat absorption belong to a class of stabilized lipid particles involved in lipid transport called plasma lipoproteins (Fig. 4). It is now customary to classify these according to their density. Thus the chylomicrons, consisting largely of triacylglycerols (about 85%) and only 2% of proteins, are the largest and least dense of the lipoproteins. Because of their size, their presence in plasma shortly after a fatty meal can easily be recognized by a cloudy appearance. They are the main carriers of fat derived from the *diet*. Very low density lipoproteins (VLDL) form a class smaller in size, with less triacylglycerol (about 50%) and about 20% each of cholesterol and phospholipid. They are the main carriers of triacylglycerols formed in the liver from carbohydrates. The third class, still

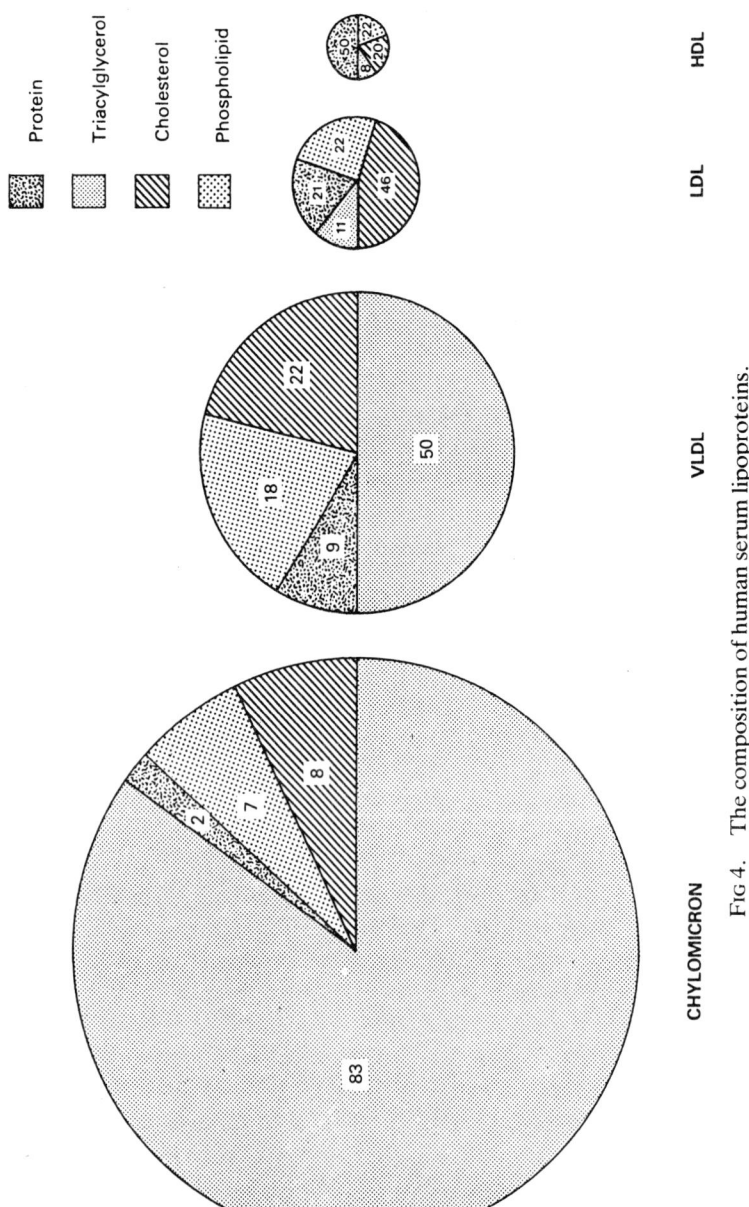

FIG 4. The composition of human serum lipoproteins.

smaller in size and with a protein content of about 20%, the low density lipoproteins (LDL), act as the main carriers of cholesterol in the plasma. Finally, the smallest of the major classes, the high density lipoproteins (HDL), have a protein content of 50% and about 20% each of phospholipids and cholesterol. Their role is to transport excess cholesterol from membranes to the liver where it can be degraded or converted into bile acids. It should be emphasized that the lipoproteins do not fit rigorously into these categories; there is a continuous spectrum of particle sizes in the plasma but it is convenient to classify them in this way, based on the centrifugation methods used to isolate them, and such a classification coincides well with their known functions or origins.

Chylomicrons disappear rapidly from the bloodstream with a half-life of about 10 min. This occurs mainly by the hydrolysis of the glyceride moiety by a lipoprotein lipase (LPL) released from the capillaries of the adipose tissue or muscle. The release of LPL is under dietary and hormonal control. After a meal containing significant amounts of fat and carbohydrate, when the supply of energy may exceed the body's immediate needs, the secretion of insulin ensures that adipose tissue LPL is active. Fatty acids are released from chylomicron lipids and taken into adipose tissue where they are re-esterified into triacylglycerols. The uptake of fatty acids may also be accompanied by the uptake of steroids and fat-soluble vitamins. A congenital deficiency of LPL can lead to a hyperlipoproteinaemia (Type I) characterized by high circulating concentrations of chylomicrons. During a fast or in starvation, adipose LPL is inactive while muscle LPL is 'switched on' and fatty acids are released from circulating lipoproteins (mainly VLDL) and taken up into muscle tissue to be used as fuel. Also during fasting, fatty acids may be mobilized from adipose tissue by the activation of an intracellular triacylglycerol lipase (activated in the fasting state by adrenalin and inhibited in the fed state by insulin). The fatty acids combine in the plasma with albumin and are transported to the liver where they are synthesized into triacylglycerols to be released into the plasma again, mainly as VLDL.

The chylomicrons are not totally degraded by lipoprotein lipase. Smaller denser lipoprotein particles, called 'chylomicron remnants', remain and are further metabolized in the liver to LDL which are depleted in triacylglycerol and enriched in cholesterol. The uptake of cholesterol by tissues can occur by non-specific mechanisms similar to pinocytosis. However, our understanding of lipid transport in the last decade has been revolutionized by the discovery that the cells of many tissues have specific receptors on their surfaces that interact with plasma lipoproteins.[19] After binding to the receptor, the lipoprotein–receptor complex is taken into the cell where the lipoprotein is broken down by lysosomal enzymes. The resulting free cholesterol interacts with intracellular membranes to inhibit a key enzyme of cholesterol biosynthesis. The intracellular production of cholesterol is thus reduced and does not become re-activated until circulating cholesterol levels (reflecting, in part, the dietary intake of cholesterol) begin to fall. When the concentration of cholesterol in the plasma is low, the number of receptors for LDL is low and the rate of endogenous cholesterol synthesis is high. When the concentration of LDL in plasma rises, the number of receptors increases. This elegant mechanism enables the body to maintain its supply of cholesterol within well-controlled limits depending on the supply of dietary cholesterol. This is necessary because cholesterol is a vital compound, yet in large concentrations may have undesirable metabolic effects. Several defects in lipoprotein metabolism leading to abnormal blood-lipid concentrations have been correlated with defects in the receptor-mediated uptake mechanism.

When the diet contains little fat, the body synthesizes fatty acids from carbohydrates in a wide range of tissues, but mainly the liver. The presence of fat positively inhibits the tissue enzymes that synthesize fatty acids and it is now known that the polyunsaturated fatty acids, especially linoleic acid, have a more potent inhibitory effect than the saturated or mono-unsaturated ones, although the mechanism of the inhibition is unknown.[20] The fatty acids synthesized by the liver from dietary carbohydrates are mainly incorporated into VLDL triacylglycerols and exported into the plasma.

For many years, HDL received little research attention but recently it has become clear that they play a vital role as scavengers for cholesterol. The cholesterol is present in HDL largely as cholesterol esters. A plasma enzyme (lecithin-cholesterol acyltransferase, LCAT) converts the free cholesterol in HDL to cholesterol ester. The cholesterol-depleted HDL thus formed interact with membranes, such as those of red blood cells, and pick up free cholesterol from the membrane. This is then transported to the liver where it is degraded and the HDL are returned to the plasma to continue the cycle.[21]

The transport processes described here have great importance in terms of the diseases of lipid metabolism described in Section 7.

4.3. Biosynthesis of Lipids

It is not the intention, in this chapter, to discuss the detailed biochemical pathways for lipid biosynthesis; for these, the reader is referred to Gurr·

THE NUTRITIONAL SIGNIFICANCE OF LIPIDS

and James,[1] and Hawke and Taylor (Chapter 2). Rather, the regulation of lipid biosynthesis in the face of changing needs and dietary intakes is discussed.

As implied in Section 4.2, the rates of lipid biosynthesis in the body are related to the intake of dietary lipids. When the concentrations of dietary lipids are very low, rates of fatty acid synthesis are high, particularly in the liver, in order to supply the needs of the body in respect of structural and storage lipids. When energy intake, as well as the proportion of dietary carbohydrate, is high, the rate of fatty acid synthesis is high in the liver, ensuring an ample supply of VLDL and the esterification and storage of fat is particularly high in adipose tissue which acts as a reservoir for storage of excess energy. However, this hardly ever occurs in western humans whose diets generally contain a high proportion of energy as fat (about 40% in the UK). This probably means that at most times the enzymes involved in fat synthesis are 'switched off' and that the needs for both storage and structural fats are satisfied from dietary intake.

So that the appropriate quantities and types of fatty acid are directed into storage, into cellular structures or burned as fuel, in spite of a constantly changing intake from the diet, a well-balanced metabolic control system must operate.

In general, metabolic control is exercised through alterations in the activity of enzymes in metabolic pathways. Rapid changes may occur when the *amount* of enzyme remains constant but its *activity* is altered by subtle changes in its structure which may occur when the protein combines with small molecules. Longer-term adaptive changes are brought about by changes in the absolute number of enzyme molecules without the catalytic activity of individual molecules being affected.

Diet may influence metabolic control in several ways. A very direct way is the obligatory supply from the diet of a coenzyme. The enzymes of fatty acid synthesis require the B-group vitamins, pantothenic acid and biotin, as coenzymes. Deficiency of these vitamins therefore leads to defects in lipid biosynthesis with profound effects on health.

A major role of diet in the control of lipid metabolism is almost certainly to bring about specific changes in the concentration of circulating hormones which induce or repress the synthesis of some enzymes of lipid metabolism. This is as yet a poorly understood area of nutritional biochemistry because of the difficulties of isolating and purifying enzymes that are usually bound very tightly to cellular membranes.

There are many hormones involved in the regulation of lipid biosynthesis but the dominant role is probably played by insulin, high levels of which characterize the fed state when ample carbohydrate fuel is available from the diet. It suppresses glucose production by the liver and encourages glycogen and fatty acid synthesis. In adipose tissue it encourages the rapid passage of glucose into the cells where it is available for lipid synthesis, and it also inhibits the breakdown of lipids in adipose tissue. More important perhaps than the concentration of a single hormone is the ratio of the concentrations of different hormones. For example, a high ratio of insulin to glucagon favours esterification of fatty acids into glycerides and a low rate of fatty acid oxidation. Another characteristic of hormones is their different modes of action in different tissues. Thyroid hormones stimulate the rate of triacylglycerol synthesis in the liver but have the opposite effect in adipose tissue. High concentrations of circulating tri-iodothyronine are associated with carbohydrate-rich diets.[22] High serum cortisol concentrations result in an elevated rate of glyceride synthesis in the liver and arise from the ingestion of excessive quantities of saturated fatty acids and sucrose.[23] Obesity is also associated with elevated plasma glucocorticoid concentrations.

Much less is known about the dietary and hormonal control of structural phospholipid synthesis because of the difficulties of studying the controlling enzymes. However, an important factor in controlling membrane lipid synthesis, and therefore composition, is the availability of essential fatty acids from the diet and the dietary intake of fatty acids that influence essential fatty acid metabolism (Section 5.3).

4.4. Breakdown of Tissue Lipids

4.4.1. Hydrolytic Release of Fatty Acids from Glycerolipids
All tissues contain lipases. Reference has already been made to adipose tissue lipase and lipoprotein lipase, enzymes that are under strict hormonal control. Membranes contain lipases that cleave fatty acids from different positions of the phospholipid molecules and also the phosphate and 'base' moieties. As a result of the continued activity of lipases and acyltransferases (enzymes catalysing esterification), there is a continual 'turnover' or replacement of all parts of lipid molecules in cells which allows a fine control of their metabolism. Especially important is the membrane phospholipase that liberates essential fatty acids for prostaglandin synthesis (Section 5.3).

4.4.2. Oxidation of Fatty Acids for Cellular Energy
Oxidation in this context refers to the controlled step-by-step breakdown of fatty acids to yield metabolic energy, not the auto-oxidation leading to

food spoilage. This is known as β-oxidation and is the chief means by which the energy locked up in fatty acids is made available to the living cell.

β-Oxidation is catalysed by a group of enzymes in the mitochondria of the cell and the complete oxidation of one molecule of palmitic acid yields eight molecules of acetyl-CoA, which are further degraded in the tricarboxylic acid (Krebs) cycle to carbon dioxide and water. The reduced pyridine nucleotides generated in this cycle are oxidized in the mitochondrial electron transport chain yielding ATP, required to power many biosynthetic reactions (including lipid synthesis).

The rate of β-oxidation may be controlled by changes in the activity of the enzymes or via the supply of precursors which are, in turn, subject to dietary and hormonal control. The supply of precursors is determined by the lipolytic release of long-chain fatty acids from triacylglycerols, mainly those stored in the adipose tissue, the regulation of which was briefly described in Section 4.3.

Diet may also influence β-oxidation by supplying carnitine, a nitrogenous base, which is required for the transport of long-chain fatty acids into mitochondria. Also, two of the enzymes of β-oxidation require a coenzyme, flavin adenine dinucleotide, derived from the B-group vitamin riboflavin. Deficiency of riboflavin in experimental animals leads to defects in the cellular oxidation of fatty acids.[24]

Although there is some controversy about the relative importance of sugars and fatty acids as fuels in some tissues, it is now accepted that fatty acids are the preferred fuels in most animal tissues. Brain is the prime example of a tissue that cannot utilize long-chain fatty acids and this is the main reason why blood glucose levels have to be so strictly maintained in animals. Blood glucose concentration is maintained at a nearly constant level by a series of reactions known as gluconeogenesis which are driven by β-oxidation in a number of ways. Firstly, β-oxidation provides two essential cofactors—ATP and NADH—and, secondly, acetyl-CoA, the end product of β-oxidation, activates a key enzyme in gluconeogenesis. Inhibitors of β-oxidation (e.g. hypoglycin) also impair gluconeogenesis so that the animal becomes hypoglycaemic.

The flow of fatty acids into pathways which result in their esterification or in their β-oxidation is under hormonal control. A high ratio of glucagon to insulin favours β-oxidation while insulin suppresses the release of fatty acids from adipose tissue and hence also β-oxidation and gluconeogenesis. In diabetes, where there is a lack of insulin, β-oxidation and gluconeogenesis are enhanced and the rate of acetyl-CoA production exceeds the rate at which it can be utilized by the tricarboxylic acid cycle. The result is an excessive accumulation of ketone bodies (ketosis) which can occur in any condition where there is a reduced supply of carbohydrates, as in starvation or nutritional imbalance.

5. ROLES OF FOOD LIPIDS

5.1. A Source of Energy

Triacylglycerols, when burned in a bomb calorimeter, supply ~38 kJ of energy per gram compared with about 17 and 16 kJ g^{-1} for proteins and carbohydrates, respectively. As storage fuels, triacylglycerols have the advantage that they can be stored in an anhydrous form, representing more energy for less bulk than complex polysaccharides such as food starches or body glycogen which are highly hydrated.

Very careful calorimetry of fats differing in chain length and unsaturation indicates that each has a slightly different energy value. There are reports that the consumption of unsaturated fats results in smaller increases in body energy than saturated fats, presumably because their metabolism is less efficient and more metabolic energy is lost as heat.[25] However, these measurements were made in very few subjects and, as there are large differences between individuals in metabolic efficiency, these results need confirmation. The energy value of medium- and short-chain fatty acids is considerably less than that of long-chain fatty acids because of their pathways of metabolism. The former are absorbed into the portal blood and carried to the liver where they are oxidized and are not deposited in adipose tissue. The energy value of dairy fats, which contain a higher proportion of short- and medium-chain acids than most other fats, may therefore be somewhat lower than average.

5.2. A Supply of Structural Lipids

Dietary fats supply structural lipids in the form of fatty acids or cholesterol. A dietary supply of cholesterol and saturated or monounsaturated fatty acids is not, however, obligatory since the body can synthesize its own lipids, with the exception of essential fatty acids, to compensate for dietary insufficiency.

5.3. Essential Fatty Acids

In 1929, Burr and Burr[26] described how acute deficiency states could be produced in rats by feeding fat-free diets and that these deficiencies could be eliminated only by adding certain specific fatty acids to the diet. It was shown that linoleic and arachidonic acids were responsible for this effect;

the term *vitamin F* was coined for them, although they are now always referred to as *essential fatty acids*.

Essential fatty acid deficiency can be produced in a variety of animals, including humans, but symptoms for the rat are the best documented (Table 6). Well-documented essential fatty acid deficiency in humans is rare, but it has been shown that children fed fat-free diets develop skin conditions similar to those produced in rats and that these disappear when linoleic acid is added to the diet.[27] Individuals who are unable to absorb

TABLE 6
THE MAJOR EFFECTS OF ESSENTIAL FATTY ACID DEFICIENCY IN LABORATORY RATS

Skin	Dermatosis; increase in water permeability; drop in sebum secretion; epithelial hyperplasia
Weight	Decrease
Circulation	Heart enlargement; decrease in capillary resistance; increase in capillary permeability
Kidney	Enlargement, intertubular haemorrhage
Lung	Cholesterol accumulation
Endocrine glands	Adrenals: weight decrease in females and increase in males Thyroid: reduction in weight
Reproduction	Males: degeneration of seminiferous tubules Females: irregular oestrus and impaired reproduction and lactation
Metabolism	Changes in fatty acid composition of most organs Increase in cholesterol levels in liver, adrenal and skin Changes in heart and liver mitochondria and uncoupling of oxidative phosphorylation Increase in triacylglycerol synthesis and release by liver

long-chain fatty acids, either because of conditions such as idiopathic steatorrhea or as a result of intestinal surgery,[28] are as close to deficiency in essential fatty acids as any group yet studied. Other vulnerable groups are hospital patients on intravenous alimentation with glucose and amino-acid mixtures.[29]

Animal tissues contain enzymes (desaturases) capable of desaturating saturated fatty acids, normally at position 9. Enzymes are also present that can produce polyunsaturated fatty acids but the additional double bonds are created between the first double bond and the carboxyl group. A

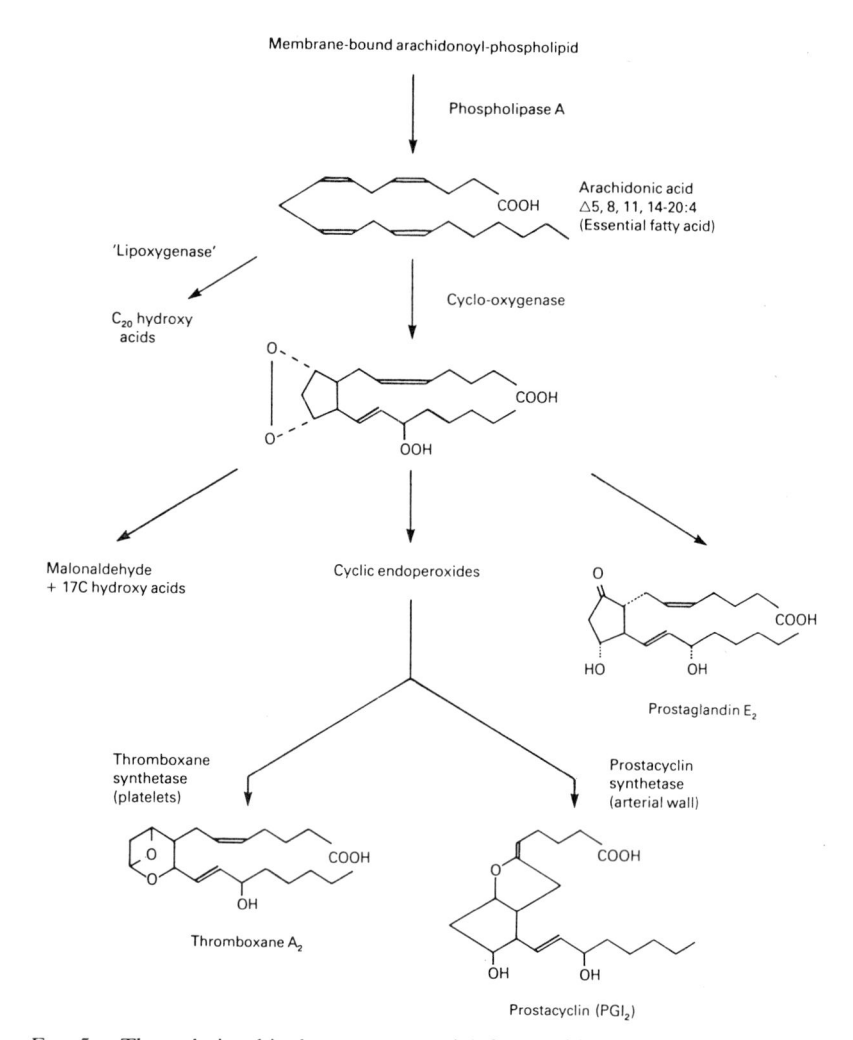

FIG. 5. The relationship between essential fatty acids and prostaglandin-like substances. In the short-hand formulae, the carbon chains are represented by zig-zag lines. At each intersection of lines there is a carbon atom, i.e. $C^{\diagdown}C^{\diagup}C^{\diagdown}C$ is represented by $\diagup\diagdown\diagup\diagdown$.

dietary source of essential fatty acids is required because animals, during the course of evolution, have lost the ability to introduce a double bond at position 12 of fatty acids. The function of the essential fatty acids and the reason why a particular double-bond structure was necessary to achieve this function remained obscure until it was demonstrated that arachidonic acid (a metabolite of linoleic acid and one of the fatty acids having the most potent essential fatty acid activity) could be converted by enzymes present in a wide variety of tissues into substances called prostaglandins which are oxygenated fatty acids with an unusual five-membered ring system (Fig. 5).[4] They and their related metabolites (Fig. 5) exert a range of profound physiological activities at concentrations down to 10^{-9} g/g tissue. These include the abilities to cause contraction of smooth muscle, to inhibit or stimulate platelet adhesion, and to cause vaso-constriction or dilation with a related influence on blood pressure. They are so potent in their action that they must be produced locally and destroyed immediately after they have produced their effect by enzymes that convert them into inactive metabolites. The excretion of these metabolites in urine has been used to estimate daily production of prostaglandins and to assess the quantities required by the body. In order to achieve local production, the essential fatty acids are released from membrane phospholipids by specific phospholipases and transferred to the enzyme that synthesizes prostaglandins, which is also located in the membrane. The structural, or membrane, phospholipids can therefore be regarded as a vast body store of essential fatty acids that are immediately available for prostaglandin synthesis when required. As they are used up, they must be replenished by dietary fatty acids (or like arachidonic acid, synthesized from dietary linoleic acid). This explains the need to maintain a dynamic 'turnover' of fatty acids in membranes.

The daily production of prostaglandins in humans has been estimated at about 1 mg day^{-1} and this has to be set against the daily intake of around 10 g of essential fatty acids. It is still a matter of controversy whether prostaglandin formation is the sole function of essential fatty acids. Certainly, the metabolic functions of membranes, such as those of the mitochondria, are impaired during essential fatty acid deficiency in ways that may not be related to prostaglandin formation and it is possible that essential fatty acids are important for maintaining the required spatial configuration of lipids in membrane architecture in ways that are currently obscure. Another point of controversy is whether the *n*-3 family of fatty acids (represented by α-linolenic acid, all-*cis*-9,12,15-octadecatrienoic acid) has an essential fatty acid role in humans that is distinct from that of the *n*-6 family based on linoleic acid. There is some evidence[29] to suggest that this is so, but more research is needed.

There is some evidence to suggest that lipids, especially those containing polyunsaturated fatty acids, have a role in the regulation of the immune system. Immune defence may involve the synthesis of specific immunoglobulin antibodies (humoral immunity) or the activation of lymphoid cells (cell-mediated immunity). Lymphocytes, for example, respond to antigens by dividing to provide new populations of lymphocytes that secrete lymphokines, powerful chemicals that are involved in triggering further immune reactions. This type of lymphocyte response may be conveniently studied with cells in culture and when certain fatty acids are added to the culture medium they can suppress the stimulation of lymphocytes by antigens. The suppressive effect is greater with polyunsaturated fatty acids such as linoleic acid and arachidonic acid than with monounsaturated or saturated fatty acids and there is some reason to believe that this effect may be mediated through the formation of prostaglandins.[30] There is still some debate about the ability of monounsaturated and even saturated fatty acids to influence lymphocyte response to antigens and their observed actions may be connected with their influence on the metabolism of polyunsaturated fatty acids and prostaglandins. It is still uncertain whether these effects on immune function are significant in the living animal. Diets rich in polyunsaturated fatty acids have been employed as adjuncts to conventional immunosuppressive therapy to reduce rejection of kidney grafts.[31] Such diets certainly prolong the survival time of skin grafts in mice.[32] Diets rich in linoleic acid also appear to be beneficial in treating patients with multiple sclerosis,[33] perhaps by suppressing an abnormal immune response to one of the body's own proteins, which is a feature of this disease. A confusing observation is that diets *deficient* in essential fatty acids also suppress immune responses and it seems likely that there is an optimal supply of polyunsaturated fatty acids for the functioning of the immune system, above and below which the response is impaired.[34] This is a subject deserving much more attention from the nutritional standpoint.

It is difficult to put an absolute figure on essential fatty acid requirements. Skin lesions seen in children and patients on intravenous alimentation disappear when the diet contains 1% of its energy as linoleic acid, and this figure is widely quoted as the requirement for essential fatty acids. Intakes of this order are also effective in preventing the accumulation of all-*cis*-5,8,11-eicosatrienoic acid which occurs during essential fatty acid deficiency (see below). However, a joint FAO/WHO report[35] suggests that for human adults a value of 3% of dietary energy is more appropriate. The average UK diet may contain 10 g linoleic acid, representing about 3% of energy intake. Many people take much less or much more than this, but there is probably also a very wide range of individual requirements.

Dietary essential fatty acids should not be considered in isolation. Many dietary constituents affect the requirement for or the utilization of essential fatty acids. Thus saturated, *cis*-monounsaturated and isomeric unsaturated acids depress the utilization and increase the requirement for linoleic acid when their proportion to linoleic acid is very high (around 10-fold). One of the reasons for this is that monounsaturated fatty acids compete with the desaturase enzyme that is the first and controlling step in the metabolism of linoleic acid to arachidonic acid. The affinity of the enzyme for linoleic acid is much greater than for oleic acid and normally linoleic acid competes effectively for the enzyme and is efficiently metabolized. When the ratio of oleic to linoleic acid is very high, however, oleic acid may be metabolized at the expense of linoleic acid. One result of this is the accumulation of a triunsaturated fatty acid of the *n*-9 family (all-*cis*-5,8,11-eicosatrienoic acid) in membranes instead of arachidonic acid.[1] This fatty acid has no essential fatty acid activity and does not give rise to a biologically active prostaglandin. The ratio of arachidonic acid to this eicosatrienoic acid in tissues is used as a sensitive biochemical index of essential fatty acid nutritional status and to assess requirements for essential fatty acids.

Because of the hydrogenation of unsaturated fatty acids in the rumen, dairy products are poor sources of essential fatty acids. Milk fat contains 2·4% of dienoic acids but the amount present as *cis*,*cis*-9,12-octadenadeienoic acid is only 1·4%. Nevertheless, since dairy products contribute about 35 g fat/day to the average UK diet, the total intake of linoleic acid from this source is about 0·5 g.

5.4. Fat-Soluble Vitamins

Dietary fat is also important in so far as it contributes several fat-soluble vitamins: A, D, E and K. Vitamin A (retinol) is found as such only in animal fats. Vegetables, such as dark-green leaves, and vegetable oils, such as palm oil, contain a precursor (provitamin A), β-carotene, which is converted into retinol in the body. Because there are losses during the absorption of carotene and its conversion into retinol, it is convenient to describe the vitamin A activity of the diet in terms of 'retinol equivalents'. Normally, 1 μg retinol equivalent is taken as corresponding to 6 μg β-carotene, although there are strong suggestions that this factor generally underestimates the retinol equivalents of most diets. Milk fat is relatively rich in retinol equivalents and because of the better absorption of β-carotene from milk fat than from many other foods, 1 μg retinol equivalents corresponds to only 2 μg β-carotene in milk. The carotene content of milk depends on the diet of the cow (higher in cows on pasture than in winter) and breed (higher in Channel Island than in Friesian cattle possibly because of the lower conversion into retinol in the gut).[36] This accounts for the yellower colour of Channel Island compared with Friesian milk. The concentration of retinol in dairy foods is proportional to the concentration of milk fat in the product. Other animal foods rich in retinol are fish-liver oils, liver and eggs. Adipose tissue fat contains no retinol although it may contain β-carotene. In the UK, all margarine for retail sale is required by law to contain about the same amount of vitamin A (added as synthetic retinol and β-carotene) as butter.

The average UK diet contains about twice the recommended intake of vitamin A, with two-thirds coming from retinol and one-third from carotene. Excessive intakes result in an accumulation in the liver with resulting toxicity.

Vitamin D occurs in significant quantities only in marine oils and in margarine (to which it is required by law to be added). Some argue that it should not be considered to be a vitamin. The generation of vitamin D from a precursor in the skin by the action of sunlight is generally the most important source for the majority of people who need little or none from food. However, a dietary source is more important in children, pregnant and lactating women, whose requirements are particularly high, and people who have little exposure to sunlight. Dairy products, especially butter, can contribute small but useful amounts to the diets of such people. It is quite important to avoid excessive intakes, especially in young children, since this causes more calcium to be absorbed than can be excreted. The excess is deposited in the kidneys and causes damage to them.

Vitamin E activity is shared by several related tocopherols, the most active being α-tocopherol. Although it is assumed, by analogy with other animals, to be essential in the human diet, this has never been proven. Most food fats contain small quantities, but it is found in the largest concentration in vegetable oils. Because one of its roles is thought to be in the inhibition of free-radical oxidation of unsaturated fatty acids (i.e. as an anti-oxidant), it is generally held that intake should be considered in relation to the polyunsaturated fatty acid content of the diet rather than in absolute amounts. A ratio of vitamin E to linoleic acid of 0·6 mg/g is generally recommended.[37] In general, those vegetable oils containing high concentrations of polyunsaturated fatty acids are sufficiently rich in vitamin E to give adequate protection.

Some fat is necessary in the diet to improve the absorption and utilization of fat-soluble vitamins.[35] However, there is little evidence that within the normal range of fat intakes in human diets, the amount of dietary fat significantly affects the utilization of fat-soluble vitamins.

5.5. Palatability of Food: Flavour, Aroma and Texture

Food is nutritious only in so far as it is good enough to eat, and the presence of fat contributes substantially to the palatability of food. A fat-free diet would not only lead to signs of essential fatty acid deficiency, but would also be extremely unpleasant to eat. Fats contribute to palatability principally in two ways: by olfactory responses in the nose or mouth to lipids or lipid breakdown products, and by responses to their texture in the mouth. This property of 'mouth-feel' may comprise attributes such as chewiness, grittiness, stickiness and oiliness.[38]

Odour normally results from a perception of low molecular weight volatile lipids. Short-chain fatty acids have a more intense taste and smell than longer-chain fatty acids because of their greater water solubility and volatility. They may influence the flavour of food in different ways according to the nature of that food. For example, the rancidity of milk is due to those same short-chain fatty acids that in cheese provide the characteristic and much-prized flavour. Butyric acid is a key flavour component of butter but there are an enormous number of straight- and branched-chain acids that in combination contribute to the flavour and palatability of dairy products.

Chemical and enzymic transformations of lipids may also give rise to alcohols, esters, carbonyl compounds and lactones each of which may contribute to the unique flavours of different dairy products. As in the case of free fatty acids, the same compounds may give rise to pleasant or unpleasant flavours depending on factors such as their concentration, the ratio of oil to water in the food and the proportion of other food constituents, the proteins and carbohydrates.

Lipids also contribute towards texture or 'mouth-feel' in different ways. To swallow a pure oil is extremely unpalatable, yet emulsions like milk, butter and cream are pleasant in different ways: the mouth-feel of milk is related to its colloidal structure; cream owes its attractiveness to the globular structure of the fat particles; and butter has a pleasant cooling effect on the tongue. The texture and palatability of butter can be changed by adding an emulsifier, such as a monoacylglycerol, prior to churning or by altering its fatty acid composition by feeding ruminants 'protected fat' (see Section 3.2.2). Fatty acid composition may also affect the crystal structure of a fat and its melting range. This is especially important in relation to chocolate fats which owe their palatability to rapid melting in the mouth, yet at normal storage temperatures are quite hard and easily handled. For a detailed review of the role of lipids in taste, aroma and palatability the reader is recommended to consult Forss.[38]

6. TOXICITY OF LIPIDS

It can be argued that any component of the diet, including water, can be toxic if consumed in inappropriate amounts. This applies no less to lipids than to other food components and the concept of a balanced diet will be developed in later sections.

The threshold level at which toxic effects occur is lower for some substances than for others, and this section will deal with several lipids that are potentially toxic on the basis of experiments with small animals. It should, however, be stressed that in no case has it been proved that they are toxic for humans in quantities which may reasonably occur in normal diets. The subject has been reviewed in more detail by Mattson,[14] and by Dhopeshwarkar.[39]

6.1. Cyclopropene Fatty Acids

By far the most important edible oil containing cyclopropene fatty acids (of which sterculic acid is an example) is cottonseed oil, in which the concentration ranges from 0·6–1·2%, although after processing, the oil as actually eaten probably contains only 0·1–0·5%. The metabolic importance of sterculic acid (I) is to inhibit the desaturation of stearic to oleic acid, the effects of which are to alter the permeability of membranes and to increase the melting point of fats. A biological result of this is illustrated by 'pink–white disease' in hen's eggs. If cyclopropene acids are present in the diet of laying hens, the permeability of the yolk-sac membrane is increased allowing the release of substances, including pigments, from the yolk. In dietary experiments with animals, the source of cyclopropenes is usually *Sterculia foetida* seed oil, which has a much higher content (up to 70%) than cottonseed oil. Rats die within a few weeks when fed diets containing 5% of dietary energy as sterculic acid and the reproductive performance of females is completely inhibited at levels as low as 2%. Humans have been eating cottonseed oil for many years in such products as margarines, cooking oils and salad dressings but the intake of cyclopropene fatty acids from these sources is very small. On this basis, it is presumed that low levels have no adverse effects, but whether prolonged ingestion by humans of larger amounts would be deleterious is not known. Possible synergistic effects in which fats containing cyclopropenes may enhance the potency of carcinogens deserve more research attention.[40] Dairy products contain no cyclopropene fatty acids.

$$CH_3(CH_2)_7\overset{\overset{\displaystyle CH_2}{\diagup\!\!\diagdown}}{C}\!\!=\!\!C(CH_2)_7COOH$$

(I)

6.2. Branched-Chain Fatty Acids

The occurrence of branched-chain fatty acids in ruminant tissues, and therefore in dairy products, has been mentioned in Section 3.2. These foods may contain a variety of structural isomers, including monomethyl iso- and anteiso-acids and polymethyl structures.[41-3] A common dietary branched-chain fatty acid is phytanic acid (II) (3,7,11,15-tetramethylhexadecanoic acid) which is formed from phytol, a universal constituent of green plants.

$$CH_3.\overset{\overset{\displaystyle CH_3}{|}}{CH}(CH_2)_3.\overset{\overset{\displaystyle CH_3}{|}}{CH}(CH_2)_3.\overset{\overset{\displaystyle CH_3}{|}}{CH}(CH_2)_3.\overset{\overset{\displaystyle CH_3}{|}}{CH}.CH_2.COOH$$

(II)

Branched-chain fatty acids appear to be generally excluded from animal membranes and are mainly deposited in adipose tissue. As a source of fuel, they are normally oxidized in the usual way. However, the normal process of β-oxidation may be blocked when methyl groups occur in certain positions along the hydrocarbon chain.[1] In these cases, an alternative oxidation mechanism (α-oxidation) comes into play, removing the block and allowing normal β-oxidation to proceed. In a rare genetic disease (Refsum's disease) the alternative oxidation pathway is lacking, and such people accumulate the branched-chain metabolic product in their tissues, resulting in chronic polyneuropathy, night blindness, narrowing of the visual field, skeletal malformation and cardiac complications. This disease is normally fatal and a phytanic-acid-free diet is the only therapeutic measure.[1,44] It has been suggested[45] that once the myelin membrane has accumulated a certain proportion of phytanic acid it becomes unstable, and that this is responsible for the demyelination.

An important precursor of branched-chain fatty acids is methylmalonyl-CoA. This metabolite does not normally accumulate in tissues because it is removed by an enzyme that depends on a supply of vitamin B_{12} for its activity. A rare metabolic abnormality has been described in children where a congenital deficiency of B_{12} metabolism results in a failure to remove methylmalonyl-CoA. This, in turn, leads to the formation of branched-chain fatty acids and their accumulation in tissues, causing some neurological disturbances.[46] However, no overt neuropathological signs were observed in young sheep and goats fed on barley-rich diets for up to two years, despite their having synthesized considerable amounts of branched-chain fatty acids in that period.[47] Equally, no neurological disorders were evident in baboons deprived of vitamin B_{12}, even after the administration of an inactive analogue which exacerbates the depletion of B_{12} and which occasions the hepatic production of enhanced amounts of branched-chain fatty acids.[48]

Hence it seems that the only reason for concern about the toxicity of lipids containing branched-chain fatty acids is in the case of those few individuals with specific metabolic disorders that prevent the normal utilization of these compounds. Indeed, there is some evidence that a branched-chain fatty acid (14-methylhexadecanoic acid) may play an essential role in protein synthesis.[49] These results need to be confirmed by other laboratories and future developments may indicate a vital importance for specific branched-chain fatty acids, occurring in dairy products, in nutrition and metabolism.

6.3. *Trans* Fatty Acids

Most unsaturated fatty acids found in nature contain double bonds in the *cis* geometrical configuration. As explained in Sections 2.2 and 3.3, fatty acids with the opposite (i.e. *trans*) geometrical configuration are found in nature, although in much less abundance. *Trans* fatty acids are found naturally either as short-lived intermediates in biochemical pathways or as stable end-products. As examples of the latter, *trans*-3-hexadecenoic acid is an ubiquitous constituent of photosynthetic tissue, although present in small amounts. Some seed oils (e.g. tung, *Aleurites fordii*) may contain considerable amounts of *trans* fatty acids, although they are not generally important dietary sources.

Trans fatty acids are also produced by the process of microbial biohydrogenation of dietary polyunsaturated fatty acids in the rumen of ruminant animals or by chemical hydrogenation in margarine manufacture (Sections 2.2 and 3.3).

In terms of human diet, *trans* fatty acids may be consumed, therefore, in green vegetables (very small amounts), some seed oils, dairy products (up to 20% of the monounsaturated fatty acids in butter fat),[43] and in margarines and processed cooking fats. Some margarines may have up to 55% of *trans* fatty acids[50] but the variation is enormous depending on the process and starting material. As explained earlier, whether originating naturally or due to industrial processing, fats containing *trans* fatty acids contain a large variety of positional and geometric isomers.

There is no evidence that the digestion, absorption and subsequent oxidation of fatty acids to yield metabolic energy is in any way disturbed by the presence of *trans* fatty acids or that the *trans* acids themselves are poorly utilized in this respect.

In regard to their influences on membrane properties, *trans* fatty acids have physico-chemical properties more akin to saturated fatty acids than to *cis*-unsaturated fatty acids and, in many respects, behave as though they were substituting for saturated fatty acids.

Trans fatty acids may influence the metabolism of other unsaturated fatty acids[51] largely as a result of competition between the different fatty acids for a single desaturase enzyme (Section 5.3). Linoleic acid (the primary dietary essential fatty acid) has the strongest affinity for this enzyme and is normally present in sufficient quantities to be used preferentially in the metabolic pathway, thus ensuring more than adequate levels of polyunsaturated fatty acids in the tissues for prostaglandin formation. However, when the intake of linoleic acid is low, the presence of large quantities of fatty acids, such as *trans* monounsaturated fatty acids that can act as alternative substrates for the desaturase, influences the direction of the metabolic pathway so that the metabolic products are non-essential fatty acids that either cannot give rise to prostaglandins or result in the formation of prostaglandins of unknown or unpredictable activity. Thus the ingestion of excessive amounts of hydrogenated polyunsaturated fatty acids can result in metabolic and nutritional disturbances in two ways: the dietary intake of polyunsaturated fatty acids is reduced, and the pathways for the metabolism of available essential fatty acids are disturbed by the imbalanced mixture of fatty acids ingested. Hill *et al.*[52] have shown that increasing the concentration of dietary *trans* fatty acids causes a decrease in the concentrations of essential fatty acids in heart muscle and that this effect is accentuated when the diet is marginal in its essential fatty acid content.

Recently, Hwang and Kinsella[53] have suggested that an important factor in the toxicity of *trans* fatty acids is their effect in limiting the availability of the essential fatty acids for prostaglandin formation, either by displacing them from tissues or by inhibiting the metabolic pathways leading from essential fatty acids to prostaglandins. They fed groups of rats one of four diets containing 5% by weight of fat which was either: (1) hydrogenated coconut oil (containing a negligible quantity of unsaturated fatty acids); (2) all-*trans*-linoleate; (3) an equal mixture of all-*cis*-linoleate and all-*trans*-linoleate; or (4) all-*cis*-linoleate. Diets containing *trans* fatty acids resulted in lower concentrations of essential fatty acids in the liver and in the blood platelets and a slower rate of production of prostaglandins by platelets. The differences were most marked when diets (1) and (2) were compared with (4); there were smaller differences between (3) and (4). The authors concluded that 'these data demonstrate that dietary *trans* fatty acids aggravate the symptoms of essential fatty acid deficiency and cause a reduction in the concentration of prostaglandins and their precursors, even when fed with *cis*-linoleic acid. The implications of these observations are significant in view of the presence of *trans* fatty acids in the American diet and are particularly pertinent because of the potent effects of prostaglandins and their intermediates on many physiological functions, particularly platelet aggregation and cardiovascular actions'. These conclusions should be treated with caution for the following reasons. Diets (1) and (2) were not simply 'essential fatty acid deficient'; they contained *no essential fatty acids at all*. It is well established that the presence of *trans* or even *cis* monounsaturated fatty acids exacerbates the effects of essential fatty acid deficiency under these conditions. Although diet (3) contained all-*cis*-linoleate, there was a slight depression in tissue essential fatty acid concentrations and a rather bigger depression of prostaglandin production. However, it is not established whether depressions of prostaglandin formation of this order are significant in the human diet or indeed whether the high rate of prostaglandin production from diet (4) *in which all the fatty acid was present as all*-cis-*linoleate* can be regarded as normal.

Worries about the adverse nutritional effects of *trans* fatty acids began with the publication by Kummerow[54] which reported that pigs fed diets enriched in *trans* fatty acids developed atherosclerosis. Many other studies support the contention that the metabolic effects are attributable not specifically to the presence of *trans* double bonds but to a relative imbalance between the different types of fatty acid ingested which then compete for important metabolic pathways.[55,56] In the author's view, there is no evidence that moderate (or even large) dietary intakes of fats containing *trans* fatty acids by humans have deleterious consequences as long as the linoleic acid content of the diet is adequate. The nutritional and metabolic aspects of *trans* fatty acids have been reviewed.[57-9]

6.4. Long-Chain Monoenoic Fatty Acids

Some natural edible oils contain, in addition to the widely occurring monounsaturated fatty acid oleic acid (*cis*-9-octadecenoic), appreciable quantities of monounsaturated fatty acids with a chain length of 22 carbon atoms. The most important of these are rape- and mustard-seed oils which contain erucic acid (*cis*-13-docosenoic acid) and herring oil which contains cetoleic acid (*cis*-11-docosenoic acid). Catalytic hydrogenation of marine oils containing polyenoic acids for margarine manufacture gives rise to a range of long-chain *cis*- and *trans*-monoenoic acid isomers.

When young rats were fed diets containing more than 5% of the energy as rapeseed oil (45% erucic acid) their heart muscles became infiltrated with fat.[60] After about a week, the hearts contained three- to four-times as much fat as normal hearts and although, with continued feeding, the size of the lipid deposits gradually decreased, other pathological changes were noticeable, such as the formation of fibrous tissue in the heart muscle. The biochemistry of the heart tissue was also adversely affected in that the rate of mitochondrial oxidation of substrates was slower and the rate of ATP synthesis impaired. The activity of lipases is lower on triacylglycerols containing erucic acid than on fatty acids of more normal chain length. These results have been repeated in many laboratories with many species of animals and reviewed by Vles.[61] Although there are wide differences between species and the response is much weaker in older than in younger animals, the results have been taken very seriously from the point of view of human nutrition since rapeseed oil is now a major edible oilseed crop on world markets.

It is not known whether similar lesions occur in humans, although in some countries rapeseed oil has been consumed for many years, albeit not in the high concentrations employed in many animal feeding experiments. Toxicity is likely to be less in a mixed diet with adequate quantities of essential fatty acids and a good balance of the more usual fatty acids.[14] Despite the lack of evidence for harmful effects in humans, it has been thought prudent to replace older varieties of rape, having high erucic acid contents, with new varieties of zero-erucic rapes. Food products containing hardened fish oils will, however, continue to contain some C_{22} monoenoic acids. Dairy fats contain negligible quantities of these long-chain monounsaturated fatty acids.

6.5. Oxidized Fats

The storage of fats containing appreciable concentrations of polyunsaturated fatty acids in the presence of oxygen at room temperature can result in the formation of hydroperoxides (see Section 3.4, Reference 18 and Richardson and Korycka-Dahl, Chapter 7). When these are ingested they are rapidly degraded in the mucosal cells of the gut to various oxyacids that are further oxidized to CO_2. There is no evidence for the absorption of unchanged hydroperoxides nor for their incorporation into tissue lipids. Although the growth of rats fed a fat with a peroxide value of 100 was reported to be normal,[62] other workers[63] have provided evidence for the potentiation by linoleic acid hydroperoxide of tumour growth in female rats. It should be mentioned that fats with peroxide values of much less than 10 are, in any case, organoleptically unacceptable. It seems that the normal levels of auto-oxidation in foodstuffs at room temperature may not be important from a toxicological point of view.

Heating unsaturated fats leads to more extensive chemical changes.[64] Heated fats contain a range of polymerized compounds rather than peroxides; the extent of polymerization may be 10–20% under normal household or commercial practice and up to 50% when the oil is severely abused. Only in the latter case have toxic effects been observed in animal feeding experiments and then of no great severity. Artman[15] concluded that fats heated in normal cooking processes are not harmful to humans and subsequent studies have tended to support this contention,[14] although Alexander[65] takes a less optimistic view.

7. THE ROLE OF DIETARY LIPIDS IN DISEASE PROCESSES

7.1. Background

In general, the affluent countries of the world have succeeded in reducing mortality due to infectious diseases to very low levels by ensuring a good food supply, improving hygiene and by the development of a sophisticated pharmaceutical industry. In these countries, the main causes of death are now diseases that for the most part cannot be attributed to a specific cause but in which various factors associated with affluence are thought to be involved. The principal diseases in terms of prevalence and contribution to increased morbidity or mortality are cardiovascular disease, cancer, diabetes and obesity.

Although there is undoubtedly a strong genetic component contributing to the risk of developing any one of these diseases, the evidence, which comes from studies of migrating populations, is for the dominance of environmental factors. Thus the prevalence of some types of cancer is low in Japanese living in Japan but increases dramatically when they migrate to the USA. Of the various environmental factors that could be involved, there is now strong evidence that diet plays an important part in each of the diseases mentioned, and, of the various dietary components, fat has received most attention. The evidence for a role for dietary fat in the aetiology of these diseases comes principally from three sorts of study. Firstly, there is epidemiological evidence gained by comparing the dietary intakes and other environmental factors in different population groups and making correlations between these observations and disease incidence. These studies have given rise to the concept of 'risk factors', i.e. characteristics, usually associated with affluence, prevalent among those who develop the disease. Such studies cannot provide information on the causes of the disease but can provide important clues from which hypotheses can be developed as a basis for more direct experimental studies. Secondly, there is evidence from studies with experimental animals. These can give good insights into the basic mechanisms underlying disease processes but it has to be emphasized that, since animals are not living under the same conditions as humans and may not be completely similar in their metabolism, the findings may not always be entirely relevant to the human disease. Thirdly, there are nutritional studies in which the aim is to modify the human diet and determine whether this can influence, beneficially, the course of the disease. Such studies are expensive, difficult to control and it is frequently difficult to accumulate large enough numbers or to continue the experiment for long enough to reach positive conclusions about the outcome. It is important to bear these limitations in mind when assessing the soundness of recommendations that are often made for modification of the dietary habits of whole populations.

It would be impossible in this section to give a detailed account of the evidence for and against the importance of dietary fats in the diseases of affluence. The aim is to give a brief summary of what are currently thought to be the major implications of lipids, either as causative agents or in the dietary management of the disease, and to indicate further reading.

7.2. Cardiovascular Disease

7.2.1. Definitions
Cardiovascular disease is a collective name for a number of diseases affecting the supply of blood to the heart. The most important in terms of mortality is ischaemic heart disease (IHD). In this disease, a reduced supply of blood (and therefore of oxygen) to the heart results in the death of a localized mass of heart muscle. The cause is usually the occlusion of a

coronary artery by a thrombus (*thrombosis*), a process that is more likely to occur when the arteries have already become narrowed by the slowly developing disease, *atherosclerosis*. This is an irregular thickening of the inner wall of the arteries that reduces the size of the lumen and is caused by the accumulation of plaque, consisting of smooth-muscle cells, connective tissue and considerable deposits of lipid, of which cholesterol esters comprise the major part.[66]

The likely sequence of events seems to be initial damage to the *vessel walls*, development of *plaque* at the site of the damage, and occlusion of the vessel by a *thrombus*. Whereas there is little evidence to suggest that dietary lipids influence the condition of the vessel wall, the possibility should not be totally discounted. The other major factors influencing IHD, namely plaque and thrombus formation, are both affected by the quantity and quality of circulating lipids.

7.2.2. Atherosclerosis: Relationship of Blood Lipids to the Disease

The most important blood lipid fractions influencing plaque formation are the low density (LDL) and high density (HDL) lipoproteins which bear an inverse relationship to each other. Population groups having low mean HDL and high mean LDL concentrations are also those with a high incidence of IHD, although there are some exceptions.[67,68] The association between blood lipid levels and incidence of IHD is much weaker in individuals within a population, implying that whereas, on average, high plasma LDL concentrations indicate increased risk of the disease, it may not be true for a particular individual. Other factors of environment or life style may exert a greater influence.

7.2.3. Atherosclerosis: Relationship of Blood Lipids to Diet

Plasma LDL concentrations are influenced by the amount and nature of dietary fat. Thus, dietary saturated fats tend to result in increased LDL concentrations while polyunsaturated fats tend to decrease them. However, dietary saturated fats are roughly twice as potent in raising plasma LDL concentrations as polyunsaturated fats are in lowering them.[69,70] Thus the ratio of polyunsaturated fats in the diet is as important, if not more so, than the total amount of dietary fat. There is no effect on the plasma cholesterol concentration when the polyunsaturated/saturated fat ratio is 0·5; a higher ratio would lower it and a lower ratio would increase it by a calculable amount.

The absolute amounts or proportions of saturated fats in the diet are not the only factors to affect blood LDL levels. It has been shown that when

THE NUTRITIONAL SIGNIFICANCE OF LIPIDS 405

the fatty acids in the triacylglycerol molecules of butterfat are randomized by interesterification (Section 3.3.3), the fat no longer has the effect of raising blood LDL concentrations when fed to humans. Butter itself has a marked tendency to cause elevated LDL levels.[71] The randomized butter was more rapidly hydrolysed by pancreatic lipase *in vitro*[72] and in feeding experiments, chylomicrons appeared more rapidly in the plasma than after a similar intake of butter.[73]

The reason why saturated fats with fatty acids esterified in a particular configuration give rise to elevated LDL concentrations whereas monounsaturated acids have no effect and polyunsaturated fatty acids have the opposite effect is still unclear. It is most likely due to subtle effects on the chemical composition of the different plasma lipoprotein fractions with a resulting influence on their physical properties which, in turn, results in changes in rates of uptake by tissues. This could be due to changes in membrane receptor activity or the activity of enzymes, such as lipoprotein lipase, involved in clearing lipoproteins from plasma. It is well established that individuals with a genetic predisposition to hyperlipoproteinaemia may have defects in the receptor uptake mechanism or a deficiency of lipoprotein lipase.[66,74] Polyunsaturated fats can accelerate removal of cholesterol from the body by increasing the secretion of sterols and bile acids.[75] However, it is not known whether this accounts for all the plasma cholesterol-lowering activity. A possibility to consider is that although cholesterol may be removed from *plasma*, it in fact 'overflows' into tissues. It could then only be removed satisfactorily if sufficient HDL were present to act as a scavenger.[21,67]

Dietary cholesterol (when allowance is made for the fact that foods rich in cholesterol also tend to be rich in saturated fats) seems to exert a much smaller effect on plasma lipoproteins. This may be partly due to a lower efficiency of absorption and partly to the efficient metabolic regulation of cholesterol synthesis by dietary cholesterol (Section 4.2). From experiments in humans, it appears that average serum cholesterol concentration increases with increasing intake of cholesterol up to about 600 mg day^{-1}, beyond which there is no additional effect, probably because the absorptive plateau has been reached.

Although LDL has been implicated as the major plasma-containing fraction to influence vascular disease, some evidence has been presented that high circulating concentrations of VLDL may also play a part.[76] Some individuals are particularly sensitive to dietary carbohydrate which may give rise to elevated concentrations of VLDL.

To summarize: studies with experimental animals and humans suggest a strong link between high circulating blood lipoprotein concentrations (LDL and possibly VLDL) and the risk of developing atherosclerosis. High concentrations of HDL have a protective effect. Among the major factors in the diet which influence the concentrations of these lipoproteins are the amounts, fatty acid composition and molecular configuration of dietary triacylglycerols. In some individuals, the rate of the sequence: dietary fat → blood lipoprotein concentration → atherosclerosis, could be influenced by changes in the diet. This is often called the 'lipid hypothesis'. The background to atherosclerosis has been clearly summarized in a series of articles in *Science*.[77-9]

7.2.4. Thrombosis

Until fairly recently, attention had been given only to the role of dietary lipids in the early stages of cardiovascular disease, namely atherosclerosis. In the final stage, namely the formation of a thrombus which may occlude

TABLE 7
OPPOSING EFFECTS OF PROSTACYCLINS AND THROMBOXANES ON THE CARDIOVASCULAR SYSTEM

Thromboxanes in platelets	Physiological effect	Prostacyclins in arterial wall
Stimulates	Platelet aggregation	Inhibits
Constricts	Arterial wall condition	Relaxes
Lowers	Platelet cAMP[a] levels	Raises
Raises	Blood pressure	Lowers

[a] cAMP = cyclic adenosine monophosphate, a chemical transmitter that regulates the activity of certain enzymes.

an artery, the adhesiveness of blood platelets is an important factor. During the 1970s, it became apparent that the prostaglandins (Section 5.3) were closely concerned with coagulation in blood vessels and the balance between the different types of prostaglandin formed may be important in thrombus formation.[80] Prostacyclins (Fig. 5) generated in vessel walls inhibit platelet aggregation and may be the biochemical mechanism whereby adhesion of platelets to vessel walls is normally prevented.[81] Thromboxane (Fig. 5) produced by platelets has the opposite effect (Table 7), being a powerful stimulator of platelet aggregation.[82]

THE NUTRITIONAL SIGNIFICANCE OF LIPIDS   407

The intake of dietary polyunsaturated fats can affect both the total daily production of prostaglandins[53] and the balance between them.[83] Of particular interest, a dietary intake of large amounts of all-*cis*-7,10,13,16,19-eicosapentaenoic acid (*n*-3 family), found in some fish oils, can shift the balance towards an anti-aggregating state and could explain the enhanced bleeding tendency in Eskimos who consume such a diet.[84] It has been suggested[83] that the popularization of the role of polyunsaturates in the dietary prevention of IHD has led to a gradual increase in the amount of linoleic acid consumed in UK diets and in the ratio of *n*-6 to *n*-3 fatty acids. This might adversely affect the coagulating properties of blood. Much more research is needed before the long-term effects of dietary fats on blood coagulation *in vivo* are understood and before this knowledge can be translated into practical dietary advice.

The relatively high ratio of *n*-3 to *n*-6 fatty acids in butter[85] may prove to be an advantage when the role of the two families of fatty acids in platelet function and thrombus formation is more clearly understood.

7.2.5. Dietary Guidelines and Implications for the Dairy Industry

Although there is clear evidence that dietary modification may influence the distribution and concentration of blood lipoproteins, there is no clear evidence that the reduction of cholesterol-rich LDL concentrations in the general population will influence the total mortality from IHD. Several so-called 'intervention studies' have indicated that replacement of dietary saturated fatty acids with polyunsaturates (usually in association with a reduction in total fat intake) could reduce the incidence of myocardial infarction, but none succeeded in reducing total mortality.[86-9]

Nevertheless, public health authorities in many countries have recommended dietary modifications aimed at reducing plasma cholesterol concentrations. Amongst those who believe that dietary modification can be effective as a preventative measure are two schools of thought. One holds that recommendations should be reserved for individuals known to be particularly at risk. The rate of development of the disease can be checked by dietary modification, for example in the small number of people who are genetically predisposed to exceptionally high cholesterol levels and who normally die prematurely of IHD.[66] The arguments against recommendations for the general population are the lack of firm evidence for the general effectiveness of such measures in reducing the incidence of IHD and the paucity of information concerning other consequences of plasma cholesterol reduction. The other view is that the problem is so severe that there is no sensible alternative to applying dietary guidelines to the general population. The implications of such a policy for agriculture and food production are enormous. The argument against specific recommendations applied to individuals is that it would require widespread and expensive screening programmes to determine precisely who is 'at risk'. The whole controversy and background to the lipid hypothesis has been admirably reviewed.[90]

The present author's view is that people in general eat food because they like it not because they perceive that it will achieve some long-term indefinable medical benefit. Those who know they have a medical problem may well take a different view and it is the role of the food industry, of which the dairy industry is a part, to satisfy consumer demand. The availability of appropriate food products is a key factor in satisfying such demands. The margarine industry has tackled this problem successfully, in part by producing margarines with a high polyunsaturate content and with a minimum of modification to the ingredient oils. Dairy products are potentially under threat because of the high proportion of total dietary fat that they supply and because of their high content of saturated fatty acids. A reduction in butter consumption (and, it should be pointed out, the consumption of hydrogenated margarines which form by far the largest proportion of margarine sales) is a convenient way to achieve a reduction in plasma cholesterol. However, if the contribution of polyunsaturated fatty acids can be increased by only a little, as suggested by Marr and Morris,[91] a substantial hypocholesterolaemic effect can be achieved with a comparatively small reduction in saturated fat intake.

The way forward for the dairy industry probably lies in the application of appropriate technologies to diversify the range of its products. The provision of a range of low-fat milk drinks (flavoured or not, fermented or not), the randomization of butter and the provision of spreads achieved by blending butterfat with vegetable oils are three such approaches. The last is simpler and more economical than the feeding of 'protected fats' to ruminants (Section 3.2.2) which has not been generally successful. The new products will, of course, have to be named differently from traditional products. However, the sanctity of traditional dairy foods can only be maintained at some cost to the industry.

7.3. Cancer

It has long been known that restricting the energy intake of rodents reduces the genesis and growth of tumours[92] and that high-fat diets increase the incidence of both spontaneous and chemically induced tumours.[93,94] For a review of this and subsequent work the reader is referred to Visek et al.[95]

Clinical evidence relating dietary fat and cancer is less clear-cut. The possible involvement of dietary fat in the aetiology of human breast cancer has been investigated in a number of epidemiological studies. Two age distributions of breast cancer can be discerned: an early group (women aged 40–44) and a later group (women aged 65–69). Hems[96] found that only the 'late' cancer was correlated with dietary fat (and also sugar) intake. Data from 22 different countries were examined and the mean per caput intakes of fat, sugar, total energy and meat were calculated from United Nations statistics. Similarly, Lea[97] found a highly significant correlation between the consumption of fats and oils in 24 different countries (UN statistics) and the death rate from malignant neoplasms of breast, ovaries and rectum in each group over age 55. A negative correlation was found with neoplasms of stomach and uterus. However, more recent large-scale epidemiological studies designed to explore the relationship between dietary lipids and IHD have provided incidental data that seem to contradict earlier findings. The work of Rose et al.[98] and several subsequent studies have shown an apparent relationship between low serum cholesterol concentrations and increased incidence of cancer, whereas the reverse might have been expected in view of the positive correlation between high fat intake and raised serum cholesterol values. Wynder[99] used data from 21 countries to demonstrate a positive correlation between daily total intake of fats and oils and age-adjusted mortality from colonic cancer.

It should be emphasized that these are correlations only and tell us nothing about causal relationships between diet and cancer or whether there is distinction between saturated fats, unsaturated fats and sterols. It has been postulated that dietary fat influences the populations of microorganisms in the lower gut, favouring the establishment of bacteria that metabolize bile acids to carcinogens or co-carcinogens.[100] This might explain, in part, the correlation between high cancer incidence and low *plasma* cholesterol, in view of the relationship between low plasma cholesterol and increased sterol secretion in the bile (Section 7.2.3). There is experimental evidence that high-fat (20% by weight) diets render animals more susceptible to chemically induced tumours under conditions in which there is increased faecal excretion of sterols and sterol degradation products.[101] Some experiments appear to have indicated that tumour development was greater when the dietary fat was unsaturated (corn oil) than when it was saturated (coconut oil).[102] There is no evidence to suggest that polyunsaturated fatty acids act directly as carcinogens. They may act as potentiators either by stimulating the secretion of sterols in bile, as discussed above, or by potentiating the effect of a known carcinogen. Mertin[103] has suggested that their effect is due to a damping down of the body's immune defences (Section 5.3) that might otherwise retard cancer development. Carroll and Hopkins[104] showed that a diet containing 3% sunflower-seed oil and 17% beef tallow fed to rats potentiated the formation of tumours induced by dimethylbenzanthracene just as much as one containing 20% sunflower-seed oil alone. Rats on these diets developed at least twice as many tumours as those fed only 3% sunflower-seed oil or 20% saturated fat. These authors considered that there is a basic requirement for polyunsaturated fats in the potentiation of mammary tumour induction but that there is also an overall requirement for a high fat intake since 3% sunflower-seed oil did not enhance tumour yield.

There is no clear-cut evidence to link polyunsaturated fat content of human diets with cancer incidence. Carroll,[105] in reviewing epidemiological data, found a significant correlation with total fat intake, a weak correlation with animal fat and no correlation with vegetable fat (which may be expected to contain more polyunsaturated fatty acids). He suggested that most human diets are likely to contain the small amounts of polyunsaturated fatty acids required for potentiation of tumour development but over and above that the quantity is more important than the quality of dietary fat.

Epidemiological studies have indicated a positive association between low intakes of vitamin A or its precursors and increased susceptibility to cancer.[106] Some experimental data for rats deprived of vitamin A lend support to this.[107-9] In affluent countries, however, few people are deficient in vitamin A and, in terms of therapy, interest is likely to centre on the pharmaceutical use of synthetic retinoids rather than natural vitamin A which in large doses is extremely toxic.[110] In developing countries, the principal result of vitamin A deficiency is xerophthalmia which will continue to be a more severe problem than cancer.

7.4. Diabetes

Diabetes is another common disease of 'affluent countries' which is slowly but gradually increasing. The term is a cause of confusion since it is applied to at least two diseases that are quite different in their associated metabolism. In *juvenile-onset diabetes* the pancreatic β-cells are deficient in the capacity to produce insulin and the patient is unable to maintain normal glucose homeostasis. The condition is associated with leanness and a reduced level of fatty acid metabolism. It is controlled by administration of appropriate amounts of insulin and control of the diet to maintain stable blood glucose concentrations.

Maturity-onset diabetes is invariably associated with obesity, hyperlipoproteinaemia and an increased tendency to develop atherosclerosis. In contrast to the juvenile type, lipid metabolism tends to be enhanced and blood glucose *and* insulin levels are elevated. The condition is associated with a decreased sensitivity of tissues, such as adipose tissue, to insulin; in other words, more hormone than normal is required to produce a given metabolic effect. The pancreas responds by producing more insulin which in turn enhances the general level of lipid synthesis. The management of the disease is predominantly by means of dietary modification.

Whereas diabetes was once regarded as a disease of carbohydrate metabolism, it is now apparent that faulty lipid metabolism is equally important. The main practical implication is a radical change in dietary management of diabetes. Extremely low carbohydrate diets are no longer emphasized to the extent they used to be, although it is recommended that any increases in carbohydrate should contain a higher proportion of non-digestible polysaccharide (dietary fibre). The chief measures are to reduce total energy intake to avoid obesity and to increase the proportion of polyunsaturated to saturated fat to control hyperlipoproteinaemia.[111]

7.5. Obesity

This is one of the predominant health problems of 'affluent' societies. The outward sign of obesity is an excessive accumulation of adipose tissue which affects body contours. Such an accumulation of stored energy results when the intake of energy is not precisely matched by energy expenditure. However, there is a very poor correlation between energy intake and the prevalence of obesity among individuals because of marked individual variations in energy expenditure. Thus although obese people consume more energy than they require, it is not true to say that they eat more than lean people; quite often the reverse is true.[112]

The potential for foods to give rise to energy deposition in adipose tissue depends on their total energy content, their energy density (energy per gram) and the efficiency with which they are metabolized in the body. Although it might be thought that the high energy density of fats would result in an earlier satiety than a similar amount of energy from carbohydrate, it is apparent that some individuals can consume very large amounts of fat[25] before feeling satiated. Laboratory rodents given increasing amounts of dietary fat adjust food intake so as to maintain their total energy intake constant, but there is no firm evidence that humans do likewise. When the concentration of fat in the rodent's body reaches much over 20% by weight, however, its ability to automatically adjust energy intake is lost and the animals become obese.[113] Palatability is another factor affecting the intake of energy from fat. Rodents fed a variety of human foods (the so-called 'cafeteria' or 'supermarket' diet[114,115]) do not effectively control their energy balance and become obese. It is certainly true that obesity is rare in countries where the diet is bland and monotonous even though energy intake is not restricted.

The assumption that humans are unable to detect changes in the energy content of foods of identical bulk forms the basis of products in which part of the fat is replaced with water (by means of appropriate emulsifier technology). These low-fat spreads will be effective in the human diet only if there is no tendency (as there is in the rodent) to compensate for the low energy density of the food by automatically adjusting total food intake over the medium- or even the long-term. Surprisingly little work has been done to test this hypothesis with human subjects and this deserves further attention. The implications for the dairy industry are profound as such research could give rise to a new generation of dairy-based low-fat spreads with excellent organoleptic properties. (It is worth pointing out that soft margarines have an image in the public mind of being of low energy density, which of course is not so.) The dairy industry is in a good position to move into the offensive by producing a range of top-quality energy-reduced products should research indicate a positive role for these products in the control of energy intake.

There is little evidence to suggest that the metabolic efficiency of polyunsaturated fats is significantly different from that of saturated fats, as has been suggested.[25] However, as discussed in Section 5.1, the energy value of medium-chain triacylglycerols is significantly lower than that of longer-chain acids and this could have application in dietetic products.[116]

Just as some types of dietary carbohydrate ('dietary fibre') are not digested by the enzymes in the human small intestine, and therefore have a low energy value, it is possible to synthesize fats that mimic the physical and organoleptic properties of natural fats but which are not digested and have little or no energy value. Examples are the glycerol polyesters.[115] Few rigorous studies on the energy value of these 'tailor-made' fats have been done and before they could be extensively used in foods their safety in use would need to be thoroughly assessed. Their potential for incorporation into foods (especially dairy-type foods) to increase the potential range and variety of low-energy diets deserves further research.

8. CONCLUSIONS

This chapter reviews and places in perspective the many and varied roles of dietary fats. The dairy industry has over several centuries achieved enormous success in marketing a substantial proportion of the fats that humans eat. This business is highly vulnerable because of its sheer size, the high cost of its products and the limited range of chemical composition of these fats. Advantage has been taken of this vulnerability by other sections of the edible-fats industry, whose products have gained ground against traditional dairy products. While one important factor has been price, clever use of some of the scientific data reviewed in Section 7.2 has enabled the promotion of some products rich in polyunsaturated fatty acids on medical grounds. Although these products form a small part of the total market, something of their 'image' has reflected on other, inferior, products whose composition is unknown to the general public.

Much of this promotion (and some of the dairy industry's attempts to combat it) has been spurious in the sense that the arguments have been about the nutritional (or antinutritional) properties of single dietary components. The important factor in good nutrition is the impact of the diet as a whole and each dietary component contributes to good nutrition in so far as it is present in an appropriate amount. It is nonsense, therefore, to speak about the good or bad nutritional qualities of 'butter' or 'polyunsaturated margarine' when considered in isolation.

The chief advantage of dairy products (at present, but for how long?) lies in their flavour and textural properties. The limited range of chemically distinct products will not allow the industry to compete successfully in the market place and it is time to abandon the notion that the composition of dairy fats is inviolate. The way forward is to extend the range and choice of dairy-based fats by the use of established or new technologies. These products will have a wide range of total fat contents compared with their traditional counterparts and will have a range of fatty acid compositions brought about by chemical modification and blending with other fats and oils.

REFERENCES

1. GURR, M. I. and JAMES, A. T., *Lipid Biochemistry: An Introduction*, 1980, Chapman & Hall, London.

2. CHRISTIE, W. W., *Lipid Analysis*, 1973, Pergamon Press, Oxford.
3. NICOLAIDES, N., *J. Am. Oil Chem. Soc.*, 1965, **42**, 708.
4. LANDS, W. E. M., *Ann. Rev. Physiol.*, 1979, **41**, 633.
5. VERNON, R. G., *Prog. in Lipid Res.*, 1980, **19**, 23.
6. MORRISON, W. R., In: *Topics in Lipid Chemistry*, F. D. Gunstone (ed.), 1970, Logos Press, London, **1**, p. 52.
7. PAUL, A. A. and SOUTHGATE, D. A. T., *McCance and Widdowson's 'The Composition of Foods*, 1980, HMSO, London.
8. MOORE, J. H., *Proc. Nutr. Soc.*, 1978, **37**, 231.
9. JENSEN, R. G., CLARK, R. M. and FERRIS, A. M., *Lipids*, 1980, **15**, 345.
10. JENNESS, R., *J. Dairy Sci.*, 1980, **63**, 1605.
11. MCDONALD, I. W. and SCOTT, T. W., *World Rev. Nutr. Dietetics*, 1977, **26**, 144.
12. GURR, M. I., *J. Dairy Res.*, 1981, **48**, 519.
13. HALL, B., *Am. J. Clin. Nutr.*, 1979, **32**, 304.
14. MATTSON, F. H., In: *Toxicants Occurring Naturally in Foods*, 1973, National Academy of Sciences, Washington D.C., p. 189.
15. ARTMAN, N. R., *Advances in Lipid Res.*, 1969, **7**, 245.
16. BRISSON, G. J., *Lipids in Human Nutrition. An Appraisal of Some Dietary Concepts*, 1982, MTP, Lancaster.
17. GALLIARD, T., In: *Recent Advances in the Chemistry and Biochemistry of Plant Lipids*, T. Galliard and E. I. Mercer (eds), 1975, Academic Press, London.
18. FRANKEL, E. N., *Prog. Lipid Res.*, 1980, **19**, 1.
19. BROWN, M. S., KOVANEN, P. T. and GOLDSTEIN, J. L., *Science*, 1981, **212**, 628.
20. JEFFCOAT, R., *Essays in Biochemistry*, 1979, **15**, 1.
21. GLOMSET, J. A., *Am. J. Clin. Nutr.*, 1970, **23**, 1129.
22. JUNG, R. T., SHETTY, P. S. and JAMES, W. P. T., *Clin. Sci.*, 1980, **58**, 183.
23. BRINDLEY, D. N. et al., In: *Lipoprotein Metabolism and Endocrine Regulation*, L. W. Hessel and H. M. J. Krans (eds), 1979, Elsevier/North Holland, Amsterdam.
24. OLPIN, S. E. and BATES, C. J., *Br. J. Nutr.*, 1982, **47**, 589.
25. KASPER, H., THIEL, H. and EHL, M., *Am. J. Clin. Nutr.*, 1973, **26**, 197.
26. BURR, G. O. and BURR, M. M., *J. Biol. Chem.*, 1929, **82**, 345.
27. SODERHJELM, L., WIESE, H. F. and HOLMAN, R. T., *Prog. Chem. Fats Other Lipids*, 1970, **IX**(4), 557.
28. PRESS, M., KIKUCHI, H., SHIMOYAMA, T. and THOMPSON, G. R., *Br. Med. J.*, 1974, **2**, 247.
29. HOLMAN, R. T., *Prog. Chem. Fats Other Lipids*, 1970, **9**, 607.
30. MEADE, C. J. and MERTIN, J., *Adv. Lipid Res.*, 1978, **16**, 127.
31. ULDALL, P. R. et al., *Lancet*, 1974, **ii**, 514.
32. MERTIN, J., *Transplantation*, 1976, **21**, 1.
33. MILLER, J. H. D. et al., *Br. Med. J.*, 1973, **1**, 765.
34. MERTIN, J. and STACKPOOLE, A., *Nature (London)*, 1981, **294**, 456.
35. FAO, *Dietary Fats and Oils in Human Nutrition*, A joint FAO/WHO Report, 1978, Rome, FAO.
36. THOMPSON, S. Y., HENRY, K. M. and KON, S. K., *J. Dairy Res.*, 1964, **31**, 1.
37. WITTING, L. A., *Prog. Chem. Fats Other Lipids*, 1970, **IX**(4), 519.
38. FORSS, D. A., *Prog. Chem. Fats Other Lipids*, 1972, **XIII**, 181.

39. DHOPESHWARKAR, G. A., *Prog. Lipid Res.*, 1981, **19**, 107.
40. LEE, D. J., WALES, J. H., AYRES, J. L. and SINNHUBER, R. O., *Cancer Res.*, 1968, **28**, 2312.
41. GARTON, G. A., In: *Metabolism and Physiological Significance of Lipids*, R. M. C. Dawson and D. N. Rhodes (eds), 1964, John Wiley, London.
42. RYHAGE, R., *J. Dairy Res.*, 1967, **34**, 115.
43. HAY, J. D. and MORRISON, W. R., *Biochim. Biophys. Acta*, 1970, **202**, 237.
44. LOUGH, K. A., *Prog. Chem. Fats Other Lipids*, 1973, **14**, 1.
45. MACBRINN, M. C. and O'BRIEN, J. S., *J. Lipid Res.*, 1965, **9**, 552.
46. KISHIMOTO, Y., WILLIAMS, M., MOSER, H. W., HIGNITE, C. and BIEMANN, K., *J. Lipid Res.*, 1973, **14**, 69.
47. GARTON, G. A., HOVELL, F. de B. and DUNCAN, W. R. H., *Br. J. Nutr.*, 1972, **28**, 409.
48. GARTON, G. A., SCARFE, J. R., SMITH, A. and SIDDONS, R. C., *Lipids*, 1975, **10**, 855.
49. HRADEC, J., DUSEK, Z. and MACH, O., *Biochem. J.*, 1974, **138**, 147.
50. ZALEWSKI, S. and KUMMEROW, F. A., *J. Am. Oil Chem. Soc.*, 1968, **45**, 87.
51. HOLMAN, R. T., *Chemistry and Industry*, 1981, 704.
52. HILL, E. G., JOHNSON, S. B. and HOLMAN, R. T., *J. Nutr.*, 1979, **109**, 1759.
53. HWANG, D. H. and KINSELLA, J. E., *Prostaglandins*, 1979, **17**, 543.
54. KUMMEROW, F. A., *J. Food Sci.*, 1975, **40**, 12.
55. LE BRETON, E. and LEMARCHAL, P., *Ann. Nutr. Aliment.*, 1967, **21**, 1.
56. ANDERSON, R. L., FULLMER, C. S. and HOLLENBACH, E. J., *J. Nutr.*, 1975, **105**, 393.
57. HOUTSMULLER, U. M. T., *Fette, Seifen, Anstrichmittel*, 1978, **80**, 162.
58. JUILLET, M. T., *Fette, Seifen, Anstrichmittel*, 1978, **80**, 445.
59. NATIONAL DAIRY COUNCIL, *Dairy Council Digest*, 1982, **53**, 7.
60. ABDELLATIF, A. M. and VLES, R. O., *Nutr. Metabol.*, 1973, **15**, 219.
61. VLES, R. O., In: *The Role of Fats in Human Nutrition*, A. J. Vergroesen (ed.), 1975, Academic Press, London.
62. ANDREWS, J. S., GRIFFITH, W. H., MEAD, J. F. and STEIN, R. A., *J. Nutr.*, 1960, **70**, 199.
63. CUTLER, M. G. and SCHNEIDER, R., *Food Cosmetics Toxicol.*, 1973, **11**, 443.
64. CHANG, S. S., PETERSON, R. J. and HO, C.-T., *J. Am. Oil Chem. Soc.*, 1978, **55**, 718.
65. ALEXANDER, J. C., *J. Am. Oil Chem. Soc.*, 1978, **55**, 711.
66. GOLDSTEIN, J. L. and BROWN, M. S., *Ann. Rev. Biochem.*, 1977, **46**, 897.
67. MILLER, G. J. and MILLER, N. E., *Lancet*, 1975, **i**, 16.
68. KANNEL, W. E., CASTELLI, W. P. and GORDON, T., *Ann. Intern. Med.*, 1979, **90**, 85.
69. KEYS, A., ANDERSON, J. T. and GRANDE, F., *Lancet*, 1957, **ii**, 959.
70. HEGSTED, D. M., MCGANDY, R. B., MYERS, M. L. and STARE, F. J., *Am. J. Clin. Nutr.*, 1965, **17**, 281.
71. CHRISTOPHE, A., MATTHYS, F., GEERS, R. and VERDONK, G., *Arch. Int. Physiol. Biochim.*, 1978, **86**, 414.
72. CHRISTOPHE, A., ILIANO, L., VERDONK, G. and LAUWERS, A., *Arch. Int. Physiol. Biochim.*, 1981, **89**, B156.
73. CHRISTOPHE, A., VERDONK, G., DECATELLE, J. and HUYCHEBAERT, A., *Arch.*

Int. Physiol. Biochim., 1982, **90**, B100.
74. BROWN, M. S. and GOLDSTEIN, J. L., *Science*, 1976, **191**, 150.
75. CONNOR, W. E., WITIAK, D. T., STONE, D. B. and ARMSTRONG, M. L., *J. Clin. Invest.*, 1969, **48**, 1363.
76. CARLSON, L. A. and BÖTTIGER, L. E., *Lancet*, 1972, **i**, 865.
77. KOLATA, G. B. and MARX, J. L., *Science*, 1976, **194**, 509.
78. KOLATA, G. B., *Science*, 1976, **194**, 592.
79. MARX, J. L., *Science*, 1976, **194**, 711, 755.
80. MONCADA, S. and VANE, J. R., *Br. Med. Bull.*, 1978, **34**, 129.
81. MONCADA, S. and VANE, J. R., In: *Advances in Prostaglandin and Thrombosis Research*, **6**, B. Samuelsson, P. W. Ramwell and R. Paoletti (eds), 1980, Raven Press, New York, p. 43.
82. HAMBERG, M., SVENSSON, J. and SAMUELSSON, B., *Proc. Natl Acad. Sci. (USA)*, 1975, **72**, 2994.
83. GIBNEY, M. J., *Biochem. Soc. Trans.*, 1982, **10**, 161.
84. DYERBERG, J. and BANG, H. O., *Lancet*, 1979, **ii**, 433.
85. TAYLOR, T. G., GIBNEY, M. J. and MORGAN, J. B., *Lancet*, 1979, **ii**, 1378.
86. RINZLER, S. H., *Bull. NY Acad. Med.*, 1968, **44**, 936.
87. DAYTON, S., PEARCE, M. L., HASIMOTO, S., DIXON, W. J. and TOMIYASU, U., *Circulation*, 1969, Supp.II, **40**, 1.
88. LEREN, P., *Circulation*, 1970, **42**, 2307.
89. MIETTINEN, M., TURPIENEN, O., KARVONEN, M. J., ELUSUO, R. and PAAVILAINEN, E., *Lancet*, 1972, **ii**, 835.
90. OLIVER, M. F., *Br. Med. Bull.*, 1981, **37**, 49.
91. MARR, J. W. and MORRIS, J. N., *Lancet*, 1982, **i**, 217.
92. TANNENBAUM, A., *Cancer Res.*, 1942, **2**, 460.
93. TANNENBAUM, A., *Cancer Res.*, 1942, **2**, 468.
94. CARROLL, K. K. and KHOR, H. T., *Lipids*, 1971, **6**, 415.
95. VISEK, W. J., CLINTON, S. K. and TRUEX, C. R., *Cornell Veterinarian*, 1978, **68**, 1.
96. HEMS, G., *Br. J. Cancer*, 1970, **24**, 226.
97. LEA, A. J., *Lancet*, 1966, **ii**, 332.
98. ROSE, G. *et al.*, *Lancet*, 1974, **i**, 181.
99. WYNDER, E. L., *Cancer Res.*, 1975, **35**, 3388.
100. HILL, M. J., *Am. J. Clin. Nutr.*, 1974, **27**, 1475.
101. REDDY, B. S., WEISBURGER, J. H. and WYNDER, E. L., *J. Natl Cancer Inst.*, 1974, **52**, 507.
102. GAMMAL, E. B., CARROLL, K. K. and PLUNKETT, E. R., *Cancer Res.*, 1967, **27**, 1737.
103. MERTIN, J., *Br. Med. J.*, 1973, **4**, 357.
104. CARROLL, K. K. and HOPKINS, G. J., *Lipids*, 1979, **14**, 155.
105. CARROLL, K. K., *Cancer Res.*, 1975, **35**, 3374.
106. BJELKE, E., *Int. J. Cancer*, 1975, **15**, 561.
107. COHEN, S. M., WITTENBERG, J. F. and BRYAN, G. T., *Fed. Proc.*, 1974, **33**, 602.
108. NETTESHEIM, P., SYNDER, C., WILLIAMS, M. L., CORE, M. V. and KIM, J. C., *Proc. Am. Assoc. Cancer Res.*, 1975, **16**, 54.

109. ROGERS, A. E., HERNDON, B. J. and NEWBERNE, P. M., *Cancer Res.*, 1973, **33**, 1003.
110. HICKS, R. M., *Proc. Nutr. Soc.*, 1983, **42**, 83.
111. LEEDS, A. R., *Proc. Nutr. Soc.*, 1979, **38**, 365.
112. GARROW, J. S., *Energy Balance and Obesity in Man*, 1978, Elsevier/North Holland, Amsterdam.
113. SCHEMMEL, R. and MICKELSEN, O., In: *The Regulation of Adipose Tissue Mass*, J. Vague and J. Boyer (eds), 1973, Excerpta Medica, Amsterdam, p. 238.
114. ROLLS, B. J., *Br. Nutr. Foundation Bull.*, 1979, **5**, 78.
115. STOCK, M. J. and ROTHWELL, N. J., In: *Animal Models of Obesity*, M. F. W. Festing (ed.), 1979, Macmillan, London, p. 141.
116. BABAYAN, V. K., *J. Am. Oil Chem. Soc.*, 1974, **51**, 260.

INDEX

Acetic acid
  mammary gland, taken up by, 40
  metabolism,
    non-ruminants, 49, 50
    ruminants, 42, 43, 47, 48, 381
Acetyl CoA
  fatty acid oxidation, 387
  synthesis, 43, 44
Acetyl CoA carboxylase, fatty acid synthesis, 43, 48, 49
Acyl CoA synthetase,
  triglyceride synthesis, 51–3
Acyl transferases, 84, 386
  see also Triglyceride synthetase
Adenosine triphosphatase,
  membranes, 107, 108, 111
Adipose tissue
  fatty acid metabolism, 370, 383
  lipids, 368–72
Agglutination (cold) of milk fat globules, 142–3, 147–8
  immunoglobulins, 143, 144, 146, 147
Agitation of milk, 126
  fat crystals, effects of, 167
  fat globules, effects on, 130, 132, 145
  lipolysis, effects on, 209–10
  MFGM, recombined milk, effects on, 137–8
Alcohols
  lipid oxidation, inhibition, 260

Alcohols—*contd.*
  lipid oxidation, termination, 272
Alkoxy radicals, 262, 270, 271, 277, 318–19
Alkyl radicals, 270, 288
Alnarp method, butter consistency, 183
Amino acids
  oxidation, lipids, 285, 325
  pro-oxidant effects, 271
Amino groups, reaction of singlet oxygen, 251
Anti-oxidants, 244, 309–27
Arachidonic acid, essential fatty acid, 277–8, 370, 374, 388
Arylesterase, blood serum and milk, 203
Ascorbic acid
  anti-oxidant, 318
  copper, oxidation, 288, 319–20, 322, 351
  copper-catalysed oxidation, lipids, 318–22, 345, 356
  riboflavin-catalysed oxidation, 331
Ascorbyl palmitate, milk, 351–2
Ascorbyl radical, 318, 319, 321

Bacteria, milk, 148, 151
  lipases, 223
    psychrotrophic bacteria, 196, 204–6, 217

Beef
  fatty acids
    muscle, 375
    suet, 369, 372
Bile salts, 368, 380, 381
  lipase, human milk, activated by, 200–2
Bixin, singlet oxygen quenching, 324
Blood, milk, fatty acids, 39–42
Brain
  fatty acids, 375, 387
  glycolipids, 366
Bregott®, 189
Buffalo, milk, 143, 147, 202
Butter
  fat crystals, 174, 175
  fatty acids, 369
  consistency of, 65, 182–9
  oxidation of, 247, 360
  rancidity, 217–18
  rheology, 176–8, 191
  structure, 174–6
Butterfat, solid:liquid ratio, 168
Butylated hydroxytoluene, synthetic anti-oxidant, 316, 378
Butyric acid, 220, 268
  milk fat, ruminants, 13
  glycerol, position 3, 24, 25, 53, 54, 69
Butyrophilin, MFGM, 94, 103–5, 111

Cancer, dietary factors, 402–3, 409–10
Carbohydrates
  digestibility, low-fat syndrome, cow, 66–7
  MFGM, 98, 99, 109–10, 128
  synthesis of fatty acids from, 384
Carbonyls
  browning reaction with amines, 325–6
  production, autoxidation of unsaturated fatty acids, 273, 277
  off-flavours, 342
Carboxylesterase, milk, 204
Cardiovascular disease, 403–4, 406–7
  blood lipids, 404–8

Carnitine, acylcarnitines, 10, 387
β-Carotene (provitamin A)
  anti-oxidant activity, 324
  milk, 393–4
  MFGM, 100
  oxidation, 284–5
  quenching singlet oxygen, 251
Carotenoids, 251, 323, 324
Caseins
  anti-oxidants, 325, 349, 358
  fat globules
    effects on, 140, 141, 147, 181
    homogenized milk, 137
    recombined milk, 316
  LPL, associated with, 198
Catalase, 295, 326
Ceramide
  fatty acids and hexosides, 21, 22, 28
  milk, 11
  MFGM, 90, 101–2
Cerebrosides, MFGM, 98
Cheese, lipolysis in ripening of, 207, 212, 221, 222–3
Chlorophyll, photosensitizer, 244, 328
Chocolate, fats, melting point, 395–6
Cholesterol, 5, 372
  diet, plasma level, 404, 405
  intracellular synthesis, 384
  membranes, 366, 367
  MFGM, 100
  oxidation, 281, 282–4
  reaction with singlet oxygen, 253
  sources for milk, 56–7
  transport in plasma, 381, 382, 383, 384
Cholesterol esters, 5, 19–20, 57
  MFGM, 100
Chromatography, fatty acids, 25–7
Chylomicrons, long-chain fatty acid transport, 381–3
Citrate cleavage pathway, mammary gland, 47
CoA esters, fatty acid metabolism, 46, 50–1
Coalescence of fat globules, 129–34
  heating, 137
  recombined milk, stability, 140
Coconut oil, 374

Coconut oil—*contd.*
  cows, 120
  human diets, 380
Cod
  fatty acids
    cod liver oil, 369
    muscle, 375
Colchicine, milk secretion suppression, 91
Colourants, foods, pro-oxidants, 249, 328
Concanavalin A, milk secretion suppression, 94–5
Cooling of milk, oxygen in liquid phase, 248
Copper, 257, 306, 307
  ascorbic acid, oxidant ability, 288, 319–20, 322, 351
  catalysis of lipid oxidation, ascorbic acid, 321–2, 345, 356
  milk, 286, 346–7
    MFGM, 287
  pro-oxidant, 287–8
    milk, 291–3
Corn oil, fatty acids, 374
Cottonseed oil, sterculic acid, 396
Cows milk
  Channel Island breeds, 122, 394
  fat content, dietary fats, effects of, 59–60, 63–5
  fatty acids, 11, 14, 72, 373
    dietary fats, effects of, 58–63, 63–5
    forms of lipid in, 19, 20, 21, 23, 27–8
    linoleic acid level, 17, 68
  low-fat syndrome, 66–7
  spontaneous lipolysis, 212–16
Cream, 150–1
  bitty, bacterial phospholipids, 208
  churning, 140, 174, 221
  consistency
    homogenization, effects of, 180–1
    rebodying, effects of, 130, 152, 181–2
  FFA concentration, rancid flavour, 218
  homogenized, 141–2, 181

Creaming, cows' milk, gravity/centrifugal, 148–51
Crystals, milk fat, 160–6
  bulk fat, 166–7
  butter, 175
  content of, methods of measuring, 168–72
  fat globules, 124–5, 167–8
    coalescence, effects of, 129–30
  oxygen, solubility, 246, 247, 360
Cultured milk, fat globules in casein framework, 141
Cyclopropane fatty acids, toxic effects, 396–7
Cytochalasin B, milk secretion, 91
Cytochromes, 297, 298, 309
  MFGM, 98, 99, 108, 357–8
  pro-oxidants, 357–8

Dairy products, future for, 408, 412, 413
Desaturases
  fatty acids, 389, 390, 395
  trans fatty acids, 399
Diabetes, faulty fat metabolism, 410–11
Differential scanning colorimetry, solid content of fat, 170–1
Digestion and absorption of food lipids, 379–81
Diglycerides, 5, 55
Dihydroxyacetone phosphate pathway, triglyceride synthesis, 50, 51, 52
Dilatometry, solid fat content, 169–70
Dioxetanes, 280
Disease, dietary lipids, 402–12
Dried milk, susceptibility to lipid oxidation, 360

Echidna, milk triglycerides, 23–5
Electronic configuration
  oxygen, ground and singlet states, 250, 251
  transition metals, effect of ligands on, 306–9
Electrons, flow of, 248–52, 271

INDEX

Emulsions (thermodynamically
   unstable), 125–6, 172, 174
Endoplasmic reticulum, mammary
   epithelial cells, MFGM
   secretion, 84–6, 90
Energy
   dietary supply, from lipids, 365, 385
   sources of, values of fatty acids, 388,
      412
Enzymes
   anti-oxidants, 326–7
   lactogenesis, 48–9
   MFGM, 107–9
   specificity, triglyceride synthesis, 52–5
Epoxidation, lipids, double bond, 280–1
Erucic acid, rapeseed oil, toxic effects,
   401
Essential fatty acids, 388–93
   membrane lipid synthesis, 386
Esterases
   milk, 203–4
   psychrotrophic bacteria, 206
Evaporated milk, heat stability, 141
Extrusion method, butter consistency,
   178–9

Fat
   content, milk, 3, 4
   diet, cow, 59–60, 63–5
      physical properties, 132, 149, 151
   fractionation, milk, 187–9
   globules, milk, 119–20
      agglutination, cold, 142–8
      air bubbles, interaction, 133–4
      coalescence, 129–33
      creaming, 148–51
      disruption, 135–6, 139–41, 181
      fat crystals, 124–5, 167–8
      homogenization clusters, 141–2,
         152, 181
      instability, 125–6
      interaction energy, 126–8
      intracellular origin/growth/
         transport, 84–91
      membrane, *see* Milk fat globule
         membrane
      rheology, 151–4

Fat—*contd.*
   globules—*contd.*
      secretion, 91–6
      sizes, 5, 120–3
      viscolization, 181
      liquid/solid mixture, milk, 168
      solid content, determination, 169–72
   Fatty acid synthetase, 43, 44–6
   Fatty acids
      branched-chain, 18, 397–8
         adipose tissue, ruminants, 370,
            372, 381
         oxidation of, genetic defects, 397,
            398
         skin, 366
      free (FFAs), 5
         blood, 39, 380, 383
         flavour
            butter, 219–20
            cheese, 395
            milk and cream, 208, 218–19
         foaming properties of milk, 122
         fresh milk, from incomplete
            synthesis, 208
         lipid autoxidation facilitation,
            267–8, 292
         lipid digestion, production, 380
         LPL, inhibition, 199
         methods of determining, 224–6
         MFGM, 98, 100
         spoilage, production, 378
      metabolism, in, 384–8
      milk lipids, 12–19
         bovine/human milk, 11
         cholesterol esters and
            phospholipids, 19–22, 101,
            102
         sources, 11–12, 38–9
            blood, 39–42
            mammary gland, 42–50
            synthesis, mammary gland, 12,
               16, 25, 45–6
         yield,
            seasons, 68–9
            stages of lactation, 67–8
      odd-numbered
         adipose tissue, ruminants, 370,
            372, 381

Fatty acids—*contd.*
  odd-numbered—*contd.*
    skin, 366
  unsaturated
    autoxidation, 261–78
    fish liver oils, 407
    hydrogenation, *see* Hydrogenation
    immune system, 392
    margarines, 376
    off-flavours from oxidation, 342
    saturated fatty acids ratio, blood, cholesterol content, 404
    several double bonds, 13, 16, 19
    plants, 374
    prostaglandins, 368, 390, 391, 399, 400
    spontaneous oxidation, 343
    singlet oxygen, reaction with, 252
    trans isomers, 11, 398–400
      margarines, 376, 377, 398
      rumen, produced in, 370, 381, 398
Fermented milk products, 332, 359–60
Ferric thiocyanate test, lipid oxidation, 333
Ferryl state of iron, 297–300
Fish liver oils, 369, 372, 394, 407
Flavin adenine dinucleotide,
    β-oxidation of fatty acids, 387
Foaming properties of milk, 222
Formate, scavenger for hydroxy radicals, 356
Free radicals
  generation,
    decomposition of hydroperoxides, 252, 265, 266, 287–9
    oxidation of unsaturated fatty acids, 261
    photochemical reactions, 328, 329
    reactions of metals with substrates, 289
  lipid autoxidation, initiation, 261–2, 265
  reaction with ground-state oxygen, 252

Galactolipids, leaves, 375

Gangliosides
  milk, 21
  MFGM, 98, 102
Glucagon, fat metabolism, 386, 387
Glucose, absorbed from blood by mammary gland, 40
  metabolism of, 42, 43, 47–9
Glycerol-3-phosphate pathway, triglyceride synthesis, 50–2
Glycerolipids, trace amounts, milk, 6, 8
Glycerophospholipids, fatty acids, 19–20
Glycocalyx, fat globules, 109–10, 128
Glycolipids, 99, 100, 366
Glycoproteins, MFGM, 103–4, 105, 106, 128
Glycosphingolipids, 11, 101–2
Goats, milk, 39, 46, 54, 202, 373
Golgi apparatus, mammary epithelial cells/MFGM, 93
Guinea-pigs, milk, 23, 202–3

Haem, pro-oxidant, 295, 296, 307–8
Haemoproteins, hydroperoxide, decomposition, 299
Heat treatment
  chemical changes
    fats, 377
    unsaturated fatty acids, 402
  cream, properties, effects of, 141, 182–5
  milk, properties, effects of, 146
  skim milk, properties, effects of, 138–9, 141
  temperature, inactivation of enzymes, 356
Hexanoic acids, glycerol, position 3, 24, 53, 54, 69
Hexose monophosphate cycle, mammary gland, ruminant/non-ruminant, 47–8
Homogenization clusters, cream, 142, 152
  agglutination clusters, compared with, 181
Homogenized cream, 141–2, 180–1

Homogenized milk, 137, 210–11
  autoxidation and lipolysis of lipids,
    140
  fat globule properties
    coalescence, 133, 140
    cold agglutination, 146
    size, 122–3, 136, 137, 140
    foaming properties, 222
    metals, redistribution, 295
Hormones, fat metabolism, 49–50,
    385–6
Human milk
  fatty acids, 11, 13, 18–19, 183, 373
  diet, effects on, 63
  lipids, 20, 23, 26–8
  linoleic acid, 63
  lipases, 200–1
Hydrocarbons, trace amounts,
    milk, 8–9
  MFGM, 98
Hydrogen peroxide, 259–60
  destruction
    catalase, 326
    metal complexes, 301–2
  production
    photochemical reactions, 329
    sulphydryl oxidase, 327
    superoxide dismutase, 256, 326
Hydrogen sulphide, released from fat
    globules on heating, 136
Hydrogenation of unsaturated fatty
    acids
  margarine manufacture, 376, 377–8
  rumen, 12, 18, 61, 370, 371
Hydroperoxides
  decomposition, 265–6
    catalysed by metal ions, 254
    producing free radicals, 252, 265,
      266, 277, 287–9
  production
    oxidation, unsaturated fatty acids,
      252, 253, 260–1, 263, 272–8
    reaction of tocopherol with peroxy
      radicals, 310, 311
    reaction, producing epoxides, 280–1
  tasteless, 342
Hydroxy radicals
  formate, scavenger for, 356

Hydroxy radicals—contd.
  production
    copper ascorbate and oxygen, 345
    hydrogen peroxide and metals/
      superoxide, 257, 260
    lipid autoxidation, 270
    water radiolysis, 260
β-Hydroxybutyric acid, 13, 40–3
Hyperlipoproteinaemia, 389, 405, 411

Ice-cream, recombined fat globules,
    140
Immune system, unsaturated fatty
    acids, 392
Immunoglobulins, milk, 143, 148
  cold agglutination, 143, 144, 146, 147
Immunosuppression therapy, linoleic-
    acid-rich diet, 392
Insulin, fat metabolism, 49, 50, 383,
    385–8
Interesterification of triglycerides, 376,
    378, 405
Intravenous feeding, essential fatty
    acids, 391, 392
Iodine value of butterfat, 185
Iron
  electronic configuration, haem, 306–8
  high oxidation (ferryl) state, 297–300
  milk, 286, 346–7
  pro-oxidant, 287–8, 291–3
*Iso*citrate cycle, mammary gland,
    ruminants, 48

Ketones, lipid oxidation termination,
    272
Ketosis, 387–8
Kreis test, lipid oxidation, 333

Lactation, stage, milk lipids, 67–8, 122,
    148, 211, 215
Lactoferrin, peroxidation inhibition,
    358
Lactones, flavour compounds, 10
Lactoperoxidase, 259, 309, 327, 354,
    358–9

Lactoperoxidase—*contd.*
  coupling,
    sulphydryl oxidase, 351
    xanthine oxidase, 345, 346, 356
Lard, fatty acids, 369, 372
Leaves, fatty acids, 16, 375
Leucocytes, milk, 151, 216–17
Light, lipid autoxidation, initiation, 264
  see also Photochemical oxidation, Photosensitizers
Linoleic acid, essential fatty acid, 388
  autoxidation, 261, 263
    catalysed by ascorbate complexes, 320
    hydroperoxide formation, 272–8, 279
  content, human milk, 63
    cow's milk, 15, 17
    cholesterol esters, 19
    dietary supplements, 263, 355
    isomers, 17–18
  singlet oxygen, reaction with, 252, 253
Linolenic acid, 16, 263, 277–8
Lipases, 386
  intracellular, 383
  milk
    cow, 197
    human, 206–7
  pancreatic, 380
  pre-gastric, new-born mammal, 223–4, 379
  psychrophilic bacteria, 196, 204–6
  see also Lipoprotein lipase
Lipid(s)
  definition, 366
  dietary
    composition, 371–5
    energy source, 365, 385, 388
    essential fatty acids, 388–93
    fat-soluble vitamins, 393–5
    modification, processing, 377–8
    palatability/texture, foods, 395–6
    spoilage, 378
    structural, 388
    toxic effects, 396–402
    see also Lipolysis, Oxidation of lipids

Lipid(s)—*contd.*
  droplets, mammary epithelial cells
    origin/growth, 84–90
    secretion, milk fat globules, 91–6
    transport, 90–1
  living organisms
    disease, 402–12
    metabolic, 368
    metabolism, 378–81, 384–6
    storage, 368–70
    structural, 366–7
    transit, 371
    transport, 381–4
Lipolysis, 5
  induced
    agitation and foaming, 208–10
    homogenization, 140, 210–11
    temperature changes, 211, 212
  spontaneous, 208, 212–16
Lipoprotein lipase (LPL), 52, 198–200
  adipose tissue, 383
  deficiency, congenital, 383
  mammary gland, 39–40, 49, 197–8
  positional specificity, 200
Lipoproteins
  apolipoproteins, activated by LPL, 199, 200
  blood, 371, 381–3
  milk, cold agglutination, 140, 141
  MFGM, 147
  receptors, cells, 383–4
Lipovesicles, 85, 86, 87, 88, 89, 90
Luminescence,
  dioxetane decomposition, 280
  peroxy radical decomposition, 334
Lysophospholipids, membrane-perturbing agents, 6, 7, 200, 380–1

Malonyl CoA, fatty acid synthesis, 43, 44
Mammary gland
  cholesterol synthesis, 57
  fatty acid synthesis, 11, 42–50, 58, 369
    short to medium length, 12, 13, 370
  triglyceride synthesis in, 54
Manganese, milk, 286
Mannitol, lipid oxidation inhibition, 260

Margarines, 186, 371, 394
  fatty acids, 375, 376
  trans forms, 376, 377, 398
  hydrogenation of fatty acids, 376, 377–8
Marine mammals, milk, 3, 4
  fatty acids, 16, 24
Mastitis, milk, effects of, 203, 204, 216
Melting behaviour, milk fat, 64–5, 69–76, 377
Membrane(s)
  cells, 366, 367, 386
  milk fat globules, *see* Milk fat globule membrane
Metal ions
  anti-oxidants, ligands, 310
  binding sites, organic compounds, 294
  ground-state oxygen, interaction, 249
  milk, lipid oxidation, 346–8
  pro-oxidant activity, 245, 286–91
    electron transfer and reduction potential, 296–7
    electronic configuration, 306–9
    ligands, 293–5
    oxygen complexes, 302–6
    redistribution, 295–6
    stabilization of oxidation state, 297–303
  reactions, yielding singlet oxygen, 254
Metallo-enzymes, 254, 393–4
Methional, off-flavours, 352
Methyl ketones, flavour compounds, 10–11
Methylmalonyl CoA, branched-chain fatty acid precursor, 397–8
Metmyoglobin, iron, 299, 309
Microfilaments, microtubules, mammary epithelial cells, 90–1
Micro-organisms, gut, food lipids modification, 380–1
  see also Bacteria, Rumen microorganisms
Milk fat globule membrane (MFGM)
  acquisition, fat globules, 84, 92–4, 95, 100

Milk fat globule membrane—*contd.*
  antibodies, milk secretion inhibition, 94–5
  changes, 136–9
  composition, 97–9
    amino acids, 106
    enzymes, 107–9
    lipids, 3, 99–102
    proteins, 102–6
    thiols, 349
    tocopherols, 316–17
  isolation, 96–7
  lipase attachment, spontaneous lipolysis, 215, 216
  molecular organization, 109–11
  pro-oxidant factors, 353–8
  weakening, butter making, 174–5
Milking machines, lipolysis, milk, 210, 214
Mink, milk, fatty acids, 20
Mitochondria, metabolic processes, 42, 47, 387
Molybdenum
  milk, 286
  xanthine oxidase, 293
Monoglyceride(s), 5, 380
  catalytic impurities, fat crystal nucleation, 161
  pathway, triglyceride synthesis, mammary gland, 50, 51, 52
Moulds
  butter, 217
  cheese ripening, 207, 223
Mouse, fatty acids, milk phosphatides, 20
Multiple sclerosis, linoleic-acid-rich diets, 392
Muscle, fatty acids, 374–5

NADH/NADPH, 42, 47–8
Nuclear magnetic resonance, solid content in fats, 171–2, 190
5′-Nucleotidase, MFGM, 107, 108, 111, 148

Obesity, 386, 411–12

Off-flavours, compounds contributing to, 342, 343, 352
Oleic acid, 59, 60, 263, 272–8
Olive oil, fatty acids, 374
Oxidation
  lipids, 242–5
    autoxidation, as chain reaction, 261–72
    bulk phase, membranes and interfaces, 314
    environmental effects on, 327–32
    measurement, 332–4
  milk lipids, 342–4, 360–1
    electrons flow, 348–52
    light and riboflavin, 352–3
    metals, 346–8
    MFGM, 353–8
    oxygen, 359–60
    serum enzymes, 358–9
    spontaneous, 344–6
β-Oxidation, fatty acids, 386–8
  branched-chain, 397
Oxidation–reduction potential of metals, ligands and, 296–7, 308–9
Oxidizability, 263
Oxygen
  active forms
    complexes with metals, 303–5
    containment, living organisms, 255
    generation, phagocytes, 360
    life span, lipids, 344
    lipid oxidation, initiation, 265, 345
    see also Hydrogen peroxide, Hydroxy radicals, Ozone, Singlet oxygen, Superoxide
  ground/triplet state, 248–50, 345
  removal from milk, and oxidative off-flavours, 359
  solubility, 245–8
  uptake, as measure of lipid oxidation, 255
Oxygenases, oxygen bound to metals, 303–5
Ozone, 260

Palm oil, fatty acids, 374

Palmitic acid, 40, 43, 53
Pasteurized milk, lipases, 199, 205, 211
Peanut oil, fatty acids, 374
Penetration method, butter consistency, 179–80
Peracyl radicals, reaction of superoxide with esters, 258
Perhydroxy radicals (protonated superoxide), 255, 258, 332
Peroxidases, 295, 297–9, 327
Peroxidation of unsaturated fatty acids, 272–8, 298, 378
Peroxide salts, as oxidants, 305
Peroxide values, as measure of lipid oxidation, 332–3, 342
Peroxides, 309–10, 311
  cyclic (endoperoxides), 278–80
Peroxy radicals
  cyclization, 276
  double bond, 280
  tocopherol radicals, 310, 311
  formation,
    autoxidation of lipids, 262, 263, 270, 272, 274
    reaction of ground-state oxygen with free radicals, 252
pH
  isoelectric, milk fat globules, 128
  lipid oxidation, 332
Phagocytes, adhering to MFGM, active oxygen production, 360
Phosphatidate phosphatase, triglyceride synthesis, 50, 53–4
Phosphatidic acid, 55, 207–8
Phosphodiesterase, MFGM/plasma membrane, 107
Phospholipases, 207–8, 380
Phospholipids, 6, 7
  anti-oxidants, 324
  cell membranes, 366
    store of essential fatty acids, 391
  chylomicrons, 381, 382, 383
  fatty acids, 19, 20, 27–8
  hydrolysis, LPL, 200
  interconversion of types, 55, 56
  MFGM, 3, 98, 99, 100, 354
    milk serum particles, 354
  synthesis, 55, 56

4'-Phosphopantetheine, 44, 45
Photochemical oxidation, 327–31
　milk, 352–3
Photosensitizers
　ground-state oxygen activation, 249, 254
　photo-oxidation, 328–31
Phytanic acid, branched-chain fatty acid, 397
Pigeons, fatty acids, cropmilk, 28
Pigs, milk
　fatty acids, 16, 19
　phospholipids, 20
　triglycerides, 23, 24, 26
Plants
　polyunsaturated fatty acids, rumen, 370, 371
　storage lipids (seed oils), 371, 373–4
　see also Leaves
Plasmalogens, 6, 8
Platelets, prostaglandins and aggregation, 406–7
Polymerization of lipid oxidation products, 242, 272
Porphyrin, pro-oxidant, 295, 299
Prolactin, 40, 49, 50, 198
Pro-oxidants, 244, 286
　see also Metal ions, Oxygen, Riboflavin
Propionic acid, ruminants, metabolism, 67, 381
Prostaglandins, 9, 10, 391
　arterial wall, inhibit aggregation of platelets, 406–7
　sources, polyunsaturated fatty acids, 368, 390, 391, 399, 400
Proteins
　MFGM, 98, 102–8, 110
　milk plasma, adsorbed onto fat globules to replace MFGM, 137

Rabbits
　enzymes, triglyceride synthesis, mammary gland, 52
　fatty acids, milk, 13, 39, 50
Rancidity
　hydrolytic, 195–6, 226–7

Rancidity—*contd.*
　oxidative, 242
Rapeseed oil, 374, 401
Rats
　fatty acids of triglycerides in milk of, 23
　syntheses in mammary gland of, 13, 47–8, 49, 54
Recombined milk, 137–9, 140, 141
Refsum's disease, 397
Rheological properties, milk products
　butter, 176–8
　fat globules, effects on, 151–4
　milk and cream, 172–4
　mixed crystals, 166
　modification
　　blending, 189–90
　　fractionation, 187–9
　　processing, 180–7
　solid/liquid fat proportions, 168
Riboflavin
　β-oxidation, fatty acids, 387
　photosensitizer, 249, 259, 328–31, 352–3
Rumen micro-organisms
　fermentation, 381
　hydrogenation, unsaturated fatty acids, 12, 18, 61, 370, 371
Ruminants
　fatty acids
　　adipose tissue, 370, 372
　　milk, 12, 41
　lipid metabolism, mammary gland, 41–2, 43, 47

Seals, fatty acids, triglycerides, milk, 23, 24
Season, cows milk, 74, 214
Secretory vesicles, fat globule release, mammary gland, 93, 94
Sectility method, butter consistency, 178
Sheep, fatty acids, milk, 373
Sialic acids, MFGM, 98, 99
Singlet oxygen, active form, 250–4
　formation, 254, 352–3, 359
　tocopherols, 310, 314, 317
Skim milk, 31, 147, 150

Skin, lipids, 366
Soybean oil
  butter consistency, 189–90
  cow's diet
    milk fat yield, 63
    milk fatty acids, 61–2, 75, 76
    fatty acids, 374
Sphingolipids, milk, 21–2, 56
Sphingomyelin, milk, 6, 7, 21, 98
  synthesis, 56, 57, 90
Spontaneous lipolysis, milk fat, 212-16
Spontaneous oxidation, milk fat, 344–6
Squalene
  milk fat, 8
  MFGM, 100
Starters, lactic, 206–7, 221
Stearic acid, 40, 41
  cow's diet, milk fatty acids, 59, 60
Sterculic acid, cottonseed oil, 396
Steroidal hormones, milk, 8
Steroids (other than cholesterol)
  milk, 8
  MFGM, 98, 100
Storage lipids
  animals, 368–71
  plants, 371, 373–4
Sulphydryl compounds, *see* Thiols
Sulphydryl oxidase, 259, 321, 327, 351
Sunflower seed oil, 374
  cow's diet
    fatty acids, milk, 64–5
    positional distribution, fatty acids, triglycerides, 70, 71, 76
Superoxide, 249, 255–9
  formation, 249, 259, 322, 329, 330
  protonation to perhydroxy radical, 255, 332
  salts, decomposition into oxygen and peroxide salts, 305
  singlet oxygen from oxidation, 254
Superoxide dismutase, 250, 257, 326, 327
  milk, 355, 359

Tallow
  blended with butter, 189
  cow's diet, fatty acids, milk, 60–1, 62

Thiobarbituric acid test, lipid oxidation, 278, 279, 333
Thiols
  anti-oxidant mixtures, 271
  anti-oxidant/pro-oxidant, 288, 322–3, 349–50, 358
  singlet oxygen reaction, 251
Thiyl radicals from autoxidation of thiols, 322–3
Thyroid hormone, fat metabolism, 386
Tocopherols
  anti-oxidant, 309, 313–14
  electrons, 348–9
  free radicals, 310
  singlet oxygen, 310, 314, 317
  pro-oxidant, 312, 313
  MFGM, 316–17
  regeneration, ascorbic acid/thiols, 312, 316
Tocopheryl radicals, 310, 311
Triglyceride synthetase, 50, 52–3, 67
Triglycerides (triacylglycerols), 2, 3
  blood lipoproteins, mammary gland, 39–40
  composition, melting behaviour of milk fat, 69–76, 161
  MGFM, 98, 99–100
  positional distribution of fatty acids, 22–7, 69–71
  synthesis, 50–5
    mammary gland, 54, 84–5
  transport, chylomicrons, 381, 382, 383
Tryptophan, pro-oxidant effects, 271–2

Ultra-violet absorption
  conjugated double bonds, lipids, 276
  measure of lipid oxidation, 333
Uronic acids, MFGM, 98, 99

Viscosity, milk and cream, 151–4, 173, 174
Vitamin A, 54–5, 224
  milk, 9, 393
  oxidation, 204–5, 378
  toxicity, excess, 310, 394

Vitamin B
  fatty acid synthesis, 385
  metabolism, congenital defect, 398
Vitamin D, 9, 285, 394
Vitamin E, 9, 285, 394
  see also Tocopherols
Vitamins, fat-soluble, 368, 380, 393–4

Water activity of foods, lipid oxidation and other processes, effect of, 291, 371
Whey, 140, 146
Whipping, cream, recombined milk fat globules, 140
Work softening of butter, 185–7

Xanthine oxidase, 94, 103–5, 107–8, 111
  amount, skim milk, fat globule damage, 209
  coupling
    lactoperoxidase, 345, 346
    generation active oxygen, 356
  denatured, pro-oxidant with copper, 357
  pro-oxidant, milk, 344–5, 355
  superoxide production, 249, 259
  tissues, 104

Yeasts, psychrotrophic, rancidity, butter, 217
Yield stress, butter consistency, 178, 191